Environmental Management

This comprehensively updated third edition explores the nature and role of environmental management and offers an introduction to this rapidly expanding and changing field. It focuses on challenges and opportunities, and core concepts including sustainable development.
    The book is divided into five parts:

- Part I (Introduction to Environmental Management): four introductory chapters cover the justification for environmental management, its theory, scope, goals and scientific background.
- Part II (Practice): explores environmental management in economics, law and business and environmental management's relation with environmentalism, international agreements and monitoring.
- Part III (Global Challenges and Opportunities): examines resources, challenges and opportunities, both natural and human-caused or human-aggravated.
- Part IV (Responses to Global Challenges and Opportunities): explores mitigation, vulnerability, resilience, adaptation and how technology, social change and politics affect responses to challenges.
- Part V (The Future): the final chapter considers the way ahead for environmental management in the future.

With its well-structured coverage, effective illustrations and foundation for further, more-focused interest, this book is easily accessible to all.
    It is an essential reference for undergraduates and postgraduates studying environmental management and sustainability, and an important resource for many students on courses including environmental science, environmental studies and human geography.

Chris Barrow is Founding Editor of the journal *Land Degradation & Development* and continues to work on it. His research and publications focus on environmental management, land degradation, water and agriculture, tropical highland environments and smallholders. He has undertaken research in Malaysia, the sub-Antarctic, highland Morocco and Amazonian Brazil (floodlands). He lectured at Hull University (1975–77), joined Swansea University as a Lecturer in 1978, and retired from a Readership in late-2011. He worked as a palaeoecologist with the British Antarctic Survey (1972–75), gained a PhD from Birmingham University (1977) and a PGCE in 1978.

T0286685

Environmental Management

Introduction, Challenges, Opportunities

Third Edition

Chris Barrow

Routledge
Taylor & Francis Group

LONDON AND NEW YORK

Designed cover image: Author photo

Third edition published 2024
by Routledge
4 Park Square, Milton Park, Abingdon, Oxon, OX14 4RN

and by Routledge
605 Third Avenue, New York, NY 10158

Routledge is an imprint of the Taylor & Francis Group, an informa business

© 2024 Chris Barrow

First edition published by Routledge 1999
Second edition published by Routledge 2006

British Library Cataloguing-in-Publication Data
A catalogue record for this book is available from the British Library

Library of Congress Cataloging-in-Publication Data
Names: Barrow, Christopher J., author.
Title: Environmental management : introduction, challenges, opportunities / Chris Barrow.
Description: Third edition. | New York : Routledge, 2024. |
Series: Routledge environmental management series | First edition published by Routledge 1999. Second edition published by Routledge 2006. | Includes bibliographical references and index. Identifiers: LCCN 2023049453 |
ISBN 9781032023717 (hbk) | ISBN 9781032039671 (pbk) | ISBN 9781003189985 (ebk)
Subjects: LCSH: Environmental management.
Classification: LCC GE300 .B375 2024 | DDC 363.7/05--dc23/eng/20231228
LC record available at https://lccn.loc.gov/2023049453

ISBN: 978-1-032-02371-7 (hbk)
ISBN: 978-1-032-03967-1 (pbk)
ISBN: 978-1-003-18998-5 (ebk)

DOI: 10.4324/9781003189985

Typeset in Sabon
by SPi Technologies India Pvt Ltd (Straive)

To Zak, Sophia and Olivia

Contents

Contents

Figures

Tables

Boxes

Preface

Environmental management (EM) is the management of the interaction of human activities on the environment and of environmental challenges and opportunities. This book explores the nature and role of EM, offering an introduction and focusing on challenges and opportunities, core concepts including sustainable development and principles and practice. Since the second edition in 2006 there has been much change and many businesses, organisations and governments now routinely use EM. In 2006 EM was part of a few teaching programmes, now full programmes are offered and there are departments of EM worldwide. There is now wide popular interest.

Part I

Introduction to environmental management

Chapter 1

Introduction

Chapter overview

- Aims and background
- Key terms and concepts
- Definition and scope of EM
- The evolution of EM
- Sustainable development (SD)
- EM problems and opportunities
- Summary
- Further reading
- www sources
- Professional bodies
- EM courses

Aims and background

EM seeks to steer development, take advantage of opportunities, help people avoid threats, mitigate problems, improve adaptability, reduce vulnerability and increase resilience. The Earth is no longer natural and humans must shift from exploiting nature to stewardship or civilisation will fail, so EM is vital. Challenges and opportunities for stewardship are interwoven: a problem can mean new opportunities and vice versa.

This book is divided into: Introduction chapters (1, 2, 3, 4) that cover: justification for EM, theory, scope, principles, concepts, goals, scientific background and social aspects; Practice chapters (5, 6, 7, 8, 9) that focus on: background, environmentalism, EM and economics, EM and business, EM and law, international agreements, EM approaches, standards, and monitoring; Challenges and Opportunities chapters (10, 11, 12) explore: resources, challenges and opportunities (natural and human-caused/aggravated); Responses chapters

DOI: 10.4324/9781003189985-2

(13, 14, 15) explore EM and mitigation, vulnerability, resilience, adaptation and how technology, social change, attitudes, fashions and politics affect responses to challenges and opportunities. Chapter 15 explores the way ahead.

Humans have increasingly caused changes which now threaten planetary stability. The risk is that:

> Within the lifespan of someone born today, our species is currently predicted to take our planet through a series of one-way doors that bring irreversible change... We, the world's better-off, live... comfortable lives in the shadow of a disaster of our own making.
>
> (Attenborough, 2020)

A new geological epoch reflects this impact: the Anthropocene (the starting date is not formally agreed but looks likely to be around 1950 AD) (Bonneuil and Fressoz, 2017; Davidson, 2019).

In 1970 few knew the term 'environment'. Acceptance that socio-economic development and environmental issues were both important spread a little between 1972 and 1992 and is today more widespread. EM demands co-ordination skills, ability to devise trade-offs, negotiation and diplomacy and foresight. Unpredictable natural disasters and human fickleness mean even the best prediction and careful observations will sometimes give little or no warning; EM must therefore be proactive and address human vulnerability and seek adaptable and resilient strategies. EM can bring together diverse stakeholders, specialists, levels of administration, sectors and nations that otherwise have little inclination to co-operate.

Many involved in EM have a specialisation, and focus on an issue, sector, country, region, environment or business. Some environmental managers work for a firm or institution but all generally profess a responsibility to a wider range of stakeholders, ultimately the global environment and world population. To some extent all people are environmental managers: individuals make choices which affect the quality of their surroundings, vulnerability and sustainability of their lifestyles.

Key terms and concepts

Throughout development from 1945 runs a Western, liberal democratic bias. This is usually short-term focused, places human needs before protection of the environment and usually seeks profit. As Figure 1.1 shows, environmental degradation by 2021 was nearing a point-of-no-return. The view among many is that there is limited time (a few decades) to set in motion development that will sustain indefinitely as many people as the Earth can give a satisfactory quality of life (which means human numbers well below present levels); en route it will be necessary to support too large a population and to cope with environmental damage, conflicts and disasters: a risky overshoot period.

Misleading data are easy to acquire; apparent causes of a problem may be symptoms, and faulty diagnosis can lead to costly mis-spending on solutions and perhaps irreversible damage (Fairhead and Leach, 1996). There should be careful questioning of received wisdom (Lomborg, 2001, 2004). Whenever possible multiple lines of evidence should be sought.

1954
World human population   2.7 billion
Carbon dioxide in atmosphere   310 ppm
Remaining wilderness   64%
1960
World human population   3.0 billion
Carbon dioxide in atmosphere   315 ppm
Remaining wilderness   62%
1968
World human population   **3.5 billion**
Carbon dioxide in atmosphere   **323 ppm**
Remaining wilderness   **59%**
1971
World human population   3.7 billion
Carbon dioxide in atmosphere   326 ppm
Remaining wilderness   58%
1978
World human population   4.3 billion
Carbon dioxide in atmosphere   335 ppm
Remaining wilderness   55%
1989
World human population   5.1 billion
Carbon dioxide in atmosphere   353 ppm
Remaining wilderness   49%
1997
World human population   5.9 billion
Carbon dioxide in atmosphere   360 ppm
Remaining wilderness   46%
2011
World human population   7.0 billion
Carbon dioxide in atmosphere   391 ppm
Remaining wilderness   39%
2020
World human population   7.8 billion
Carbon dioxide in atmosphere   415 ppm
Remaining wilderness   35%

Figure 1.1 Trends in less than one human lifetime (approximate values for three key statistics). Source of data: Attenborough, D. (2020) *A Life on Our Planet: my witness statement and vision for the future.* Grand Central Publishing, New York, NY, USA.

Definition and scope of EM

EM values prudence and stewardship pursued via a multidisciplinary, interdisciplinary or holistic approach. There is no single universal definition of EM; Box 1.1 offers a selection.

EM is usually biased towards the anthropocentric, i.e. environmental issues are considered after human development objectives have been set. There are other (non-mainstream) approaches; e.g. ecocentrism (giving the environment priority over human needs). In the coming decades the environment may have to take priority. In the future EM will have to work with nature and control greed and anthropocentric outlooks.

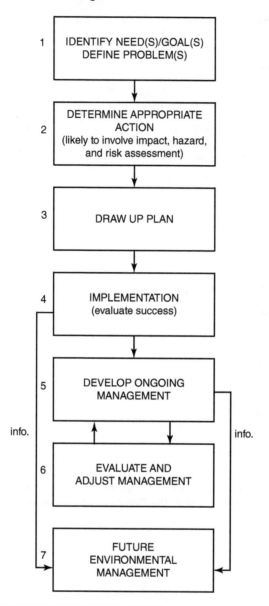

Figure 1.2 A typical scheme of practice adopted for EM.
Note: Increasingly, stages 1, 2 and 3 are influenced by broad strategic policies, and are accountable to public scrutiny (as is stage 5). Ideally, lessons learned at every stage should be passed on to improve future EM – the evaluation of stages 4 and 5 is especially helpful in future management. At stage 1 the public or a developer may not have a clear idea of needs or goals, so the environmental manager may need to help establish these.

> ### Box 1.1   Some definitions of environmental management
>
> - Formulation of environmentally sound development strategies.
> - The process of allocating natural and artificial resources so as to make optimum use of the environment in satisfying human needs at the minimum, and more if possible.
> - Seeking the best possible environmental option (BPEO), using the best available techniques not entailing excessive cost (BATNEEC).
>
> - Management of the environmental performance of organisations, bodies and companies.
> - EM – a generic description of a process undertaken by systems-oriented professionals with a natural science, social science, or engineering, law or design background, tackling problems of the human-altered environment on an interdisciplinary basis from a quantitative and/or futuristic viewpoint.

The evolution of EM

Some past societies sustained reasonable lifestyles for a few centuries before collapse. Small and scattered populations of non-sedentary and resilient people could move and adapt. Modern populations are huge, less mobile and far less adaptable, and so are more vulnerable in spite of technology.

In Western societies the belief gained hold that human welfare could be improved through work, the appliance of technology and moral development. Natural resources were to be used to these ends, and some believed that humans would conquer nature and control it. Technological optimism faltered by the 1960s as awareness of environmental problems grew. Between 1945 and the 1980s development effort was distorted or sidelined by concern for and spending on the Cold War. Development from the 1960s was seen to be concerned with the reduction of poverty; environmental concern was deemed irrelevant, a luxury that poor countries could not afford, or it was suspected to be part of a conspiracy by the rich to hold back the less-developed. It was not until 1987 that it started to be accepted that development needed effective EM. The shift to environmental concern was prompted by obvious pollution, global warming, ozone depletion, loss of biodiversity, soil degradation, deforestation, etc.

Natural resources management evolved before EM (in the 1960s) and deals more with specific components of the Earth: resources, which have utility – EM stresses stewardship rather than exploitation. Natural resources management responses to problems tend to be reactive, and to seek a quick-fix technological solution delivered with a project-by-project approach. Pre-1990s natural resources managers had limited sociological and environmental expertise, were usually authoritarian and did not involve the public; they also tended to miss off-site and delayed impacts.

Sustainable development (SD)

Short-term gain from resources often yielding environmental degradation is not a route to future security; as this has been realized SD has spread and is a concept that prepares the way for stewardship that co-evolves with the Earth. Before the 1980s few questioned 'business-as-usual' development. Environmental degradation sparked sporadic interest between the 1700s and 1940s. SD appeared around 1970 and seemed to offer a way to heed environmental limits *and* develop without slowing to zero growth (Meadows et al., 1972; Schumacher, 1973). *The Brundtland Report* helped establish the idea of SD (World Commission on Environment and Development, 1987); the *World Development Report 2003* also promoted SD (UN, 1992).

Sustainability and SD are not the same, but are often used as if they were. The former is the ongoing function of an ecosystem or use of a resource, and implies steady demands; the latter implies increasing demands for wellbeing and lifestyles and, in the foreseeable future, a growing population. Sustainability is the quantification of status and progress (environmental or social) and the goal of the SD process (Kopnina and Shoreman-Ouimet, 2015a). Even if SD is achieved in only a limited way, it may be valuable as a prompt and guide rail for development. There are parallels with seeking justice, liberty or truth; meanings and practice vary but efforts to reach them continue. A pessimistic view is that SD marks the end of the West's faith in progress, a post-industrial loss of confidence (Lovelock, 2006).

SD seeks to meet the needs of the present without compromising the ability of future generations to meet their own needs (World Commission on Environment and Development, 1987). Inter-generational equity (passing to future generations as much as the present enjoys) is added to intra-generational equity (sharing what there is between groups, including biota). Ecologists developed measures such as carrying capacity and maximum sustainable yield: the idea that an ecosystem can sustain a given level of demand. It should be noted that, even if demand is sustainable, unexpected environmental changes may upset things. Perhaps SD will be easier if new technology, altered tastes and substitution of resources enable increasing demand to be met without greater environmental impact.

Environmental economists often split SD into two (unsatisfactory) extremes:

- *Strong* – the existing stock of natural capital should be maintained or improved. Human capital (infrastructure, labour, knowledge) and natural capital (environmental assets or ecosystem services; biodiversity, natural systems, etc.) are complementary not interchangeable. The rejection of strategies such as substitution. The same amount of natural capital is passed on to future generations. Human misery is acceptable as a cost of reaching SD. SD must be based on natural capital that can be regenerated.
- *Weak* – the costs of attaining SD are weighed and unpleasant impacts are resisted, even if SD is delayed or endangered. Human capital can be substituted for natural capital: i.e. if need be it is permissible to trade natural capital through substitution (future generations receive the same total capital, but it may have changed form). What cannot yet be substituted is protected. This viewpoint concedes that existing economics and development strategies may be used.

Variants of the weak interpretation are currently mainstream, and few hold to strong SD. An example of weak SD: a state invests some profit from finite petroleum resources into (hopefully secure) long-term investments that fund things to maintain livelihoods and/or environment. SD may be pursued at local, regional, national and supranational levels, using 'top-down' or participatory approaches (Fiscus and Fath, 2018). A body, region, city or country might win a false SD at the expense of somewhere else. For example, a poor nation may dispose of waste sent from rich consumers.

SD seeks ongoing economic growth without exceeding global environmental limits: a conflict within the concept (Redclift, 1987). The question is: can the goals of SD be achieved in real-world situations? Also, it might be better to set sights lower and pursue *survivability* rather than SD during overshoot?

EM can support SD by:

- identifying key issues;
- clarifying threats, opportunities and limits;
- establishing feasible boundaries and strategies;
- monitoring to reduce the chance of surprises and supporting adaptability.

The goal of SD is to *stretch* what nature provides to the optimum using knowledge, effort and technology, and *maintain that expansion indefinitely* without environmental breakdown (in spite of unexpected social, economic or environmental changes or disasters). Thereby maximising human wellbeing, security and adaptability. This demands the ability to recognise and avoid, mitigate or adapt to socio-economic and physical threats. It is a step toward working with nature not against it. 'Business-as-usual' approaches to development will not suffice (Sachs, 2015) (see Box 1.2).

In 2001 the Millennium Ecosystem Assessment tried to stocktake the environment and in 2005 a set of Millennium Development Goals were adopted by the UN. In 2015 the UN General Assembly set 17 Sustainable Development Goals (SDGs), to be achieved by 2030 (UN, 2015; *UN 2030 Agenda on SD*: https://sustainabledevelopment.un.org/post2015/transformingourworld/publication – accessed 01/02/21). It looks *very* unlikely in 2023 that they will be fully achieved.

SD is not a costless strategy; it involves trade-off of short-term against long-term, and local/individual against other beneficiaries. SD can be misused or poorly practised and the trade-offs can be a cruel choice for individuals, groups or countries (Washington, 2015).

The question is whether SD can generate workable strategies to improve human wellbeing and prevent environmental degradation. It is possible to recognise four scenarios: 1) Over-exploitation of resources and environmental degradation (and possibly socio-economic instability and vulnerability) – *unsustainable mal-development*. 2) Usage of resources so environment is in an equilibrium state (hopefully socio-economic stability) – *steady-state development*. 3) Environmental and resource demands relaxing (perhaps a resilient and less vulnerable socio-economic system), environment and human conditions improve indefinitely – *SD*. 4) *Sustainable intensification* – somehow new skills and technology allow more crops, energy etc. with lessening degradation – *SD+*. Currently the world is in state 1) but might with effort reach 2) by 2072; state 3) is a challenge, and 4) may be a dream.

Box 1.2   Some definitions of sustainable development

- Improving the quality of human life while living within the carrying capacity of supporting ecosystems.
- Development based on the principle of inter-generational, inter-species and inter-group equity.
- Development that meets the needs of the present without compromising the ability of future generations to meet their own needs.
- A change in consumption patterns towards more benign products, and a shift in investment patterns towards augmenting environmental capital.

It is crucial to be able to measure potential for SD, progress achieved and its security. Indicators can help police and analyse situations (Bell and Morse, 2020). Composite indicators may be more use than single-dimension indicators. The following are some of the indicators so far applied to SD:

- The environmental sustainability index.
- The index of sustainable economic welfare – a socio-political measure.
- Net primary productivity – derived from the ecological concept of carrying capacity, which is the maximum population of a given species that an area can support without reducing its ability to support the same species indefinitely.
- An extension of the human development index (HDI) – the HDI was proposed by the UN Development Programme in 1990 and has become a widely used multidimensional measure of development. It has been considerably modified and now makes some (often criticised) provision for assessing sustainability.
- Composite index of intensity of environmental exploitation.
- Eco-footprint (or ecological footprint) – measures how much land is required to supply a particular city, region, organisation, country, island, sector, service (e.g. a transport system) company, group, supply of a commodity, activity or individual with all needs. It is a measure of load imposed on the environment to sustain consumption and dispose of waste. It does this by calculating how much biologically productive land and water (as area or area per capita) is needed with current demand and technology to provide inputs and dispose of outputs (wastes) safely. It is a measure of human aggregate ecological demand (inputs and outputs) and it can only be temporarily exceeded or the productive and assimilative capacity of the biosphere is weakened. The average North American had an eco-footprint in 1995 c. 4.5 times greater than in 1905. Eco-footprint is a useful ecological accounting method for assessing the demands made by humans. It can aid evaluations of what could be done to improve sustainability, and provide a framework for SD planning. However, it has been suggested that it underestimates human impacts and shows only the minimum needs for SD, without a margin for error, which the precautionary principle demands. It allows comparison of the impact of different components on the same aggregated scale and aids assessment of eco-efficiency (i.e. whether an organisation is making the most of its resources and waste disposal opportunities). There are other faults, including its two-dimensional interpretation of complex ecological and economic systems and failure to recognise issues such as consumer preference or property rights, and it offers a temporally limited snapshot view; natural systems are seldom stable and consumption per capita, fashions and settlement patterns change. It does not help to find the most appropriate path for human activities. It offers a graphic and easily communicated image; it is a relatively transparent tool; and it may stimulate creative thinking about environmental and developmental issues.

EM problems and opportunities

Some dismiss EM as 'environmental managerialism' or claim it has become institutionalised. Nevertheless, EM is widely used to address real-world problems and it is a step in evolving a way of civilisation surviving into the Anthropocene. EM is changing and is far from fixed in form. The focus tends to be more on how humans affect the environment, less on how environment affects humans – both are needed.

Many assume that living standards and technological progress will continue and even improve without upheaval. This is unwise; few nations have had 200 years without famine, less without epidemic diseases or warfare. There is only a thin veneer of technology and governance offering limited protection from disaster but it can be built upon.

It is highly unlikely that all constraints and challenges will be assessed in advance, so resilience and adaptability are crucial but have had little attention. Awareness of the past helps in forecasting but the future will still present new challenges and opportunities (Halvorsen et al., 2019). Some assume the worst: cassandras (pessimists). Others are over-optimistic ('cornucopians'); the majority in rich and poor countries do not think about threats and are disinterested or apathetic. Many adopt unhelpful perceptions based on fashion, ignorance, intolerant faith or false news. EM should offer carefully weighed warnings in a persuasive manner and also solutions. Should a problem flagged by EM not materialise or if it develops in unexpected ways there will be accusations of 'crying wolf', and future warnings might be ignored. There is no way for EM to avoid risk taking, but reliance on sound data from more than one source, careful checking and seeking win–win solutions helps. Win–win solutions aim for beneficial results, even if the problem fails to develop as expected. Unfortunately, decisions sometimes have to be rapid and based on inadequate information (crisis management).

EM may need to modify the activities and attitudes of individuals, groups and societies to achieve its goals. There are approaches which can be adopted:

(1) *Advisory*:
   - through education;
   - through demonstration (e.g. model farms or factories);
   - through the media (overt informative publications and advertisements or covert approaches – the latter includes 'messages' incorporated in entertainment);
   - through advice (e.g. leaflets, drop-in shops, etc.).
(2) *Economic or fiscal*:
   - through taxation;
   - through grants, loans, aid;
   - through subsidies;
   - through quotas or trade agreements.
(3) *Regulatory*:
   - through standards and laws;
   - through restrictions and monitoring;
   - through licensing;
   - through zoning (restricting activities to a given area).

Attempts to address a problem may themselves present challenges:

(1) Ethical dilemmas – e.g. what to conserve: indigenous hunters or game?
(2) Efficiency dilemmas – e.g. how much damage is acceptable?
(3) Equity dilemmas – e.g. who benefits from EM and who pays?
(4) Liberty dilemmas – e.g. to what degree must people be restricted to protect the environment?
(5) Uncertainty dilemmas – e.g. how to choose a course of action without adequate knowledge or data.
(6) Evaluation dilemmas – e.g. how to compare effects of options or actions.

Hasty ad hoc responses are not good, better to try to spot problems early. Politicians, NGOs, movements, lobby groups and individuals may get away with advocacy, but EM has to produce the goods. Even if the environmental manager is objective, powerful special-interest groups such as the rich, officials, lobby groups, NGOs, industry, the military, etc., may not be and may overrule advice. Enforcement is also a problem. Where EM has only advisory powers it is easy side-step it. Sovereignty, political, cultural or strategic needs threaten common-sense decisions and public attitudes shift so EM must be flexible and perceptive and if need be global.

For most of human history worries have been concerned with the acquisition of inputs: food, water, fuel. Additional worries, outputs, now have to be addressed: pollution and waste. EM faces unexpected and rapid changes and situations which develop so slowly and insidiously that long-term, inter-generational approaches are required to identify and address them. Change may be driven by factors 'over the horizon' making them difficult to address. Development may be decided before EM is involved, making meaningful steward-ship difficult. Science adopts a reductionist approach with disciplinary specialists studying components of a problem, but now wide multidisciplinary approaches are often required.

Given that EM has appeared only in the past 40 years or so, there has been progress. Tools and methodology are evolving, but environmental and social knowledge is often inadequate.

EM frequently faces:

- a poorly researched threat;
- transboundary or global challenges;
- problems demanding rapid decisions;
- acquisition of information from diverse sources.

With something as ambitious as EM, criticism is inevitable. One frequently voiced is that it is prescriptive and insufficiently analytical, and so is not reliable science. Or enforcement efforts can be branded eco-fascism. Critics flag environmental managerialism, symptoms of which include: the consideration of the environment *after* development objectives have been set; becoming lost in techniques regardless of whether they are needed; applying 'one size fits all' protocols, etc. EM can be poorly pursued as a reactive, piecemeal approach but hopefully it will be better practiced.

Encouraging EM

EM frequently has to encourage environmentally desirable practices. This demands skills of presentation and persuasion, if not cunning; there is a need for forming alliances, negotiation, networking and perhaps manipulation of others. EM can be supported by suitable standards and systems. Treaties, agreements, protocols and conferences play a role in EM. Support is also provided by the internet, which facilitates exchange of information, debate, lobbying, protest and watchfulness for problems. EM guidelines began to be issued by businesses, states and agencies in the 1980s. Today many insist on their application before supporting developments.

Summary

- EM is evolving and spreading. It has to be adapted to suit all conditions.
- EM demands a proactive approach and must integrate closely with other fields.
- Without EM, development is unlikely to be environmentally wise, adaptable and sustainable and people will be more vulnerable.

- EM must promote adaptability, resilience and awareness of vulnerabilities and focus on more than global warming.
- EM should ensure opportunities are grasped and challenges are dealt with.
- The Anthropocene Epoch has arrived and humans must change from exploiting nature to working with nature.

Further reading

EM books

Antweiler, W. (2014) *Elements of Environmental Management*. University of Toronto Press, Toronto, ON, CA.

Belcham, A. (2014) *Manual of Environmental Management*. Routledge, London, UK.

Bradey, J., Ebbage, A. and Lunn, R. (Eds) (2011) *Environmental Management in Organizations: The IEMA handbook* 2nd edn. Routledge (Earthscan), London, UK.

Broniewicz, E. (Ed) (2011) *Environmental Management in Practice*. InTech, Rijeka, HR.

Friedman, F.B. (2011) *Practical Guide to Environmental Management* 11th edn. Environmental Law Institute, Washington, DC, USA.

Gregory, R., Failing, L., Hartstone, M., Long, G., McDaniels, T. and Ohlson, D. (2012) *Structured Decision Making: a practical guide to environmental management choices*. Wiley-Blackwell, Oxford, UK.

Hyde, P. and Reeve, P. (2011) *Essentials of Environmental Management* 2nd edn. Routledge, London, UK.

Jørgensen, S.E. (Ed) (2015) *Encyclopedia of Environmental Management*. CRC Press, Boca Raton, FL, USA.

Krishnamoorthy, B. (2017) *Environmental Management: text and cases* 2nd edn. PHI, New Delhi, IN.

Lovett, J.C. and Ockwell, D.G. (Eds) (2010) *A Handbook of Environmental Management*. Edward Elgar, Cheltenham, UK.

Mitchell, B. (Ed) (2015) *Resource and Environmental Management in Canada*. Oxford University Press, Don Mills, CA, USA.

Modak, P. (2018) *Environmental Management: toward sustainability*. CRC Press, Boca Raton, FL, USA.

Murali Krishna, I.V. and Manickam, V. (2017) *Environmental Management: science and engineering for industry*. Butterworth-Heinemann, Oxford, UK.

Skinner, G., Crafer, K. and Turner, M. (2017) *Cambridge IGCSE and O-Level Environmental Management Coursebook*. Cambridge University Press, Cambridge, UK.

Theodore, M.K. and Theodore, L. (Eds) (2009) *Introduction to Environmental Management*. CRC Press, Boca Raton, FL, USA.

Thomas, I.G. (2011) *Environmental Management*. The Federation Press, Alexandria, NSW, AU.

Uberoi, N.K. (2012) *Environmental Management*. Excel Books, New Delhi, IN.

Watters, B. (2013) *Introduction to Environmental Management: for the NEBOSH Certificate in Environmental Management*. Routledge, London, UK.

SD books

Baker, S. (2015) *The Politics of Sustainable Development* 2nd edn. Routledge, London, UK.

Blewitt, J. (2017) *Understanding Sustainable Development* 3rd edn. Routledge, London, UK.

Dalby, S., Horton, S., Mahon, R. and Thomas, D. (Eds) (2020) *Achieving the Sustainable Development Goals: global governance challenges.* Routledge, London, UK.

Elliott, J.A. (2012) *An Introduction to Sustainable Development* 4th edn. Routledge, London, UK.

Kopnina, H. and Shoreman-Ouinet, E. (2015) *Sustainability: key issues.* Routledge (Earthscan), Abingdon, UK.

Robertson, M. (2021) *Sustainability Principles and Practice* 3rd edn. Routledge, London, UK.

Rooda, N. (2020) *Fundamentals of Sustainable Development* 3rd edn. Earthscan, London, UK.

Rubic, F. and von Raggamby, A. (Eds) (2012) *Sustainable Development Evaluation and Policy Making: theory, practise and quality assurance.* Edward Elgar, Cheltenham, UK.

Sachs, J.D. (2015) *The Age of Sustainable Development.* Columbia University Press, New York, NY, USA.

EM journals

- *Journal of Environmental Management* (Elsevier) – www.journals.elsevier.com/journal-of-environmental-management (accessed 12/01/21).
- *Environmental Management Journal* (Springer) – www.springer.com/journal/267 (accessed 12/01/21).
- *Journal of Environmental Management* (Academic Press) – www.scimagojr.com/journalsearch.php?q=23371&tip=sid (accessed 12/01/21).
- *Management of Environmental Quality* (Emerald) – www.emerald.com/insight/publication/issn/1477-7835 (accessed 12/01/21).
- *Australian Journal of Environmental Management* (Taylor & Francis) – www.tandfonline.com/toc/tjem19/current (accessed 12/01/21).
- *Australasian Environmental Management* (Taylor & Francis) – www.tandfonline.com/toc/tjem20/current (accessed 12/01/21).
- *Journal of Environmental Planning and Management* (Taylor & Francis) – www.tandfonline.com/toc/cjep20/current (accessed 12/01/21).
- *Journal of Environmental Economics and Management* (Elsevier) – www.journals.elsevier.com/journal-of-environmental-economics-and-management (accessed 12/01/21).
- *Journal of Environmental Assessment Policy and Management* (World Scientific Publishing) – www.worldscientific.com/worldscinet/jeapm (accessed 12/02/21).
- *Current Environmental Management* (Bentham Science) – www.eurekaselect.com/ (accessed 12/01/21).
- *Corporate Social Responsibility and Environmental Management* (Wiley) – https://onlinelibrary.wiley.com/journal/15353966 (accessed 12/01/21).
- *Global Journal of Environmental Science and Management* (GJSEM) – www.gjesm.net/ (accessed 12/01/21).
- *Journal of Environmental Accounting and Management* (L & H Scientific Publishing) – www.lhscientificpublishing.com/Default.aspx (accessed 12/01/21).
- *Public Policy & Environmental Management* (Emerald) – www.emeraldgrouppublishing.com/archived/products/collections/ppem.htm (accessed 12/01/21).

SD journals

- *Challenges in Sustainability* – www.librelloph.com/challengesinsustainability (accessed 11/10/22).

- *International Journal of Sustainable Environmental Management* (Inderscience) – www.inderscience.com/jhome.php?jcode=ijesd (accessed 12/01/21).
- *Environmental Management and Sustainable Development* (Macrothink) – www.macrothink.org/journal/index.php/emsd (accessed 12/01/21).
- *Journal of Cleaner Production* (Elsevier) – www.journals.elsevier.com/journal-of-cleaner-production (accessed 12/01/21).
- *Sustainable Development* (Wiley) – https://onlinelibrary.wiley.com/journal/10991719 (accessed 12/01/21).
- *International Journal of Sustainable Development & World Ecology* (Taylor & Francis) – www.tandfonline.com/toc/tsdw20/current (accessed 12/01/21).
- *International Journal* of *Development* and *Sustainability* (ISDS) – www.isdsnet.com/ijds.html (accessed 12/01/21).
- *Journal of Education for Sustainable Development* (Sage) – https://us.sagepub.com/en-us/nam/journal/journal-education-sustainable-development (accessed 29/11/23).
- *International Journal of Sustainable Development* (Inderscience) – www.inderscience.com/jhome.php?jcode=ijsd (accessed 12/01/21).
- *European Journal of Sustainable Development* (ECSDEV) – https://ecsdev.org/ojs/index.php/ejsd/index (accessed 12/01/21).
- *Sustainability* (MDPI) – www.mdpi.com/journal/sustainability (accessed 12/01/21).
- *Nature Sustainability* (Springer Nature) – www.nature.com/natsustain/ (accessed 12/01/21).
- *International Journal of Sustainable Development & Policy* (Ideas) – https://ideas.repec.org/s/pkp/ijsdwp.html (accessed 12/01/21).
- *Indian Journal of Sustainable Development* (Publishing India Group) – www.i-scholar.in/index.php/IJSD (accessed 12/01/21).
- *Agronomy for Sustainable Development* (Springer) – www.springer.com/journal/13593?gclid=Cj0KCQiArvX_BRCyARIsAKsnTxOz3Wk4C48D8u74eL_yhPrGUEVueq3QfQrJWYOYfc7DWxwb9i9o50oaAgIuEALw_wcB (accessed 12/01/21).
- *International Journal of Sustainable Economy* (Inderscience) – www.inderscience.com/jhome.php?jcode=ijse (accessed 12/01/21).

www sources

Note: these sources may change or disappear, and are less reliable than peer-reviewed journals.

- UK Centre for Ecology & Hydrology, Environmental Information Platform – https://eip.ceh.ac.uk/ (accessed 13/01/21).
- European Environmental Agency – www.eea.europa.eu/ (accessed 12/01/21).
- International Institute for Environment and Development – www.iied.org (accessed 12/01/21).
- International Union for Conservation of Nature – www.iucn.org (accessed 12/01/21).
- Social Science Information Gateway: social science and EM sources – www.ariadne.ac.uk/issue/2/sosig/ (accessed 12/01/21).
- WWF International, Knowledge Hub – https://wwf.panda.org/discover/knowledge_hub/ (accessed 12/01/21).
- UN Environment Programme (UNEP) – www.unep.org/ (accessed 12/01/21).

- ❏ International Institute for Sustainable Development – www.iisd.org/ (accessed 14/01/21).
- ❏ UN Department of Economic and Social Affairs, Division for Sustainable Development – www.u.org/esa/sustdev (accessed 14/01/21).
- ❏ UN Sustainable Development Knowledge Platform – https://sdgs.un.org/ (accessed 14/01/21).
- ❏ Sustainable Development Communication Network – http://sdgateway.net (accessed 14/01/21).

Professional bodies

- International Network for Environmental Management (INEM) – www.inem.org/ (accessed 12/01/21).
- Chartered Institute of Ecology and Environmental Management, UK (CIEEM) – https://cieem.net/ (accessed 12/01/21).
- Institution of Environmental Sciences (UK) – www.the-ies.org/ (accessed 12/02/23).
- Institute of Environmental Management and Assessment (IEMA) (UK) – www.iema.net/ (accessed 12/01/21).
- Chartered Institution of Water and Environmental Management (UK) – www.ciwem.org/ (accessed 12/01/21).
- Environmental management systems (EMS) (USA) – www.fedcenter.gov/programs/EMS/index.cfm?&printable=1 (accessed 12/01/21).
- Environmental professional associations, Yale School of the Environment (US environmental bodies) – https://environment.yale.edu/careers/resources/env-professional-associations/ (accessed 13/01/21).
- International Association of Impact Assessment (USA) – www.iaia.org/ (accessed 12/01/21).

EM courses

There are many EM school, undergraduate, postgraduate and short courses. Professional bodies also offer courses – some 2023 examples:

School exams:

- IGCSE Cambridge Board (UK) O-Level (with international uptake);
- IBO – International Baccalaureate – has EM courses/exams.

Other courses:

- NEBOSH Certificate in EM (UK);
- National Examination Board in Occupational Safety and Health (UK-based independent examination board);
- Open University (UK) short EM courses;
- IEMA Certificate (UK).

Chapter 2

Environmental management
Character and goals

Chapter overview

- Character and goals of EM
- Concept of limits to development
- Polluter-pays principle
- Precautionary principle
- EM challenges
- Need to be adaptable and resilient and to seek to reduce human vulnerability
- EM ethics and institutions
- Summary
- Further reading
- www sources

Character and goals of EM

EM promotes stewardship of the environment on behalf of citizens largely by experts ('citizens' includes future generations). Management is difficult to define and is often poorly conducted: it is a dynamic process which can include reduction of uncertainty, leadership, governance and encouraging motivation. Before 2000 EM was often linked to waste and pollution but is now wider. EM usually blends 'top-down' and technocratic ('trust me, I am a professional') approaches with public accountability and consultation.

Natural environmental changes have affected humans: history, archaeology and palaeoecology show this. Hopefully, EM can guide humans to use, repair and maintain a global environment that enables their long-term wellbeing and will adequately protect nature (Carey, 2015). A material, consumerist outlook is currently mainstream and has been likened to 'the ideology of the cancer cell'. Attenborough (2020) stressed *all* nations are developing countries: the richer have to strive to sustain living standards and reduce environmental impacts; the poorer struggle to improve standards and can often do little to control environmental damage.

DOI: 10.4324/9781003189985-3

The acceptance that human fortunes could be improved through material rather than religious work owes much to the appearance of scientific enquiry and rationalism after 1700. Benchmarks used to judge success of development today pay attention to economic (gross domestic product – GDP) or engineering criteria, and give little weight to environmental, cultural or social issues. Development is conducted for profit or against the clock to achieve already decided goals before a government runs out of its term of office or because there is a sense of haste. Often there is a desire to win public support/votes, not do what is wise for the longer term. Hindsight experience is often not considered because it is politically or commercially embarrassing and information tends to be restricted to limited-circulation reports; post-development appraisals are poor because there is scarce funding, and those involved may not want to highlight shortcomings. Consequently mistakes are repeated.

Development management evolved before EM as the manipulation of interventions and had little environmental concern before roughly 1990. Adopting a theatrical analogy, it has largely been as if only the actors existed and the stage, play and audience attracted little attention. *Laissez-faire* strategies with limited intervention have been common but are unwise in a crowded and fast-changing world. Today fewer question the importance of caring for the environment, yet governments commonly fail to give it priority or spend on it. People resist changing environmentally damaging lifestyles through attitude and/or because poverty renders them unable. Many citizens and bodies follow slow or no change ways ('business-as-usual') (Faure, 2018).

It is widely claimed that a turning point has been reached, and there is limited time to avert disaster (Davidson, 2019). The long-term goal of EM is to sustain a population the Earth can support with a satisfactory quality of life and civilisation. En route it will be necessary to 'overshoot' and support too large a population until numbers fall while coping with environmental damage and change, loss of biodiversity, conflicts and disasters. Overshoot began around 1987 when the human population passed what the Earth could support and continue to regenerate. By 2020 humans used over 1.7 times what the planet can sustain; some say three Earths are needed to meet current demands, some say 11. Strangely there is little concern about population at present.

There are various reactions to the idea a crisis is approaching: (1) ignore the threat; (2) promote abandonment of technology and a return to simple ways; (3) use all tools available, including technology, economic and social change to support civilisations' survival. The first is foolhardy; the second would mean disaster for most of the world's population; there seems to be no choice other than to pursue (3) and use EM to steer it. The idea that the world faces a crisis often provokes emotive debate and fire-fighting solutions that divert attention from important tasks. A crisis attitude may make things worse. If causes and treatments are not carefully researched, little will be achieved (Agyeman, 2013; Sachs, 2015).

EM goals include:

- sustaining and, if possible, improving existing resources;
- preventing and resolving environmental problems;
- reducing human vulnerability and improving resilience and adaptability;
- establishing boundaries/limits and monitoring them;
- founding and nurturing institutions that support environmental research, monitoring and management;
- warning of threats and identifying opportunities;
- where possible improving quality of life;
- identifying new technology or policies that are useful.

EM needs a focus but must come with challenges stretching from local and short-term to global and long-term; must avoid fragmented decision making; and must prioritise urgent tasks. Effective EM should scope (decide goals and set limits on efforts) before starting to act; however, this is often neglected. It has been suggested that at its core EM asks two questions: 1) What kind of planet do we want? 2) What kind of planet can we get?

The approach to EM may take different paths (Box 2.1). Reliable indicators and effective monitoring and forecasting techniques are vital but EM relies on more than measurement.

Box 2.1 Approaches to environmental management

There may be some overlap between groupings and within categories. Environmental managers are usually anthropocentric but could be ecocentric, more or less 'green', more or less supportive of technology. There is a wide spectrum of political and philosophical stances:

1 *Ad hoc approach*: approach developed in reaction to a specific situation.
2 *Problem-solving approach*: a series of logical steps to identify problems and needs and to implement solutions, e.g. China's Grain-for-Green Program.
3 *Systems approach*:
    - ecosystem (mountain; savannah; island; lake, etc.);[†]
    - agroecosystem;
    - socio-economic system;
    - global systems (atmospheric CO_2, etc.).
4 *Regional approach*: mainly ecological zones or biogeophysical units, which may be international (e.g. an internationally shared river basin). For example:
    - watershed;[†]
    - river basin;[†]
    - coastal zone;[†]
    - island;
    - command area development authority (irrigation-related);

    - administrative region;
    - sea (e.g. Mediterranean; Baltic; Aral Sea, etc.).[†]
5 *Specialist discipline approach*: often adopted by professionals. For example:
    - air quality management;
    - water quality management;
    - land management;
    - environmental health;
    - urban management;
    - ocean management;
    - human ecology approach;
    - tourism management/ ecotourism;
    - conservation area management.
6 *Strategic EM approach*.
7 *Voluntary sector approach*: EM by, or encouraged and supported by, NGOs. For example:
    - debt-for-nature swaps;
    - private reserves;
    - 'ginger groups' which try to prompt EM;
    - private funding for research or EM.
8 *Commercial approach*: EM for business/public bodies.
9 *Political economy or political ecology approach*.
10 *Human ecology approach*.

Note: [†] = biogeophysical systems

Discrete and straightforward problems are more likely to attract public support than slow-onset, often complex and insidious ones. It also helps if EM can point to benefits from its actions and not just flag threats and aims.

Concept of limits to development

The environment before 1800 was seen as something to be tamed and unspoilt lands were 'wastelands'. Romantics and proto-environmentalists bemoaned the rape of nature by industry, deforestation and hunting; the English novelist Mary Shelley warned humans could become extinct through careless science and greed (see *The Last Man* and *Frankenstein*).

In the eighteenth century in the UK Thomas Malthus offered the thesis that human population growth pressures the means of subsistence, throwing it out of balance with the environment, resulting in famine. Malthusianism and eugenics (selective breeding and perhaps culling, a subject developed by Francis Galton and others) have influenced Western policies for two centuries often in unpleasant ways. During the 1930s the frontiers were closing, forests were being cut, ocean fisheries were stressed, unsettled areas were becoming difficult to find, and Midwestern USA, eastern Amazonia and other parts of the world were suffering soil degradation. In the 1940s came a few warnings of resource and population problems (Osborn, 1948; Vogt, 1948). In the 1970s, ecologists, systems analysts, neo-eugenicists and environmentalists (neo-Malthusians) argued that, for a given species and situation, population may grow until it encounters a critical resource limit or controlling factor, whereupon there is a gradual or sudden, limited or catastrophic decline in numbers, or a shift to a cyclic boom-and-bust pattern (Ehrlich et al., 1970). Neo-Malthusians also focused on the threat from careless technology (Farvar and Milton, 1972). Hardin (1968) published a 'tragedy of the commons' essay (and unpopular work on lifeboat theory) suggesting commonly owned natural resources under *laissez-faire* conditions would be damaged because each user would seek to maximise their short-term interests (Ostrom, 2015).

In 1972 the Club of Rome (a group concerned about the predicament of humanity) reported on a systems dynamics computer world model they had developed (Meadows et al., 1972, 2004). This tried to determine future scenarios, using forecasts of industrialisation; population growth; rates of malnutrition; depletion of non-renewable resources; and a deteriorating environment. The aim was to understand the world system, and warn of possible outcomes. *The Limits to Growth* concluded: 'If present growth trends... continue unchanged, the limits to growth on this planet will be reached within the next hundred years'.

Some environmentalists called for reduced or even 'zero growth'. Around 1970 a much more palatable alternative was proposed: SD. This offered a way for growth to avoid conflict with limits, using technology, altering people's demands, finding resource substitutes. Overshoot could then be survived (perhaps for decades) until SD was achieved. In a sequel to *The Limits to Growth* in 1992 the same authors had refined their world model and fed in much-improved data: *Beyond the Limits* (Meadows et al., 1992) argued that the 1972 warnings were broadly correct, some limits had by 1992 been exceeded, and that, if trends continued, there was virtual certainty of a global collapse within the lifetime of children alive in 1992. However, it looked possible to survive overshoot and avoid collapse, provided environmental impacts were cut and there was an increase in efficiency of materials and energy use. This threw down a challenge and indicated a timescale for action.

Simon (1981) argued substitution and technology would overcome resource scarcity given a free market. Boserüp (1965) reported how, provided it is not too fast and overwhelms the adaptive ability of people, population increase may stimulate social and technological changes leading to less environmental damage, improved quality of life and decline in births. Tiffen et al. (1994) researched situations where population growth led to innovation that improved quality of life and reduced environmental degradation. However, such examples are not very widespread and in general a high population means stress on resources and environment.

There was little general awareness of the vulnerability and limitations of the Earth's eco-systems until 1965 when Adlai Stevenson popularised the catchphrase 'Spaceship Earth'. It portrayed the world as a fragile, closed environment in which 'first-class passengers' (the greedy rich) and the more numerous 'lower-class passengers' (the poor), were stressing Earth's life-support capacity. Neither first- nor lower-class passengers were in control of the 'Ship' (the mechanics were not understood) and both were vulnerable to each other and increasingly to natural hazards: a finite interdependent world.

To what level should human population fall? More people can be supported at a lower standard. Also, technology change or disaster could invalidate an estimated number. Today (2023) the world population is around 7.8 billion; in 1800 it was about 1 billion. In 2017 the UN projected a 2100 population of 10.9 billion with a growth rate of +0.1 (medium prediction). The peak growth rate was in 1968 (+2.1) and in 2020 it had fallen to about +1.1 and slowing. Recent estimates by the Club of Rome suggest 8.5 billion in 2040 and a fall to 6 billion by 2100. Whatever the overall population, some countries have young growing

Figure 2.1 'Spaceship Earth': a finite, vulnerable system. Source: Photo taken by Apollo 17 astronauts 07/12/1972.

populations and some older and declining number. Some fear population decline and aging more than overpopulation (Angus and Butler, 2011; Webb, 2020). Reduction may happen naturally or it may have to be encouraged. Overshoot of population above a desired number will probably be an environmentally unstable period.

Attempts to establish what population Earth could sustain indefinitely with an adequate lifestyle vary, with 3 billion a possible figure. The effects of crowding might be problematic even at a viable environmental carrying capacity. The lower the target population, the more resilient and less vulnerable they become (provided numbers do not fall too low), and the better lifestyle might be (in theory). There is a need to calculate an ideal and desirable global population human population for security of lifestyle and for civilisation to survive the Anthropocene. How nations can agree what their component of the overall global population should be is yet to be seen. Studies suggest sustainable numbers need to be achieved within 100 years hopefully, through contraception, education (especially), late marriage, fashion, taxation, etc.

Demographic transition will hopefully happen. Japan followed such a path from the 1950s to early 1970s and by 2023 had zero growth. Italy has had decline since 2010, as has Eastern Europe. China pursued a one-child policy from 1970 to 2013. In nations with decline there might be labour shortages, pension and tax problems; automation or perhaps migrants may take on tasks. In 2011 the richest 1 billion consumed per head 32 times the average person. Numbers have to be linked to consumption.

Demographers suggest a peak human population in 2050 of 10 billion then decline (Webb, 2020). The Green Revolution(s) have so far fed people in a non-sustainable way perhaps delaying problems: land degradation, water shortages, unrest and climate change could disrupt improved agriculture.

The planetary boundaries model was proposed in 2009 by Johan Rockström from the Stockholm Resilience Centre and Will Steffen from the Australian National University (Rockström et al., 2009a; https://sustainabilityguide.eu/sustainability/planetary-boundaries/ – accessed 20/03/21). This identified **nine planetary boundaries (thresholds)**:

1) Climate change
2) Change in biosphere integrity (biodiversity loss and species extinction)
3) Stratospheric ozone depletion
4) Ocean acidification
5) Biogeochemical flows (phosphorus and nitrogen cycles)
6) Land-system change (e.g. deforestation)
7) Freshwater use
8) Atmospheric aerosol loading (microscopic particles in the atmosphere that affect climate and living organisms)
9) Introduction of novel entities.

If a boundary/threshold is crossed, abrupt, perhaps irreversible environmental changes are likely, with serious consequences for humankind. Researchers report that four were effectively crossed by 2015: 2), 6), 1) and 5).

Polluter-pays principle

The polluter-pays principle is a shift in attitude away from develop now, and if there is a problem evade blame, seek abatement and cleanup. A shift from the burden being borne by those affected to the public in general, or better still to making the polluter pay (the developer). If forced to pay, the developer is, hopefully, less likely to cause problems. Bystanders, consumers or workers should not pay for developers' mistakes. Some problems once they happen are not curable. Penalties for pollution are often hard to enforce and relatively light; consequently, individuals and organisations motivated by profit are tempted to risk sometimes getting caught. Ideally EM educates and motivates potential polluters (and other mal-developers) to genuinely avoid polluting.

Pollution may be so huge that a company or even a state struggle to resolve or pay for them (e.g. Chernobyl or the BP *Deepwater Horizon* accidents). Sometimes damage is done, and it only becomes evident or accepted years after; in the meantime, the body responsible has closed down or it is too late to use the law to claim damages. Not all support the principle, fearing it hinders pollution managers' decision making (Faure, 2018).

Precautionary principle

The precautionary principle has been described as 'institutionalised caution', and is constructed around the goal of preventing, rather than reacting to problems. Critics see it as bureaucrats covering their backs and fear it slows or discourages action. Problems include: precautionary (proactive) planning can be side-stepped; it is applied too late to select the best development option; it is misused or neglected; there is no way to forecast impacts. The precautionary principle has four core components:

(1) taking preventive action in the face of uncertainty;
(2) shifting the burden of proof to proponents of a development;
(3) exploring a wide range of alternatives to try and avoid unwanted impacts;
(4) increasing public participation in decision making.

Acceptance of the precautionary principle means that regulatory action can precede full scientific certainty about an issue: lack of evidence is no reason for inaction which has scientific and legal implications (Harremoës et al., 2002). It risks costs which may not be justified, inappropriate responses, etc. Because there is no agreement on a definition its status is one of a broad approach, rather than a firm and precise principle of law. There have been appraisals of the precautionary principle; a few suggest it be discarded for all but general policies, although most accept that it is valuable when serious, possibly irreversible, impacts are likely. There is also the question of how much a society can afford to pay to support a precautionary principle approach.

Some key tools support the principle; for example: environmental impact assessment (EIA) and social impact assessment (SIA). Impact assessment effectively forces developers to look before they leap and, if problems are anticipated, to delay acting until there can be effective avoidance or mitigation.

EM challenges

Whatever the approach, EM involves a myriad of individual and collective decisions by persons, groups, and organisations, and together these decisions and interactions constitute a process; a process that can hopefully be managed. Of the challenges that beset EM, inadequate data is common. The ideal is data that allows a scenario to be observed as it changes.

With improved computers, software and tools such as geographical information systems (GIS), this is often possible, but sometimes all that is available is a 'snapshot' view (i.e. limited in time and space). Predictions can be difficult with stable environments, but many are unstable and some processes are becoming uncertain with global change; once stability has been upset there may be unexpected and sudden feedbacks or shifts to different states. The behaviour of economic systems is equally challenging to predict, and human behaviour is fickle, with politics, tastes and attitudes often suddenly altering.

Need to be adaptable and resilient and to seek to reduce human vulnerability

Since roughly 11,700 BP humans have enjoyed relatively benign conditions, apart from a few challenges like the Little Ice Age *c.* AD 1500 to 1750. Things now look likely to become more unstable.

Some threats are random and difficult, if not impossible, to recognise in advance; others develop in an insidious way. Worse, a problem may have indirect and cumulative causes, unrelated factors suddenly conspire to cause trouble, or a process develops a positive or negative feedback which (respectively) quickly accelerates or slows down change. Not all problems will be seen or mitigated, so EM must reduce vulnerability and promote adaptability and resilience.

EM ethics and institutions

For effective EM there must be the means of ensuring proper conduct, this is assisted by laws, ethical codes and EM systems. Ethics are a system of cultural values motivating people's behaviour. They draw upon human reasoning, morals, knowledge of nature and goals to act as a sort of plumb-line for development and shaping a worldview. Ethics operates at the level of individuals, institutions, societies and internationally. From the sixteenth century in Western countries the Protestant ethic is said to have encouraged individuals to be responsible for self-improvement through good acts and hard work (Weber, 1958). Generally there were *laissez-faire* controls until the 1930s when the Soviet Union, fascist regimes, and even the USA (with developments like the Tennessee Valley Authority in 1933) explored state manipulation of development.

Environmental ethics can be divided into four groups:

(1) *Technocratic* environmental ethics = resource-exploitative, growth-oriented
(2) *Managerial* environmental ethics = resource-conservationist, oriented to sustainable growth
(3) *Communalist* environmental ethics = resource-preservationist, oriented to limited or zero growth
(4) *Bioethicist* or *deep ecology* environmental ethics = extreme preservationist, antigrowth

Group (1) is anthropocentric and places faith in the capacity of technology to overcome problems. Group (4) is unlikely to attract support enough to be viable, and offers little guidance to EM. The ethics of groups (2) and (3) are more likely to support SD and provide guidance for EM. Another grouping of environmental ethics is:

i)   *Anthropocentric* – human welfare is placed before environment or biota
ii) *Ecocentric* – focused on ecosystem conservation (holistic outlook)
iii) *Biocentric* – organisms are seen to have value *per se*

Ethic i) predominates.

EM might use education and media to alter social attitudes so that there is awareness of issues and an acceptance of a new ethics. It will have to develop effective institutions. Manuals, conventions, agreements and decision tools can help guide the identification of goals and preparation of action plans. The reductionist approach of splitting problems into component parts for study and solution lies at the core of rational, scientific study (which the world owes a great deal to). But some feel a holistic 'overall view' approach should replace 'compartmentalised and inflexible science'; a mistake, as there is a need for both.

Human institutions (structures of rules and norms that shape and constrain individual behaviour) and social capital (a set of shared values and knowledge that allows individuals to work together in a group to effectively achieve a common purpose) are valuable but change over time and can be difficult to understand and control, and building new ones may be hard. A key international institution charged with EM is the United Nations Environment Programme (UNEP – founded in 1973). The UNEP was located in Nairobi to broaden representation, with mixed results. The location is peripheral and the UNEP has relatively limited funding. Institutions such as the UNEP rely on the quality of their arguments to convince countries, business and citizens to accept a strategy, and have been given little in the way of sanctions to enforce policies. Governments often sign environmental agreements and then fail to pay towards or do what they promised.

Summary

- The world and its resources are finite, yet human demands increase. EM must manage consumption as much as population.
- Modern humans are more numerous than in the past and are less adaptable and more vulnerable.
- There is no one single approach to EM, but there are key concepts.
- A precautionary and proactive approach is wise.
- EM should promote resilience and adaptability.

Further reading

Attenborough, D. (2020) *A Life on Our Planet: my witness statement and vision for the future*. Grand Central Publishing, New York, NY, USA.

Bardi, U. (2011) *The Limits to Growth Revisited*. Springer, New York, NY, USA.

Lomborg, B. (Ed) (2007) *Solutions for the World's Biggest Problems: costs and benefits*. Cambridge University Press, Cambridge, UK, USA.

Pearce, F. (2011) *The Coming Population Crash: and our planet's surprising future*. Beacon Press, Boston, MA, USA.

Randers, J. (2012) *2052: a global forecast*. Chelsea Green Press, White River, VT, USA.

Scudder, T. (2010) *Global Threats, Global Futures: living with declining living standards*. Edward Elgar, Cheltenham, UK.

www sources

- Carey, J. (2015) The 9 limits of our planet… and how we've raced past 4 of them. Ideas. TED.com 05/03/2015 – https://ideas.ted.com/the-9-limits-of-our-planet-and-how-weve-raced-past-them/ (accessed 27/01/21).
- Limits to Growth: the 30-year update – http://donellameadows.org/archives/a-synopsis-limits-to-growth-the-30-year-update/ (accessed 02/02/21).

Chapter 3

Environmental management and science

Chapter overview

- Environment and environmental science
- Structure and function of the environment
- Biosphere cyclic processes
- How stable and resilient are environments?
- Threatening environmental events
- Biodiversity
- Environmental limits and resources
- Environmental and ecosystems modelling, the ecosystem concept, environmental systems and ecosystem management
- Applying the ecosystem concept to tourism, conservation and heritage management
- Environmental systems and ecosystem planning and management – biogeophysical units
- Summary
- Further reading
- www sources

Environment and environmental science

Knowledge about the Earth, its organisms and human affairs is incomplete, so forecasting and decision making are imperfect. Nevertheless, compared with the situation before the International Geophysical Year (1957–1958), the International Biological Program (1964–1974) and the Millennium Ecosystem Assessment (2001–2005), there is better understanding of the structure and function of nature. Environment is the sum total of all the living and non-living elements and their effects that influence life. An ecosystem is a community of life forms in concurrence with non-living components, interacting with each other. Ecology is the study of the relationships between living organisms, including humans, and their physical environment.

DOI: 10.4324/9781003189985-4

When EM relates to science it can adopt one of two approaches: *multidisciplinary*, which involves communication between various fields but without much breakdown of discipline boundaries; *interdisciplinary*, where the various fields are closely linked in a coherent way. The interdisciplinary (or even a holistic) approach is advocated as a cure for the fragmentation of science; multidisciplinary is easier.

Interaction between non-living (abiotic) and living (biotic) components generates the natural environment. Little human interference is an increasingly rare thing. Humans are part of nature and in a way our alterations are 'natural'. Many organisms alter the environment and the change may be slow or rapid, localised or global. Environmental change by humans is deliberate or unwitting and usually degradation, rather than remediation or improvement. EM must understand human–environment interactions.

Before the 1990s social studies and science found it difficult to communicate because they had different traditions and languages. Today there are stronger links.

Many environmentalists listen to scientific reason, some take little heed, oppose it and/or do not understand it. EM may be confronted by irrational environmentalists and politicians who try to present their views as scientifically sound. There will be demands for firm answers that may be difficult to come by and citizens who switch-off from crucial issues. Popular pressure can distort support for research: some fields are attractive to citizens and politicians, and others (even if vital) are not.

Figure 3.1 South Georgia, sub-Antarctic. A simple flora and fauna, which, excepting larger marine mammals, has had limited human disturbance and offers opportunities for ecosystem studies. Reindeer, cats and rats were introduced causing damage to vegetation and birds. In the 2020s deer, cat and rat control was applied (opposed by some environmentalists). Once a base for sealing and whaling, those industries had gone by 1967. Foreground a gentoo penguin (*Pygoscelis papua*) nest site. Source: Photo by author, 1973.

Those involved in EM have worldviews and company/group policy affects how they proceed. Nevertheless, there are common problems: 1) data inadequacy; 2) modelling difficulties; 3) analytical difficulties; 4) insufficient time for research; 5) pressures from various stakeholders; 6) funding limitations. The problem under study may be complex and there may be a request for answers before data is available.

Reconstruction of past scenarios (through palaeoecology, archaeology, environmental history, etc.) can warn of change and hazards establish trends, suggest possible future scenarios and human reactions. Backcasting (a planning method that starts with defining a desirable future/goal and then works backwards to identify policies and programmes that will hopefully achieve that specified future working from the present) can advise scoping and goal setting. It is also possible to imagine possible scenarios and use cautiously (Kolbert, 2006, 2021; Vogelaar et al., 2019).

Developments which aid environmental science and EM include:

- growing international co-operation;
- standardisation of measurements and definitions;
- remote sensing, telecommunications and computing/data-processing advances;
- the internet which facilitates exchange of information and makes it harder for individuals, companies or national authorities to hide environmental problems;
- improved communications between environmental science and social studies.

There has been interest in a holistic approach since the 1920s (Smuts, 1926), yet this is still poorly defined; it implies acceptance that 'the whole is greater than the sum of the parts'. It can be argued science relies on reductionism (the view that everything is explainable from the basic principles), focused, objective research, empiricism (use of data to prove a case) and is compartmentalised (isolation of fields of study from each other). Holistic research seeks to understand the totality of problems; it puts study of the whole before that of the parts. The danger is that the strengths of science are undervalued and, in all but the simplest environments, problems tend to be so complex that a holistic approach is very difficult (Berkes and Bierkes, 2009; Behrens, 2010), so there are situations where holism will not work, and there are dangers in over-enthusiastic use.

Structure and function of the environment

Popular 'laws of ecology' have been published (EM implications in brackets):

1 Any intrusion into nature has numerous effects, many of which are unpredictable (EM must cope with the unexpected).
2 Everything is connected; humans and nature are bound together; what one person does affects others and a wider world (EM must consider chains of causation, looking beyond the local and short term).
3 Care must be taken that substances produced by humans do not interfere with any of the Earth's biogeochemical processes (EM must monitor natural processes and human activities).

In 1927 Charles Elton described ecology as 'scientific natural history'. Modern definitions include: the study of the structure and function of nature; or the study of interactions between organisms and their non-living environment (Odum, 1975). Synecology is the study of individual species–environment linkages. Autecology is the study of community–environment linkages. Ecology is a guide for EM, offering concepts and techniques. A word seldom seen before 1970, ecology has also come to mean a viewpoint: 'concern for the environment'.

Human ecology developed to facilitate the study of interrelationships among people, other organisms and environment (Dyball and Newell, 2015). Human ecology supports many practical applications including the planning of communities and regions, resource use, etc.

Political ecology is an approach to understanding the function of the human-altered environment. It has largely displaced cultural ecology, marking a move to more ecosystems focus (hopefully environment is given adequate consideration). It tries to assess environmental problems using concepts and methods mainly derived from political economy (Bryant, 2015; Perreault et al., 2015). A central premise is that ecological change cannot be understood without consideration of the political and economic structures and institutions with which it interacts (Blaikie, 1985; Peet et al., 2011; Bridge et al., 2015; Robbins, 2020; Roberts, 2020).

The global complex of living and dead organisms forms a relatively thin layer: the biosphere. The term ecosphere is used to signify the biosphere interacting with the non-living environment, biological activity being capable of affecting physical conditions even at the global scale. The global ecosphere can be divided into various climates. Climate may be affected by:

- variation in incoming solar energy due to fluctuations in the Sun's output or possibly dust in the atmosphere or space;
- variation in the Earth's orbit or change in its inclination about its axis;
- variation in the composition of the atmosphere: alterations in the quantity of dust, gases (e.g. CO_2; methane), water vapour and cloud present (which may be affected by biological activity, human pollution, volcanicity, impacts of large comets or asteroids);
- altered distribution of continents, mountain uplift/loss, changes in oceanic currents, or fluctuation of sea level that may expose or submerge continental shelves;
- EM must not assume that climate is fixed and stable.

Trophic level and organic productivity

Organisms in an ecosystem may be grouped by function according to their trophic level (position in the food chain/web at which they gain energy/nourishment). The first trophic level, primary producers (or autotrophs), in all but a few cases convert solar radiation (sunlight) into chemical energy. The exceptions include deep ocean hydrothermal-vent communities and some microorganisms below ground. Seldom are there more than four or five trophic levels because organisms expend energy living, moving, and in some cases generating body heat; transfer of energy from one trophic level to the next is unlikely to be better than 10% efficient. Given these losses in energy transfer, it is possible to feed more people if they eat at a low rather than high trophic level. Thus grain supports a bigger population than would be possible if it were used to feed animals for meat, eggs or milk.

The sum total of biomass (organism mass expressed as live weight, dry weight, ash-free dry weight or carbon weight) produced at each trophic level at a given point in time is termed the standing crop. This needs to be treated with caution; if taken at the end of an optimum growing period it indicates full potential; if taken during a drought, cool season,

Figure 3.2 South Georgia, sub-Antarctic. Island ice and vegetation shifted with climatic change long before human activity.   Source: Photo by author, 1973.

period of agricultural neglect or insect damage, it is an underestimate. Primary productivity is the rate at which organic matter is created (usually by photosynthesis, although in some situations by other metabolic processes) at the first tropic level. The total energy fixed at the first trophic level is termed gross primary production; minus the estimated respiration losses, this gives net primary productivity (in g m^{-2} d^{-1} or g m^{-2} yr^{-1}). Net primary productivity gives a measure of the total amount of usable organic material produced per unit of time. Most cultivation (agroecosystem) functions well below the net primary production of more productive natural ecosystems. Thus there should be potential for improvement of agriculture (Struik and Kuyper, 2017; Martin-Guay et al., 2018; Pretty and Bharucha, 2018).

Widely used concepts are: maximum sustainable yield and carrying capacity (Box 3.1). These should be treated with caution. Maximum sustainable yield may be calculated, but if the environment changes a reasonable resource exploitation strategy based on it could lead to over-exploitation. Maximum sustainable yield calculations can give a false sense of security. A given ecosystem usually has more than one carrying capacity, depending on factors such as the intensity of use and the technology available. Some organisms adjust to their environment through boom and bust, feeding and multiplying during good times, and in bad suffering population decline, migrating or hibernating; calculating carrying capacities for such situations can be difficult. Biogeophysical carrying capacity may need to allow for behavioural carrying capacity – a population could be fed and otherwise sustained but feel crowded and stressed to a degree that limits their survival.

The carrying capacity of an ecosystem may be stretched by trade, human labour and ingenuity, technology, military power (tribute from domination). Net primary productivity often increases at the cost of species diversity and vulnerability. The timing of resource use

> **Box 3.1 Ecological concepts and parameters which are useful for environmental management**
>
> ● *Maximum sustainable yield*
>
>   The fraction of primary production (as organic matter) in excess of what is used for metabolism (net primary production) that it is feasible to remove on an ongoing basis without destroying the primary productivity (i.e. safe harvest). Under US Law, maximum sustainable yield would be defined as: maintenance in perpetuity of a high level of annual or regular periodic output of renewable resources.
>
> ● *Carrying capacity*
>
>   Definitions vary and can be imprecise. Examples include: the maximum number of individuals that can be supported in a given environment (often expressed in kg live weight per km²); the amount of biological matter a system can yield, for consumption by organisms, over a given period of time without impairing its ability to continue producing; the maximum population of a given species that can be supported indefinitely in a particular region by a system, allowing for seasonal and random changes, without any degradation of the natural resource base.
>
> ● *Assimilative capacity*
>
>   The limiting resource may not be an input such as food or water, it may be inability to deal with outputs (waste products). A given environment has some capacity to purify pollutants up to a point where the pollutant(s) hinder or wholly destroy that capacity; this is termed the assimilative capacity.

may be crucial: for example, rangeland might feed a certain population of livestock, provided that grazing is restricted during a few critical weeks (at times when plants are setting seed, becoming established or are otherwise temporarily vulnerable). If this is not done, or a disaster like a wild fire strikes, degradation may occur and fewer or no resources can be extracted in the future. Within the simplest ecosystems there can be complex relationships among organisms and between organisms and environment. There are often convoluted food webs; complex pathways along which energy (food) and perhaps pollutants are passed; interdependencies for pollination, seed dispersal, etc.

Certain pesticides, radioactive isotopes, heavy metals and other pollutants can become concentrated in organisms feeding at successively higher trophic levels, so that apparently harmless background contamination could, through such biological magnification (bio-accumulation), prove harmful to humans and other organisms without assimilative capacity having obviously broken down.

The ecosystem

The biosphere is composed of many interacting ecosystems, the boundaries between which are often indistinct, taking the form of transition zones (ecotones), where organisms from adjoining ecosystems may be present. It is possible for some organisms to be restricted to an ecotone only. Large land ecosystems or biomes (biotic areas) provided they are not disturbed have a prevailing regional climax vegetation and its associated animal life, in effect regional-scale ecosystems. Biomes, such as desert biomes or grassland biomes, often mainly reflect climate, but can also be shaped by fire, drainage, soil characteristics, grazing, etc. So a biome may change in character over time.

Ecosystems are recognised as environmental or landscape units. The ecosystem has become the basic functional unit of ecology and useful for defining management boundaries (Tansley, 1935). There are various definitions, which include: an energy-driven complex of a community of organisms and its controlling environment; a community of organisms and their physical environment interacting as an ecological unit; an integration of all the living and non-living factors of an environment for a defined segment of space and time. Ecosystems have six major features: interdependence, diversity, resilience, adaptability, unpredictability and limits. They also have a set of linked components, although the linkages may not be direct – a network or web with organisms as nodes within it (Table 3.1). An ecosystem boundary (distinct or transition zone) may be defined at organism, population or community level, the crucial point being that biotic processes are sustainable within that boundary. It is also possible to have different physical and functional boundaries to an ecosystem. No two ecosystems are exactly the same, but one may recognise general rules and similarities. There are two ways of viewing ecosystems: 1) as populations – the community (biotic) approach, in which research may be conducted by individuals; 2) as processes – the functional approach (studying energy flows or materials transfers), best investigated by a multidisciplinary team. Once understood, ecosystems can often be modelled, allowing prediction of future behaviour. There are three broad types of ecosystem: i) isolated systems – boundaries recognisable and more or less closed to input and output of materials and energy; ii) closed systems – boundaries prevent input/output of materials, but not energy; iii) open systems – boundaries may be difficult to recognise and these allow free input/output of materials and energy. Many of the Earth's ecosystems are type-(iii) and are often interdependent. Alternatively, ecosystems may be classified as: A) natural – unaffected by humans; B) modified – some change due to humans; C) controlled – whether by accident or design humans play a dominant role.

Table 3.1 **Classifications of environmental systems**

(A) BY FUNCTION
Isolated systems Boundaries are closed to import and export of material and energy.
Closed systems Boundaries prevent import and export of material, but not energy. For example, Biosphere-2 is supposed to function with no other exchanges. The Earth is largely a closed system, although it receives dust, meteorites and solar radiation.
Open systems Boundaries allow free exchange of material and energy. Many of the Earth's ecosystems are of this form and may actually be interdependent.
(B) BY DEGREE OF HUMAN DISTURBANCE
Natural systems Unaffected by human interference.
Modified systems Affected to some extent by human interference.
Control systems Human interference, by accident or design, plays a major role in function (includes most agricultural systems).

Note: Biosphere-2 is a 1.27 ha enclosed environment experiment near Tucson, AZ, USA. It uses sunlight and was sealed with eight people inside and planted with several ecosystems from 1991–1993 forming Earth's 2nd biosphere. One role was to test ideas for a Mars base (Allen, 1991) (ownership and function has changed). In 2014 China ran a similar experiment, Yuegong-1, for 200 days, a c. 98% sealed system using LEDs for plant growth.

A naturalist might map the ecosystem of an animal, say a bear, by reference to the resources it uses (i.e. as a function of the organism), so the area may alter with the seasons and differ according to the age or sex of the animal. Alternatively, ecosystem delineation could be by function (i.e. as a sort of landscape unit).

Ecosystems cover a great range of scales: one 10,000 km^2, another less than 1 km^2; or the *c*. half-litre of water in a pitcher-plant. In a stable ecosystem each species will have found a position, primarily in relation to its functional needs for food, shelter and so on. This position, or niche, is where a given organism can survive most effectively. Some organisms have very specialised demands and so occupy very restricted niches; others can exist in a very wide niche. Niche demands are not always simple: in some situations a species may be using only a portion of its potential niche, and alteration of a single environmental parameter may suddenly open, restrict or deny a niche for an organism. Species usually compete for a niche with other organisms.

Ecosystems can be subdivided, according to local physical conditions, into habitats (places where an organism or group of organisms live), populated by characteristic mixes of plants and animals (e.g. a pond ecosystem may have a gravel bottom habitat and a mud bottom habitat). Within an ecosystem change in one variable may affect one or more, perhaps all other, variables.

Even something as well-defined as a cave may exchange water and nutrients with regional groundwater or receive debris from outside. To simplify study, ecologists have enclosed small natural ecosystems, created laboratory versions (e.g. phytotrons – sealed greenhouses and growth chambers) and study very simple ecosystems. Controlled environment experiments are valuable for exploring what effects changing global climate and CO_2 may have on crops and biota.

From the 1940s, systems diagrams have been constructed to show energy flows between components of ecosystems. Similar approaches were adopted by social scientists and business managers as frameworks for study and as means of prediction. Applied systems theory and systems modelling have been steadily improving and are used in EM (Odum, 1983). While the ecosystem approach may not give precise modelling results, it can often provide a framework for analysis. However, it can be difficult to recognise boundaries and measure what goes in and comes out. In addition, the assumption that an ecosystem will behave in a linear, predictable manner is over-optimistic, because some of the processes that are operating work at random. According to systems theory, changes in one component of a system will promote changes in other, possibly all, components.

Given time, natural, undisturbed ecosystems theoretically reach a state of dynamic equilibrium or steady state. Regulatory mechanisms (checks and balances) counter changes within and outside the ecosystem to maintain the steady state (climax). However, since each ecosystem has developed under a different set of variables, each has a different capacity to resist stresses and to recover. Humans often upset regulatory mechanisms, so response may be distorted. When ecosystems are exposed to stress, some responses may be immediate and others gradual or delayed. To manage ecosystems it is necessary to know longer-term behaviour as well as short-term response.

Ecosystems adjust to perturbation through regulatory mechanisms. When the relationship between input and output to the system is inverse (e.g., increased sunlight causes more cloud which reduces the impact of that sunlight on the surface), it is termed a negative feedback. The opposite is a positive feedback, whereby an effect is magnified. There is a risk that a feedback may result in a runaway reaction, which is especially dangerous if it damages a crucial biogeochemical or biogeophysical cycle (a fear with global warming).

Biosphere cyclic processes

Within the biosphere, numerous cyclic processes move and renew supplies of energy, water, chemical elements and atmospheric gases. These cycles affect the physical environment and organisms, and some are greatly affected by life forms. Upset by occasional catastrophic events, such as volcanic eruptions or asteroid strikes, biogeochemical and biogeophysical cycles are assumed to ultimately reach dynamic stability. Nevertheless, EM must not assume an unchanging natural environment.

Organisms play a vital part in some cycles, of which the most critical include the maintenance of the atmospheric gas mix and ensuring global temperature remains within acceptable limits. There are over 30 known biogeochemical cycles; some have a turnover of as little as a few days and others hundreds or millions of years; the latter mean that some materials are non-renewable as far as humans are concerned. Biogeochemical and biogeophysical cycles are not fully understood; for example, there is much to learn about the cycling of carbon, phosphorus, sulphur and many other elements. Without better insight, accurate modelling and prediction of global change is difficult. Cycles may be classified as 1) natural, 2) upset by humans, and 3) recycling (managed by man and perhaps sustainable). Many of group 1) have already been converted to 2); conversion of some type 2) to 3) is an important goal for EM.

How stable and resilient are environments?

Some environments are less prone to change and/or recover from damage better than others.

Stability

Stability (constancy) is invoked by those interested in establishing whether conditions will remain steady or will return via a predictable path to something similar to the initial steady state after disturbance. It is widely held that ecosystem stability is related to biological diversity (biodiversity): the greater the variety of organisms there are the less likely there is to be change in biomass production, although population fluctuations of various species may still occur. However, it is possible that a change in some parameter could have an effect on all organisms. Thus, diversity may help ensure stability, but does not guarantee it. An ecosystem may not have become stabilised when disturbed: it may have been close to a starting point, or it could be undergoing cyclic, more or less constant or erratic change. An ecosystem may return to stability after several disturbances but fail to return after a subsequent similar upset for various reasons. The manner of return may vary. Some ecosystems are in constant non-equilibrium or frequent flux, rather than in a stable state at or near carrying capacity. Return to a pre-disturbance state is therefore uncertain.

Stability can have a number of meanings, including: lack of change in the structure of an ecosystem; resistance to perturbations; or a speedy return to steady state after disturbance. EM is likely to want to know whether an ecosystem is stable, and what would happen if it were disturbed. Natural ecosystems are rarely static: the best EM can expect is a dynamic equilibrium, not fixed stability.

Equilibrium is in part a function of sensitivity and resilience to change. Sensitivity may be defined as the degree to which a given ecosystem undergoes change as a consequence of natural or human actions. Resilience refers to the way in which an ecosystem can withstand change. Originally it was proposed as a measure of the ability of an ecosystem to adapt to a continuously changing environment without breakdown.

There is debate as to whether an ecosystem evolves through slow and steady evolution of species (phyletic gradualism) or experiences generally steady, slight and slow evolution punctuated by occasional sudden extinctions, after which there may be rapid and considerable biotic change (the whole has been called punctuated equilibrium) (Gould, 1984). The implication is that survival might be more down to luck than fitness, although adaptability is a very useful trait. Whatever the process, a climax stage is reached via more or less transient successional stages, at any of which succession might be halted by some limiting factor. The concept of ecological succession, pioneered by Clements (1916), is complex. According to this concept, organisms occupying an environment may modify it, sometimes assisting others: a birch wood may act as a nursery for a pine forest, which ultimately replaces the birch: thus birch is a successional stage (sere) en route to a pine stage. Two types of succession are recognised: 1) primary succession (pioneer stage) and 2) secondary succession. The former is the sequential development of communities from a bare, lifeless area (e.g. the site of a fire, volcanic ash or newly deglaciated land). The latter is the sequential development of biotic communities from an area where the environment has been altered but has not had all life destroyed (e.g. cut forest, abandoned farmland, etc.). Many communities do not reach maturity before being disturbed by natural forces or humans so type 2) situations are common.

Forests may maintain maturity, not become senile and degenerate, through patch-and-gap dynamics; clearings caused by lightning or storms, etc., allow regeneration. Pioneer communities usually have a high proportion of plants and animals that are hardy, have catholic niche demands, and disperse well. Mature, climax communities are supposed to have more species diversity, recycle dead matter better and be more stable.

Resilience

Resilience is displayed by many things; for example: organisms, ecosystems, communities, regions, individuals, societies, institutions and nations. Resilience may be defined in many ways:

- the ability to return to maintain a steady-state ecosystem;
- the facilitation of adaptive behaviour;
- the speed of recovery of a disturbed ecosystem;
- the number of times a recovery may occur if disturbance is repeated.

The concept of resilience is important to human ecologists and EM as well as ecology. Some societies absorb or resist social change and continue with traditional skills and land uses or develop satisfactory new ones; other societies fail, and their resource use and livelihood strategies degenerate. In humans, resilience and vulnerability are not fixed or predetermined; they vary as a consequence of environmental factors, institutions, attitudes (individuals and government), innovation, etc.

Based on sensitivity and resilience, a classification of land, which may be modified to apply to ecosystems, is:

1. Ecosystems of low sensitivity and high resilience. These suffer degradation only under conditions of poor management, human or natural catastrophe. Generally these are the best ecosystems to stretch to improve production of food or other commodities.

2. Ecosystems of high sensitivity and high resilience. These suffer degradation easily but respond well to management and rehabilitation efforts.
3. Ecosystems of low sensitivity and low resilience. These initially resist degradation but, once a threshold is passed, it is difficult for management and restoration efforts to save things.
4. Ecosystems of high sensitivity and low resilience. These degrade easily and do not respond readily to management and rehabilitation efforts. It is probably best to leave such ecosystems alone or to alter them radically.

Managers or researchers often wish to establish in advance, or sometimes after a disturbance, what the consequences will be:

  i)   Will the ecosystem re-establish its initial state?
 ii)   Will there be a shift to a new state?
iii)   If (i) takes place, how rapid will the recovery be and how complete?
 iv)   What path does the recovery take?
  v)   How often can recovery occur?
 vi)   Will the same recovery path always be followed?
vii)   Will successive, similar disturbance have the same effect?
viii) Would change still occur if there were no disturbances?

So, in a given situation the path of recovery/succession can be unpredictable.

In a resilient/stable ecosystem each species is assumed to have found a position (niche), primarily in relation to its functional needs: food, shelter, etc. This niche is where a given organism can operate most effectively. Some organisms have specialised demands and so occupy very restricted niches, while others can exist in a wide range of niches. A species may be using only a portion of its potential niche. Alteration of a single parameter affecting competition with other organisms may suddenly open, restrict or deny a niche for an organism.

Threatening environmental events

Environmental threats/challenges may have a predictable pattern of recurrence or be impossible to predict with current knowledge, although some of the latter may give warning signs as they start to manifest. Threats may be unmistakable and sudden, less obvious and sudden, or they may unfold gradually and obviously or gradually and insidiously. Slow developing threats may remain unnoticed for generations. For example, widespread gradual soil loss may only be apparent if a religious site protects an area or some marker has been installed. Sometimes a system changes virtually imperceptibly until a threshold (tipping-point/planetary boundary) is reached, whereupon there suddenly appear obvious and possibly drastic effects. Given long enough, chance events probably affect the survival of organisms as much as evolution; this has been described as 'contingency'. Events which give insufficient time for adaptation may allow some organisms to prevail for quite fortuitous reasons, rather than survival of the fittest/wisest; *adaptability* (and luck) count.

There is evidence of prehistoric mass extinctions: at least 15, with five especially severe: *c.* 440, *c.* 390, *c.* 345, *c.* 220 and *c.* 65 million years BP. There is plenty of evidence of volcanic, tsunami and asteroid/comet impacts. The Toba eruption (Sumatra *c.* 74,000 BP) might have threatened human survival. Recent moderate eruptions had weak global impact: El Chichon (Mexico 1982) and Mt Pinatubo (Philippines 1991) reduced global temperatures. But these events failed to scare governments enough to prepare for the future (possible loss of one or more harvests worldwide). EM should seriously consider infrequent natural threats.

Repeated natural global greenhouse events have happened (one *c.* 680 million years BP). Cold glacial phases (icehouse/glacial events) alternating with warmer (interglacial) or less cold (interstadial) phases have occurred through Earth's history. During glacials ice expanded to lower latitudes from the poles and worldwide to lower altitudes from high ground, and there were marked shifts in vegetation globally. The most recent glacial-interglacial phase began perhaps 40 million BP and became pronounced from about 15 million BP. The Quaternary Period (*c.* 2.6 million BP – ongoing) has had over 20 major glacial–interglacial oscillations. The interglacials each lasted between 10,000 and 20,000 years and each glacial spanned about 120,000 years. The last interglacial was about 132,000 to 120,000 BP and the last glacial maximum was about 18,000 BP. Ice retreated to broadly present limits world-wide by *c.* 10,000 BP. Between 7,000 and 3,000 BP average conditions may have been 2°C warmer than now. Today we worry about human-caused global warming but climate has clearly not been stable.

There are well-established links between climate and CO_2 in the atmosphere (approximately 25% reduction during glacials compared with the present), along with low levels of methane in the atmosphere, and low sea levels (perhaps 140 m below present). During inter-glacials, carbon dioxide and methane in the atmosphere were higher than currently, and sea levels 40 m or more above present.

During the past 12,000 years meltwater flows into the Atlantic suddenly affected oceanic circulation for decades, one reduced the Gulf Stream, dropping average annual temperatures in Western Europe and eastern USA by several degrees. Natural ongoing global warming may have cool spells and not be warm for all. Change may be sudden and humans have made things less predictable.

Biodiversity

Ecological diversity is the range of biological communities that interact with each other in a given environment. Biodiversity (biological diversity) refers to species diversity plus genetic diversity within those species. Loss of biodiversity occurs when species extinction exceeds the rate of species creation. Extinction is sometimes sudden, occasionally catastrophic, but also an ongoing, gradual process. Humans have greatly accelerated the rate of extinction (Kolbert, 2014). Loss of biodiversity is one of the most serious challenges facing EM. The consequences, in addition to the immorality of loss, are reduction of potential for new crops, pharmaceuti-cals, etc., and possibly a less stable and resilient environment. Biodiversity is irreplaceable and losses often happen before effective action can be taken, or measures are inadequate.

Environmental limits and resources

Humans have exceeded some limits and are in overshoot or a crisis situation with only a few decades to resolve things.

Environmental limits

Von Liebig's law of the minimum states that whichever resource or factor necessary for sur-vival is in short supply is the critical or limiting one which restricts population growth (e.g.

water, space, nutrients, recurrent fires, predators, etc.). The population reaching a limit may suffer gradual or sudden, limited or catastrophic collapse in numbers, a vacillation, or a cyclic boom-and-bust growth pattern. Solar energy drives most of the Earth's ecosystems. Solar radiation receipts are thus a main limiting factor and it is possible to estimate maximum potential global food production by mapping available surfaces and factoring in photosynthetic potential. Climate change debates assume fairly constant solar input. Few of the world's agricultural strategies function at anything like potential maximum photosynthetic efficiency; improvement of food and commodity production without further expansion of farmland is thus theoretically possible. Estimates of the population the Earth might sustain vary. It may be possible to produce 40 tonnes of food per person for the 1990 global population, but there might be land degradation, falling investment, environmental and social hindrances, etc. Agricultural improvement may not happen. There may be a global population of 6, 10 or 30 billion in 2100. Attempts have been made to estimate a comfortable *sustainable* global population; one (low) assessment is 200 million, another 3 billion, both far below the present global population.

Resources

A resource is something which meets perceived needs or wants. Resources become available through a combination of increased knowledge, improving technology, ability to access sources, and changing individual and social objectives. Resources are defined by perceptions and attitudes, wants, technical skills, legal, functional and institutional arrangements, as well as by political customs, utility (usefulness or enjoyment something provides), aesthetic quality, etc. Resource demands change as human perceptions alter, new technology is developed, fashions vary, the environment alters and new materials are substituted. It is unwise to wait and see what 'business-as-usual' will bring and better to seek stronger controls on key resources.

A crude division is into:

- those that can be safely and easily stretched;
- those that can be stretched with care;
- those that cannot or should not be stretched.

Resources may be grouped as: non-renewable (finite or exhaustible and can be used only once); renewable (if well managed, and there is no natural disaster, these can be used indefinitely); inexhaustible (e.g. sunlight, gravity, wind power).

The Gaia hypothesis

The Gaia hypothesis was proposed in 1965 by James Lovelock (with later input from Lynn Margulis). James Hutton and Pierre Teilhard de Chardin expressed similar views earlier. Lovelock suggested the biosphere acts as a self-evolving homeostatic system. The Gaia hypothesis (less contentious terms are Earth system science or co-evolution) received limited support before the 1990s; since then it has been better accepted. The hypothesis was seen as pseudoscience until Goldsmith and Margulis published the Daisy World model in 1983 which offered a rational mechanism.

There are several ways of interpreting the Gaia hypothesis (Lovelock and Margulis, 1973); whichever is accepted, it challenges the attitude that humans can exploit the Earth (Lovelock and Margulis, 1973, 1979, 1988, 1992; Lovelock, 2009). Whether or not it is valid it has stimulated thought about environment, biogeochemical cycles and development. It provoked research into the global carbon cycle and nature-based solutions (N-BS). The hypothesis suggests that life has not simply adapted to the conditions it encountered, but also alters and controls the global environment to keep it habitable. The hypothesis seeks to explain the survival of life on Earth by treating the organic and physical environment as parts of a single system ('Gaia') in which biotic components act as regulators enabling control and repair (not a conscious process, nor is there implied a design; the idea that nature is benign is wishful). The view is that Earth is to some degree a self-regulating system. Temperature and the composition of the Earth's atmosphere, according to the hypothesis, are regulated by its biota, the evolution of which is influenced by the factors regulated.

Without Gaian regulation, supporters argue, global temperatures would be inhospitable to higher life forms, and atmospheric oxygen would probably be locked up in rocks. Earth is a sort of superorganism, a homeostatic system with feedbacks controlling global temperature, atmospheric gases and availability of nutrients. The controls involve a number of biogeochemical cycles, notably those of carbon dioxide, nitrogen, methane, ammonia, ocean nitrates and salts, oxygen, sulphur, carbon and phosphorus. The system functions unconsciously in the interests of the physical environment and biota. If true, humans are part of a fragile complex system: upset Gaian mechanisms, and there could be catastrophic, runaway environmental changes (Lippert, 2014).

Two Gaian models emerged by the 2020s: 1) Life influences Earth's processes (and substantially affects abiotic processes) – the widely supported *weak* Gaia hypothesis. 2) The original *strong* Gaia hypothesis, that life controls Earth's processes. Not widely accepted.

Environmental crisis

Warnings that the Earth faces crisis or is in crisis have appeared since the 1970s (Ehrlich, 1970). Crisis is a turning point (threshold or tipping point), a last chance to avoid, mitigate or adapt. Perception of crisis is subject to changing beliefs, fashion, technological ability, etc. Not all agree on what constitutes a crisis, the term is subject to emotive usage. Some on the political left suggest the idea of crisis serves as a cover-up to divert attention from doing anything about problems. Others feel that environmental problems are mainly due to unsound concepts of development and modernisation, a social or ethical fault or result of commerce, globalisation or capitalism (Lomborg, 2001; Diamond, 2005).

Environmental problems have often been interpreted as indicating a progressive loss of ecological stability, but it may partly reflect more research and awareness. There are serious local, regional and global environmental and socio-economic 'hotspots'. But, a crisis-fighting, rushed and short-term focus to EM and development planning is not wise.

Identification of a large-scale crisis may be a mistaken response to a patchy, localised problem (hotspots), reflecting inadequate observation. Hotspots may be real or reflect researchers' tendency to: view roadside areas and miss the less accessible; conduct studies during dry seasons; interview unrepresentative groups of people; do research that is too short term, etc. There is also researcher bias and 'helicopter research': outsiders without local insight make the study. Blaikie and Unwin (1988) cited an example of a gully erosion area in Zimbabwe identified as a crisis, yet the study revealed only 13% of total soil loss was from the spectacular gullies, while 87% was from insidious inter-gully sheet erosion (missed). Funds could have been misspent treating gullying.

Cumulative causation: things interacting might lead unexpectedly to a crisis. With any complex system the breakdown of one component might be insignificant, but sooner or later a particular failure or combination, or other factors, contribute to collapse. EM has to recognise significant thresholds and monitor (Hamilton et al., 2015). Ultimate environmental threshold assessment, derived from threshold analysis, is based on the assumption that there are boundaries which may be broken by direct or indirect, including cumulative, impacts. There have been regional disasters which ultimate environmental thresholds assessment might have helped avoid, such as the 1970s to 1990s ruination of the Aral Sea, or recent wildfires.

Global warming is seen as *the* crisis, but Lawson (2008) and many others argue it is a politico-pseudo-scientific construct. A wise view is that global warming *is* a threat but other challenges also deserve attention, carbon emission control is not enough and adaptability and resilience are vital.

Africa is frequently singled out as having an environmental or poverty crisis, or both, especially in the *Maghreb* and sub-Saharan margins. Africa will roughly double its population between 2020 and 2050 (Oba, 2021). Drought is often cited as the cause of difficulties, yet there is no conclusive evidence that rainfall receipts have diminished (some African lakes are rising). Possibly drought in Africa is an exposure of other weaknesses: a 'litmus of development'. Often it is possible to recognise a problem, but tracing why it happens can be difficult. Population growth cannot simply be blamed; there are situations where, despite very low settlement density, there has been severe damage. And there are densely settled regions with challenging environments where people have sustained themselves for centuries. Population growth projections alone are not a reliable indicator.

The *Millennium Ecosystem Assessment* (2005), a worldwide ecosystem stocktaking, indicated serious degradation of the Earth's life-support systems, lack of sustainable resources use and a growing risk of abrupt and drastic environmental change. It prompted thinking and offers some hope that with proactive approaches disaster may be averted (millennium assessment.org/en/index.html – accessed 29/06/21).

Environmental and ecosystems modelling, the ecosystem concept, environmental systems and ecosystem management

Once understood and monitored, environmental systems or ecosystems may be modelled using a variety of approaches, including theoretical, physical, analogue or computer models. Environmental and ecosystem modelling are used for anything from sediment transport to hydrology, groundwater, climate change, carbon sequestration, ocean–atmosphere energy and chemical flux (Grant and Swannak, 2008; Jakeman et al., 2008; Jørgensen, 2009, 2011a, 2011b, 2016a, 2016b; Geary et al., 2020). The ecosystem concept was adopted by the International Biological Program.

The approach focuses on energy flows or nutrient transformations (Chapin et al., 2011). Biotic activity within an ecosystem can be divided into that of producers, consumers and decomposers, so study needs data on population dynamics, productivity, predator–prey relations, parasitism, etc. Study of non-biotic aspects of an ecosystem may focus on estimation of biomass or micrometeorology. There has been a shift from description of the structure of ecosystems to a focus on trying to understand function, processes, mechanisms and systems behaviour. The ecosystem boundary is adopted as the spatial and temporal limit to an environmental task, and the ecosystem concept allows EM to look at portions of complex nature as an integrated system (Marques et al., 2015). The concept may be applied to cities, agriculture and many other situations, although these are not actually true, discrete units in terms of energy flows or function.

An ecosystems approach allows an interdisciplinary or even holistic view of how complex components work together, and it can enable the incorporation of human dimensions into assessments (Odum, 1983). This requires multidisciplinary or interdisciplinary teamwork that includes consideration of science and social science issues (Fish, 2011). It is important that planners and analysts have a clearly thought out interpretation of what an ecosystem approach means before using it. EM may treat an ecosystem rather like a factory: seek to improve and sustain output and reduce cost. And there may be several different products (ecosystem services) such as agricultural produce, tourism, water supply, conservation, etc.

Sustainable ecosystem management seeks to maintain ecosystem integrity and, if possible, produce food and other commodities on a sustained basis. Many of the principles used by ecosystem management are normative (i.e. moral and ethical rather than strictly scientific), which has attracted criticism. Concern has been voiced over the lack of satisfactory established principles for ecosystem management, and that it may lead to a broad and possibly superficial approach. Experience gained in one ecosystem may be of limited value for other, even similar, ecosystems. The character of natural ecosystems may be difficult to establish where there has been disturbance, so it is difficult to agree what conservation or land restoration should aim for.

The ecosystem approach could be seen as methodology (with models to simulate the ecosystem) and mindset (with a focus on function and properties of ecosystems), the strength of this approach being synthesis of the complexity of problems faced, enabling assessment of consequences. In practice there has been specialisation; e.g. ecosystem studies of risk; ecosystem quality management; assessment of ecosystem potential; and ecosystem conservation. It is not only ecologists and EM who have adopted an ecosystems approach: other disciplines do so, including human ecology, cultural anthropology, planning, management and urban studies.

The decision to adopt an ecosystems approach will usually be based on an assessment of whether its advantages outweigh its disadvantages (Box 3.2). A commodity or service orientation may be fine if the goal is to maximise production of a single product or service; it is less satisfactory where the ecosystem yields several products, and it is important to know hazards, limits and opportunities (Box 3.3).

Sometimes ecosystem boundaries are clear physical features but often they are less well delineated. And it may be necessary to define an ecosystem in 3D, not just mapping an area. The quest is for an eco-socio-economic planning unit, which is stable, clearly defined and likely to support SD or sustainable intensification.

Ecosystems (and environmental systems) may be analysed using systems theory, which enables complex, changing situations to be understood and predictions made. Systems theory assumes that measurable causes produce measurable effects. There have been attempts to combine ecological and economic models in systems analysis. For example, a systems analysis approach to environmental assessment and management was used in the Oetzertal (Austria) from 1971 for some years, as part of the UNESCO Man and Biosphere Program. This alpine valley ecosystem experienced change through tourism, especially skiing, and, with the help of the modelling, managers now have a clear idea of what is needed to sustain tourism and maintain environmental quality. In the early 1990s the USA established a nation-wide EM and Monitoring Program (EMAP) to aid ecological risk analysis by assessing trends in condition of ecosystems – a controversial and expensive exercise.

The Millennium Ecosystem Assessment (MEA) completed in 2005 provided information for the Convention on Biological Diversity, the Convention to Combat Desertification, the Ramsar Convention on Wetlands and the Convention on Migratory Species.

Box 3.2   Advantages and disadvantages of
          the ecosystem approach

Advantages	Disadvantages
Comprehensive, holistic approach for understanding whole systems.	May neglect sociocultural issues such as politics, power and equity.
Different view of science that recognises diversity of cause and effect, uncertainty, and probabilistic nature of ecosystems.	Ecological determinism: danger of generalising from biophysical to socio-economic systems.
Draws on theory and methods from different fields to generate models and hypotheses.	Nebulous: a vague, superorganismic theory of poor empirical foundation that relies on analogy and comparison.
Contributes to understanding limits, complexity, stresses and dynamics.	Non-standard definition of 'ecosystem'.
Encourages preventive thinking by placing people within nature.	Reification of analytical systems; in some approaches linked to reductionist and equilibrium views.
Facilitates locally appropriate, self-reliant, sustainable action.	Narrow spatial focus on local ecosystem structures and processes.
Facilitates co-operation, conflict reduction, institutional integration.	Functionalist and/or energy analysis are overemphasised.
Requires recognition of mutual dependence on all parts of a system (e.g. natural/cultural, person/family).	Duplicates and/or overlaps other disciplines without a special contribution of its own.
Results in criteria for management actions.	If ecosystem approaches can apply to everything they may be meaningless.
Facilities studies that integrate a range of disciplines (holistic).	

Source: Slocombe (1993: 298, Table 3 (with modification))

Applying the ecosystem concept to tourism, conservation and heritage management

Tourism commonly takes place in sensitive ecosystems: coastal zones; alpine areas; etc. The value of the ecosystems approach is that it can highlight vulnerable features and threatening human behaviour, which may be easily overlooked. Heritage sites can be established to conserve cultural and natural features, including wildlife and old crop varieties in arboreta and the gardens of large estates. Ecosystem studies can determine how such areas may be sustained and augmented.

Applying the ecosystem concept to urban and periurban management

An ecosystem approach can help to identify strategies to reduce urban and periurban pollution, develop local production of food, conserve biodiversity and provide employment,

Box 3.3 How the ecosystem approach can advise the environmental manager

Three selected situations:

Range management

- what type of stock;
- stocking rate;
- the state of the range;
- how to manage grazing rotation;
- whether to augment with seeding or fertiliser;
- potential threats;
- parallel usage opportunities/ecosystem services (e.g. recreation, conservation, forestry).

Forest management

- whether the forest trees are healthy and regenerating;
- whether the mix of species is steady or in decline;
- whether there are threats;

- what harvesting is possible and how;
- parallel usage opportunities/ecosystem services (e.g. forest products, conservation, tourism);
- whether forest can be established/ re-established in currently unforested areas.

Conservation management

- whether conservation is viable in the long term;
- what mix and number of species can be carried;
- whether a cull or improvement in breeding is needed;
- whether there are threats;
- what parallel uses/ecosystem services are possible (e.g. ecotourism);
- whether there are alternative ecosystems to provide backup.

assess other ecosystem services, etc. (McDonnell et al., 2009; Mostafavi and Doherty, 2010; Douglas et al., 2011; Guntenspergen et al., 2011; Itard et al., 2011; Chadwick and Francis, 2013; Elmqvist et al., 2013; Tanner and Adler, 2013; Forman, 2014; Hagan, 2015; Genletti et al., 2019).

Applying the ecosystem concept to conservation management

Forest management and wildlife conservation make use of the ecosystem approach (Bailey, 2009). Nature reserves are islands in a sea of disturbance, so the study of island ecosystems provides information on rates of extinction and evolution and minimum size of habitat, and suggests linkages between habitats necessary for sustained conservation: likely impacts of climate change or acid deposition; clarification of vital pollination and seed dispersal needs; information on predator–prey relationships, etc. (MacArthur and Wilson, 1967; Mueller-Dombois et al., 1981). A conservation area may fail to sustain biodiversity because disruptive effects penetrate towards its core, an ecosystem approach can forewarn of this. Caution is needed; island biogeographic theory has been little tested.

Studies have been made in Amazonian Brazil to improve understanding of the impact of various intensities of disturbance and ecosystem fragmentation on biodiversity using different sizes and patterns of forest reserves. This Minimum Critical Size of Ecosystem Study, and similar work elsewhere, adds to island biogeography for establishing what are viable locations, size and pattern of conservation areas.

Environmental systems and ecosystem planning and management – biogeophysical units

An early step by planners and managers is to determine the limits of their task (scoping), part of which is establishing a suitable sized and stable unit which reflects the structure and function of nature, and goes further to facilitate consideration and management of social, economic, cultural and other aspects of human–environment interaction. Diamond (2005) examined past societies which have sustained themselves and those which have failed, noting social factors tend to dominate natural factors in determining success; also, a number of small regional units with adaptable bottom-up organisation fared well. Some sort of biogeophysical canton or county might be a promising route to EM.

Ecozones, ecoregions and ecodistricts

There have been attempts to divide the Earth into ecozones or life zones for study, planning and management (Schultz, 2005). One of the best-known land use classifications is the Holdridge life zone model. This is based on the relationship of current vegetation biomes to three parameters: annual temperature, annual precipitation and potential evapotranspiration (Holdridge, 1967, 1971). It is a simple system requiring few empirical data and giving objective mapping criteria, and has a spatial resolution of one-half degree latitude/longitude; 37 life zones and over 100 life-zone classes are recognised. The model has been applied to soil carbon sequestration research (Jungkunst et al., 2021). The model is used in land use classification and predicts ecoclimatic areas but does not directly model actual vegetation or land cover distribution (Van Dyke, 2008). Zoning can be done with an eye for dovetailing mutually supportive activities and encouraging co-operation between sectors, agencies, etc. Areas can also be zoned according to their biodiversity, conservation needs, vulnerability, resilience, susceptibility to hazards, etc. GIS techniques allow EM to zone with virtually any variable that suits needs.

The Netherlands National Institute of Public Health and Environmental Protection have developed a framework for hierarchical ecosystem classification to try to overcome the confusion resulting from the use of many different geographical regionalisations by various bodies; it is used for regional environmental policy. It ties in with GIS, is useful for state-of-environment reporting and has been quite successful. Similar approaches have been tried in several countries, such as Canada, the USA and Belgium. Eco zones/districts/regions are used for research, as management units and routes to SD (Bailey, 2009).

Coastal zone and marine ecosystem planning and management

It is in the coastal zone that most human activity is concentrated (Moksness et al., 2009). With the threat of global warming and rising sea levels, coastal zone management and integrated coastal zone management are important (Ahlhorn, 2017; Ramkumar et al., 2018; Krishnamurthy et al., 2019).

An ecosystems approach has been explored for managing the Baltic Sea, the Mediterranean, the Aegean, the North Sea and the Japanese Inland Sea; also for the Great Lakes of North

America, the Aral, Caspian and Black Seas, and Lake Baikal. Some of these involve several countries, and in order to control pollution management must extend inland and ensure coverage of the hydrology of the whole basins. Marine protected areas, pollution control, aquaculture management, fisheries management, marine acidification monitoring and concern for coastal wetlands and reefs have an interest in ecosystems approaches (Kidd et al., 2011; Wondolleck and Yaffee, 2017).

River basin planning and management

In a river basin flowing water acts as an integrative element and is something to use to seek development. Watersheds offer a similar management/governance/study landscape unit, but the focus is more on moisture and soil conservation. River basins have long been used for integrated or comprehensive regional development planning and management, flood management, pollution management, environmental research and EM (Xiangzheng Deng et al., 2014; ADB, 2016; Smith et al., 2017; Komatina, 2018; Fusheng Li et al., 2022). The river basin is suitable for applying an integrated or holistic ecosystem approach, and is useful when several states share a river system (Brebbia, 2011; Peden et al., 2012; Squires et al., 2014; Ramkumar et al., 2015; Bucur, 2016; Xiangzheng Deng and Gibson, 2019; Collins et al., 2020; Kittikhoun and Schmeier, 2021; Xenarios et al., 2021). The shared management of rivers traversing more than one state is often dealt with through a basin approach (Schmeier, 2013).

There is probably more experience with the use of river basins as a means for integrated environmental–socio-economic planning and management than with any other ecosystem approach. Some concerned with ecosystem services have also used a basin approach (Chicharo et al., 2015).

Watershed/catchment planning and management

A watershed (catchment) offers a biogeophysical unit usually with well-defined boundaries within which agroecosystem use, human activity and water resources are interrelated. Watersheds are component parts of a basin. Researchers and environmental managers have made use of watersheds or subdivisions (micro-watersheds) to study how land use changes have affected hydrology, soil conservation and human welfare since the 1930s (starting with the US Forest Service Coweeta Experimental Forest). Watershed experiments seek to establish the effects of disturbing vegetation or soil, monitoring inputs to the basin (e.g. sunlight and rainfall) and outputs (flows, nutrients, sediment) by measuring quantity and quality of flows from streams or material removed by humans as produce. One of the best known of these is the Hubbard Brook Experimental Watershed (USA). Watersheds are useful for forestry, agricultural development, erosion control, water supply, pollution, fisheries management, etc. (Ferrier and Jenkins, 2021) and as units for integrated biophysical and socio-economic management (Ferrier and Jenkins, 2010; Smith et al., 2017; Dinesh Kumar, 2019).

Bioregionalism

Bioregionalism argues for human self-sufficiency at a local scale and support for natural, rather than political or administrative units, for managing development (Evanoff, 2014).

Bioregionalism is usually seen as a sense of place, adopting a life region and a social unit, and places emphasis on local adaptation to environment: citizen awareness of the ecology, economy and culture. It is often possible to nest a series of bioregions within each other. Bioregionalists generally seek community development in environmentally friendly ways, and strive for self-sufficient sustainable units (Lockyer and Veteto, 2013). Supporters of permaculture, a form of organic farming, often advocate bioregionalism and re-tribalisation of society. Bioregionalism has been promoted by biogeographers such as Peter Berg (Lynch et al., 2012; Cato, 2013; Glotfelty and Quesnel, 2014). The bioregion can be seen as intermediate between biogeographical provinces and ecosystems or groups of ecosystems.

Agroecosystem analysis and management

An agroecosystem differs from a natural ecosystem in that it is modified by humans to obtain food or other agricultural produce, usually they are simpler than natural ecosystems and perhaps less stable. An agroecosystem approach requires integration of ecology, economics and social studies and can be used to reduce the impacts of farming or other activities on the surrounding environment and improve the function of the exploited agroecosystem (Krishna, 2013). The approach can help reduce land degradation, improve soil management and may support ecosystem services like carbon sequestration, etc. (Benkeblia, 2019; Lemaire et al., 2019; Wezel et al., 2020). Examples of agroecosystems include wheatfields, rice paddys, and aquaculture.

Four agroecosystem properties can be recognised:

1 *Productivity* – output, yield or net income from a valued product per unit of resource input. This may be measured as yield or income per hectare, total production per household or farm, or at a regional or even national scale. Alternatively, it may be expressed as calories.
2 *Stability* – the constancy of productivity in the face of climatic fluctuations, market demand, etc.
3 *Sustainability* – the capacity of an agroecosystem to maintain productivity in the face of environmental challenges and degradation arising from its exploitation.
4 *Equitability* – the evenness of distribution of the productivity benefits among humans.

Most of the efforts to modernise agriculture focus on 1) above. This demands an understanding of ecosystems and of how natural processes are modified by agricultural objectives (Gliessman, 2014, 2019).

Telecoupling

Telecoupling refers to socio-economic and environmental interactions between distant coupled human and natural systems (CHANSs), which have become more extensive and intensive in the globalised era (Kapsar et al., 2019). The integrated framework of telecoupling examines global flows of information, energy, matter, people, organisms, diseases, financial capital and goods and products, etc. For example, demand for soya in the EU impacts on distant forests (Hull and Liu, 2018; Fris and Nielsen, 2019). The concept/framework is applied in land use change studies and may help EM formulate new types of governance focused on SD.

Landscape ecology approach

Landscape ecology focuses on spatial patterns at the landscape scale (e.g. hedges, fields, etc.), and how their distribution determines the flow of energy and materials and affects organisms (MacArthur and Wilson, 1967; Falk et al., 2011; Almusaed, 2016; With, 2019). The response of an ecosystem to disturbance frequently depends on its neighbouring ecosystems; organisms may escape if there are suitable nearby ecosystems and recolonise after disturbance ceases; also, energy or materials may be transferred between ecosystems. An ecosystem seldom functions in isolation and its ability to withstand stress may depend on how a nearby ecosystem is being managed, or on whether the boundaries are altered; a road or cleared area of forest may prevent animal or plant dispersal to an alternative site. The landscape ecology approach extends ecosystem management to a group of more or less neighbouring or linked ecosystems (could be urban ecosystems), and problems are generally dealt with in a holistic way (Coulson and Tchakerian, 2010; Bojie Fu and Bruce Jones, 2013; Rego et al., 2019). The approach is useful for resource and biodiversity management and predicting impacts of land use change and for supporting SD (Hong et al., 2010; Francis et al., 2016; Lopez and Frohn, 2017; Nobukazu Nakagoshi and Sun-Kee Hong, 2018).

Ekistics

Ekistics is the science of human settlements: it draws upon human ecology and regional planning and treats urban territory as a living organism, adopting an interdisciplinary, problem-solving approach; similar to an ecosystems approach, especially in its focus on networks. The approach seeks harmony between the inhabitants of a settlement and their physical and socio-cultural environments (Doxiadis, 1977; Rukhsana et al., 2021).

Summary

- EM must look carefully at the physical, social and economic factors involved in each situation before drawing conclusions.
- Lessons from the past are valuable to EM, but caution is needed because history rarely repeats itself exactly and future conditions may be novel.
- The environment is subject to natural change, sometimes sudden and unexpected. Humans have made such shifts more likely and prediction less easy.
- Ecosystems are widely used as study, planning and management units.
- Few ecosystems are natural; many have altered drastically and must therefore be managed to avoid degradation.
- The carrying capacity of an ecosystem may be stretched by means of trade, human labour and ingenuity, technology, etc.
- Net primary productivity often increases at the cost of species diversity (biodiversity loss).

Further reading

Beeston, M. (2019) *Environmental Populism: the politics of survival in the Anthropocene.* Palgrave Macmillan, London, UK.

Lovelock, J. (2006) *The Revenge of Gaia: why the Earth is fighting back – and how we can still save humanity.* Allen Lane, London, UK.

Roberts, J. (2020) Political ecology. In: F. Stein, S. Lazar, M. Candea, H. Diemberger, J. Robbins, A. Sanchez and R. Stasch (Eds), *The Cambridge Encyclopaedia of Anthropology* (pp. 1–17). Cambridge University Press, Cambridge, UK.

Tyrell, T. (2013) *On Gaia: a critical investigation of the relationship between life and Earth.* Princeton University Press, Princeton, NJ, USA.

www sources

❏ Gaia hypothesis outline: https://courses.seas.harvard.edu/climate/eli/Courses/EPS281r/Sources/Gaia/Gaia-hypothesis-wikipedia.pdf (accessed 07/12/22).

❏ Ecosystem management (Science Direct): www.sciencedirect.com/topics/agricultural-and-biological-sciences/ecosystem-management (accessed 07/12/22).

❏ Ecosystem Management (Colorado State University): www.nrel.colostate.edu/research/ecosystem-management/ (accessed 08/12/22).

❏ A Guide to Ecosystem Models (Nature): https://docs.google.com/document/u/0/ (accessed 07/12/22).

Chapter 4

Environmental management background

Chapter overview

- Environmental concern 1750 to 1960
- Environmental concern 1960 to 1980
- Environmental concern 1980 to the present
- Environmentalism, ecologism and the Green Movement
- Social sciences and environmentalism
- The greening of economics
- Environmental accounts
- Estimating the value of the environment and natural resources
- Paying for and encouraging EM
- Green aid
- Natural capital and ecosystem services
- Debt, structural adjustment and the environment
- Debt-for-nature/environment swaps
- Trade and EM
- Summary
- Further reading
- www sources

Environmental concern 1750 to 1960

Some societies protect plants and animals for culture, religion or local economy; e.g. baobab in parts of Africa. Here and there rulers established reserves for hunting and recreation long before the twentieth century. By the 1760s some colonial powers legislated to try to protect soil and forests, as in Tobago, Mauritius, St Helena, and some other countries. In the nineteenth century European and American leisured classes took an interest in natural history but had little awareness of the structure and function of the environment.

DOI: 10.4324/9781003189985-5

Two broad groupings of those interested in nature evolved in Europe and America:

1) *Utilitarian environmentalists*

In the late nineteenth century the British sought assistance from German foresters to sustain timber production in Burma (now Myanmar) and India. Political theorists like Kropotkin (in Russia), argued for small, decentralised communities close to nature and avoiding industrialisation and the division of labour. The UK had utopian liberals and proto-socialists such as William Morris (Morris, 1891), John Ruskin and Robert Owen; the latter founded utopian settlements, with limited success, in the UK, Ireland and the USA in the 1820s. In South Africa and a few other countries, legislation tried to reduce soil erosion, control hunting and conserve forests and areas of beauty. By 1900, game reserves had been established in Kenya and South Africa, some by ex-hunters. In North America by the 1850s only a few feared frontiers were closing; one who clearly saw this was George Perkins Marsh, who in 1864 published *Man and Nature*. This and publications by others prompted two utilitarian groups concerned for the American environment: 'preservationists' and 'conservationists'. The former included John Muir, who wished to maintain unspoilt wilderness areas; the latter included Gifford Pinchot and were prepared to see protection combined with careful land use. EM often faces this preservation/conservation dilemma today.

Travels by Alexander von Humboldt (between 1799 and 1804) and his book *Cosmos*, and by Charles Darwin (between 1831–1836) and his books, especially *The Voyage of the Beagle* and *On the Origin of Species*, stimulated interest in nature. The British established conservancies in India. John Muir in the USA founded the Sierra Club in 1892, still an influential NGO; it has played a role in environmental concern since the 1960s, and helped establish Friends of the Earth.

After 1917 divergence of development paths between Russia, other socialist economies, developing nations and the West made little difference: all blocs had and have serious environmental problems (Komarov, 1981; Smil, 1983), and have contributed to developing EM.

2) *Romantic environmentalists*

The eighteenth- and nineteenth-century Industrial Revolution led in Europe and North America to filthy cities, damaged countryside, loss of commons, disease and misery. Various intellectuals questioned capitalism, agricultural modernisation and industrial growth. These 'romantics' saw the environment as a source of inspiration, and advocated a less damaging relationship with nature. They included poets like Wordsworth and Coleridge, writers like Henry Thoreau (1854), and artists like Holman Hunt and John Turner, but their contribution is more escapist than practical or visionary.

Drought, depression and aggressive farming in the US Midwest, between 1932 and 1938, caused crop loss and soil erosion. The wind-blown dust was visible in Chicago and Washington, DC. Many farmers were ruined and displaced and the folk singer Woody Guthrie and novelist John Steinbeck protested at the degradation and misery; at first seen as subversives, they provoked public and government concern. To lessen the problems, President Franklin D. Roosevelt promoted integrated development of natural resources and in 1933 established the US Soil Erosion Service and in 1935 its successor, the US Soil Conservation Service (Montgomery, 2007; Cory et al., 2015).

The Second World War cut concern for the environment, accelerated development of resources and led to new threats such as atomic weapons. After 1945 efforts focused on economic and industrial reconstruction, on raising agricultural production, and until roughly 30 years ago on the Cold War. A few publications on the environment did appear from the late 1940s (Osborn, 1948; Vogt, 1948). It was probably Aldo Leopold (1949) who stimulated environmentalist interest in the late 1960s. In 1949 the UN held one of the first international environmental meetings, the Conservation Conference at Lake Success (USA), and during the early 1950s helped establish the International Union for the Protection of Nature, which in 1956 changed its name to the International Union for Conservation of Nature and Natural Resources (IUCN).

Environmental concern 1960 to 1980

By the 1970s there had developed what has been variously called an environmental(ist) movement, environmentalism, ecology movement, an environmental revolution, and the conservation movement. In the 1960s and 1970s, particularly in California, public-interest law groups (e.g. the Environmental Defense Fund or the Natural Resources Defense Fund), supported by grants or foundations, acted on behalf of groups of citizens (previously legal action had to be by individuals) to protect the environment. Understanding of the structure and function of the environment was improved by the International Geophysical Year (1957–1958), the International Biological Program (1964–1975) and the International Hydrological Decade (1965–1974), plus expanding research institutions. The USA Civil Rights movement, 'hippies', the anti-Vietnam War movement, European anti-nuclear protests and 1960s to 1970s pop culture encouraged people to ask questions including about environment and development. However, after a peak in the mid-1970s, media and public interest declined until the mid-1980s. Some credit for raising environmental concern also goes to broadcasters like Peter Scott, David Attenborough (UK) and Jacques Cousteau (FR), and to Princes Rainier (MC) and Phillip (UK).

In the 1970s, the environmentally concerned, although active in publication, litigation and protest, were relatively non-political (in New Zealand, Germany and the UK environmental political activity appeared) (Bahro, 1986). Environmentalists at the time were prone to apocalyptic advocacy so became known as prophets of doom, ecocatastrophists or cassandras.

In 1965 Adlai Stevenson used 'Spaceship Earth' in a speech; the idea that the world was a vulnerable, finite, closed system spread. The International Biological Program, and the UNESCO Man and Biosphere Program, helped establish that global-scale problems were real and the Earth's resources were finite. In the 1970s many identified population growth as the primary cause of environment and development problems. The more extreme neo-Malthusians even discussed the possibility of triage (withholding aid for over-populated countries). Hardin (1968, 1974a, 1974b) published essays on common property and on 'lifeboat theory'; his arguments were criticised along with the views of the neo-Malthusians.

Other publications shook complacency: especially *Blueprint for Survival* (Goldsmith et al., 1972) and *The Limits to Growth* (Meadows et al., 1972). Debate developed between those advocating slow or even zero economic growth and opponents to 'limits' like Simon (1981). Those too optimistic about limits have been called cornucopian. Schumacher (1973) argued large organisations led to specialisation, economic inefficiency, environmental damage and inhumane working conditions. The remedies he offered included Buddhist (or Gandhian) economics, intermediate technology (smaller working units, local labour and resources) and respect for renewable resources.

Around 1970 the USA passed the National Environmental Policy Act (NEPA). NEPA established environmental impact assessment (EIA) and proactive environmental legislation; within ten years EIA had spread worldwide to become an important input to EM.

Environmental concern 1980 to the present

Two publications: the *World Conservation Strategy* (IUCN, UNEP, WWF, 1980) and the *Brandt Report* (Independent Commission on International Development Issues, 1980) triggered interest. The *Brandt Report* stressed problems would be solved only if it was recognised that rich and poor countries had a mutual interest: aid was not a question of charity but of global interdependence. The *World Conservation Strategy* promoted conservation for sustainable development (SD). The *Brundtland Report* (World Commission on Environment and Development, 1987) again highlighted SD and urged a marriage of economics and ecology.

By the 1980s the World Bank was adjusting its policies to give greater support to EM and a Green Movement had emerged, particularly in Europe, that made politicians of all persuasions aware of environmental issues. Green politics peaked in the 1980s, but other parties increased their environmental activity. From the 1990s in many countries, popular environmentalism and environmental justice movements appeared.

Roughly from 2015 popular environmental concern again blossomed, mainly focused on climate change. By 2021 environmental activism in the West, some of it in the form of eco-disruption/disobedience (e.g. Extinction Rebellion), was booming. One controversial citizen/youth voice was Greta Thunberg (2019).

Environmentalism, ecologism and the Green Movement

Many greens see economic growth and consumerism as tainting both capitalist economies and socialist states. Less radical green philosophy can be embraced by existing politics, from liberal to dictatorship; radical *deep*-green beliefs demand fundamental changes in politics, worldviews and ethics.

Environmentalism

Environmentalist is a label applied to those involved in environmental matters (Pepper, 1996). Environmentalism is a moral code or a set of mediating values to manage human conduct; environmentalists aim at improving the environment. A diverse group with varying ideologies and objectives; some environmentalists embrace technology and the free market; some reject technology and/or capitalism. Mixed with moderate environmentalism are silly, alarmist, even dangerous, but appealing, views (Zubrin, 2012; Jensen et al., 2021). There is some backlash against environmentalism (Horner, 2007; Rogers, 2010; Lomborg, 2001, 2020; Shellenberger, 2020). Social media has made campaigning and lobbying wider, faster and easier, but the downside is that falsehoods can be easily propagated (Delingpole, 2012; Zehner, 2012; Seymour, 2018). Environmentalism includes alternative views/romantics (Heynen et al., 2008), neo-liberalism (Islam, 2013; Buscher et al., 2014; Apostolopoulou and Cortes-Vasquez, 2018), eco-socialism (Huan Quingzhi, 2010; Okereke, 2010), eco-authoritarianism (Yifei Li and Shapiro, 2020) and eco-fascism (Biehl and Staudenmaier, 2001). There are also faith/religious environmentalists.

Box 4.1   Some common green characteristics

The 'four pillars of green':

1 ecology;
2 social responsibility;
3 grassroots democracy;
4 non-violence.

The 'six values of green':

1 decentralisation;
2 community-based economics;
3 post-patriarchal principles;
4 respect for diversity;
5 global responsibility;
6 future focus.

Green characteristics:

- holistic approach;
- disillusionment with modern unsustainable development paths;
- non-violence;
- a shift in emphasis away from philosophy of means to ends;
- a shift away from growth economics;
- a shift towards human development goals;
- a shift from quantitative to qualitative values and goods;
- a shift from impersonal and organisational to interpersonal and personal;
- commonly a feminist interest;
- a decentralised approach – 'think globally, act locally'.

Some environmentalists seek an end to modernism (fulfilling human needs through the development of technology and the creation of wealth), calling for postmodern alternatives. Postmodern as a concept is confused, although a postmodern period might be recognised from the 1960s ongoing (Kallis and Bliss, 2019).

Ecologism

Ecologism is an ideology that argues for care of the environment and a radical change in the human relationship with nature to get it (Baxter, 1999). Ecologism lies at the radical or fundamentalist end of the environmentalist spectrum with a strong spiritual component. Unlike most environmentalists (and EM) they are unwilling to manipulate and alter the environment even if human needs are pressing.

Green spirituality

Environmentalists include those seeking reconnection with nature and those who focus on faiths and their relationship with nature; examples include Pierre Teilhard de Chardin (1959, 1964) and Matthew Fox (Fox, 1989). Green spirituality grew after 2000 (Gottlieb, 2010; Taylor, 2010; Van Eyk McCain, 2010; Mattson et al., 2011; Philpott, 2011; Kelson et al., 2016). Interest comes from Christians, Jews, Muslims, Hindus, Buddhists, Jains and those looking to pre-Christian or pre-Islamic religions or philosophy for inspiration to transform humans so they improve reverence for nature (Cooper and James, 2017; Steiner, 2008; Taylor, 2009; Scruton, 2012; Sponsel, 2012; Macey, 2013). Islamic interest in environment, climate change and SD has expanded (Foltz et al., 2003; Abdul-Matin, 2010; Al-Jayyousi, 2012; Gade, 2019; Khalid, 2019).

Environmental problems are sometimes blamed on a supposed Judaeo-Christian world-view of human dominion over nature. In 1986 the World Wide Fund for Nature held a meeting at Assisi, where leaders of Buddhist, Christian, Hindu, Islamic, Judaic and other faiths established an International Network on Conservation and Religion, and published the Assisi Declarations on Man and Nature. Interest is ongoing (Francis, 2015).

The Green Movement

The Green Movement is a diverse social or cultural movement that shares an environmental concern and which embarks on advocacy and/or political action. Green means environmentally friendly; greening means environmental improvement (Doherty, 2016). The movement draws upon the writings of Henry Thoreau, Theodor Roszak, Ivan Illich, Aldo Leopold, Martin Luther King, etc. Although green implies politicised environmentalism, a few groups are not politically active, and even eschew politics.

There is little about green philosophy that is new. Greens may be socialists, conservatives, intellectuals or uneducated, poor or rich, Buddhists, Christians, Muslims or secular. Most fear that industrial nations are pursuing an unsustainable, dangerous development path. The Green Movement has spread worldwide (Day, 2005; Dabashi, 2011; Geall, 2013; Glaeser, 2013; Shiyi Chen, 2013; Zhang, and Barr, 2013; Finamore, 2018; Yanzhong Huang, 2020; Pourmokhtari, 2021; https://greeninitiatives.cn/the-birth-of-chinas-green-movement/ (accessed 21/05/21).

Some European environmentalists became politically active in the 1970s; a leading role was played by the Hamburg Greens (die Grunen). The Greens' popularity in Europe faltered after the mid-1980s, partly because established political parties hijacked the cause. In the USA environmentalists concentrated more on getting supportive legislation and advocacy, and green politicisation barely took off.

Dark- deep- light- and bright-greens Greens may be subdivided into romantic, anarchistic, and utopian; or simply into 'light-' or 'dark-' (and recently 'bright-green') (Grasso and Giugni, 2022). The light-green (or shallow) and deep-green (or deep) division was largely initiated by the Norwegian philosopher and founder of deep ecology Arne Naess (1988; Drengson and Devall, 2016). Deep-green/deep ecology seeks to replace the existing social, political and economic status quo with new environmentally appropriate bioethics and politics; it adopts a biocentric (ecocentric) outlook, granting all life intrinsic value (Devall, 2021). Deep ecology includes social ecology and eco-feminism, and some advocates incorporate Taoist or Gandhian philosophy. Deep ecology can give non-scientific input similar importance to (if not greater than) scientific, and is sometimes hostile to science. Some radical, deep ecologists act as eco-warriors such as Earth First! or Extinction Rebellion, and climate or animal rights activists may use violent protest, monkeywrenching or ecotaging (Schreurs and Papadakis, 2007; Scarce, 2016; Extinction Rebellion, 2019). There has been significant backlash against the more extreme Green Movement.

Social ecology is sometimes seen as a deep-green stance; it was largely initiated (in the USA) by the anarcho-socialist Murray Bookchin (1990). Social ecology supporters see environmental problems to be the result of social issues, and adopt an anthropocentric, decentralised, co-operative approach, a sort of eco-anarchy. Another difference from mainstream deep ecology is that social ecology is humanist rather than ecocentric. There is ongoing evolution (Wright et al., 2013; Eiglad, 2015; Fischer-Kowalski et al., 2016; Zebich-Knos and Grichting, 2017; Stokols, 2018), including groups interested in resilience-building (Ungar, 2012).

Light-green or shallow ecology seeks to apply ecological principles to ensure better management and control of the environment for human benefit; it is usually anthropocentric. There is less rejection of established science characteristic of deep-green/deep ecologists. Shallow ecology is more inclined to try to work with existing economics and ethics (Waddington, 2010). Light-greens see environmental protection as a personal responsibility; deep-greens believe damage is an unavoidable part of industrialised civilisation, and seek radical political change.

Bright-green environmentalism dates from about 2003, the term coined by Alex Steffen (Steffen, 2006). These greens feel that changes are needed in the economic and political operation of society, but that better designs, new technologies and social innovations are the means to make those changes and that society cannot protest its way to survival. Bright-greens focus on tools, models and ideas that are available and realistic (Jensen et al., 2021; Schwartz and Cohen, 2021).

Social sciences and environmentalism

Social sciences are vital in addressing the human–environment interrelationship (Moran, 2010; Walker and Schand, 2017). The potential inputs to EM from the social sciences are:

- to provide information on social development needs and aspirations to explain, present and predict future human attitudes, ethics and behaviour;
- to study and develop ways of focusing the activities of social institutions, non-governmental organisations, groups of consumers and so on to achieve better EM;
- to show EM social constraints and opportunities;
- to unravel the often complex and indirect social causes of environmental problems;
- the articulation and fulfilment of the shared interests of people;
- national governments have mainly been reactive rather than forward-looking: social science will be needed to clarify how people think, nations relate to each other and institutions behave if a more proactive approach is the goal.

There has been exchange of concepts and jargon by social sciences from and to the sciences, but sometimes things become distorted because not all social scientists derive concepts by a process of logic or experiment and may seek bolt-on justification for values they already hold (Vaccaro et al., 2010; Scoones, 2015; Mullen, 2018).

Anthropology can inform EM about human behaviour, attitudes and beliefs, institutions, and organisational capacity, etc. (Dove and Carpenter, 2007; Brightman and Lewis, 2017). Anthropological input to EM has been strong in the fields of relocation and resettlement, pre-development appraisal, impact assessment, conservation area management planning, studies of resource use, SD, hazard perception, bioregionalism (Lockyer and Veteto, 2013), and survival strategies adopted by land users (Kopnina and Shoreman-Ouimet, 2015a, 2015b, 2017; Wang Tianjin, 2018). Ethnobotany involves anthropologists and ethnographers in identifying useful crops, pharmaceuticals and so on (Schultes and von Reis, 1995; Vajravelu, 2009; Albuquerque et al., 2017; Martinez et al., 2019). Environmental sociology and ecosociology explore the interactions between society and the environment; these fields have eclipsed the human ecology approach (Gross and Heinrichs, 2010; Bell, 2012; Young, 2015; Burns and Caniglia, 2017; Carolan, 2020; Legun et al., 2020).

A late twentieth-century paradigm shift?

One can argue there was a worldwide paradigm shift in the late twentieth century (ongoing), whereby diverse political groups, religious persuasions and citizens old and young share concern for the environment to a far greater extent than in the past. Still citizens and politicians make Faustian bargains: decisions that sacrifice long-term wellbeing for short-term gain, and/or, worse, act in their own self-interest (Gatt, 2018). But there are dangers in adopting green policies without carefully researching their likely impacts; Sri Lanka supported organic agriculture and much reduced fertiliser use after 2019; soon the country slipped from rice self-sufficiency to importing, other export crops declined and the economy had collapsed by 2022.

Ethics for EM

Ethics are the non-legal rules and principles which order human existence. Ethics are related to values – things which people hold dear and wish to support. Worldviews, the perceptions a person or group have of their surroundings, overlap with ethics and values. Even within one family, perceptions vary so generalisation must be cautious. Environmental ethics and laws are evolving but that process is incomplete and there are often inadequacies (King-Tak Ip, 2009; Keller, 2010; Rolston III, 2012; Hourdequin, 2015; Nolt, 2015; Thompson and Gardiner, 2015; Attfield, 2016, 2018; Iannone, 2017; Sandler, 2017; Franks et al., 2018; Chemhuru, 2019; Bassham, 2020; Traer 2020). EM ethics range from eco-friendly utilitarianism to authoritarian eco-fascism.

Women and the environment

Environment-human interrelationships are gender-sensitive (gender being a set of roles) (Detraz, 2017). Women have played a key, if not the major, role in establishing environmentalism and green politics and many of the world's conservation bodies (Carson, 1962; Carrol, 2019). Eco-feminism explores women and environment issues in a wide diversity of ways, some a form of feminist political ecology, some spiritualism, some cultural/literary (Harcourt and Nelson, 2015; Phillips and Rumens, 2015; Adams and Gruen, 2022).

It has been argued that women view environment and development differently from men because of their reproductive role and gender differences in employment, income, freedom and perception of resources. Women influence future behaviour and can support environmental care when educating children. The role, income and influence of women are generally growing. Women and children frequently have different diets and exposure to pollution and other threats. Whether rural or urban, women and men are likely to have different livelihood activities and dissimilar access to resources (Hawley, 2015; Le Masson and Buckingham-Hatfield, 2017; MacGregor, 2017; Vakoch and Mickey, 2018).

Rural societies often experience out-migration of men to work in cities, mines or overseas; there are also areas where more women than men find overseas employment or become skilled local workers. Women are often sidelined from inheritance and form the poorest sector of society. Where men are dissolute and lazy, women may initiate pollution control, conservation, community forestry/tree planting, and improvement of water supplies. But it is common for just men to be consulted, while women work the fields. Women are resilient and inventive because they have to be when migrant males fail to send back remittances, etc.

Debates on women and development can be subdivided into women in development (WID), women and development (WAD), and gender and development (GAD). WID focuses more on improvement of women's welfare and their role in economic development. WAD explores relationships between women and the development process, not just strategies to improve the integration of women. GAD looks more at the roles of sexes, their needs and interests, and ways in which each can actively participate in development.

Eco-feminism sees parallels between the oppression of women and exploitation of the natural world: a gender-neutral approach is inadequate and masculine control may have to be opposed. Romantic environmentalists flag the Earth goddess, sensitivity-to-nature aspects of femininity. Radical eco-feminists include Anita Roddick who used her *Body Shop*® stores in the 1980s in the UK to support fair trade and environmentally sound marketing.

Social aspects of resource use

An understanding of people's attitudes, capacities and needs is vital for managing resources, biodiversity conservation, pastoral development, etc. New socially appropriate and workable ways may have to be found, and new managing institutions built and maintained for EM.

Social forestry and community forestry deal with the establishment and management of forest, woodlots and hedges by and for local people. The focus is on planting and maintaining trees, in the most appropriate manner, with minimal dependency on the outside, making use of local knowledge and with awareness of land rights and traditions (Lacuna-Richman, 2011; Arts et al., 2012; Hanna and Bullock, 2012; Joseph et al., 2018). Forestry may be linked to farming (agro-forestry) (Divya, 2008; Schwab, 2009). Social foresters can help identify why people destroy trees and find ways of countering such behaviour.

Conservation efforts (biodiversity and heritage conservation) have often been insensitive to locals, which can alienate them, sometimes triggering poaching and other destructive activities (McMahon, 2010; Porter-Bolland et al., 2013; Chitty, 2017; Berkes, 2021). However, there are strong criticisms of some community-based conservation. When the cost of failure is the extinction of biodiversity, a top-down approach can be excusable until workable alternatives are found. Chico Mendes (assassinated 1988) and others fought clearances and promoted the concept of the extractive reserve in Brazil as a way of protecting flora and fauna and allowing local people to collect products with limited forest damage (Cardoso, 2018). Similar conservation results can be obtained from tolerant forest management strategies: the thinning of understorey species and encouragement of useful trees, although disturbance prevents total biodiversity protection.

The greening of economics

Economists have long invoked the 'invisible hand of the market' as a mechanism which supposedly ensures that it becomes uneconomic to exploit a potentially renewable resource until it is badly damaged. Unfortunately, the market has not been an effective control: there are numerous examples of ruined soil, fisheries and lost forests. The reason for this is that 'the free market' does not provide consumers with proper information, because the social and environmental costs of production have not been part of economic models. Private profits are being made at public costs in the deterioration of the environment and the general quality of life, and at the expense of future generations. Another problem is the difficulty in valuing many resources; for example, it is not easy to assign a value simply because a species is rare, and some things are presently valueless since a use has yet to be found.

Resources and environment may be used to give outputs like crops or benefits (ecosystem services such as recreation) or there may be non-use (intrinsic) value (e.g. conservation provides material for future pharmaceutical use or crop breeding). When a resource or the environment has current utility (i.e. can give 'satisfaction'), this may be gained directly, say by the use of land for recreation or tourism, or indirectly through manufacturing.

Many attempts to define economics mention resources or environment; e.g. economics is essentially the stewardship of resources; or economics is concerned with the allocation, distribution and use of environmental resources (Harris and Roach, 2013). It is thus puzzling why, before 1990, there was little contact between economics and environmental studies. Given the difficulties involved in effectively valuing nature, and in dealing with human use of the environment and resources, such criticisms are slightly unfair. Also, before the 1980s few economists recognised that the Earth was finite and most encouraged expansion.

One of the first to publish on resource and conservation economics was Ciracy-Wantrup (1952). Environmental economics is now mainstream (Kennet et al., 2012a, 2012b, 2015; Schmelev, 2016; Haab and Whitehead, 2014; Martinez-Alier and Muradian, 2015; Managi and Kuriyama, 2017; Spash, 2017; Endres and Radke, 2018; Acar and Yeldan, 2019; Anderson, 2019; Buchholz and Rübbelke, 2019; Tietenberg and Lewis 2020). Still many companies (especially asset management firms) seek maximum profits for their shareholders, which does little to encourage investment in the environment.

Green economics stresses quality of life and is less concerned with capital accumulation. There is now pollution control economics (Eloi, 2020) and environmental taxes. Banks and trade unions promote green awareness (Räthzel and Uzzel, 2013). The two widely stated goals of green/environmental economics are: 1) to cut extravagant resource exploitation; 2) to seek SD. Green economists argue that environmental care should stimulate economic growth by improving the health of the workforce, making it more productive and creating employment in the green sector (pollution control, carbon sequestration, etc.).

A portion of the world's population is not directly affected by economics, due to subsistence lifestyles and remote situations. Barefoot (Gandhian) economists advocate small-scale enterprises (Max-Neef, 1992a, 1992b). Innovation does not solely happen in rich countries: developing nations have world-class economists, ideas also flow South to North.

Economists have tried to improve the environmental sensitivity of a key tool: cost–benefit analysis (CBA), and there have been attempts to incorporate economic evaluation into EIA (Pearce et al., 2006; Hanley and Barbier, 2009; Nyborg, 2012; Atkinson et al., 2018). Ecosystem valuation is vital for EM seeking to resolve conflicts, achieve trade-offs and develop green accounting (Kumar and Thiaw, 2013; Trudgill, 2014; Ryan, 2016). There has been greening of accountancy (Bartelmus and Seifert, 2019).

Human capital / social capital / cultural capital / built capital

Human capital includes: education; mental wellbeing; health; communication skills; and management skills. It is possible to invest in human capital and improve it by educating or training. It can be lost and be very difficult or impossible to regain. Social capital is the networks of relationships among people who live and work in a particular society, enabling that society to function effectively. As with economic capital there can be a subdivision (social bonding, bridging and linking capitals). Sociologists recognise cultural capital: individuals' social assets (education, intellect, etc.). Human capital is about skills and knowledge; social capital more about linkages/interactions between people. Built capital is any pre-existing or planned formation that is constructed or retro-fitted to suit community needs (any human-made environment): networks of infrastructure and economic development.

Global environmental problems and economics

EM must manage transboundary issues (Brauch et al., 2008). Of these transboundary/global issues, climate change attracts huge attention (Hawken, 2017; Abunnasr et al., 2018; UNCTAD, 2020). Globalisation greatly affects the environment and society but could offer opportunities for EM (Ho-Won Jeong, 2005; F. Harris, 2012; Newell, 2012; Christoff and Eckersley, 2013; Newell and Timmons Roberts, 2016; Verma, 2018; Eastwood, 2019).

Environmental accounts

There are many ecosystem/environmental auditing approaches: eco-audits, environmental stocktaking, eco-review, eco-survey, etc. (Aronsson and Löfgren, 2012; Gray et al., 2014; Debnath, 2019; Bebbington et al., 2021). State-of-the-environment accounts, environmental quality evaluations and environmental accounts systems collect data on the environment and resources to try and show the state of a land or marine area (Meyer and Newman, 2022). These accounting procedures treat the environment as natural capital and try to measure its depletion or enhancement. Techniques such as eco-footprinting seek to trace and value flows of resources and activities associated with discrete areas or activities. Land use and land cover (LULC) classification (and assessment of change) using satellite, UAV (un-manned aerial vehicle) or aircraft aerial remote sensing or ground assessments is important for EM seeking to monitor and account for degradation and improvement of land, and various ecosystems (Helming et al., 2008; Giri, 2012; Mölders, 2012).

The foundation for these procedures has often been the UN model of standard national accounts, usually with 'satellite accounts' added for environmental items; some call these 'environmentally adjusted national accounts' (UN, 1993; UN et al., 2014). Such accounts seek to establish the stocks of resources, value of environmental features and their use over time. National environmental accounts systems (new systems of national accounts, green accounts, patrimonial accounts or state-of-the-environment accounts) have been developed to assist with data gathering and storage and to value environment and natural resources. Canada, Denmark, Norway, France, Japan, USA, The Netherlands and the World Bank have developed national state-of-the-environment accounts and the UNEP has been promoting them in developing countries. Some follow The Netherlands' model, comparing output of each sector of the economy with how much it depletes finite resources such as fossil fuels.

A natural resource accounting system can help EM establish what percentage of, say, mineral exploitation profits to invest in long-term SD so that a region or country does not suffer boom and decline (Giampietro et al., 2014; UN, 2017). In practice, being able to make such investments depends on the type of government, people's attitudes and the persuasiveness of EM. Natural resource accounts can show the linkages between the environment and the economy, may be useful for forecasting, and can establish which habitats are of importance. They should make land use more rational, but stop short of encouraging a crucial change in people's and administrators' attitudes towards environmentally sound development.

Estimating the value of the environment and natural resources

Resource inputs have been divided by assessors into: *renewable* ('stock resources') which are robust enough to withstand poor management; *potentially renewable* (dependent on wise management); and *non-renewable*. Some renewable resources can be degraded to non-renewable through poor management, environmental change or natural disaster. Certain resources cannot be remade if damaged or exhausted (e.g. biodiversity). The absorptive capacity of the environment, its ability to absorb and neutralise damaging compounds or activities, should be assessed by economists: it tends to be undervalued. There may be opportunities to substitute for a given resource, using labour, capital or alternative materials.

Cost–benefit analysis

Cost–benefit analysis (CBA) seeks to identify the impact of development on each person affected at various points in time, and so estimate the aggregate value which each person gains or loses. Dissatisfaction with CBA effectiveness in valuing environmental facets led to suggested improvements, some favouring quantitative approaches, and others qualitative (Graves, 2007). There has been progress, including CBA application to global environmental issues (Pearce et al., 2006; Hanley and Barbier, 2009; Nyborg, 2012; Livermore and Revesz, 2013; Meisel and Puttaswamaiah, 2017; OECD, 2018).

BATNEEC and BPEO

Two principles have been widely used when assessing EM costs:

(1) Best available techniques not entailing excessive costs (BATNEEC) introduced by the EEC in 1984. This places the onus on developers to adopt the best techniques available, with only excessive cost as an excuse for not doing so.
(2) Best practicable environmental option (BPEO) is where a choice exists for a process, technique or substance; it is the best option, with respect to environmental protection (having carefully considered local conditions, individual circumstances, currently available technology and financial implications).

Shadow prices

The difficulty of establishing the value of externalities, including environmental factors, in monetary units has been addressed in several ways: one is to use shadow prices. Applied to currently unknowable or difficult-to-calculate costs in the absence of available market prices, a shadow price is a value that reflects the true opportunity cost of a resource or service. Opportunity cost is the value of the next-best alternative when a decision is made, i.e. what is given up. The real value of something reflects the most desirable alternative use for it. For example, the opportunity cost of producing an extra unit of manufactured goods is the lost output of childcare, food production, etc., forgone as a result of transferring resources to manufacturing activities. In consumption, opportunity cost is the amount of one commodity that must be forgone in order to consume more of another.

Paying for and encouraging EM

EM may be funded by national and international taxation and by business (e.g. through sponsorship) (Ramutsindela et al., 2017). Some of the funding may be made available to developing countries for EM tasks via provisions such as the Global Environmental Facility (GEF) (World Bank, 2016). The GEF is also a funding mechanism for treaties and conventions like the Convention on Biological Diversity; the UN Framework Convention on Climate Change; and the UN Convention to Combat Desertification. In addition to national and international (UN) bodies, NGOs/charities funds are available. Money can be raised from profits on recreational activity such as lotteries (some cultures would be unhappy to receive money from gambling/usury) or a levy on other things. The generation of funds from investment into green activities is expanding with some life assurance companies, banks and pensions schemes investing only in environmentally or socially beneficial activities; international bonds similar to those issued in wartime by the UK and USA may be a possibility (Ramiah and Gregoriou, 2015).

Environmental, social and governance (ESG) investments have become established; they support SD, environmental and social goals but returns to the investor take priority. Global impact investments are a more recent development more focused on addressing world problems, e.g. investing in carbon reduction.

Fair trade

Many small farmers in developing countries have transitioned from subsistence to production of export crops. They become vulnerable to market trends, controls on transportation, marketing and so on. For example, coffee provided a useful income until the late 1990s when world prices fell; the consequences in countries like Peru were small farmer migration to cities, narcotics production, and in some cases a resort to shifting cultivation. Attempts to even out market fluctuations through coffee cartels such as the International Coffee Agreement (set up in the 1960s) have not been enough. Escape from poverty and reduction of land degradation in areas with smallholder coffee or cocoa producers now lies with fair trade initiatives through companies such as *Cafédirect®*. Fair trade arrangements seek to improve the revenue going to producers by cutting out the middleman and aiding transport/marketing (Zaccaï, 2007; DeCarlo, 2011; Hudson et al., 2013). Many local co-operatives track market prices. They may also be able to offer produce as something with local character and so carve out a market. There are critics of fair trade (Ndongo, 2014; Ehrlich, 2018; Fridell et al., 2021).

Contract farming

Small producers can be vulnerable to swings in market prices and have difficulties acquiring inputs such as improved seeds, fertilisers and irrigation equipment. They also face problems in packing and transporting produce to market, and in advertising and selling it. A number of large companies, especially supermarkets, have developed contract farming approaches. The supermarket finds a reliable supplier which it can control, and it knows the price will not fluctuate too much. The farmer is insulated from very low prices but misses out on high prices, but enjoys stable medium prices in return for signing a contract. The seller provides input and advice, transport and marketing, and insists on quality control and work practices (which sometimes include environmental care and employee welfare measures) (Leung Ping Sun and Setboonsarng, 2014; Kuzilwa et al., 2017). In some cases production would be unlikely to start without the support of company patronage. The alternative might be degenerating subsistence agriculture, land damage, poverty and out-migration. There are downsides: farmers become dependent on others; reliance on a distant market entails risks; an unscrupulous company could in the long term drive down producer prices once farmers become locked-in, rather like sharecroppers it is a sort of globalisation debt-peonage. Contract farming needs to be monitored to ensure that it improves livelihoods, does not cause exploitation or environmental damage and perhaps aids EM.

Green taxes

Taxation may be used to discourage undesirable activities; reward beneficial activities through rebates or reduced taxation; encourage growth; and make issues public through the release of accounts and profit data (Ekins and Speck, 2011; Kreiser et al., 2012; Pirlot, 2017). Eco-taxes can be added to products, energy, services and materials to reflect true environmental costs. But these measures mean the consumer pays. While there have been national measures for some time, green taxation on an international scale has grown since about 2000, triggered by global warming, transboundary pollution, competition for internationally shared resources, etc.

A tax on say CFCs reflects their impact on the ozone. Green taxes counter the pursuit of lower prices by externalising the true costs. It is important that attempts to integrate external costs of production into prices do not burden the poor or punish the middle classes. The aim should be to give people and companies incentives to invent, innovate and respond to environmental challenges. Green taxation ideally encourages manufacturers to reduce waste and environmental damage to keep down their costs and thus prices to the purchaser.

Pigouvian taxes

These are intended to be levied on external costs, e.g. pollution to ensure that a manufacturer pays all costs from raw material and energy provision to final collection and recycling. Pigouvian taxes may present problems: large companies can make sufficient profits to afford fines, but small companies could be crippled. Thus the polluter-pays principle can be a virtual licence to pollute if the fines are not set high enough.

Carbon emissions taxes and incentives

Interest in climate change/carbon pricing and taxation has expanded dramatically (Parry et al., 2012; Farid et al., 2016; Crampton et al., 2017; Hsu, 2017; Weishaar et al., 2017; Rabe, 2018; Metcalf, 2019). There are taxation approaches that have potential for controlling global climate change: tradable emission permits/quotas; carbon (emissions) tax; energy use tax; taxation associated with technology transfers; reduced taxation for providing carbon sinks, etc.

Tradable energy quotas

Tradable energy quotas (TEQs), also called marketable or auctionable permits or tradable emissions permits, have been adopted by a number of countries (TEQs is also used for toxic equivalent concentrations by ecotoxicologists). There is interest in TEQs for dealing with transboundary atmospheric pollution, especially CO_2 emissions (Tietenberg, 2006). The

1997 Kyoto Convention established a TEQ 'club', which trades emissions permits among its members (Oberthur and Ott, 1999). From 2005 it committed signatory nations to reducing greenhouse gas emissions by around 5.2% below 1990 levels by 2012. Revised and extended from 2012 to 2020, its key elements are:

- a system of TEQs;
- tough verification of emissions;
- financing arrangements to aid poorer countries to comply.

The Kyoto Convention (2016) was superseded by the Paris Agreement, within the United Nations Framework Convention on Climate Change (UNFCCC), on climate change mitigation, adaptation and finance (Victor, 2004; Klein et al., 2017; Figueres and Rivett-Carnac, 2020; Martinez Romera and Broberg, 2021; Van Calster and Reins, 2021).

Energy use taxes

Energy taxes (e.g. carbon taxes) discourage pollution by increasing costs: for example, burning low-grade coal would attract a higher tax per BTU (1 kW = 3412 BTU) than the use of oil or gas, which emit less carbon. Tax on vehicle fuel, domestic power supplies and household heating fuel can be used to discourage excessive consumption. Energy taxes (and subsidies or other rewards for adopting desired practices) encourage efficient use and change to non-polluting alternatives, but may be unfair to countries with less scope for the latter, such as those lacking hydroelectricity or already committed to coal or oil.

REDD and REDD+

In 2008 the framework on Reducing Emissions from Deforestation and Forest Degradation (REDD) was launched to create an incentive for developing countries to protect, better manage and wisely use their forest resources, contributing to the fight against climate change. REDD+, the '+' referring to additional roles of conservation, sustainable management of forests and enhancement of forest carbon stocks in developing countries, is a climate change mitigation solution. REDD+ aims to achieve climate change mitigation by incentivising forest conservation (both ongoing in 2023) (Angelsen et al., 2012, 2018; Lyster et al., 2013; Arts et al., 2020; Dehm, 2021).

Green aid

Environmental care is now usually a condition of aid (Hicks et al., 2008; Garcia, 2014). Funding and aid agencies have increasingly focused on EM and SD, and they also check for risks before supporting developments.

One aid initiative is the GEF (Global Environment Facility – a multilateral environmental fund that provides grants and funding for projects related to biodiversity, climate change, international waters, land degradation), established in 1992 as a venture between governments of industrialised and developing countries. The GEF is jointly managed by the World Bank, UNDP and UNEP to assist developing countries to tackle globally relevant environmental problems such as climate change; loss of biodiversity; management of international waters; and stratospheric ozone depletion (Betzold and Weiler, 2019; www.thegef.org/

– accessed 28/06/21). The GEF is targeted at poorer countries and involves NGOs in identifying, monitoring and implementing projects. There are criticisms; one is that donors to the GEF might simply cut back on other aid; another, that participation is not open or wide enough; and that some developing countries want poverty alleviation included with environmental care support.

There is a wide diversity of aid approaches: recipients may be governments, bodies and groups of people or individuals. Aid may be in the form of grants, loans, equipment, training, secondment of skilled staff and so on. Donors can be international agencies, NGOs, individuals, groupings of governments or national governments. Sometimes donors contribute aid directly to recipients, or it can be via an intermediary such as an NGO or a UN body. When aid is government to government it is termed bilateral aid; when several governments or an international organisation have contributed it is multilateral aid. Frequently aid is tied, that is, conditional: a recipient may have to behave in a particular way or a percentage must be used to buy goods and services from the donor company/nation, sign trade agreements, or a donor military base is required. Aid may be in the form of funding, foodstuffs or other supplies, sometimes training or secondment of skilled manpower rather than donation of goods or funds.

Green aid is conditional (tied): it depends on the exercise of environmental care and/or seeks environmental improvement. A risk is this is perceived as neo-protectionism or neo-colonialism, intrusion into sovereignty, an extra cost, a risk that support could be ended.

Aid may be well intended, but even providing something 'harmless' like roads or wells can cause serious environmental problems. Environmentally benign aid is not easy to achieve and problems are often not intentional. What to a donor seems like sensible safeguards to avoid environmental and socio-economic damage may appear to a recipient to be excuses for conditionality, delay and there may be loss of a portion of funding to pay for appraisals, safeguards and remedial measures. Nations may be reluctant to commit themselves to the GEF, either for fear it could slow their economies or because they wish to ensure tight control over how aid is spent.

Crowdfunding is a means of financing things that fail to attract official support and/or which demand rapid response; citizens see online publicity and give small amounts that accrue. Another 2020s funding source for the environment is donation by very rich individuals.

Natural capital and ecosystem services

Capital (economic capital) is wealth in the form of money *or other assets* owned by a person, group, organisation, society, etc., available for a purpose such as starting a company or investing. Alternatively, it is anything that confers value or benefit to its owner, such as infrastructure or intellectual property. Since the 1960s other forms of capital have been recognised (social, human, built, cultural).

Natural capital

Natural capital is the world's stock of natural resources, which includes sunlight, habitat, plants and animals, atmosphere, geology, soils, water and all living organisms. Natural capital often cannot be substituted by economic capital (Helm, 2015; Henry and Tubiana, 2018). There may be ways to replace/substitute some natural resources, but it is unlikely there will ever be ways to replace many ecosystem services, such as the protection provided by the ozone layer, or the climate moderation effect of tropical forests. If lost, natural capital can be difficult or impossible to restore. Consumption/degradation of natural and social capital may have insidious effects until a threshold or tipping-point is reached and then impacts may appear suddenly.

The value of biodiversity conservation in economic terms is considerable, but not adequately acknowledged. A 1990 assessment suggested the contribution of plant and animal species to the US economy was about 4.5% of GDP (*c.* US$87 billion). If the value were better known, conservation might be given more support.

Degradation of natural and social capital has important consequences but action to alleviate it is not a priority. It has been pointed out that market failure is the explanation, with four possibilities: 1) The benefits of natural or social capital depletion can usually be privatised, while the costs are often externalised (i.e. they are borne not by the party responsible but by society in general). 2) Natural capital is often undervalued by a society that is not fully aware of the real cost of the depletion. 3) Information asymmetry is possibly a third reason; the link between cause and effect is obscured, making it difficult for actors to make informed choices. 4) Many exploiters are not good optimisers because they are caught in a business-as-usual mentality.

Natural capital accounting is the process of calculating and valuing the total physical stocks and flows of environmental goods and services in an ecosystem or region. The natural capital approach (NCA) was promoted (in 2008) by the International Institute for Sustainable Development for identifying and quantifying natural resources and associated ecosystem goods and services that can help integrate ecosystem-oriented management with economic decision making and development (www.iisd.org/system/files/publications/natural_capital_approach.pdf – accessed 07/05/2021). The Natural Capital Protocol offers a decision-making framework for identification, measurement and valuation of impacts and dependencies on natural (and social) capital, and was launched in 2016 (there is also a Social & Human Protocol) (https://capitalscoalition.org/capitals-approach/ – accessed 03/05/21). Ecosystem valuation and natural capital accounting are evolving (Barbier, 2011; Maxwell, 2015).

Valuing natural capital focuses stock (income is a flow, a return on wealth) (Kareiva et al., 2011; Islam et al., 2019). For resources traded in markets – oil, land, timber, crops, etc. – the value of small quantities of market goods can be measured by their observed price. In competitive markets, prices reflect both the marginal cost of producing the good to suppliers and the marginal value to consumers. The natural capital asset index (NCAI) is a composite index promoted in 2020 by the World Bank which tracks changes in the capacity of terrestrial ecosystems to provide benefits to people. The NCAI is intended to complement the World Bank's Human Capital Index (HCI) and does not include monetary values but is composed in a way which reflects the relative contribution of habitats to human wellbeing (Czúcz et al., 2008). The extent to which natural capital can be substituted with manufactured or human capital in production is a key determinant of the possibility of long-run sustainable economic development (Cohen et al., 2019).

Ecosystem services

Healthy natural ecosystems (capital assets) provide people with free goods and services – ecosystem services such as wild foods, fuelwood, insects for pollinating surrounding crops, conservation, clean air, storm protection, aesthetic inspiration, mental and physical wellbeing (Wallace, 2007; Tallis et al., 2011; Potschin et al., 2016; Everard, 2017). Many ecosystem services are undervalued, or have no value assessed in current decision-making frameworks, although they are often crucial, for example in supplying clean drinking water, decontaminating wastes, and helping ensure resilience and productivity of agro-ecosystems (pollination, etc.).

Following the Millennium Ecosystem Assessment (2005), ecosystem services have been studied and grouped into categories: 1) provisioning, such as the production of food and water; 2) regulating, such as the moderation of climate; 3) supporting things like nutrient cycles and oxygen production; and 4) cultural, spiritual and recreational benefits. Many ecosystem services have been given values to enable comparisons with human infrastructure and services (Kareiva et al., 2011; Bouma and van Beukering, 2015; Berkhard and Maes, 2017; Muddiman, 2019; Costanza et al., 2020; www.ceeweb.org/work-areas/priority-areas/ecosystem-services/how-to-value-ecosystem-services/ – accessed 03/05/2021; *Ecosystem Services*, Elsevier – ISSN 2212-0416).

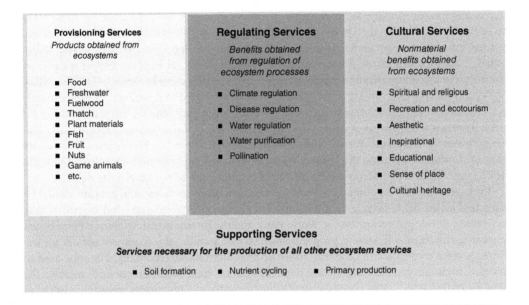

Figure 4.1 Ecosystem services. Falkenmark, M., Finlayson, M., and Gordon, L.J. (2007) Agriculture, water, and ecosystems: avoiding the costs of going too far. In: D. Molden (Ed), *Water for Food. Water for Life: a comprehensive assessment of water management in agriculture* (Figure 6.2, p. 241; Chapter 6, pp. 233–277). Copyright 2007 by Routledge (Earthscan). Reproduced by permission of Taylor & Francis Group.

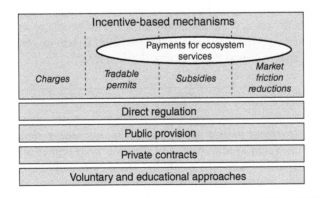

Figure 4.2 Paying for ecosystem services.

There is a need to find effective ways to value the ecosystem or environment for EM to help assess impacts and conduct better CBA (Wratten et al., 2013; Ninan, 2014; Perrings, 2014; Hufnagel, 2018; Mace et al., 2018; van Beukering et al., 2013). However, an ecosystem service value is an estimate with inherent quantitative uncertainty (Grunewald and Bastian, 2012; Jacobs et al., 2013; Managi, 2013; Patton, 2019; Schröter et al., 2019).

Debt, structural adjustment and the environment

When countries acquire debts they are bound into interest repayments that may cause poor EM and environmental damage. If the situation becomes serious it may make sense for those debts to be written off. During the 1970s many developing countries financed their economies or improved infrastructure by taking loans; some have done so since the 1990s (Macaes, 2020). Falling prices for exports of primary produce, rising costs for oil imports, and in some cases disorder and mal-administration led to escalating debt; the debt crisis broke in 1982 (UN, 1992). Various impacts of debt have been recognised: 1) money diverted to servicing debt is unavailable for EM; 2) resources are put under pressure to earn foreign exchange for interest or to pay off debt; 3) means to combat debt cause difficulty, notably structural adjustment measures; 4) sovereignty may be weakened (perhaps debt-trap diplomacy); 5) funding bodies insist on austerity measures, which may cause the breakdown of established land use and land degradation. In 1991 under the Trinidad Terms the Paris Club of creditors agreed to cancel some debts.

Linkages between economics and environment are often complex, and caution should be exercised when debt–damage relationships are recognised. When recession began to take hold in the developed countries, the World Bank and the International Monetary Fund began imposing structural adjustment programmes to try to stabilise the economies of debtor nations, protect creditors and generally shore up the international economy. The tool used to try and stimulate growth, and ensure debt repayment, fight inefficiencies and improve the flow of traded goods, was the structural adjustment loan. These measures varied in detail from country to country, but were always granted on condition the recipient deregulated their economy, reducing state expenditure and freeing exchange rates. The goal was to give priority to export earnings, make the economy more efficient by cutting spending on wages and welfare, and reduce state controls to boost productive sectors. There was limited success, and significant or marked ill-effects: reduced household incomes, increased unemployment, inflation, cutback in support for environment, welfare and public services (e.g. subsidies for fertiliser cut). Less spending on EM led to land abandonment, rural-urban migration, etc. Structural adjustment may impact on the environment through progressive disempowerment of the poor. At its worst, structural adjustment caused people to sacrifice environmental assets for short-term survival (Bigger and Webber, 2021). Debt can be generated by those seeking to counter environmental deterioration (Larkin, 2013).

Debt-for-nature/environment swaps

Debt-for-nature/environment swaps: the former focus on conservation, the latter address other issues. They are widely seen to provide a way for the recipient to pay off some or all foreign debt with less loss of face than would be caused by defaulting (Kokenes, 2008). The debtor country retains control over conservation or environmental activities, and banks write off the expense involved. The earliest debt swaps were negotiated by Ecuador and Bolivia in 1987, and subsequently by many other countries (www.oecd.org/env/outreach/

debt-for-environmentswaps.htm; https://green-bri.org/debt-for-nature-swaps-in-the-belt-and-road-initiative-bri/; www.eurekalert.org/pub_releases/2021-01/aaft-arh012521.php – accessed 08/05/2021).

Debt swaps mostly involve conversion of hard currency debt to local currency debt. When a lender realises it will probably never recoup, it sells the debt, at a discount, to another who then releases cash to the debtor country in local currency, and the donation supports EM or conservation. Some debt swaps are bond-based (a central bank pays interest on a bond created, usually for an NGO, over a few years); others are government policy programmes (under which the recipient government pledges to implement a policy or initiatives aimed at improving the environment or conservation). Climate change mitigation deals, such as for carbon sequestration, are sometimes linked to debt-for-nature swaps (debt-for-climate swaps). Through these a developed country or company establishes tree plantations to lock up CO_2 to compensate for emissions elsewhere. The developing country land is usually cheap and has better tree-growing conditions than colder climates.

There are opponents of debt swaps, especially in the recipient countries and on the staff of some NGOs. Criticisms are that:

- they offer limited potential to pay off debts (because they are tiny compared with typical national indebtedness);
- they may be used to 'smear' indigenous environmental groups' efforts, i.e. opponents of environmental protection spread rumours of foreign interference to divert attention from other issues;
- there may be difficulties in adopting them in some countries due to different accounting and regulatory systems;
- there is no guarantee of ongoing protection or care;
- they may be seen as an erosion of a developing country's sovereignty;
- if operated through NGOs, swaps may not assist or train local agencies;
- they do little to change commercial forces that damage the environment;
- they have so far been applied to a limited range of activities, mainly park and reserve establishment and maintenance;
- the main beneficiaries, it has been argued, are the debt-seller banks.

Trade and EM

Trade affects the environment and socio-economic conditions in negative and positive ways (Baughen, 2007; Gallagher, 2008). Some forms of trade may be less damaging: e.g. export of renewable forest products should be less damaging than logging, and may discourage deforestation if it is carefully controlled and local people benefit. To combat logging the *Body Shop*® store chain tried to encourage environmentally benign forest product trade by minimising middleman profits. Falling commodity prices on the world market mean farmers get poor returns on crops, yet, committed to purchasing inputs, they are forced to expand the area farmed, or intensify production, or practise shifting cultivation and the extraction of other resources to supplement their farming activities, leading to environmental degradation. Going back to a pre-cashcrop economy cannot solve the problem. Through trade, countries can obtain materials and continue to expand production. It may mean that production impacts (pollution due to manufacturing and problems associated with consumption of goods) are felt over a wider area.

Box 4.2 The positive and negative effects of free trade on environmental management

Free trade might help EM through:

- ending tariff barriers that raise produce prices, causing farmers to overstress land for profit;
- reducing the dumping of cheap US and European food surpluses, which, by making it difficult for developing country producers to get a fair price, discourage them from leaving land fallow or investing in land improvement and erosion control.
- removing restrictions that make it difficult for developing countries to produce and sell finished wood products to other developing countries. This should yield much better profits and reduce logging;
- harmonising standards and co-ordinating trade impacts on the environment on a global scale.

Free trade might harm EM because:

- much existing or proposed environmental legislation could be interpreted as illegal non-tariff trade barriers. There is thus a reduction in controls which discouraged logging, trade in endangered species, use of cattle growth hormones such as bovine somatotropin (bST);
- trade liberalisation may lead to increased specialisation of production that may overstress a resource or environment;
- the struggle to keep down costs to be competitive may mean exports are expanded to compensate and resources or the environment are put under stress;
- reduced import restrictions will remove opportunities to counter trade in hardwoods and endangered species;
- there may be increased opportunities to sell commodities (e.g. beef, sugar),

and this might encourage increased forest clearance and poor land management in countries that are keen to boost production;

- producers may think twice about spending money on pollution control or other forms of EM if another country does not, and they are competing with it to sell similar goods, on otherwise equal terms;
- it may be less easy, without the threat of trade restrictions, to get countries to reduce CO_2 emissions or other pollution;
- poor countries reduce domestic food prices, import grain, and raise more export crops like soya;
- any domestic support for the peasantry in developing countries or poorer farmers in developed countries could be interpreted as unfair protection. Small farmers might become marginalised and then damage the land trying to survive;
- larger farmers, encouraged by free trade to practise industrial (agrochemical-using) agriculture to produce export crops, may damage the land;
- there is a risk that foreign inputs and MNC controls will increase, leading to more dependency;
- if free trade leads to reduced home production there is a risk of problems if overseas supplies fail;
- it could be difficult to pass and enforce national environment and resource management or health protection laws;
- reducing the dumping of cheap US and European food surpluses, which, by making it difficult for developing country producers to get a fair price, discourage them from leaving land fallow or investing in land improvement, erosion control, etc.

In 1974 the Group of 77 (G77), a coalition of mainly developing countries, demanded a New International Economic Order (NIEO). The NIEO included plans for new commodity agreements, alteration of what were seen as unfair patent laws and general North–South economic reform, especially expanded free trade as a way of creating employment and wealth. Some advocate a new protectionism: a reduction in the volume of trade, as an alternative to free trade, to cure the market problems that led to demands for NIEO.

The main vehicle for reform has been the General Agreement on Tariffs and Trade (GATT) (Watson, 2013; McKenzie, 2020). There are other multilateral trade agreements (e.g. the North American Free Trade Agreement (NAFTA), which became the United States-Mexico-Canada Agreement (USMCA) in 2018; the Asia Pacific Economic Cooperation (APEC); the Common Agricultural Policy (CAP)).

Debate on the global and local environmental impacts of World Trade Organization (WTO) policies can be fierce (Najam and Meléndez-Ortiz, 2007; Copeland, 2014; Teehankee, 2020). Ideally, the WTO could be harnessed to support SD (Matsushita and Schoenbaum, 2016).

When the Roman Empire adopted 'free trade' (c. AD 100) grain prices fell, prompting landowners with slaves to practise ruthless farming that caused soil degradation and smaller farmers were forced out of business. Richard Cobden was aware of the environmental implications for the UK of freeing up trade by the Repeal of the Corn Laws (1846) (ending legislation which protected farmers from falling wheat prices); with free trade farmers drained and cleared land, intensified land use and damaged farmland.

One problem is WTO signatory countries have less control over their imports because quotas and restrictions can be interpreted as trade barriers, which are outlawed. There are worries that free trade favours developed countries' biotechnology.

Bodies such as the OECD are keen to harmonise free trade and the environment. It would be wise to seek greater co-ordination between free trade organisations and the UN Commission on Trade and Development (UNCTAD).

In 1985 GATT undertook to try to restrict the export of hazardous materials. Measures were taken to improve controls; e.g. in 1986 the FAO issued an International Code of Conduct on the Distribution and Use of Pesticides. The FAO and WHO set up the Codex Alimentarius Commission to establish food standards, including acceptable pesticide levels, and this publishes standards annually. But any nation that already has, or is setting, standards higher than the Codex may be deemed to be putting up trade barriers (www.iisd.org/system/files/publications/envirotrade_handbook_2005.pdf – accessed 11/05/2021).

Summary

- After 1970 the 'Spaceship Earth' icon spread: people realised the world was a vulnerable, closed system.
- When the history of the twentieth century is written, the most important social movement of the period will be environmentalism.
- Today a diversity of political groups, religious persuasions, old and young, share concern for the environment to a far greater extent than has been the case in the past.
- The failure to weave environmental sensitivity into economics has been flagged as a cause of many of the world's problems.
- For the last 40 years there has been an effort to develop green economics.

Further reading

Endres, A. (2011) *Environmental Economics: theory and policy* 3rd edn (English trans). Cambridge University Press (Kohlhammer Verlag), Cambridge, UK.

Hussen, A. (2019) *Principles of Environmental Economics and Sustainability: an integrated economic and ecological approach* 4th edn. Routledge, Abingdon, UK.

Jackson, T. (2017) *Prosperity without Growth: foundations for the economy of tomorrow* 2nd edn. Routledge, London, UK.

Laurent, E. (2020) *The New Environmental Economics: sustainability and justice*. Polity Press, Cambridge, UK.

Lomborg, B. (2020) *False Alarm: how climate change panic costs us trillions, hurts the poor, and fails to fix the Planet*. Basic Books, New York, NY, USA.

Smith, S. (2011) *Environmental Economics: a very short introduction*. Oxford University Press, Oxford, UK.

Tietenberg, T. and Lewis, L. (2020) *Environmental Economics: the essentials*. Routledge, New York, NY, USA.

www sources

❏ Green economics site – www.greeneconomics.net/what2f.htm (accessed 06/06/21).
❏ Social Science Information Gateway (SOSIG): environment section – www.ariadne.ac.uk/issue/19/planet-sosig/ (accessed 06/06/21).
❏ Social Science Information Gateway – www.worldcat.org/identities/lccn-no95-47014/ (accessed 06/06/21).

Part II

Practice

Chapter 5

Environmental management, business and law

Chapter overview

- The US National Environmental Policy Act (NEPA) – a 1970 environmental Magna Carta?
- EM and business/organisations
- Corporate visions of stewardship – a paradigm shift to EM ethics?
- Approaches adopted to promote EM in business/organisations
- EM and law
- Summary
- Further reading
- www sources

The US National Environmental Policy Act (NEPA) – a 1970 environmental Magna Carta?

Before NEPA the USA had little effective federal control over the environment. NEPA was signed into US Law on 1 January 1970, to reform policy, force action and influence the private sector to reorientate values (Anderson, 2011; Eccleston, 2011, 2014, 2020; Greenberg, 2012). NEPA requires environmental impact assessment (EIA) prior to federally funded projects that might 'significantly' affect the environment. Section 102 (2)c of NEPA requires US federal agencies to prepare an EIA statement (EIS) (bearing the costs against taxes, and sending copies to federal and state agencies and the public), prior to development.

There are three main elements in NEPA:

1) It announced a US national policy for the environment.
2) It outlined procedures for achieving the objectives of that policy.
3) Before supporting or funding any development likely to significantly affect the environment NEPA required federal agencies to conduct an EIA.

DOI: 10.4324/9781003189985-7

Provision was made for the establishment of a US Council on Environmental Quality (CEQ) to advise the US President on the environment, review the EIA process, review draft environmental impact statements (EISs), and see that NEPA was followed. Also, in 1970 the US Government created the US Environmental Protection Agency (EPA), its brief to co-ordinate the attack on environmental pollution and to be responsible for the EIA process.

NEPA was the first time US law had allowed for development to be delayed or abandoned for the long-term good of the environment, and for efforts to be made to co-ordinate public, state, federal and local activities. Effectively, NEPA put environmental quality on a level with economic growth, a revolution in values in a country where state intrusion was anathema; so, many see it as a sort of Magna Carta, although it stopped short of making a healthy environment a constitutional right.

NEPA is statutory law: it was written after deliberation, and did not evolve from custom, practice or tradition. Consequently, it was imperfect; there were problems, especially delay, as litigation took place over various issues. Many felt NEPA had been abducted by lawyers and could become a bureaucratic delaying tactic. These teething problems have largely been resolved, although some feel NEPA should be strengthened.

Effective implementation of EIA demands legislation and law enforcement to ensure that:

- there are no loopholes, so that no activity likely to cause impacts escapes EIA;
- the assessment is adequate;
- the assessment is heeded;
- the public are kept informed or, ideally, involved in assessment.

EM and business/organisations

It is not enough for a developer to obey law; there is also a need to at least appear concerned for public relations and to avoid negligence charges. Other factors also prompt business greening:

- globalisation;
- increasing public demand for access to information;
- activity of green business groups;
- trade union and NGO concern for environmental issues;
- a wish by companies to reduce inspection by regulatory bodies;
- insistence by funding, insurance and licensing bodies that required environmental impact assessment (EIA) and eco-audit be conducted;
- ethical (green) investment policies;
- genuine sense of responsibility;
- avoidance of litigation;
- the establishment of increasingly powerful environmental ministries in most countries;
- formation of professional bodies;
- promotion of the Integrated Systems for EM and the Business Charter for SD;
- provision of courses on EM by university business schools;
- the UN Center on Transnational Corporations has promoted SD.

Satisfying investors and shareholders dominates business; the adoption of EM implies concern for a wider range of stakeholders: the public, bystanders, employees, consumers, and the environment. EM must address its objectives within the context of company practices. Larger businesses, government departments and institutions adopt corporate approaches to EM, hopefully numerous SMEs (small and medium enterprises) will follow (Cramer and Karabell, 2010; Hollender and Breen, 2010; Lovett and Ockwell, 2010; Belcham, 2014; Ajith Sankar, 2015; Pritwani, 2016; Wagner, 2021).

Accidents like Bhopal in 1984 and *Exxon Valdez* 1989 (which cost Exxon over US$2 billion) and in 2010 the *Deepwater Horizon* explosion/spill have prompted business EM (Konrad and Shroder, 2011).

Corporate visions of stewardship – a paradigm shift to EM ethics?

'Fordism' from the 1920s has emphasised mass production, mass consumption, corporate control and resource exploitation (Link, 2020). The problem is *how* will people (consumers) and business (supplying the consumers) shift to something more supportive of environmental goals (Adoue, 2011; Spencer, 2018). Fordism (the system of mass production and social interference pioneered in the early twentieth century by the Ford Motor Co., and the typical post-1945 mode of economic growth and its associated political and social order) is still dominant. From the 1960s 'barefoot economists' and environmentalists have questioned consumerism (i.e. excessive consumption stimulated through marketing). There is a risk that greening of business is merely the adoption of EM tools to improve profits and public relations.

One significant prompt was the publication in 1991 of a Business Charter for Sustainable Development by the International Chamber of Commerce (ICC) at the 1991 World Industry Conference on Environmental Management (Box 5.1). Another has been the pressure for carbon neutral production or carbon negative activity (sequestration) (Welford, 2013).

Box 5.1 Business Charter for Sustainable Development: principles for environmental management

1 Corporate priority – recognise EM as among the highest corporate priorities and as a key determinant of SD; to establish policies, programmes and practices for conducting operations in an environmentally sound manner.

2 Integrated management – integrate these policies, programmes and practices fully into each business as an essential element of management.

3 Process of improvement – continue to improve corporate policies, programmes and environmental performance, taking into account technical developments, scientific understanding, consumer needs and community expectations, with legal regulations as a starting point; and to apply the same environmental criteria internationally.

4 Employee education – educate, train and motivate employees to conduct their activities in an environmentally responsible manner.

5 Prior assessment – assess environmental impacts before starting a new activity or project, and before decommissioning a facility or leaving a site.

6 Products and services – develop and provide products or services that have no undue environmental impact and are safe in their intended use, that are efficient in their consumption of

energy and natural resources, and that may be recycled, reused or disposed of safely.

7   Customer advice – advise, and where relevant educate, customers, distributors and the public in the safe use, transportation, storage and disposal of products provided; and to apply similar considerations to the provision of services.

8   Facilities and operations – develop, design and operate facilities and conduct activities, taking into consideration the efficient use of energy and materials, the sustainable use of renewable resources, the minimisation of adverse environmental impact and waste generation, and the safe and responsible disposal of waste.

9   Research – conduct or support research on the environmental impacts of raw materials, products, processes, emissions and wastes associated with the enterprise, and on the means of minimising any adverse impacts.

10 Precautionary approach – modify the manufacture, marketing or use of products or services or the conduct of activities, consistent with scientific and technical understanding, to prevent serious or irreversible environmental degradation. The 1991 Second World Industry Conference on Environmental Management promoted the precautionary principle. One problem for those proposing a development is how much proof of a risk they need before taking possibly expensive precautions; what seems to be widely followed is to establish whether there is a 'reasonably foreseeable risk'.

11 Contractors and suppliers – promote the adoption of these principles by contractors acting on behalf of the enterprise, encouraging and, where appropriate, requiring improvements in their practices to make them consistent with those of the enterprise; and to encourage the widest adoption of these principles by suppliers.

12 Emergency preparedness – develop and maintain, where significant hazards exist, emergency preparedness plans in conjunction with the emergency services, relevant authorities and the local community, recognising potential transboundary impacts.

13 Transfer of technology – contribute to the transfer of environmentally sound technology and management methods throughout the industrial and public sectors.

14 Contributing to the common effort – contribute to the development of public policy and to business, governmental and intergovernmental programmes and educational initiatives that will enhance environmental awareness and protection.

15 Openness of concerns – foster openness and dialogue with employees and the public, anticipating and responding to their concerns about the potential hazards and impacts of operations, products, wastes or services, including those of transboundary or global significance.

16 Compliance and reporting – measure environmental performance; to conduct regular environmental audits and assessments of compliance with company requirements, legal requirements and these principles; and periodically to provide appropriate information to the board of directors, shareholders, employees, the authorities and the public.

Corporate social responsibility (CSR)

CSR is concern for the relationship a body or company has with society; in part it is self-regulation. It is a concept encouraging integration of social and environmental concerns into business operations and interactions with stakeholders (Crowther and Seifi, 2021). It has the potential to help an organisation boost sales, cut costs, improve customer relations and help attract and retain employees. It should benefit society, environment, EM and SD. It may be conducted through business activity or by an organisation making donations to a cause. It shapes the supply chain, employee benefits, investment, etc. Standards have been developed to measure and certify CSR performance and it has spread beyond Western business with developments of CSR being promoted in Islamic nations by NGOs like the Islamic Reporting Initiative (http://islamicreporting.org/ – accessed 29/09/21) (Lee and Kotler, 2005; Beal, 2013; Moon, 2014; Bhinekawati, 2017; Camilleri, 2017; McIntyre et al., 2018; Frache et al., 2020).

CSR may be recent but similar ideas have been around since Victorian times when some employers provided employees and others with housing, public facilities and services (e.g. Bourneville, UK – Cadbury Ltd.; Port Sunlight, UK – Lever Bros., Ltd.).

Figure 5.1 Corporate social responsibility(ies) (CSR). Source: Jermisittiparsert, K., Siam, M.R.A., Rashid Issa, M., Ahmed, U. and Palii, M.H. (2019) Do consumers expect companies to be socially responsible? The impact of corporate responsibility on buying behaviour. *Uncertain Supply Chain Management* 7 (January), 741–752. Figure 2, p. 743.

The triple bottom line

The triple bottom line accounting framework helps business balance economic goals with reducing damage to environment and society and cutting waste. It seeks to incorporate three dimensions of performance: social, environmental and financial (or people, planet and profit – the '3 Ps'). It includes ecological (or environmental) and social measures that can be difficult to assign appropriate means of measurement (Bals and Tate, 2017).

Environmental, social and corporate governance (ESG)

Business is increasingly including non-financial issues when conducting analyses: environmental, social and governance. ESG appeared in the 2020s and is more finance focused than CSR, but overlaps much of the latter.

Approaches adopted to promote EM in business/organisations

Some of the shift towards EM in business and law is driven from within: sometimes management has become enlightened; staff may take the initiative; consumers may demand it; insurers may reduce premiums if measures lower the risk of accidents and costly claims; other companies, retailers or consumers may force it by refusing components or products from environmentally unsound suppliers; government or international regulations may prompt greening; and NGOs, lobby groups, media, and funding bodies can also encourage the shift. Increasingly businesses insist that their component suppliers and other subsidiaries meet environmental (and other CSR) criteria. A crucial question is: do businesses just seek to comply with regulations, avoid liability, consumer criticism, etc., or does it go beyond compliance, constituting real, motivated, proactive EM support?

A tax per unit of pollution, or other environmental damage, does not really discourage discharges that may be difficult to monitor. It may be better to adopt regulation which seeks to ensure that the capacity of the environment to cope with damage is not exceeded (Kreiser et al., 2012). Money can be saved when businesses recover wastes and use less energy or raw materials. In practice, savings may be less clear-cut recouped over long periods or in ways that are not easy to measure.

Businesses may have conservative habits and adopt slow gradual change. But, some companies innovate (Gallagher et al., 2016). Breakthroughs in clean energy and pollution control could come from companies, not UN and state bodies (Epstein and Buhovac, 2014; Berners-Lee, 2021). Private sector space technology companies in the last decade have substituted NASA in various capabilities.

Industrial ecology

This approach examines industrial, economic and resource activities from a biological and environmental, rather than a monetary point of view (Clift and Druckman, 2016). Industrial ecology regards waste and pollution as uneconomic and harmful, and seeks to dovetail them with demands for raw materials. This industrial symbiosis means that wherever possible industry should be aware of supply-chains, use by-products and go beyond the reduction of wastes to make use of what remains from the producer or other bodies (Bouchery et al., 2016; Xiaohong Li., 2018). Effectively, the environmental price of a product is included in its retail price. Industrial ecology tries to understand how the industrial system works and interacts with the biosphere, and then uses this knowledge to develop a systems approach aimed at making industrial activity compatible with healthy ecosystem function. Industrial ecology is most easily practised in systems with clear boundaries (Boons and Howard-Grenville, 2009; Fiksel, 2009, 2022).

Some groupings of companies and settlements already dovetail; e.g., Kalundborg (DK) has a power station, oil refinery, pharmaceutical companies, concrete producer, sulphuric acid producer, fish farms, horticultural greenhouses and district heating which utilise wastes, heat and CO_2, etc. This happened more or less spontaneously, as companies sought to minimise costs of energy and raw materials and cut the output of waste. Finland has embraced industrial ecology, as have Belgium, Japan, The Netherlands and Sweden.

Ecological engineering

Ecological engineering is the design, creation and management of technologies and implementation of processes and systems to prevent and control environmental risks, and often also to restore and reverse environmental damage and seek SD. It can be concerned with processing by-products and waste or recovery of minerals from effluent, mine spoil (often using biotechnology), assessing technological innovations, avoidance of emissions causing climate change, provision of clean water, improvement of air quality, clean energy supply, sewage treatment, adaptation to sea level rise, etc. There is overlap with wastewater engineering and other fields. The goal is to create sustainable ecosystems which integrate human society and natural environment for the benefit of both (Matlock and Morgan, 2010; Mines, 2014; Spellman, 2015; Kutz, 2018; Prasad, 2018; Desai and Mital, 2020; Nazaroff and Alvarez-Cohen, 2021; wwwaesociety.org – accessed 17/06/21).

Green marketing

This is the presentation of products to the consumer as environmentally benign/green. Green marketing raised its profile as CSR developed, to become part of advertising and marketing (McKenzie, 2011; Martin and Schouten, 2011; Belz and Peattie, 2014; Esakki, 2017; Iannuzzi, 2017; Ottman, 2017; Quoquab et al., 2017; Verma and Naidoo, 2019; Grant, 2020; Carvill et al., 2021; Mukonza et al., 2021). However, less enlightened companies make erroneous or false claims (greenwashing).

Marketing influences people and can play a key role in gaining support for EM.

Green consumerism and consumer protection bodies

Consumer protection bodies have been promoting environmental concern since the 1960s worldwide. Consumer protection organisations are advocacy, public awareness or lobbying groups that seek to protect people from unsafe products, predatory lending, fraud, false advertising, pollution, environmentally damaging developments, etc. Some test products, others have played a vital role spreading and establishing environmental and other standards. Some are national and some international: e.g. CAP, Malaysia; VOICE and CGSI, India; BEUC, the EU (Mansvelt, 2011; Irfan, 2015; Singh Malyan and Duhan, 2019).

By the 2020s consumers were increasingly questioning products, e.g. the consumption of meat, leather, etc. Many consumers are now willing to pay more for green products, including organic food.

Eco-labelling

The marking of goods to indicate to consumers that they are environmentally friendly has become widespread (Boström et al., 2011; van der Ven, 2019; for a listing of ecolabels see: www.ecolabelindex.com/ecolabels/ – accessed 25/08/21). In most cases the product is judged against similar goods by an independent agency (not formal eco-auditing). Eco-labelling is a way of influencing the behaviour of consumers, helping them identify the environmental impacts of products, and encourages manufacturers to reduce these impacts.

Eco-labelling focuses on the product and often nothing is said about the process of production or distribution. So, an 'environmentally friendly' product may come from a factory which causes pollution, or presents a disposal problem after use. There is a need for standardisation and policing of eco-labelling worldwide (Teisl, 2007; Joshi, 2008).

Total quality management and environment

Total quality management, also called company-wide quality management, aims to provide assurance of adherence to policy and specifications through a structured management system, and to enable demonstration of it to third parties through documentation and record keeping. It aims for continuous improvement of work processes to enhance an organisation's ability to deliver quality products/services in a cost-effective and, hopefully, environmentally sympathetic manner (Sharma, 2018). One of its goals is customer satisfaction leading to long term management success.

EM systems

An EM system (EMS) demonstrates and enables adherence to a suitable environmental policy, the meeting of appropriate environmental objectives (equivalent to specifications in quality management) and shows interested parties that the system requirements and objectives are met. EMSs usually require that, following assessment/audit, a company/body publishes and regularly updates an environmental policy statement. An EMS provides an organisational structure, procedures and resources for implementing environmental policy (Dupont, 2012; Price, 2014). It also provides a language of performance and quality that can be understood by management (Salomone et al., 2013; Price, 2014). So far, adoption of EMSs has been mainly voluntary with rapid growth of interest and continuing modification and improvement.

EMSs are well established, although there are critics who argue it is possible to rig them by setting easy to achieve targets and that it is more important (and difficult) to nurture satisfactory environmental ethics.

Green and sustainable supply chain management

Businesses/other organisations have realised that they need to look beyond their immediate operations if they are to practice effective EM and SD and consider inbound (procurement of materials, energy, etc.) and outbound (services, packaging, storage, transport, end-of-life recycling, etc.) supply chains (McKinnon et al., 2010; Sarkis and Dou, 2013; Fahimnia et al., 2015; Khan et al., 2017; Paksoy et al., 2018; Abdul and Khan, 2019; Achillas et al., 2019; Ali et al., 2019; Khan, 2019; Sarkis, 2019; Bak, 2021). Supply chains can be complex, vulnerable and widespread, so green supply chain management demands close integration with EM and life cycle assessment, green engineering and green management (Falatoonitoosi et al., 2016). Adopting a profitable and secure '4R' approach (reduce, redesign, remanufacture, reuse) presents challenges (Emmett and Sood, 2010; Wang and Gupta, 2011).

Worldwide there is adoption of 'supply just in time': a component (e.g. a drug or transformer part) is sourced from one or a limited number of producers just prior to need. The motivation seems to be cost-saving: less storage space, reduced risk of obsolete or damaged parts, etc. The result is growing global dependence on easily disrupted supplies and less diversification. This increases human vulnerability.

Life-cycle assessment

Many activities have different stages (e.g. designing and manufacturing, running, end-of-life recycling or disposal, plant construction, distribution, etc.) (Hauschild et al., 2018; Vezzoli, 2018; Cays, 2020). Equipment is usually subject to corrosion, wear and tear, and so varies in performance and presents different risks as it ages and as management acquires experience (or becomes complacent). Industrial and power generation sites often accumulate contamination, and so the environmental threat is not constant. It is therefore undesirable to assess impacts or develop EM policies by simply taking a time-limited snapshot view. Life-cycle assessment (LCA) has been developed to try to consider the whole of an activity, which may extend beyond the time horizon of a single owner. It is usually a cradle-to-grave study of an activity or company.

LCA can suggest efficiencies and warn in advance of problems that need to be avoided or mitigated (Finkbeiner, 2011; Schenck and White, 2014; Reddy et al., 2015; Jolliet et al., 2016; Ren and Toniolo, 2019; Muthu, 2021a, 2021b; Passarini and Ciacci, 2021).

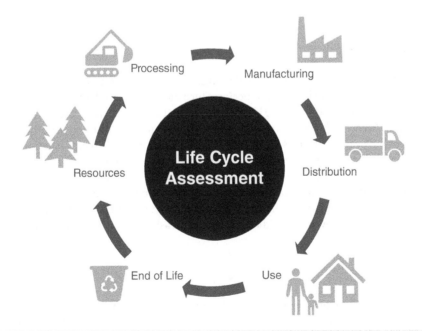

Figure 5.2 Life cycle assessment (LCA). Increasingly 'resources' includes recycled materials. Source: Khan, A., for Textile Learner (2022) *Life Cycle Assessment (LCA): the cradle to grave of your clothes*. Departmental paper, Bangalore, IN.

Covenants

A government or other regulatory body can provide companies with a more stable regulatory environment and encourage development of better pollution control plans or adoption of an EMS through a covenant. This is a written, voluntary agreement signed by the company or other body and the government or agency seeking regulation. The Netherlands has made extensive use of covenants as part of an integrated approach to national EM. A Dutch company signing a covenant would be expected to produce a development plan every four years, to be reviewed by local authorising bodies. The plan coverage includes pollution control and energy conservation and it is seen as a way of getting national policies implemented at local level. The covenanting approach can be quite effective, particularly in cutting pollution. However, some NGOs are not keen, viewing it as not sufficiently open to third parties to check. There are also some worries that it may lead to a softening of enforcement controls. Nevertheless, it is an approach which encourages company self-regulation.

Small and medium enterprises/businesses and the environment

There are numerous small and medium enterprises/businesses (SMEs) worldwide so their approach to the environment has huge impact. Small bodies may not have much money and personnel for environmental care but they may prove adaptable and willing to experiment compared with large bodies; EM and SD must not be just for large corporations (Cohen et al., 2013; Hillary, 2017). SMEs are often poorly regulated and among/in dwellings so these are further reasons why it is important they adopt EM. A key point is that for SMEs the EM/SD efforts will probably be the task of one person and need to be easy and cheap. SMEs can over time become big businesses.

Greenwashing

There are companies/bodies which see EM and/or SD as a cost and a burden. There is a temptation to lie about environmental performance and hide behind a façade of greenwashing: false green publicity (Pearse, 2012; Pierre-Louis, 2012; Johansen, 2015; Berrone, 2016; Grant, 2020). Some make calculated 'fig leaf' claims to hide or distract from bigger problems, or pass-off self-interest as moral awareness. Occasionally greenwashers are simply mistaken because some environmental issues are new, unproven and complex (Bowen, 2014). Hopefully, NGOs and citizens now check any claims made (Miller, 2017a).

Improving media and internet communications helps counteract greenwashing by making it easier to find and exchange information and expose/attack offenders. Various NGOs discourage greenwashing by public ridicule and greenwashing awards (e.g. www.corpwatch.org accessed 05/08/21).

EM and business: the current situation

Only in about the last ten years has the idea of ethical investment become practical and popular, allowing investors to direct their money to green activities, those acceptable to pacifists, etc. (Bril et al., 2020). There are signs that shareholders may shoulder more of the costs of environmental damage and insist on greener activities. Some shareholders have started to ask questions on environmental policies at board meetings or annual reporting.

Alas, environmental controls are still seen by some as 'green tape', slowing down development and raising costs.

EM and law

Law is complex, diverse and evolving; most law schools worldwide now offer environmental law courses; few did 30 years ago (Scotford, 2017; Eisma-Osorio et al., 2020). Law should provide a vital framework for regulating the use of the environment, but further development of law is needed (Burnett-Hall and Jones, 2012; Krämer, 2012; Fisher et al., 2013).

Law may be used to both control and encourage actions, and must be able to deal with citizens, business, and transboundary/global issues and future developments. EM seeks proactive measures and may deal with individually owned (private) resources; national resources; shared resources; open-access resources; common property resources; global resources, etc. Law covers some of these better than others. The world has many different legal systems: Roman Law, Customary Laws, Islamic Law, Tribal Law, etc. EM may have to work with unfamiliar national or local laws, or may have to seek agreement between parties with different legal systems. Some countries have legal systems that combine more than one system; say, indigenous and state law. Areas may be subject to state, federal, secular and religious laws. In most countries statutory law is written by politicians and passed by national legislature; and judges compile common law (with reference to past cases and prior statutory law). Legislation largely evolves in response to problems, so there is delay between need and the establishment of satisfactory law.

Developing and implementing environmental law is pointless if enforcement is weak (Macrory, 2014; Martin and Kennedy, 2015; Krämer, 2016b; Paddock et al., 2017; Hedemann-Robinson, 2019). Weak policing, war and strategic competition also cause environmental law problems. Law if enforced can encourage satisfactory performance, enable authorities to punish those who infringe legislation, or confiscate equipment that is misused or faulty, or close a company; it may also be possible for employees, bystanders and product or service users to sue for damages. Existing laws are predominantly anthropocentric: putting human needs before environment; there are signs this may change.

Some countries have been active in developing EM law, notably Sweden, The Netherlands, the USA, China (Martin et al., 2012), Canada, Australia and New Zealand (Nolan, 2020).

Box 5.2 Forms of regulation or legislation

Principles, standards, guidelines, etc., which are not firm laws, but which help lawmakers (definitions are not rigidly fixed).

Principle – A step towards establishing a law. Once established, tested and working, it can be incorporated into law.

Standard – Levels of pollution, energy efficiency and so on that are desirable or required. They provide a benchmark so that different individuals, bodies and countries are as far as possible dealing with the same values. A treaty may incorporate standards.

Guideline – Suggestions as to how to proceed, usually without real force of law.

Directive – Documents that set out a desired outcome, but to some extent

---

leave the ways of reaching it to companies, states or countries.

Licence – A right granted to a body, which agrees to terms or pays, which requires adherence to strict practice and does not give any guarantee of permanent ownership or usufruct.

Law (and statutes) – Require certain actions or standards, and may punish failure to achieve them.

Treaty – A solemn, binding agreement between international entities, especially states. Treaties can lay down rules or treaty constraints. Stricter, more precise treaties are likely to involve fewer states. It can be a slow process and EM often needs rapid action. Vague treaties are quicker and easier to get signed. Few multilateral treaties are adopted in less than five years: the UN Law of the Sea Convention took nine years (1973–1982) and some take much

more. Treaties can be difficult to enforce – often enforcement is attempted by an international organisation (e.g. the International Whaling Commission, the International Atomic Energy Agency). Some are largely ignored.

Declaration – A general statement of intent or drafting of guidelines to follow. Softer than the obligations of a treaty.

Convention – Multilateral instrument signed by many states or international institutions. Conventions can be vague, which ensures that countries are not afraid of signing, but this can undermine effectiveness.

Protocol – Less formal agreement, often subsidiary or ancillary to a convention.

Contingency agreement – A way of dealing with uncertainty surrounding many global EM issues. Agreement of what to do if something happens.

Most laws, whether civil or criminal, punish wrongdoers and deter others from infringing rules and agreements.

Three developments have had particular impact upon environmental legislation:

1   *The precautionary principle* has evolved to deal with risks and uncertainties faced by EM. The meaning is still not firmly established by law. Western law demands that a misdemeanour has actually been committed, or is clearly planned; adopting the precautionary principle may demand legal action *before* something happens.

2   *The polluter-pays principle* holds that the polluter pays for damage caused by a development and implies that potential polluters pay for monitoring and policing. A problem with this approach is that fines may bankrupt SMEs, yet be low enough for a large company to write them off, which does little for pollution control. There is debate as to whether the principle should be retrospective (e.g. today in many countries, a purchaser who acquires a contaminated site is often forced to clean up the mess others have left. How long back should liability stretch?). Developing countries are seeking to have developed countries pay more for CO_2 controls, arguing the latter polluted over centuries.

3   *Freedom of information* is needed because if the public, NGOs or official bodies are unable to obtain information, EM may be hindered. The USA led with a Freedom of Information Act, followed by the EU in 2004. Few countries have well developed disclosure. Some governments and multinational corporations fear industrial secrets will leak to competitors if there is too much disclosure, and there are situations where authorities declare strategic needs and suspend disclosure.

In many countries, court actions, even if they were fought in the public interest, had to be brought by an individual, who, if they lost, paid costs. This acted as a deterrent to anyone tackling government, a large company or powerful individual wrongdoers. It is desirable that NGOs and bodies be allowed to bring legal actions to protect the environment, possibly as group cases (class actions). Since the 1990s the internet has also been a way to spread information and counter secretive development.

European law and EM

The European Community (EC)/European Union probably has the largest set of environmental laws on the planet. A UN agency that acts as a pan-European forum is the UN Commission for Europe (UNECE), which supports SD, environmental research and has launched or serviced several agreements dealing with issues such as pollution. The European Environmental Agency has not got as much enforcement power as the US Environmental Protection Agency, and serves mainly to gather information on the European environment. The EU has also established a European Environmental Information and Observation Network (Krämer, 2016a; Kingston et al., 2017; van Zeben and Rowell, 2020).

Efforts to develop an overall European environmental policy resulted in the publication of an Environmental Action Programme; this proposes principles to which EU environmental legislation should adhere (Farmer, 2012; Maljean-Dubois, 2017; Van Calster and Reins, 2017; Rowell and van Zeben, 2020).

International law and EM

International law seeks to help relations between states (Sands and Peel, 2012; Koivurova, 2013; Fisher, 2017; Freestone, 2018; Hey, 2018; Anton, 2021; Boyle and Redgwell, 2021). It is difficult to force a sovereign state to sign, and then honour, a treaty. International law thus depends a great deal on voluntary agreements by governments and international bodies (the Brussels and Lugano Conventions on Environmental Law cover this issue of compliance) (Sand, 2015). When negotiation fails a possibility is to refer the case to the International Court of Justice (The Hague), or set up an International Joint Commission. International law tends to be *laissez-faire* and ad hoc.

Before the 1950s, co-operation, exchange of information, agreement and international guidelines or rules were generally initiated by international public unions (e.g. the International Postal Union, or the International Telegraphic Union). Nowadays, UN agencies (e.g. FAO, WHO, UNEP, etc.) often initiate the development of international environmental law. NGOs frequently lobby for environmental legislation.

Observers note the UN-supported system of environmental treaty making is valuable but needs strengthening (the UN General Assembly can only recommend, not insist, that law be made). Some countries have complained that international law is too US- or Euro-centric and there is a wish in Muslim nations to see more application of Islamic Law (Idllalène, 2021). Most of the UN-prompted multilateral treaties have been developed by a two-step process: a relatively vague framework convention which acknowledges a problem (most countries are happy to sign); that step prompts action, especially data collection, discussion and propaganda, which reduces opposition and raises interest so that a protocol may be introduced and agreed to.

International law faces challenges. One of the greatest is the management of global commons, relating especially to climate change (Park, 2013; Robinson et al., 2013; Bodansky et al., 2014; Carlarne et al., 2016; Craik et al., 2018; Mayer, 2018), as well as stratospheric ozone, etc. Many resources and pests migrate or move, so that effective management of ocean fisheries, disease or locust control, etc., needs to be through multilateral agreement (Sands and Peel, 2012).

International law and sovereignty issues

Sovereignty is often a constraint on EM. Countries are usually reluctant to sign any agreement which affects their sovereign powers; some environmentalists thus oppose sovereignty (Smith, 2011). Growing transboundary and global environmental problems make it vital to get cooperation (Núñez, 2017).

The 1977 Stockholm Declaration on the Human Environment affirmed the sovereign right of states to exploit their own resources and their responsibility to ensure that activities within their jurisdiction or control do not cause damage to the environment beyond the limits of their national jurisdiction (Stockholm Principle 21). This has had considerable influence on subsequent international environmental law making. International trade agreements, notably the GATT/WTO provisions, mean that if a country has environmental protection laws, say, controlling timber cut in an environmentally unsound fashion, or fish caught using nets that kill dolphins, these measures may be problematic because they are seen to impair free trade. Conversely, there may be situations where globalisation helps countries adopt and enforce better standards. Globalisation of patent rights generates concern; MNCs and TNCs seek to recoup research costs and control markets; poor countries suffer biopiracy. The patenting and control of sales of crop seeds and pharmaceutical products cause much friction.

Protection and extension of sovereignty can lead to wars; the testing and storage of weapons; and territorial claims. Hostile environmental modification is covered by the 1977 Environmental Modification Convention, and there are international controls on nuclear, chemical and biological weapons. The Arctic is becoming more accessible as Earth warms, triggering sovereignty and resource conflicts (Conde and Sánchez, 2016) and globally competition for water is increasing, with shared rivers a particular challenge (Szwedo, 2018).

A number of trends are apparent: there has been a move towards the precautionary principle; obtaining damages for, or penalising, transnational pollution has been patchy; there has been limited progress in establishing 'environmental rights' (i.e. rights for the

---

Box 5.3 A selection of treaties, agreements and so on relating to environmental management

Internationally shared resources

In 1972 the USA and Canada signed the Great Lakes Transboundary Agreement for the comprehensive management of the water quality of the Great Lakes.

Protection of endangered species

The 1946 International Convention for the Regulation of Whaling; the 1973 Convention on International Trade in Endangered Species (CITES); the 1979 Bern Convention on the Conservation of European Wildlife.

Protection of environmentally important areas

There are many areas agreed by scientists, social scientists and other specialists to be in need of protection. Protection may be supported by a state; privately funded by a group or individual; or by an international body or bodies. For example, there is a worldwide scatter of Biosphere Reserves; the UK has state-protected Sites of Special Scientific Interest; many countries have reserves and national parks. Some conservation areas are established and watched over by international treaty. The 1971 Ramsar Convention provides a framework for protection of wetland habitats, especially those used by migrating birds. The UN Educational, Scientific and Cultural Organization (UNESCO) supports and oversees many sites of special cultural value.

The Antarctic

In Antarctica territorial claims have been set aside (but not eliminated) under the Antarctic Treaty which came into force in 1960 (see Figure 5.3 on page 91). Basically this is an international treaty by which signatories have agreed to keep Antarctica and its surrounding seas open for scientific research by all nations deemed to be pursuing scientific exploration south of 60°S. The treaty requires demilitarisation, no nuclear weapons and a commitment to conservation (Beck, 1986; Triggs, 1988; Holdgate, 1990).

While it has been quite a flexible treaty, modified as need arose, it has been put under some pressure as interest in resource development (notably oil, minerals, krill, squid and fish) comes into conflict with its conservation requirements. There are also demands from non-treaty nations (those which have not maintained a significant research presence there) and some NGOs for there to be changes to give the whole world (probably through the UN), not just signatory nations, control of Antarctica (a coalition of over 200 NGOs and non-treaty nations – the Atlantic and Southern Ocean Coalition – has been seeking such a goal). There have been some moves which in theory could allow mineral resources to be used – the 1988 Convention on the

Regulation of Antarctic Mineral Resource Activities allows exploitation only if very stringent environmental assessments are made and accepted by treaty nations.

Transboundary pollution

In 1965 Canada and the USA became involved in the Trail Smelter pollution case. The outcome was acceptance that no state has the right to permit use of its territory in such a way as to injure another territory. The 1972 UN Conference on the Human Environment in Stockholm was in part called for by Sweden, because of concern about acid deposition generated by other countries. In 1979 the Geneva Convention on Long-Range Transboundary Air Pollution addressed the problem of transboundary sulphur dioxide atmospheric emissions. The resolution of transboundary impacts had become an active field of diplomacy. The 1991 UN Economic Commission for Europe Convention on Environmental Impact Assessment in a Transboundary Context obliged signatory states to act to avoid transboundary environmental impacts.

Controls on global warming

The UN Framework Convention on Climate Change (1992) obliged signatories to stabilise CO_2 emissions at 1990 levels by AD 2000. The 1997 Kyoto Conference was intended to settle details of CO_2 reduction and to see that targets were enforced. However, some industries opposed any limit on greenhouse gas emissions, and hindered agreements. In 2023 there is far from firm and complete control.

Ozone damage controls

Efforts to phase out and if possible ban the use of CFCs were made at the 1985 Vienna Convention for the Protection of the Ozone

Layer. The 1987 Montreal Protocol on Substances that Deplete the Ozone Layer – revised 1990 – aimed for a 50% cut in CFCs over a short period. The Protocol is a landmark in that for the first time nations agreed to impose significant costs on their economies in order to protect the global environment. By 2023 ozone thinning had reduced, but especially in the Southern Hemisphere, there was still concern.

The Law of the Sea

In 1954 the International Convention for the Prevention of Pollution from Ships was established to try to reduce the discharge of waste oil from oil-tankers and other ship-related discharges (with limited success). For ocean pollution control to be effective, agreements that cover rivers, effluent outfalls, air pollution and so on are also required, because pollutants arrive in the sea from such sources. In 1959 the UN established the International Maritime Organization to deal with marine safety, law, pollution control, etc.

From the 1970s some nations with coastlines began to declare extensions of their territorial waters from three to twelve, or even 200 nautical miles. The 1950 Continental Shelf Convention was largely behind this trend towards extension of exclusive sovereign rights to continental shelf or seabed resources. To try to formalise these trends the Third Conference on the Law of the Sea was held in 1974. The UNEP's Regional Seas Programme brought together coastal states of a number of marine regions, resulting in several regional seas treaties, covering: the Mediterranean; the Gulf; West Africa; Southeast Pacific; Red Sea; Caribbean; East Africa; and the South Pacific. In 1977 the North Sea ceased to be high seas as far as fish and mineral exploitation were concerned, when the EC established zones laying claim to the continental shelf.

Meeting in Jamaica in 1982, the UN launched the Convention on the Law of the Sea (effective to 2500 m depth from the shore). Controls over damage to ocean fisheries are still woefully inadequate at the time of writing (2023).

environment, natural objects or organisms). Various agreements have reaffirmed and extended state sovereignty over natural resources (especially in respect to ocean territorial limits).

Natural resources are often under no clear and enforceable single ownership or national sovereignty. Sea fishing is so poorly controlled that some argue 'the final roundup' is currently taking place. There were indications in 2003 that nine of the world's 17 largest ocean fisheries were over-harvested. In 2021 remote fisheries were reporting aggressive foreign trawlers. Unilever, Europe's largest fish trader, established a Marine Stewardship Council to support an approved trader label which is supposed to indicate that fishing companies catch their products in a sustainable way. Unfortunately, it is nowhere near enough, and soon, it seems, staple fish such as Atlantic cod may be on the CITES Endangered Species List.

Currently there is a sort of 'enclosure' or privatisation of common resources (e.g. genetic material via patent law and claims of ownership of intellectual property rights). This disadvantages developing countries. There have been attempts by business to uncover and control traditional knowledge (in common ownership) using ethnobotanists and social scientists. There have been 'biopiracy' protests over attempts to patent products based on folk remedies associated with India's neem tree (*Azadirachta indica*); the raw material and ideas for its use are taken for product development and then the results are sold, and attempts are made to protect the trade by patent. As biotechnology develops, similar issues are likely to increase.

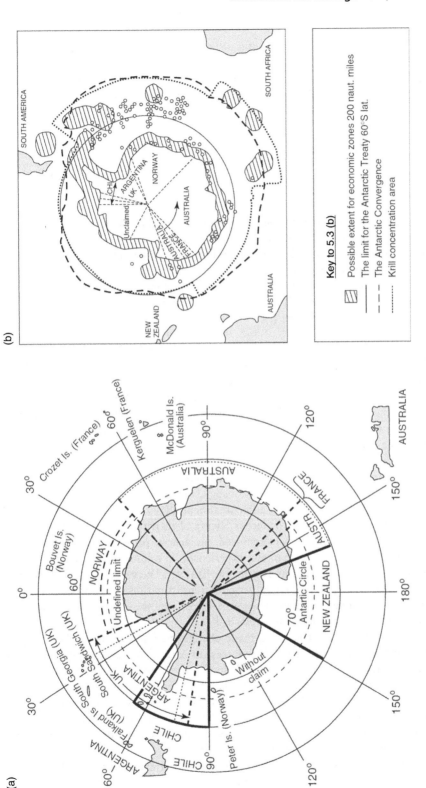

Figure 5.3 The Antarctic: a) territorial claims; b) possible economic zones to 200 nautical miles, and limit of Antarctic Treaty (60°S lat.).

Key to 5.3 (b)

Possible extent for economic zones 200 naut. miles
— — — The limit for the Antarctic Treaty 60°S lat.
– – – The Antarctic Convergence
········· Krill concentration area

Conflict management and EM

The process of managing environmental conflicts runs from identifying issues and stakeholders to developing compromises and enforcing agreements; communication is thus a central part of things. Sometimes the focus is local and relatively simple and sometimes international and highly complex (Maser and Pollio, 2012; Peterson and Clarke, 2016; Olsson and Gooch, 2019; Swain and Ejendal, 2020).

Indigenous peoples and environmental law

It is likely there are presently over 250 million indigenous people who interact with environmental law with respect to:

1) protection of natural environment together with indigenous people;
2) rights of indigenous people over natural resources;
3) rights over traditional knowledge (e.g. to prevent ethnobotany becoming biopiracy (gathering indigenous knowledge which is patented and sold));
4) damages to indigenous people for past environmental wrongs by outsiders;
5) views of indigenous people, which could be fed into environmental law making;
6) impacts of climate change/sea level rise.

Indigenous people often retain knowledge, skills and beliefs that relate closely to the environment. The protection of the environment is often vital to their physical and cultural survival, and they have insights which may aid wider EM, SD and law making. The rights of indigenous peoples are recognised by the UN Commission on Economic Development (UNCED) 1992 Convention on Biological Diversity and by the 1994 Draft UN Declaration of the Rights of Indigenous Peoples (Lennox and Short, 2018). Nevertheless, indigenous people often still have no written land tenure, making them vulnerable to abuse or resettlement if there are natural resources to be exploited.

Worldwide, governments and companies involved in resource exploitation come into contact with local people. In the past the relationship was seldom beneficial for the latter. Nowadays, laws often insist that there is fairer treatment, NGOs and the media are watching, and the peoples are increasingly vociferous, aware of their rights, own mobile phones, are likely to hire lawyers and are better organised. Indigenous peoples often network with similar groups around the world exchanging experiences. In a number of countries indigenous people have helped put in place more effective EIA/SIA. The values placed on environmental resources by modern society may be very different from those of indigenous peoples. State governments and foreign companies may not share those views. In some governments local tradition and values are scorned as 'backward'.

In recent decades many countries have made changes to improve indigenous peoples' control of their environment and natural resources. In Australia, New Zealand, the USA, Canada, Finland, Amazonian Brazil and elsewhere aboriginal people have fought for their sovereign rights to control and manage, or at least share in, resources (Abate and Warner, 2013; Westra, 2013; Tobin, 2016; Devere et al., 2017; Gilio-Whitaker, 2020; Jarratt-Snider and Nielsen, 2020; Turner, 2020). The Australian High Court has ruled that Australia's indigenous people enjoy native title and access rights to land leased by Euro-Australian farmers, which means two land users should legally coexist.

Who should bear the costs after resources exploitation? For example, the Pacific island of Nauru provided phosphates for some 90 years. Does it have any claim on the past colonial power to remedy damage? Nauru claimed through the International Court of Justice for damage done before its independence in 1967. Similar actions have arisen in Australia, and Pacific islands, over nuclear weapons test sites, and in Papua New Guinea concerning mining. Strategies to conserve biodiversity and control carbon emissions (like REDD+) may affect indigenous peoples so they need legal protection (Jodoin, 2017; Tehan et al., 2017).

International conferences and agreements

Keeping up with environmental agreements and gaining an overview is far from easy, although there are reviews (Chambers, 2008; Andresen et al., 2012; Desai, 2014; Huggins, 2017; Kayalica et al., 2017; Papantoniou and Fitzmaurice, 2017; Sand, 2019). It can be easier to focus on agreements relating to particular issues such as control of global warming, oceans, acid deposition, ozone damage, etc.

Alternative dispute resolution

Disputes about resource exploitation and EM can be addressed in a number of ways: 1) through legal measures (judicial); 2) through political measures; 3) through administrative measures; 4) through alternative dispute resolution measures (which may not use existing law but can also complement it) (Fiadjoe, 2004; Christie, 2008; Ogaji, 2019). Alternative dispute resolution 4) may be by:

- negotiation;
- mediation;
- arbitration;
- public consultation.

Negotiation is a process whereby two or more groups agree to meet to explore solutions, in the hope of reaching consensus. Mediation is similar to negotiation, but involves a mutually accepted neutral third party that finds facts and tries to facilitate agreement. The mediator may act with groups that are unwilling to meet face to face, if necessary filtering the exchanges to help reach agreement. Arbitration involves a third party like mediation, but at the outset the parties involved agree to give the arbitrator power to make decisions (which may or may not be binding). Public consultation may be by meeting(s) or phone survey, even referendum.

Summary

- The world is increasingly globalised and shaped by business and consumerism. The greening of business plays a key role in EM and SD.
- There is too little business and law awareness of the need to improve human adaptability, reduce vulnerability and increase resilience.
- There is concern for too narrow a range of issues, and policies are not sufficiently proactive.
- Some businesses genuinely embrace green approaches, some make half-hearted efforts, and others have exploited greening and practise greenwashing.
- Business is increasingly a nurturing ground for new EM and SD ideas and tools.

Practice

- Law is crucial for EM and SD, aiding regulation of resource use; protection of the environment and biodiversity; mediation, conflict resolution and conciliation; formulation of stable, unambiguous undertakings and agreements, etc.
- Enforcement of law is vital.
- Law tends to lag behind in meeting EM and SD needs.
- Law and business must increasingly embrace non-Western ways.

Further reading

Brady, J. (Ed) (2005) *Environmental Management in Organizations: the IEMA handbook*. Routledge (Earthscan), London, UK.

Etsy, D. and Winston, A. (2009) *Green to Gold: how smart companies use environmental strategy to innovate, create value, and build competitive advantage*. Wiley, New York, NY, USA.

Fisher, E. (2017) *Environmental Law: a very short introduction*. Oxford University Press, Oxford, UK.

Sand, P. (Ed) (2019) *International Environmental Agreements*. Edward Elgar, Cheltenham, UK.

Tinsley, S. and Pillai, I. (2006) *Environmental Management Systems: understanding organizational drivers and barriers*. Routledge (Earthscan), London, UK.

www sources

- Business for Social Responsibility – www.bsr.org/ (accessed 14/04/22).
- Green ethics and investment: numerous pension funds, investments, banks, etc., profess to be environmentally friendly. See *Green Money Journal* – https://greenmoney.com/ (accessed 14/04/22).
- Greening of Industry Network (UK) – www.greeningofindustry.org (accessed 14/04/22).
- *OECD Guidelines for Multinational Enterprises* (2011) – www.oecd.org/ corporate/ mne/ (accessed 14/04/22).
- International Business Ethics Institute – www.business-ethics.org/ (accessed 14/04/22).
- International Network for Environmental Management (INEM): seeks to develop and apply principles of EM; non-profit organisation – www.inem.org/ (accessed 14/04/22).

Chapter 6

Participants in environmental management

Chapter overview

- Learning from past peoples
- Stakeholders
- Participatory EM
- Funding and research bodies
- Communications
- Controllers
- Accreditation
- International bodies and agreements
- Unions
- Summary
- Further reading
- www sources

Learning from past peoples

EM can benefit from the past (and current traditional knowledge):

- *Environmental history* – hindsight on how past stakeholders reacted to challenges and opportunities (Diamond, 2020). Environmental history has explored ENSO, sea levels, the Little Ice Age, etc. (Fagan, 2000, 2008, 2013, 2019). This gives insight into how present landscapes and traditions evolved, and highlights threats which recur over a long time span.
- *Archaeology and palaeoecology* – supplement historical data. Some past cultures and surviving traditional cultures offer ways which, with or without improvement, could help (Diamond, 2005, 2012, 2020, 2021; Sveiby and Skuthorpe, 2006; Faulseit, 2016; Fagan and Durrani, 2021), one example being biochar (Woods et al., 2009; Scheub et al., 2016).

DOI: 10.4324/9781003189985-8

Use of information from the past must be cautious (Ewert et al., 2004).

Box 6.1   Stakeholders/participants in environmental management

- Existing land or resource users (males and females, age groups and class groups may make different demands): there may be multiple users.

- Groups seeking change: government (there may be conflicting demands from various ministries, federal states or policy makers), commerce (national, MNCs/TNCs), international agencies, NGOs, media, etc.

- Individuals seeking gain or change.

- Groups pressed into making changes: the poor with no option but to overexploit what is available without investing in improvement; refugees, migrants, relocatees (forced to move or marginalised so that they change the environment to survive), workers in industry, mining and so on, who face health and safety challenges.

- Public (may not be directly involved): may be affected as bystanders; may wish to develop, conserve or change.

- Facilitators: funding bodies, consultants, planners, workers, researchers, etc.

- Controllers: government and international agencies, traditional rulers and religions, planners, law, consumer protection bodies and NGOs, trade organisations, media, concerned individuals, academics, global opinion and the environmental manager.

Note: There is usually more than one stakeholder, some involved at different points in time and with varying degrees of involvement. A group/individual may become more aware of developments and/or is empowered and acts more effectively. There are subtle differences between 'involvement' and 'participation': the former may imply simply telling people what is happening. Participation means that there is some degree of involvement (often short of influencing how a development takes place).

Stakeholders

The point is frequently made that people cause environmental problems; therefore they should be involved in efforts to resolve or avoid them. It is valuable to conduct a stakeholder analysis before undertaking EM to establish who is involved and how to harness their support and reduce problems.

Stakeholders may not participate; there are voiceless groups: 'the dumb' and the 'blind'. The dumb include those informed of the implications of development but unable to adequately promote their views and affect change. The blind include the disinterested and those unaware of developments. The latter include scientists who fail to perceive a problem or opportunity because the issue is un-researched or unfamiliar, runs counter to their worldview, differs from the established viewpoint, happens too slowly to register, or fast and unexpectedly.

Stakeholder analysis and stakeholder management

It is important to work out power relations, interests, capabilities, needs, etc. Usually the focus is on key individuals or groups and their likelihood to support or oppose development. Stakeholder analysis and management are widely used in business to assist in achieving goals. Using these techniques, EM can gain insight into views, abilities, interests and relationships (Orr, 2013; Dukes and Hirsch, 2014; Gray et al., 2017; Haddaway et al., 2017; Vogler et al., 2017).

Facilitators

There are many bodies and individuals that promote and assist EM, for example: lawyers; the media; NGOs; international agencies (notably the UNEP); standards organisations; research bodies, etc. Facilitators assist change but do not invent or initiate it.

Citizens

Whatever a nation's politics, EM needs citizen support or even input. There has been expansion of citizen science. This can be the collection of knowledge/data, monitoring, processing data, developing and testing procedures, all conducted as fun and assisting a wide variety of science and other disciplines (Dickensen and Bonney, 2012; Cooper, 2016; Piera and Ceccaroni, 2016; Wynn, 2017; Hecker et al., 2018; Cavalier et al., 2020; Lepczyk et al., 2020). Citizen science is a progression (helped by computers and the internet) beyond whistle-blowing and protest to *meaningful* participation. Citizens can check pollution on beaches, monitor conservation issues and much more (USA Government crowdsourcing and citizen science site: www.citizenscience.gov/about/# – accessed 08/09/21). Other benefits are that the assistance is 'costless' and often rapid.

Indigenous groups

The study of current indigenous groups can assist EM. Dominant societies commonly overlook, ignore, scorn, exploit or persecute indigenous people. Indigenous people have often been driven from their lands or other resources, or relocated (Tidwell and Scott Zellen, 2017; Cahir et al., 2018). Recently groups have been strengthening their control over their lives and access to resources aided by new laws (Nelson and Shilling, 2018; Boiral et al., 2020; Thornton and Bhagwat, 2021).

There is a growing practice of respecting, consulting, learning from and involving indigenous people in EM, and understanding and making wider use of their knowledge (Grim, 2001; Edington, 2017; Head et al., 2020; Parsons et al., 2021). Geertz (1971) tried to understand the process of exploitation and ecological change, focusing on Indonesian farmers. Indigenous people can manage/co-manage and police conservation and things like water management (Ross et al., 2010; Stevens, 2014; Parsons et al., 2021; Spee et al., 2021). Care is needed to ensure outsiders do not exploit or misuse traditional knowledge or indigenous rights.

It is a myth that pre-modern folk are always in balance with nature. There are cases where indigenous peoples with traditional rights form joint ventures with outsiders, accept fees for waste disposal, build casinos, or trap and sell wildlife. Nevertheless, some indigenous people have developed sustainable and environmentally friendly livelihoods (Rodgers et al., 2011; Bollier and Helfrich, 2012; Linebaugh, 2014; De Moor, 2015; Shiva, 2020). Traditional knowledge has been studied by various fields (O'Faircheallaigh and Ali, 2008; Abate and Warner, 2013; Westra, 2013; Federici, 2019; Roothaan, 2019; Gilio-Whitaker, 2020).

Women

Women and development can be subdivided into: women, environment and development (WED) (Hawley, 2018); gender and development (GAD) (Arora-Jonsson, 2013; Leach,

Figure 6.1 Cultivation on land cleared from montane tropical forest. Indigenous people here have traditional rights and sometimes work with commercial farmers who would not themselves be permitted to clear.   Source: Photo by author, 2008.

2016; Detraz, 2017; MacGregor, 2017); and women in development (WID). Male and female are likely to respond to opportunities differently, so no single citizen social group is uniform. To get women's participation in soil conservation, biodiversity conservation, etc. it is necessary to ensure they enjoy the fruits of their labour (Buechler and Hanson, 2016; Wehrmeyer, 2017; Pham and Doane, 2021).

Eco-feminism argues other environmentalists have gender-neutral attitudes that are not enough to control male domination of women and nature (Phillips and Rumens, 2015; Vakoch and Mickey, 2018; Isaacs, 2020; Karpf, 2021).

Individuals and groups seeking change

Powerful individuals, businesses and special-interest groups seek to control policy making and development; few do so with the aim of improving environmental care rather than profit or merit. Philanthropy does happen; John D. Rockefeller jnr. supported conservation before the 1960s and recently environmental funding has come from Bill Gates, Jeff Bezos (a US$10 billion fund for climate action in 2021) and Elon Musk (US$ 100 million to support CCT in 2021). A recent review identified around 130 billionaire eco-philanthropists, who might rapidly support something and be unlikely to care about established ideas or need popular support and who may take risks when others will not. In the 2020s such people catalysed space technology which is of great value for understanding and managing the Earth (Greenspan et al., 2012).

Individuals and groups with little power

Marginalisation – treating a person, group, or concept as peripheral – is common. With media and social media it may mean discarding and blocking sensible ideas because they simply do not appeal to those who have closed minds and can bully. The causes of marginalisation and associated environmental degradation may lie with hidden external/distant policies, fashions and trade. Marginalisation can cause people to migrate, swell city populations and form refugee camps (McLeman et al., 2016). External aid for the marginalised might not always be needed, change might come from within according to (Marxist) theorists who argue the weak are able to assess their plight, learn and rise above it (Freire, 1970).

Displaced people People relocate for many reasons, some chose to move, some are paid to do so; others are displaced – driven. The move may be expected or sudden and unexpected; the latter group usually have no funds or tools, and are badly disorientated. Those driven by aspiration may have dreams prompted by media. Eco-refugees are people displaced by natural or human-induced environmental disaster or environmental degradation. In 2021 the displaced may have numbered one billion – estimates are often inaccurate (Fiddian-Qasmiyeh et al., 2016; Gemenne et al., 2016; Mokhnacheva et al., 2016; DeHaas et al., 2020). The world is slow to address practical and legal issues (Oliver-Smith and Shen Xiaomeng, 2009; Westra, 2009; Mayer and Crépeau, 2017; Behrman and Kent, 2018; Ahmad and Jolly, 2019; Sciaccaluga, 2020).

Much attention is focused on climate refugees anticipated to be displaced by global warming, sea level rise or related land degradation/disease (McAdam, 2010; Piguet et al., 2011; White, 2011; Piguet and Laczko, 2013; Mayer, 2016; Manou et al., 2017; Robbins and Wennerstein, 2017). Many states already have peacekeeping, finance and governance problems which eco-refugees might increase (McLeman, 2013; Boas, 2015; Miller, 2017a, 2017b; Krieger et al., 2020a, 2020b).

Displacement can also be within a nation (relocatees), e.g. due to dam construction, flooding, drought, land degradation, tsunami, resettlement or unrest (van der Land, 2017). Displaced people, even when aided, may have difficulty in establishing new or restarting their old livelihoods (Jäger and Afifi, 2010). At the time of writing stateless Rohingya people were resettled in refugee camps in India, Bangladesh (estimated 1.3 million), Myanmar (estimated 1.4.million), Saudi Arabia and China. (UNHCR, 1992; www.hrw.org/tag/rohingya – accessed 30/08/21). Relocatees may re-migrate after damaging an area and repeat the cycle; they may even be used to clear forest or establish sovereignty.

One of the best-researched aspects of human displacement is dam-related resettlement. Studies have been undertaken since the 1950s and agencies have developed guidelines; with decades of hindsight it should be possible to reduce problems, but real-world situations show otherwise. For example, there is a tendency to consider only the people in the area flooded by a reservoir, yet downstream of a dam many more suffer changes in livelihood or are forced to relocate unassisted.

In Malaysia, the Federal Land Development Authority (FELDA) opened up large areas of forest for resettling small farmers from land-hungry states, growing oil palm and rubber; Indonesia has an ongoing transmigration programme settling people from Java on less populous islands. Similar state-supported land development and resettlement programmes may be found in Bolivia, Brazil, Columbia, Ecuador, Kenya, Sudan, Ethiopia and several other

Figure 6.2 Tucuruí Dam, Tocantins River, Amazonian Brazil, c. three years before completion (early 1980s). The reservoir flooded about 2,300 km² and led to the relocation of many smallholders, some settled in the region only a decade or so earlier by official land development programmes. Source: Photo by author, 1981.

countries. These schemes cost a lot, move relatively few and might fail to sustain the settlers and prevent damage to the environment.

Lack of land reform has often been identified as cause of displacement; if insecure or denied enough land people move. Environmental causes may be merely a final trigger for relocation because the poor have already become vulnerable or less able to recover through other factors.

A number of agencies are trying to predict future eco-refugee scenarios (Warner et al., 2011; Richards and Bradshaw, 2017; Gemenne and McLeman, 2018; Nash, 2019; Ajibade and Siders, 2021; Betts, 2021; Ginty, 2021).

Public

The public consists of many stakeholders with different, perhaps conflicting, views and goals which vary through time (Berry, 2018). Powerful groups tend to dominate and the weaker get marginalised, so EM has to establish the needs and potential and try to work with this. Politics usually influences public perceptions (Rich, 2019). Disclosure does not guarantee a wise and supportive public and can cause panic or speculation.

Participatory EM

Once stakeholders are known EM may try to involve them. One way of ensuring that the weaker have some say in what should be done is to try to empower them so that they state their viewpoints. There is a choice of top-down or bottom-up approaches. Some countries prefer to tell people what to do and others encourage participation, although there can be a mix; e.g. China may adopt a top-down policy but also involve coordinated local participation (Hewitt et al., 2017). Participation and empowerment have become important; less clear is how effective it is. People may prefer to have the state co-ordinate firmly, and authorities sometimes feel their public is not ready for participation and can hinder effective EM.

Co-investment is an approach whereby locals make efforts to improve environmental care, such as soil and water conservation works, and an aid agency or government provides help in the form of funds, advisors, materials, machinery or whatever is not locally available. Where environmental damage is caused by insecurity of tenure or a weak legal claim to resource use providing better tenure and documentation may resolve problems.

EM needs local knowledge (Berry and Mollard, 2010; Hasan, 2018). Participatory approaches to data gathering, problem solving and implementation are used by the social sciences, agencies, NGOs, marketing, etc. (Chambers, 1994a, 1994b; Kochskämper et al., 2019). Participatory monitoring and evaluation of projects or programmes (community monitoring and evaluation or participatory monitoring and evaluation) aims to establish what stakeholders want, need, do, and could adopt. There is no single standard procedure but using a multidisciplinary team is important. For consultations, it is better to hold a number of sessions with few participants than a few with large numbers. Typical methods include focus groups, group discussions, observation and asking locals to draw maps or diagrams. It is important that methods and objectives are clearly explained and those consulted should be selected to be representative and the data cautiously interpreted. Those consulted may need anonymity. Assessors must also be prepared for various viewpoints, not all will support proposals. In addition, it is desirable to assess impacts upon neighbouring communities and to gather information on social capital. Social capital helps determine how vulnerable, resilient and innovative a group is.

Participation can be invaluable, but it can be slow, costly and sometimes ineffective. Received wisdom is not enough; local knowledge and objective multidisciplinary or interdisciplinary study are needed (Fairhead and Leach, 1996; Leach and Mearns, 1996). And an 'outsider's' viewpoint is also needed as well as that of locals.

Aarhus Convention

The Aarhus Convention (1998) was promoted by the United Nations Economic Commission for Europe (UNECE) and supports the right of every citizen to receive environmental information that is held by public authorities; and seeks 'to enhance public access to information through the establishment of coherent, nationwide pollutant release and transfer registers (PRTRs)'. The PRTRs place indirect obligations on private enterprises to report annually to their national governments on their releases and transfers (https://ec.europa.eu/environment/aarhus/; https://ironline.american.edu/blog/beginners-guide-environmental-agreements/; www.academia.edu/43335715/Environmental_Conventions_and_Protocols_Notes – accessed 03/10/21).

Transition Towns Movement

This participatory movement started about 2006 in Kinsale, IE and Totnes, UK and promotes grassroots community projects to increase self-sufficiency to reduce the potential effects of energy, climate change and economic instability (Transition Network: https://transitionnetwork.org/ – accessed 08/09/21; Kenis and Mathijs, 2014). The founders – Rob Hopkins, Peter Lipmann and Ben Bronwyn – drew some inspiration from permaculture. The movement has spread to many countries including the USA and encourages *local* action and 'skilling-up', so that people have more control, are less dependent and can prompt changes (Hopkins, 2008, 2013, 2019; Kenis and Mathijs, 2014).

Funding and research bodies

Funding bodies can support and focus desirable developments. Many funding bodies have developed EM units, guidelines and manuals. Regional development banks like the Asian Development Bank have commissioned EM studies and training. There have been cases where failure to carry out EM measures has led to withdrawal of funding from even large projects already underway (e.g. the Narmada Dams, India). Diverse bodies fund EM: universities, private research companies, independent international research institutes, and UN or other international agencies.

Commercial research and some universities are reluctant to focus on untested and unprofitable fields; there is thus a risk that some environmental and social issues will be missed. Since about 2010 more EM funding sources have appeared, including lotteries, cloudfunding, crowdfunding and green bonds. Investment and banking have also become greener (Oulton, 2009; Ramiah and Gregoriou, 2015; Lehner, 2017; Swiss Sustainable Finance, 2017; Schoenmaker and Schramade, 2018; Osório, 2020; Thompson, 2021; Tobin-de la Puente and Mitchell, 2021).

Communications

Communication, and linked to it education, are important to convey ideas, develop and change attitudes, and monitor developments (Kääpä, 2018; Sklair, 2021). Some active in EM should seek media training to help inform citizens and engage them in environmental challenges. From Victorian times newspapers have prompted environmental action: in 2023 UK media were lobbying to clean rivers and coastal seas (Jurin et al., 2000; Das, 2019; Armstrong et al., 2021). But media are sometimes not objective or accurate and may focus more on entertaining than informing (Abbati, 2019; Harris, 2019; Yusuf (Wie) and Saint John III, 2021; Freedman et al., 2022).

For EM the internet is an important source of information and means for dissemination, discussion, steering attitudes and countering backlash against actions (Haberlein, 2012; Brereton, 2019). Improved telecommunications make environmental monitoring easier as remote instruments can send information back (often in real time) via satellite and mobile phone links. Development of computers, software, GIS, trail cameras and drones make data collection, handling and analysis far more powerful than was dreamed possible even ten years ago.

Controllers

Stakeholders who control include government officials, business executives, the rich, large investors, religious officials, military staff, etc. These may not be easy to identify. There are various ways in which use of the environment and attitudes towards it are controlled; in most societies traditions evolve, and in some cases develop into taboos or laws.

The vast diversity of attitudes of controllers can be split into stances: development; conservation; preservation, etc. Moral and religious beliefs influence how people deal with the environment; sometimes the influence on EM is positive, sometimes it is damaging (Berry, 2018; Malik, 2018). For example, Diamond (2005) examined how Easter Islanders had *perhaps* been controlled by religion, and felled trees causing environmental collapse. In any society charismatic individuals and influential books can alter environmental outlooks and establish organisations (Gallagher et al., 2016; Schneider-Myerson, 2018).

Traditional controls and laws break down or become outmoded. In a region of Amazonia traditional fishing conserved stocks through taboos; then outsider commercial fishing flouted them with no ill-effect, so locals ignored controls and soon stocks crashed. Psychology, education and marketing have been researching attitudes and controllers (Clayton and Myers, 2015).

Traditions and spirituality

Societies commonly control resource use through leaders who exercise secular or religious power (James, 2016; Cooper and James, 2017; Kaza, 2019). Religious bodies have been active supporting indigenous peoples, in improvements to slum areas and in poverty alleviation (Abdul-Matin, 2010; Gottlieb, 2010; Mattson et al., 2011; Morris, 2012; Sponsel, 2012; Grim and Tucker, 2014; Jenkins et al., 2016; Gade, 2018, 2019; Hancock, 2018; Rankin, 2020). Recent developments in Australia are bringing Aboriginal traditions and beliefs and Western worldviews together generating new ways of using resources and viewing the environment. Some worldviews and governments oppose birth control and some traditions drive biodiversity loss (e.g. traditional medicine demanding tiger bones, rhino horn, pangolin scales, etc.).

Climate change and SD have been addressed by faiths (Iqbal, 2005; Loy et al., 2009; Gerten and Bergmann, 2011; Al-Jayyousi, 2012; Bandarage, 2013; Khalid, 2019).

Accreditation

Those involved in EM before 2010 were subject to little professional oversight. There is growing adoption of quality assurance and control by professional bodies and, thankfully, a dwindling opportunity for consultants and EM professionals to practise if not accredited. Accreditation is the process whereby a state department or professional body registers and certifies practitioners if they meet set standards. The body then (hopefully) monitors their performance, demands improvements if needed, calls for updated skills when necessary, supports new skills acquisition and warns or even strikes-off (i.e. removes credentials for practice) any who do not adequately comply. Accreditation is an important shift to policing EM, and helps set standards (e.g. ISO 14001) and improves accountability of EM staff.

International bodies and agreements

Few international bodies or agreements shaped EM before the 1970s, now they are highly influential.

NGOs and EM

International bodies, including NGOs, have become watchdogs and influencers of corporate, government and special-interest group activities. They have a multifaceted role: a crude division is into 1) 'environmental institutions' and 2) environmental activist groups. There is overlap but the latter tend to be NGOs and more concerned with activism. NGOs lobby, media, fund-raise, educate, research, benchmark and help establish standards, act as ginger groups, and much more (Kaiser and Meyer, 2016). Between 1909 and 1988 international bodies (e.g. IUCN or UNEP) increased from around 37 to about 309, and NGOs (e.g. Oxfam, Friends of the Earth, etc.) expanded from about 176 to about 4,518 (Lewis et al., 2021).

An important role for NGOs is to act as a link between the local, national and international (Fowler and Malunga, 2010; Yekini et al., 2021). Many NGOs have a tiered local-to-international structure and command huge resources in terms of funding and expertise. NGOs are very diverse: some are catalysts, some are politically orientated and some apolitical.

NGOs staff can be very dedicated; while most do sound work, in some cases they can be aggressive, or bent on ill-advised crusades. EM may need to work around this. (Listings of non-profit organisations/NGOs active in environmental protection in 2021 are provided by https://greendreamer.com/journal/environmental-organizations-nonprofits-for-a-sustainable-future; www.gov.uk/renew-driving-licence-at-70; https://guides.lib.berkeley.edu/c.php?g= 496970&p=3427176 – accessed 01/10/21.)

Often NGOs are swifter to respond to environmental problems and challenges than other organisations or governments, and grow from the grassroots in response to issues (Lyon, 2010; Lewis, 2014). The risk is that NGOs' supporters may expect to see 'magic bullet' solutions and sometimes lose interest or withdraw support if these are not quickly forthcoming.

Millennium and ongoing development goals

International bodies and NGOs set development goals in 2000. These cover human development and environment (for a list of internationally agreed environmental goals and objectives see: www.unep.org/resources/report/compilation-internationally-agreed-environmental-goals-and-objectives - accessed 02/10/21). UN COP26 (2021) climate change goals may be found at https://ukcop26.org/cop26-goals/ (accessed 01/10/21). Environmental agreements are listed/outlined in the following online sources: https://blog.studyiq.com/important-environmental-conventions-protocols-part-1-free-pdf-download/; https://byjusexamprep.com/liveData/f/2019/5/environmental-conventions-and-protocols-notes-pdf-in-english-72.pdf; https://mail.google.com/mail/u/0/#inbox?compose=new (accessed 03/10/21).

In 2016 the UN launched the 2030 Agenda for Sustainable Development which called for efforts to achieve 17 Sustainable Development Goals (SDGs) by 2030; for example SDG 13 seeks 'climate action', others address land degradation, biodiversity, etc. (https://sdgs.un.org/goals – accessed 01/10/21).

Intergovernmental bodies shape development: the G20 has recently (2022) eclipsed the G8/G7 as the primary forum. The G20 contains a larger, less ideologically coherent group of major powers which have met annually since 2008: Argentina, Australia, Brazil, Canada, China, France, Germany, India, Indonesia, Italy, Japan, Republic of Korea, Mexico, Russia, Saudi Arabia, South Africa, Turkey, UK, USA and EU. The G20 addresses major issues related to the global economy, e.g. international financial stability, climate change mitigation, and SD – www.g20.org/about-the-g20.html (accessed 02/10/21).

Calls have been made to simplify environmental and EM agreements, protocols and conventions (Young, 2014). Environmental agreements are discussed by a number of sources (UNEP, 2005; Boasson et al., 2012; Sand, 2019; Sjostedt, 2020).

Unions

Unions (labour or employees' unions) in rich and poorer countries have become involved with environmental matters and some have international influence (Nissen, 2002; Räthzel and Uzzel, 2013; Hampton, 2015; Slatin, 2017). They may encourage or pressure governments to act on issues of health, safety and environment, which could prevent disasters like Bhopal. Large unions have considerable investment funds and can steer their members towards ethical and green insurance and pension companies. Unions may oppose special-interest groups, crime and unsympathetic government. In Brazil co-operatives and unions of poor farmers and rubber tappers oppose large ranchers bent on clearing land and evicting poorer people. There can be negative effects: loggers' unions in western North America oppose environmentalist resistance to clearance of old forests fearing loss of members' livelihoods; and petrochemical, car manufacturing and coal-mining unions may lobby against carbon emission control agreements to protect members' jobs.

Summary

- Stakeholders can be individuals, groups, institutions, organisations, nations, etc.
- Increasingly, developers show some degree of environmental and social responsibility, which often means informing citizens.
- Communication is important and may necessitate EM using media professionals.
- Effective EM deals with people at the local or community level. Adoption of community/ participatory approaches should not be automatic; the benefits and contribution to environmental goals must be assessed carefully.
- Accreditation is spreading and offers a way to achieve and maintain quality of EM practitioners.

Further reading

Chambers, R.H. (2013) *Rural Development: putting the last first* (first published 1983). Routledge, Abingdon, UK.

Diamond, J.M. (2020) *Upheaval: how nations cope with crisis and change*. Penguin (Allen Lane), London, UK.

Ewert, A.W., Baker, D.C. and Bissix, G.C. (2004) *Integrated Resource and Environmental Management: human dimensions*. CABI Publishing, Wallingford, UK.

Practice

Fagan, B.M. and Durrani, N. (2021) *Climate Chaos: lessons on survival from our ancestors.* Hachette, London, UK.

Wehrmeyer, W. (Ed) (2017) *Greening People: human resources and environmental management* 2nd edn. Routledge, Abingdon, UK.

www sources

❑ Article: Public participation and the environment: do we know what works? – https://pubs.acs.org/doi/10.1021/es980500g (accessed 18/10/21).

❑ Environmental Change Through Participation (*ESCAP Report*, 2019 downloadable) – www.unescap.org/resources/environmental-change-through-participation (accessed 18/10/21).

Chapter 7

Environmental management approaches

Chapter overview

- EM focus and stance
- Local, community, regional and sectoral EM
- Adaptive EM
- Tools, expert systems and decision support for EM
- Systems and network approaches for EM
- Environmental management systems
- The state and EM
- Non-Western EM
- Transboundary and global EM
- Integrated EM
- Strategic EM
- Summary
- Further reading
- www sources

EM focus and stance

There are frameworks to shape EM application, guides to policy and procedures, and standards and systems. EM systems (EMSs) include the widely adopted ISO14001-2015 framework (Croner Publications, 1997; www.iso.org/files/live/sites/isoorg/files/store/en/PUB100371.pdf – accessed 26/10/21; https://asq.org/quality-resources/iso-14001 – accessed 29/10/21).

DOI: 10.4324/9781003189985-9

Practice

The EM approach adopted should reflect the situation needs, time and funding available, state policies, etc. (Broniewicz, 2011; Mulvihill and Harris Ali, 2016; Healy et al., 2020). For example:

- ad hoc (for a specific problem/situation);
- problem solving (resolve problems and develop solutions);
- systems approaches (e.g. ecosystem, agroecosystem);
- priority to environment or to human needs;
- top-down (authoritarian);
- bottom-up (grassroots – inclusive/participatory);
- centralised;
- decentralised;
- socialist;
- free market;
- company focus;
- non-business focus;
- light-green/bright-green (technology accepted);
- dark-green (technology unwelcome);
- giving priority to social development (poverty alleviation);
- giving priority to environment before human welfare.

The above may be combined in various ways. The goal is to maintain and, if possible, improve environmental quality and human welfare (and adaptability). There is generally more than one route to goals: perhaps one is the best all-round solution, one the best (environmentally) practical option (BPEO), one is that favoured by the government, another by a company, etc.

The Netherlands adopted a National Environmental Policy Plan (NEPP) in 1989, an integrated environmental policy based on explicit control principles and clearly formulated long-term objectives (Bennett, 1991). This is in contrast to the more usual incrementalist (step-by-step) approach adopted by much environmental planning and management.

A functional grouping approach is common: e.g. management by a pollution control agency or a conservation body, and this may hinder multidisciplinarity. Sometimes a quantitative approach is used for EM (Zhen Chen and Heng Li, 2016) and ecosystems approaches appear to offer ways forward. It may also have to proceed in a piecemeal manner, with inadequate funds, jurisdiction and time. Add public and administrative mood swings and things become difficult (Wright, 2017). EM may be faced with a crisis-management (reactive, short-term response) situation even though the ideal is anticipatory planning.

EM may be subdivided into the following components:

1 Advisory
- advice, leaflets, phone helpline;
- media information (which can be covert, i.e. hidden in entertainment or open);
- education;
- demonstration (e.g. model farm).
2 Economic
- taxes;
- grants, loans, aid;
- subsidies;
- quotas.

3   Regulatory/control
  ● standards;
  ● restrictions;
  ● licensing of potentially damaging activities.

EM can adopt one of three stances: i) preventive management – which aims to preclude adverse environmental impacts; ii) reactive or punitive management – which aims at damage limitation or control; iii) compensatory management – mitigation of adverse impacts through trade-offs: e.g. protect some habitats of conservation or aesthetic value, and develop other localities.

EM might focus on: A) modifying anthropogenic inputs (input management – controlling use); B) responding to ecosystem attributes (output management – driven by assessment of resources). Ideally an EM framework will integrate A) and B) to control environmental degradation most effectively. While co-ordination of EM approaches is desirable, it is difficult to see how too rigid a framework can help, given that each situation is unique. EMSs are increasingly used. These, together with eco-auditing, help provide internationally recognised foundations for EM to draw upon. An environmental manager can follow a textbook approach, but will be influenced by politics and his/her own outlook.

It is possible to say that whatever approach is selected a crucial element is to be sensitive to key issues, people's needs and fears, environmental limits, etc. To some extent the ends are more important than the means; the approach may not matter when problems threaten human survival or biodiversity. However, for many challenges it should be possible to attain goals in a cost-effective way without resorting to draconian measures.

In the past a command-and-control (top-down) approach to EM was usual, relying on regulations, fines, inspections, etc. That has been giving way to a more hands-off approach,

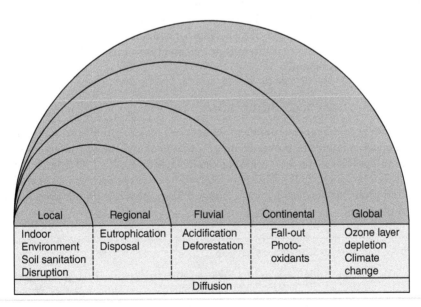

Figure 7.1 Linkages of levels in sustainable development tasks (based on the approach adopted by the National Environmental Policy Plan of The Netherlands). Source: Carley, M. and Christie, I. (1992) *Managing Sustainable Development*. Earthscan, London, UK, Figure 9.2, p. 199.

often bottom-up management, and relying on rewards rather than punishments to obtain results. For dangerous activities there have to be strict controls.

EM can be centralised or decentralised; technocratic or appropriate/human-based; sensitive to local needs (of people and environment) or insensitive. The level of activity is also diverse; EM may operate at the:

- local or even micro level (e.g. involving individual farmers);
- sectoral level (groups of villagers, farmers or bodies);
- regional level (watershed, river basin, island, etc.);
- state or national level;
- global level (bodies like the World Bank, OECD, etc., have established EM departments, policies and guidelines).

Alternatively a special-interest approach may be used, possibly combined with one of the above. This includes powerful groups, cartels, NGOs demanding adherence to a given agenda.

Political ecology focus and stance

Political ecology is the study of the relationship between society and nature; or the application of ecology to politics and study of political competition for control of natural resources (Bryant, 2015; Perreault et al., 2015; Robbins, 2020; Benjaminsen and Svarstad, 2021). Political ecology brings together cultural ecology and political economy. Cultural ecology is interpreted in different ways by geographers and anthropologists; the former view it as exploration of how society and humans affect the environment; the latter see it as study of how the natural environment affects socially organised behaviour. Political ecology holds that radical changes in human social habits and practices are required in order to counter environmental degradation and achieve SD. The approach implies an interest in cause–effect relationships, study of the different interest groups involved in using the environment, their economy, habits and livelihoods.

Political economy focus and stance

An understanding of human–environment interactions may be gained through examination of how the social relations of power relate to the control and use of resources and nature. Economics and politics shape the world (Blaikie, 1985; Stilwell, 2011; O'Brian and Williams, 2020; Clift, 2021; Cohen, 2022). There are likely to be different perceptions of environmental needs and problems between planners, policy makers, government ministers and various sections of the public.

Human ecology focus and stance

Human ecology is the study of relations between humans or society and nature through a multidisciplinary approach, or the study of ecosystems that involve humans. Wide ranging, a little vague, the scale of approach may be local to global, and it supports a systems approach and even holistic study (Savory and Butterfield, 2016; Devi, 2019; Diesendorf and Hamilton, 2020). It influences architecture and urban planning (Steiner, 2016), land use planners, resource planners and those with wider focus seeking to understand the relationship of humans with nature (Dyball and Newell, 2015; Campbell, 2017).

Participatory focus and stance

It is now common to inform or involve citizens in EM (Depoe et al., 2011). These efforts vary from tokenism to well-researched and effective measures (Coenen et al., 2012; Armeni, 2016; Murray, 2016). Participation can be demanding in resources, staffing and time (Chompunth, 2013; De Vente et al., 2016; Haddaway et al., 2017; Huong Ha and Rose, 2017; Kochskämper et al., 2019; Jager et al., 2020; Yun Tong et al., 2021). Participation has not always yielded good results (Tapia et al., 2019). Participation efforts should be aware there are usually more than one group of stakeholders involved in EM (Koohi et al., 2014; UNIDO, 2015; Nasreen et al., 2019).

Participatory appraisal There are a number of participatory appraisal approaches, notably participatory rural appraisal (PRA) and rapid rural appraisal (RRA) (Chambers, 1994a, 1994b, 2013). These evolved to support agricultural development, EIA, healthcare, sanitation, control of land degradation, adaptation to environmental change, decentralised planning, livelihoods and resource assessment, etc. PRA and RRA are used by NGOs and development agencies in rural and urban situations. Market research helped shape some of the methods (Shukla, 2016; Sandham et al., 2019; Lelo et al., 2021). In non-democratic societies village communes or people's discussion groups have offered routes to participation. Recent developments of approaches include community-based participatory research and participatory action research (PAR). Participatory approaches are often used for modelling and in initiatives seeking SD. PRA and RRA rely on multidisciplinary study and close contact with people to get a picture of their needs, capabilities, limiting factors, opportunities and threats. Often the appraisal is as much a learning process for EM as a data gathering exercise.

Local, community, regional and sectoral EM

EM is frequently tackled on a local, community, bioregional or regional scale (Birkes et al., 2007; Margerum, 2008; Wilmsen et al., 2008; Mmbali, 2013; Child, 2019). The social sciences offer experience on community development aspects of natural resource management which EM may draw upon. The bioregional concept can provide a framework: river basins (and sub-basins), watersheds, small islands and other suitable functional landscape units. A number of sectors have developed EM approaches, standards and pools of expertise; e.g. conservation, tourism, mining, oil, etc.

Adaptive EM

Whenever possible an adaptable strategy should be used to cope with unforeseen challenges and opportunities (Allan and Stankey, 2009; Robinson, 2022). Adaptive EM means different things to various people: to some it is a tool or approach that can be quickly modified to suit a particular situation; systems modellers see it as meaning the ability to explore various 'what if?' scenarios; or it can be an approach that is flexible and able to respond to new challenges as they arise, yielding know-how (learning by doing). The latter is the most common interpretation. Adaptive approaches have been applied to rangeland management and conservation areas to control pollution in the Great Lakes of North America, the Baltic, etc.

EIA practitioners developed adaptive environmental assessment and management (AEAM) (Holling, 1978, 1986) which seeks to integrate and deal with economic, social, environmental and other issues. AEAM recognises diverse stakeholders, and it addresses uncertainty. 'Hedging' and 'flexing' strategies for decisions are useful where there is uncertainty. Hedging is a process of trying to avoid the worst consequences, and flexing is a continuing search for other possible options even after a decision has been taken.

However, while adaptive EM is widely promoted, it is commonly not applied fully or successfully (Rist et al., 2013; Williams and Brown, 2013; Macleod et al., 2016).

Tools, expert systems and decision support for EM

EM has developed tools but also draws on many developed by other fields.

Tools for EM

For EM tools see: Thomas and Murfitt (2011), Thompson (2014) and Katz and Thornton (2019); *Environmental Management Tools for SMEs: a handbook* (1998), www.eea.europa.eu/publications/GH-14-98-065-EN-C/file (accessed 18/03/23).

For EMSs software see: Sheldon and Yoxon (2012); www.capterra.com/environmental/free-list (accessed 11/11/22).

Expert systems

Expert systems are computer programs that rely on a body of knowledge to cope with a difficult task usually performed only by an experienced human expert (Elling, 2010). Where there is a shortage of skilled EM staff an expert system could be used with caution as a guide. Costly to establish, an expert system should improve with use. Some draw upon fuzzy logic (Senthil Kumar and Kalpana, 2019; Stefanescu et al., 2011). Some EMSs make use of expert systems to aid organisations and businesses with the initial assessment for certification.

Decision support for EM

Problems faced in decision making include complexity, limited time, difficult-to-trace webs of interrelationships, etc. Decision support systems are derived from operational research and management studies; they deal with complexity by 'playing to learn fast'. Usually they take the form of interactive computer-based systems, which help the decision maker to model and solve problems. Some argue anything which aids decision making, like GIS, diagrams, or life cycle assessment, is a decision support tool. Whether complicated aids like the multiple criteria method are of practical value is debatable. The need is for easy, reliable and effective approaches that can help EM weigh goals against costs and risks, and help structure strategies in the best way possible (Frederic and Humphreys, 2008; McIntosh et al., 2011; Gregory et al., 2012; Janssen, 2013; Massei et al., 2014; Reynolds et al., 2014a; Hewitt and Macleod, 2017).

There is use of computer-based systems in support of EM decision making (Sànchez-Marrè et al., 2008; Zasada et al., 2017). Some systems integrate the use of GIS and modelling as well as aiding decision making, for example the ecosystem management decision support (EMDS) system (Reynolds et al., 2014b).

Systems and network approaches for EM

Systems analysis and network approaches are applied by EM to establish causal relationships between organisms and between organisms and system components (O'Higgins and Al-Kalbani, 2015; Kahraman and Sari, 2017; Jørgensen, 2019a; Fath and Jørgensen, 2021). The key premise is that everything in a system is inter-related and interdependent. A systems approach can be demanding of research and slow to perfect, but when established is useful for ongoing management of particular situations, and as a way of making sense of complexity.

Often a system diagram is constructed to provide a simple description of a (frequently complex) system that exists or needs to be built. This hopefully shows relationships and system structure and aids understanding, modelling, quantification and problem solving.

Ecosystem approaches

Ecosystems offer a practical, discrete, well defined and often stable functional unit to work with (Jørgensen, 2016a, 2016b, 2019b; Ramkumar et al., 2016; Zhifeng Yang, 2017; Schmutz and Sendzimir, 2018; De Lucia, 2020; Nazrul Islam and Jørgensen, 2020; Fischer, 2021). The approach has advantages for those seeking resilience in development (Garbolino and Voiron-Canicio, 2020), or with an interest in ecosystem services and ecosystem valuation.

Ecosystem services Humans are overexploiting the Earth's ecosystems but measures of progress like gross domestic product (GDP) fail to account for ecosystem values (ecosystem services). Ecosystem services (or ecological services) are the direct and indirect contributions of ecosystems to human wellbeing and resilience, and have an impact on our survival and quality of life. They may be grouped into:

- ❑ *Provisioning services* – the ability of humans to obtain products from ecosystems, such as water, food and other resources, including wood, oil, genetic resources and medicines.
- ❑ *Regulating services* – any benefit obtained from the natural processes and functioning of ecosystems. For example: climate regulation, natural hazard regulation, countering pollination, water purification, pollination.
- ❑ *Cultural services* – non-material benefits: spiritual enrichment, intellectual development, recreation and aesthetic values.
- ❑ *Supporting services* – those which relate to habitat functioning and therefore influence survival. For example: photosynthesis, the water cycle, nutrient cycles, biodiversity.

Ecosystem service values can be factored into decision making (ecosystem services assessment and environmental accounting) (Hester and Harrison, 2010a; Grunewald and Bastian, 2015; Biggs et al., 2015; Birkhofer et al., 2015; Bouma and van Beukering, 2015; Martin-Ortega et al., 2015; Potschin et al., 2016; Everard, 2017; Maes and Burkhard, 2017; Hale et al., 2018; Hufnagel, 2018; Kumar, 2019; Arcidiacono and Ronchi, 2021).

Bioregional approaches

Bioregional approaches are ecosystem-based and delineated by natural processes rather than politics, jurisdiction or regionalism (Johnson et al., 1999). Bioregional approaches have some links with deep ecology and seek harmony with nature. Bioregions are informed by nature, defined by physical qualities and cultural/social conditions. Some green political groups seek to establish bioregional units managed by assemblies (Sale, 2000; Glotfelty and Quesnel, 2014). Bioregionalism has been applied in non-urban, urban, terrestrial, aquatic and marine situations and has been explored by economists and cultural studies (Lynch et al., 2012; Scott Cato, 2012; Fanfani and Ruiz, 2020a, 2020b; Skipton et al., 2020; Devall, 2021).

Agroecosystem approaches

Agroecosystems are spatially and functionally coherent units of agricultural activity (or aquaculture), which include the living and non-living components involved and considers their interactions. They are used by humans to produce food and commodities (fuel, fibre, wood, oils, etc.). Agroecosystem approaches are widely used (Lal and Stewart, 2013; Mendez, 2013; Obermann, 2014; Vandermeer and Perfecto, 2017; Altieri, 2018; Benkeblia, 2019; Gliessman, 2019; Kumar, 2019; Lemaire et al., 2019; Caldwell and Wang, 2020; Chabbi, 2020; Harper, 2020; Larramendy and Solonesky, 2021).

Urban ecosystem approaches

Urban areas are among the most modified of any ecosystems used by humans. Urban ecosystem management involves preserving or recovering habitat remnants, restoring damaged processes such as the water cycle and managing ongoing functions (Alberti, 2015). Urban areas impact on much more than their immediate vicinity and periurban surroundings, city air and water pollution can be felt over considerable distances. Supplying an urban area with power, water, food and so forth can affect a huge region. If urban living can be made less sprawling and damaging and more attractive and sustainable by EM then there may be an opportunity to reduce pressures on other ecosystems by concentrating people where things can be best managed. Japan is an example, where urbanisation is dense in a narrow coastal belt and other areas have been spared much settlement and disruption. Developments in technology offer opportunities for low impact SD in urban areas (e.g. vertical farming; reuse of wastewater). Problems caused by a legacy of past rapid and negligent urbanisation are still a challenge and the problems are still being created in some cities.

The application of an ecosystems approach to urban EM has been developing (Adler and Tanner, 2013; Elmqvist et al., 2013; UNEP, 2013; Parris, 2016; Ossola et al., 2018; Aitkenhead-Peterson and Volder, 2020; Breuste et al., 2021) (*Urban Ecosystems* – Springer Journals: print ISSN 1083-8155; *Sustainable Cities and Society* – Elsevier Journals: print ISSN 2210-670; IUCN Commission on Ecosystem Management – www.iucn.org/our-union/commissions/commission-ecosystem-management – accessed 30/11/23).

SMART cities or ecosystem cities approaches Some cities have begun to address pollution, water use, transport, etc., by being more proactive using data collection to reduce impacts and wastage. The field is termed SMART cities or 'ecosystem cities'. Infrastructure is designed to reduce resource demand (efficient heating, cooling, water supply, etc.) and waste reduction and better monitoring are developed. One key aspect is data collection from CCTVs, mobile phones, household meters, traffic sensors, thermostats, etc. These digital (big data) developments may have great potential to assist EM and support urban SD and are also applied in other fields (Townsend, 2013; Sookoor et al., 2017; Böhm et al., 2019; Visvizi and Miltiadis, 2019; Halegoua, 2020; Inkinen et al., 2021; Saravanan and Sakthinathan, 2022).

River basin ecosystem approaches

Interest in river basin approaches declined slightly between the 1980s and 2005. It is again increasing and river systems look likely to suffer as global warming reduces snow accumulation in mountains cutting summer flows and extreme weather events occur. All this when demand for water is rising and availability is falling (in quality, quantity and year-round reliability). Water inputs and polluted water outputs can be managed in a basin. More nations are competing for shared river basin resources (Pearce, 2019). Co-ordination and oversight are needed for EM and river basins offer a discrete bio, geo, physical, ecological, mappable and to some extent socio-legal-economic unit (Squires et al., 2014; Xiangzheng Deng et al., 2014; Boer, 2015; Ramkumar et al., 2016; Smith, 2017; Smith et al., 2017; Khan and Adams III, 2019; Bharati et al., 2020; Lautze, 2020; Metcalfe et al., 2020; Do Carmo, 2021; Young, 2021). Most basins can be divided into sub-basins to aid management at a more local scale.

Watershed and catchment systems approaches

The watershed (catchment) component of a river basin system is widely used as a study, monitoring or management unit in rural and urban situations. A watershed or catchment is the ecosystem which contributes moisture to a stream and groundwater, more-or-less separated from the next watershed by a divide. Activity in a watershed can impact the greater area of a basin; for example, soil erosion from poorly regulated farming may cause siltation and floods for many kilometres downstream. Managing basin watershed/catchments to improve infiltration and snow retention can benefit the rest of the basin by improving water availability.

Watershed/catchment management can be top down or bottom-up (Sabatier et al., 2005). Erosion control, land rehabilitation, ecosystem restoration, fisheries, wildlife conservation, flood control, pollution control, agricultural innovation, or SD may be the focus, or an integrated approach may be favoured (Kumar, 2010; Vaughn, 2011; Haigh et al., 2012; Menon and Pillai, 2012; Mohan and Saikia, 2012; Kuhn and Emery, 2013; Lannon, 2013; Gonenc et al., 2015; Lotus, 2016; Zhang et al., 2019; Reddy et al., 2019; Hart et al., 2020; Maliwal, 2020; Ferrier and Jenkins, 2021; Yousuf and Singh, 2021).

Socio-economic and socio-economic-environmental systems

EM deals with environmental-socio-economic systems so social and behavioural networks have to be researched and managed as well as environmental (Umetsu and Sakai, 2014; Sato et al., 2018; Behnassi et al., 2021; Biggs et al., 2022). Social and behavioural networking tools include social network analysis (Glaser et al., 2012; Borgatti et al., 2013; Scott, 2017; Perz, 2019), stakeholder analysis, qualitative network models, fuzzy cognitive maps and Bayesian belief networks. These help show social structure and how stakeholders participate and act; for example, in waste management (Virapongse et al., 2016; de Ohveira, 2017; Salpeteur et al., 2017; Sayles et al., 2019; Cerqueti et al., 2021).

Environmental management systems

An EMS is an organised approach to managing the environmental effects (and health and safety and SD) of an organisation's operations; it involves integrating environmental respect and awareness with economy and quality of production. Health and safety and EMSs overlap (Shematek et al., 2016; Will, 2019). EMSs are widely used by large and some smaller businesses and a range of institutions (Sheldon and Yoxon, 2006, 2012; Tinsley and Pillai, 2006; Welford, 2013; Price, 2016).

Adopting an EMS enables an organisation to set goals, monitor performance against them, and it shows when to take corrective action or make improvements and become more competitive; also, it supports the development of a reflective outlook which seeks to be environmentally sound (Emmanuel, 2014; Price, 2016). As well as achieving cost savings through environmental initiatives, an EMS allows an organisation to integrate EM into overall management. Sometimes the EMS is conducted in-house (hopefully using environmental scientists), but it is most likely to be undertaken by an accrediting body or a subcontractor. EMSs can help ensure adequate standards through checks and certification, with pressure to maintain standards and a possibility of de-certification if standards fall (Bentlage and Weiß, 2006).

The EMS process should be one of continuous, ongoing improvement, with a cycle of goals set, checks conducted and results published. Thus, the process takes a body beyond mere compliance and encourages it to become proactive and stimulates good practice. Use of an EMS should also help keep a body aware of changes in knowledge, legislation, etc. An EMS can help ensure a structured, standardised and balanced approach to EM, and improve the organisations' image, attractiveness to employees, etc. Regulators are more likely to treat bodies using EMS with a soft touch, and management may enjoy greater peace of mind and pride. Insurance companies may offer better terms to a body with an EMS accreditation (perhaps insist on it) because it shows the body has considered risks.

The first EMS was the British Standards BS7750 released in 1992. Another EMS, the European Union Eco-Management and Audit Scheme (EMAS), is a site-specific and proactive approach promoted since 1995. The very widely used ISO14000 series includes over 60 certification systems: ISO14001 to 14061. These apply to eco-audit, life-cycle assessment and so on (Dentch, 2016; Briggs, 2017). These are derived from the ISO14000 series, and relate to the ISO9000 quality management series. Some clients use ISO9002, but ISO14001 (launched in 1996) provides what is virtually the world standard framework and guidelines which support the voluntary development of assessment and environmental practices. It also indicates what is needed for an EMS (scoping), provides the format, objectives, targets and implementation, review procedures, correction measures, and so on. ISO14001 and its future derivatives will probably be the world standard for EM. ISO14001 may help ensure that issues are investigated because an objective international assessor is involved. The ISO14001 system is often adopted by organisations with little EM experience.

Increasingly businesses insist on suppliers or subcontractors having EMS certification, and failure to do so can be a hindrance. EMSs only certify each client, not the level of actual environmental performance. A problem is that an organisation or government joining an EMS may have insufficient resources to address problems revealed. Critics argue that EMSs may be a substitute for adequate EM, and are bureaucratic, mechanistic and insufficiently flexible; consequently they lead to mere compliance, not a real will to improve. In addition, the cost of an EMS could deter bodies with limited funds and for some clients efforts outweigh benefits. It is difficult at present to de-certify a body if it is granted EMS status and then becomes careless. Bodies adopting an EMS submit to a cycle of periodic audit and review. An EMS may be applied to a body's single site or a number of sites around the world.

An EMS usually requires a participating body to prepare an environmental policy statement and regularly update it. The standards used to eco-audit typically test whether the body:

- has identified overall aims;
- understands constraints on achieving aims;

- identifies who is responsible for what;
- sets an overall timetable for achieving aims;
- has determined resource needs;
- has selected a project management approach;
- has a progress monitoring system.

EMSs can be difficult to pursue effectively as a result of institutional politics, funding problems, data shortages, need for industrial secrecy, etc. EMSs are a lucrative field of consultancy activity. Most EMSs nowadays are undertaken by subcontracting to a team accredited by and using the approaches developed by one of the main global bodies, notably ISO.

The state and EM

The process of EM can be managed by the state, international bodies, citizen bodies, business, NGOs, international agencies, etc. (Hausknost and Hammond, 2020). While much EM is in state or agency hands, civil organisations and NGOs may act as catalysts (Peters et al., 2009; Newell, 2012). A decentralised approach might prove less robust against special-interest groups, large companies and so on than a centralised, state-supported approach. It is not uncommon for states in a federal system to come into conflict among themselves or with central government over environmental issues. One reason for the formation of the EPA in the USA was to co-ordinate and integrate efforts under a federal system.

Non-Western EM

EM has to work in all countries and international situations (Pinto et al., 2018). In addition to developing EM practices to meet their own needs some countries also have to deal with overseas companies extracting resources, manufacturing goods and sending waste to them for disposal/processing (Andeobu et al., 2021; Sikdar, 2021). There is a need to adapt EM to tropical conditions where environmental and ecological processes differ from temperate or Mediterranean situations (Baker, 2020).

China

China has progressed with EM; Chinese businesses invest on a world scale and the country is increasingly influential so EM developed in the PRC is likely to spread (Watts, 2010; Geall, 2013; Shapiro, 2015). The roots of China's EM lie in the Environmental Protection Law passed in 1989 (Ma, 2017; Yuhong Zhao, 2021). In 2015 another Environmental Protection Law was passed and there has been much subsequent development of EM (Steger, 2003; Carter and Mol, 2007; Jia Wang and Junhui Liu, 2010; Kuo et al., 2011; Guizhen He et al., 2012; McLaren, 2013; Schulze, 2015; Jiahua Pan, 2016; Angang Hu, 2017; Jianming Yang, 2017; Sternfeld, 2017; Di Zhou, 2020; www.fao.org/3/p4150e/p4150e01.htm – accessed 26/10/21; Jing Wu and I-Shin Chang, 2021). China has also been developing SD approaches (Day, 2005; Kristen, 2005; P.G. Harris, 2012; Yang Xiaojun and Jiang Shijun, 2019).

India

EM has been developing in India with both rural and urban focuses (Kulkami and Ramachandra, 2006; Maliwal, 2006; Xuemei Bai, 2010; Jayamani, 2012; Ajith Sankar, 2015;

Mohanraj et al., 2015; Krishnamoorthy, 2017; Kandpal, 2019; Fadly, 2020; Singh et al., 2020; WEPA Outlook on Water Environmental Management in Asia 2018, http://wepa-db. net/en/publication/2018_outlook/index.html – accessed 30/10/21).

Southeast Asia

Southeast Asian nations have been developing environmental legislation and EM since the early1980s. EM applications to fields like small and medium enterprises (SMEs), urban growth and pollution, urban traffic, logging, rubber and palm oil processing, may be of interest to other tropical regions (Abdullah, 2007; Nitivattananon and Noonin, 2008; Setthasakko, 2010; Maizatun, 2011; Abu Bakar et al., 2016; Mokthsim and Salleh, 2016; Lopez and Suryomenggolo, 2018; Panya et al., 2018; Ueasangkomsate and Wongsupathai, 2018; Hoang et al., 2019; Nguyen Hoang Tien, 2020; Rapiah et al., 2020; Sunil, 2020; Loh, 2021; Samah and Kamarudin, 2022; Chowdhary et al., 2023; www.ehso.com/IntlSoutheastAsia.htm; https://iclg.com/practice-areas/environment-and-climate-change-laws-and-regulations/ indonesia – accessed 30/10/21; *Indonesian Journal of Environmental Management and Sustainability* – ISSN: 2598-6260, e-ISSN: 2598-6279; Handbook BIO3800 – Tropical EM – Monash University Handbook, https://handbook.monash.edu/2020/units/BIO3800htt – accessed 30/10/21).

Indonesia passed an Environmental Management Act (No. 23 of 1997); Vietnam updated its EM legislation in 2022. Singapore is one of the world leaders in EM, urban development and waste management (Ngee-Choon Chia and Sock-Yong Phang, 2006; Lin-Heng Lye, 2007; Lin-Heng Lee et al., 2010). EM in the Philippines has been outlined by Guia-Pedrosa (2016).

West Asia, Middle East and Northern Africa

Some nations, especially those with petroleum industries, have wealth, expanding higher education and are urbanised but there is sparse literature on EM practice, environmental politics, SD and environmental laws for other parts of these regions (Nhamo and Inyang, 2011; Khurshid et al., 2014; Habbash, 2016; Gerged et al., 2017; Verhoeven, 2018; Asiri et al., 2020; Khdair and Abu-Rumman, 2020). Much of the activity is undertaken by consultants and little in the way of reports is widely disseminated.

Israel, with population growth in a narrow territory with water shortages, generates EM challenges. So far results have been mixed (Llan et al., 2013). To progress there is a need for co-operation with neighbouring countries, for example to manage the Jordan River or the Dead Sea.

North Africa (the Maghreb) and the Middle East (the MENA or WANA region) has challenges from environmental degradation dating back to pre-Roman times (a history which might offer EM lessons). There are droughts, water shortages, soil erosion, political unrest and migration from these regions (Shaw, 1995; Mikhail, 2012; Samimia et al., 2012; Farmer and Barnes, 2018; Swearingen and Bencherifa, 2021; Williams, 2021). In 2021, prompted by the UNEP Regional Office for West Asia, the countries of the region met to set environmental priorities for the following five years. The Asian Development Bank outlined some regional sustainable EM of projects in Central and West Asia in 2019 (www.adb.org/projects/48109-001/ main – accessed 07/11/21), Nomani (2015), Ebadi et al. (2020) and Marsden (2017) have reviewed environmental issues, EM and SD in the region and east to China. Dust storms, salinisation/alkali problems and other forms of land degradation, and a desire for sustainable agricultural development, are among the challenges (Hui Cao et al., 2015).

EM has been used by businesses and rural developers in Iraq and Iran (Hutham et al., 2021; Fatemi et al., 2018; Thabit and Ibraheem, 2019).

Africa

Given NGO activity and establishment of the UNEP in Nairobi there has been less adoption of EM than might be expected (Strydom et al., 2009; Nhamo and Inyang, 2011; Kwashirai, 2013; Maware, 2013, 2014; Scales, 2014; Strydom and King, 2017; Hugo, 2018; Musavengane et al., 2019; Tauringana, 2019; *African Journal of Environmental and Waste Management* – ISSN 2375-1266). There has been a limited use and adaptation of ISO14001, especially in more industrialised South Africa (Tene et al., 2021), but much of the rest of Africa is rural and poor, making some EM unsuitable (Spee et al., 2021).

South and Central America and the Caribbean

EM has been adopted in Latin America by business, international or regional banks, power generating authorities and the military, reflecting resource extraction, city growth and investment, narcotics production, etc. (de Vries et al., 2001; Ruthenberg, 2001; Scavone, 2005, 2006; Espinosa and Walker, 2006; Livermann and Vilas, 2006; Lostarnau et al., 2011; World Bank, 2013a; Macpherson and O'Donnell, 2015; Minaverry and Caceres, 2016; Sattler et al., 2016; Edelman et al., 2017; Fernández, 2017; Criollo et al., 2018; Vizeu Pinheiro et al., 2020; *Latin American Journal of Management for Sustainable Development* – ISSN 2052-0336; *Resources* 2005(1) issue on EM in Colombia: www.resources.org/common-resources/environmental-management-in-colombia/ – accessed 07/11/21). Thomas-Hope (2013) provides information on policy and practice in the Caribbean.

In Brazil EM is applied by industry and used in urban settings, for waste management and for some conservation areas, hydroelectric developments and mining (da Silva and de Medeiros, 2004; Jabbour et al., 2010, 2012, 2013; Berardi and de Brito, 2015; Jabbour, 2015; Belal et al., 2017; Broietti et al., 2018; Beuron et al., 2020; Burgos-Ayala et al., 2020; D'Amico et al., 2020; da Costa de Sousa et al., 2021; *Revista Brasileira de Gestão Ambiental e Sustentabilidade* – ISSN: 2359-1412).

Transboundary and global EM

Climate change, pollution, conservation, disease control, water resources, etc. have to be managed at the supra-national level. There is the question of what body should foster international co-operation to search for solutions to transboundary and global problems, oversee implementation and, if EM is to be anticipatory, identify potential problems and conflicts (Fernandez and Carson, 2003; Earle et al., 2015; Warner and Marsden, 2016; Brooks and Olive, 2018; Yeung et al., 2021). Some see UN bodies as able to fulfil these roles (agreements on UN Sustainable Development Goals have helped prompt EM interest since 2015); others suggest it should lie with internationally respected research centres, NGOs or regional groupings of nations like ASEAN (Koh, 2012). At present overall co-ordination and enforcement is too weak.

Transboundary EIA and international law give some support for cross-border EM (Marsden and Brandon, 2015). Also, many shared river basins have gained experience. With the spread of free trade since the WTO, the North American Free Trade Agreement (NAFTA)

1994 and similar undertakings, EM must cope with problems caused if environmental controls are interpreted as a trade barrier.

The Basel Convention on the Control of Transboundary Movements of Hazardous Wastes and their Disposal (1992) has helped with cross-border waste/pollution EM. If resources like water and food become stressed by climate change and demand, transboundary co-operation will be increasingly challenged.

Integrated EM

Much EM has been reactive, narrow in focus, piecemeal and poorly co-ordinated. Integrated approaches (IEM) have been explored to try to counter these problems (Jørgensen et al., 2019a; Integrated *Environmental Assessment and Management* – ISSN: 1551-377, eISSN: 1551-3793). But there is disagreement as to what exactly IEM is (Kumar, 2009; Sarkar et al., 2016; Wang et al., 2021). Terminology is vague; for those involved in corporate EM 'integrated' means the development of an EMS that combines health, safety and environmental quality issues. Alternatively, the Dutch Government, concerned with the EM of the North Sea, would see 'integrated' as the assessment of all relevant environmental factors: pollution, fisheries, erosion, etc. Yet another interpretation is integration of EM with environmental engineering or social and economic issues (Gupta et al., 2021).

With a number of approaches, there is a risk that academics and professionals become too engrossed in techniques and forget they are a means to an end (Ewert et al., 2004). To add to the confusion some EMS bodies offer 'integrated' EMS which blends two or more standards from different disciplines into one (e.g. mixing ISO 9001, ISO 14001 and ISO 50001). The US EPA has an IEM system (IEMS) implementation guide: www.epa.gov/saferchoice/integrated-environmental-management-systems-iems-implementation-guide – accessed 13/11/21.

Strategic EM

Strategic EM (SEM), strategic environmental assessment and strategic planning overlap and link with fields like health and safety, EIA, strategic risk management, strategic innovation management, the quest for SD and strategic business management and planning. Although adopted by business, institutions, regional authorities/planners, it has been argued that there are situations where SEM may not be the best option and that it is seeking to do too much to be achievable (Göktepea et al., 2014; Rosenberg, 2015; Ruokonen and Temmes, 2019).

Strategic environmental assessment

The formalised, proactive, systematic and comprehensive process of evaluating the environmental effects of a policy, programme or plan and alternatives is known as strategic environmental assessment (Dalal-Clayton and Sadler, 2005; Schmidt et al., 2005; Short et al., 2005; Fischer, 2007; Therivel, 2010; Sadler et al., 2011; Epstein and Buhovac, 2014; Fischer and González, 2021). This has developed from EIA, assists EM and is widely applied to aid programmes, structural adjustment, changes in public transport policy, urban planning, and so on, usually at a level above a single project (Campeol, 2020).

Summary

- Each situation faced by EM is unique; the approach adopted to deal with it reflect the attitudes and background of those involved, time and funding available, etc.
- There is generally more than one route to an EM goal: perhaps one is the best all-round solution, one the BPEO, one is that favoured by the government, another is favoured by a company, etc.
- As if it is not enough to have to deal with complexity, incomplete knowledge and poor data, EM often has to cope with situations where the development objectives and strategy have already been decided by politicians, special-interest groups, aid agencies, etc.
- Whenever possible an adaptable strategy should be adopted to cope with unforeseen problems and opportunities.
- Whatever approach is selected an essential element is to be sensitive to issues, people's needs and fears, environmental limits, etc.
- EM will frequently be shaped by an adopted EMS, typically ISO14001-2015 series.

Further reading

Allan, C. and Stankey, G.H. (2009) *Adaptive Environmental Management: a practitioners guide*. Springer, Dordrecht, NL.

Everard, M. (2017) *Ecosystem Services: key issue*. Routledge (Earthscan), Abingdon, UK.

Grunewald, K. and Bastian, O. (Eds) (2015) *Ecosystem Services: concept, methods and case studies*. Springer-Verlag, Berlin, GR.

Inkinen, T., Yigitcanlar, T. and Wilson, M. (Eds) (2021) *SMART Cities and Innovative Urban Technologies*. Routledge, Abingdon, UK.

Jørgensen, S.E., Marques, J.C. and Nielsen, S.N. (2021) *Integrated Environmental Management: a transdisciplinary approach*. CRC Press, Boca Raton, FL, USA.

O'Higgins, T. and Al-Kalbani, M.S. (2015) *Systems Approach to Environmental Management: it's not easy being green*. Dunedin Academic Press, Dunedin, NZ.

Robbins, P. (2020) *Political Ecology: a critical introduction* 3rd edn. Wiley, Hoboken, NJ, USA.

Therivel, R. (2010) *Strategic Environmental Assessment in Action*. Routledge (Earthscan), Abingdon, UK.

Sheldon, C. and Yoxon, M. (2012) *Environmental Management Systems: a step-by-step guide to implementation and maintenance* 3rd edn. Taylor & Francis, London, UK.

www sources

- EU Science Hub – best EM practices – https://ec.europa.eu/jrc/en/research-topic/best-environmental-management-practice (accessed 01/12/21).
- Journal: *Ecosystem Services* ISSN 2212-0416. www.sciencedirect.com/journal/ecosystem-services (accessed 01/12/21).
- Journal: *Ecosystems and People* ISSN 26395908, 26395916. www.tandfonline.com/toc/tbsm22/current (accessed 01/12/21).

Chapter 8

Data, standards, indicators, benchmarks, goal setting and objectives, monitoring, surveillance, models and auditing

Chapter overview

- Tools, data, data analysis, statistics and interpretation
- Indicators, standards and benchmarks
- Setting goals and objectives and getting an overall view
- Monitoring
- Surveillance
- Modelling
- Environmental auditing/assessment, eco-auditing, environmental accounting, SD auditing and environmental compliance auditing
- Summary
- Further reading
- www sources

Tools, data, data analysis, statistics and interpretation

EM usually has to assess the *current* situation or progress, although goal setting and modelling look forward (and use may be made of backcasting). Useful knowledge and warnings come from studies of historical documents and prehistoric evidence. There is some overlap between the focus in this chapter and the future-looking *proactive* approaches explored in the next chapter. EM methods, techniques, tools, etc., are mainly borrowed and adapted from other disciplines. If a tool or specialist is not available EM has to assemble a toolkit and devise an approach and, if the problem is novel, research is needed to develop something. Whether tried and tested or new, the method, technique or tool usually needs to be checked, adapted and focused. Pilot studies and test runs should be a key part of EM, but they are frequently dispensed with, due to the pressure to address issues fast and at reasonable cost.

DOI: 10.4324/9781003189985-10

There are some disadvantages in using consultants, off-the-shelf techniques or software:

● consultants tend to be hired for as short a time as possible and may have moved on and be unavailable when problems develop;
● sometimes consultants seek timely completion bonuses and this might reduce care;
● the specialist may report what they feel the commissioning body wants to hear to avoid friction and help secure future contracts;
● consultants may be unfamiliar with the local environment, the socio-economic or cultural situation;
● there may be a focus on getting results which do not slow development, rather than offer a BPEO;
● off-the-shelf techniques and software are likely to need adaptation and results should be treated with caution.

Whenever possible EM methods, techniques and tools should be standardised so that the results can be checked and compared with past and future studies or results from elsewhere, although there will usually be modification needed for each case.

Common EM tasks and tools include:

● Developing terms of reference; scoping/setting limits to the exercise; desk research; test models; checking nearby and other relevant cases.
● Setting goals and objectives; scoping; strategic planning; stakeholder assessment; brainstorming, SWOT assessment.
● Selecting methods: consulting guidelines/benchmarks/standards; strategic management and planning tools.
● Assessing what funding is required, checking cost-effectiveness, reviewing spending.
● Pilot studies, trials, market research studies.
● Informing the public and authorities and seeking reactions.
● Collecting relevant data: surveys; focus groups, etc.; historical – records, palaeoecology, archaeology; GIS; establishing monitoring instruments/observers; desk research.
● Providing support for EM and standardisation: EM systems (EMSs).
● Monitoring ongoing measurement to provide data on progress (may give early warning of problems).
● Modelling: checking present situation (perhaps simplifying complexity to understand what is happening).
● Evaluation/auditing: processing data; eco-audit; sustainability audit; project/programme evaluation tools during the exercise.
● Public relations, public consultation, communication with company or government decision makers; written reports/presentations; briefings; press reports; drop-in shops; websites, chat sites.

EM should progress with goal setting, scoping and selection of approaches; pilot studies; data acquisition, modelling/assessment/analysis and finally interpretation. Scoping is the identification of the important issues to be considered, boundaries of study and state of knowledge, it is ideally a rapid exercise that offers a roadmap. This may take the form of a meeting of all parties involved yielding a report or a workshop or some form of systematic review (Moustafaev, 2015).

Data may be divided into quantitative and qualitative (Ashley and Boyd, 2006; British Ecological Society, 2014; Bergin, 2018). These each come in two forms: reliable and unreliable. A statistician might prefer to divide data into parametric and non-parametric. Whenever possible data should come from more than one source and more than one tool should be used, so that decisions are based on more than one line of evidence. The trend is towards multidisciplinary teams dealing with EM problems. Ideally, data leads via assessment/analysis and interpretation to knowledge and wisdom.

In the past EM was dominated by natural scientists who dealt mainly with quantitative data (Zhen Chen and Heng Li, 2016; Cameletti and Finazzi, 2018; Jiaping Wu et al., 2022). Qualitative data was once disparaged as subjective and unreliable but it is very important for EM (Grbich, 2012; Merriam and Tisdell, 2015; Macura et al., 2019; Miles et al., 2019; Bazeley, 2020; Bercht, 2021). Qualitative data are valuable provided they are collected properly and the interpretation is careful. There can still be a quantitative-qualitative (objective-subjective) divide between the differing traditions. Some social studies researchers are still suspicious of empirical study and what they see as scientists' failure to engage with social reality (Burnette, 2022).

Data are often wanted in a hurry and cheaply; thus tools which are 'quick and dirty' (i.e. fast but perhaps not very accurate) are used. Techniques and tools seeking to be rapid have been widely used in rural situations, but may be employed for assessing urban areas. Where data are needed fast the skills of market research and sociology may be tapped. Tools employed by market researchers include: focus groups, questionnaire surveys, observational studies, phone interviews, etc. A recent and very useful development is citizen science, the recruitment of those willing and capable of collecting, monitoring data and helping assess as an altruistic pastime (Roy et al., 2012).

Fuzzy data

Fuzzy models or sets are mathematical means of representing vagueness and using imprecise information (hence the term fuzzy). These models have the capability of recognising, representing, manipulating, interpreting and utilising data and information that might be vague and lack certainty (Viertl, 2011; Chmielowski, 2016; Kahraman and Sari, 2017). Fuzzy logic is used in computing programs that yield results in degrees of truth (on a scale of say 1 to 10) rather than 100% true or false. Often it is used in multi-criteria risk assessment, modelling and decision making (Jingzheng Ren, 2021a).

Big data

Originating in the 1990s big data approaches seek to analyse, systematically extract information from, or otherwise deal with datasets that were until recently too large or complex to be dealt with by traditional data-processing software (R. Sharma et al., 2021). The hope (big data is far from tested and established) is to reveal patterns, trends and associations, especially relating to human behaviour and interactions (Sun and Scanlon, 2019). It is key to approaches like SMART cities. There is huge investment in gathering and managing big data (Kitchin, 2014). Data is harvested from web activity, web-connected equipment (e.g. house thermostats and meters, computers, media use, mobile phones, CCTV, healthcare and genetic databases, traffic controls, etc.). Governments and business are using big data to provide terrorism alerts and monitor various other things; for example, by the National Security Agency (NSA) in the USA. China plans to use it to give all its citizens a personal social credit score affecting benefits sometime in the 2020s. Mass surveillance using big data analysis technology could be used to steer people in SMART cities and allocate services and welfare.

The benefits may include better control of citizen response to EM; the disadvantages are data misuse and excessive surveillance.

Big data could improve modelling of climate change and soil carbon sequestration, support SD, monitor SMART cities, help understand and manage pollution, weather, pandemic disease and genetics (Keeso, 2014; Woodward et al., 2014; Etzion and Aragon-Correa, 2016; Jia Wang and Moriarty, 2018; Blair et al., 2019; Pedersen and Wilkinson, 2019; Beier et al., 2020; Lucivero, 2020; Yuan Su Yanni and Yu Ning Zhang, 2020).

Open data

Open data are data which are shared and distributed without restrictions (see: Google Public Data Explorer; European Union Open Data Portal, etc.). Open data are often part of big data by virtue of the fact that they entail reuse of information that may be collected for other purposes. Open data are assumed to be held by governments (UN, 2014; N. Sharma et al., 2021). Open data should not be confused with open source software (software available free or at low cost) (Schmidt et al., 2016). Earth monitoring satellite data is partly open access as is most meteorological data (Borowitz, 2017).

Data assessment/analysis

Environmental data analysis has developed rapidly (Miller et al., 2005; Zuur et al., 2007; Bryan and Manly, 2008; Yue Rong, 2011; Acevedo, 2012; David, 2017; Wildi, 2017; Zhihua Zhang, 2017; Dormann, 2020; Emetere, 2020; Emetere and Akinlabi, 2020; Von Frese, 2020) (modelling overlaps with data analysis). Interpretation of data is a crucial stage in any approach which may be assisted by statistical tests which can also counter assessor's conscious or unconscious bias or show patterns where they are difficult to discern.

Indicators, standards and benchmarks

EM has to be able to judge if anything is changing, whether data is reliable and what significance a change has.

Indicators

An index is a scale on which can be shown value, quantity or position. An indicator is something that shows unambiguously the state of something. A standard may be a point on an index scale of pollution: e.g. so many ppm of a compound is unacceptable. The right indicator(s) need(s) to be selected. Some indicators are precise and reliable, others less so. Ecologists and others have explored critical indicators; e.g. a single parameter (or parameters) which determines whether an ecosystem or livelihood can flourish. The concept of carrying capacity is based on the belief that an ecosystem can sustain only a certain density of particular organisms, and if that is exceeded, predator–prey balance, nutrient supply or waste disposal will break down. It is risky to assume that because pressure on an ecosystem is below some threshold, all is well; a change of climate, arrival of a new species, or other unforeseen development may topple the balance. EM has to build in margins of error around any indicator or critical threshold that is being monitored.

Sometimes when a broader focus is needed, or the process to be monitored is complex, a composite index may be devised which is the sum of a number of different measurements (e.g. the Human Development Index (OECD, 1991; UNDP, 1991)) or various composite SD indicators. Something like soil acidity can be simply measured with a chemical colour change kit and interpreted using a colour chart (index) to generate a pH value, or with an electronic probe.

Standards often rely upon indicators. These should be things that can be relatively easily measured, are reliable and which have specific meaning and point out something: the stage reached, quality, stability, vulnerability, etc. Indicators are widely used to try and assess whether things are getting better or worse (Dada et al., 2013; Bell and Morse, 2020; *Ecological Indicators* – ISSN 1470-160X). Living species with known sensitivities may be used to indicate heavy metal pollution, acid deposition, frost occurrence, soil qualities, etc. The chances are that if an indicator is needed there is already at least one developed by ecology, economics, healthcare, pollution control, biodiversity assessment and conservation, social development, etc.

One task is to establish/formalise indicators so they are always measured the same way, and another is to validate them. Ideally an indicator should be sensitive, but not so much so that it triggers false alarms; it should respond fast, reliably and unambiguously, and if possible be cheap and easy to use. An indicator may be a single object or event, such as a distinctive tree which shows the soil is fertile, or lichen showing there is little acid deposition, etc. A hormonal test kit may show a colour change in the presence of some compound – like a pregnancy test. Bioindicators and biomarkers (based on indicator species) are often used for monitoring but reliance should not be placed on a single bioindicator/biomarker.

SD indicators In 1992 *Agenda 21* called for indicators of SD; if they highlight the underlying causes of environmental damage it will help prevent wasted efforts treating symptoms or pursuing cosmetic cures. Because there is no single established definition of SD, and there are different strategies for pursuing it, and the starting point and challenges differ from situation to situation, it is difficult to develop a universally accepted index to measure it (Lawn, 2006; Mieila, 2013; Bell, 2017; Muthu, 2019; Spangenberg, 2019; Bell and Morse, 2020; Nielsen, 2020; Thore and Tarverdyan, 2021) (see Box 8.1).

Box 8.1 Measuring sustainable development

- Index of sustainable economic welfare – this is a socio-political measure.

- Net primary productivity is derived from the concept of carrying capacity, which is the maximum population of a given species that an area can support without reducing its ability to support the same species indefinitely.

- Environmental space – is the amount of any particular resource that can be consumed by a country without threatening the continued availability of that resource, assuming that everyone in the world is entitled to an equal share.

- An extension of the human development index (HDI) – the HDI has been considerably modified and now makes some provision for assessing SD. However, it has a long way to go before it measures SD and environmental issues adequately, and analysts suggest a 'green index' to complement the HDI, rather than further greening the HDI.

- Factor X concept – this asks by what factor can/should the use of energy/ resources be reduced and still have the same utility? This is a flexible way of monitoring and modelling how to extract more from resources being used. It can be modified to ask by what factor must resource flows to affluent societies be reduced to allow the poorer societies to improve their living conditions?

- Composite index of intensity of environmental exploitation – similar to the HDI.

- Less general, more focused sustainability indices – these have also been developed for specific ecosystems and sectors of activity; for example, a sustainable land management index, and a sustainable agriculture index. These have been prompted by doubts about the long-term viability of modern agriculture as a consequence of soil degradation, pollution by agrochemicals and heavy use of petrochemical energy inputs.

- Indicators of farm level sustainability – might prove useful for highlighting the key inputs and practices, which hinder sustainability.

- Eco-footprint – measures how much land is required to supply a particular city, region, country, sector, activity or individual with all needs usually expressed in hectares per capita. It is a measure of load imposed on the environment to sustain consumption and dispose of waste. The concept is based on the idea that each individual uses a share of the productive capacity of the Earth's biosphere both for resources and disposal of wastes. It is a measure of human aggregate ecological demand and it can only be temporarily exceeded or the productive and assimilative capacity of the biosphere is weakened. Eco-footprinting is thus a useful ecological accounting method for assessing the demands made by humans on various productive areas. It can help show where human demands are problematic, aid evaluations of what could be done to improve sustainability, and provide a framework for SD planning. However, it has been suggested that it underestimates human impacts and shows only the minimum needs for SD, without a healthy margin for error, which the precautionary principle seeks.

Eco-footprinting may be pursued using a number of different models, for a company, island, region, sector of production, city, etc. A valuable tool for use in the quest for sustainable development, it allows comparison of the impact of different components on the same aggregated scale. It also aids the assessment of eco-efficiency (i.e. whether an organisation is making the most of its resources and waste disposal opportunities). Eco-footprinting has been used by regional planners seeking SD (e.g. for cities, regions or small islands). It may also be used by businesses, cities and other bodies to measure their environmental performance. Eco-footprinting offers a two-dimensional interpretation of complex ecological and economic systems; fails to recognise issues such as consumer preference or property rights; and offers a temporally limited snapshot view. Natural systems are seldom stable and it may fail to allow for this, and consumption per capita, fashions and settlement patterns change and this may not be registered. Also, it does not help to find the most appropriate path for human activities. However, it does offer a graphic and easily communicated image; it is a relatively transparent tool; and it may stimulate creative thinking about environmental and developmental issues. Eco-footprinting should be used with care, and in combination with other tools.

127

Judging progress towards SD demands prediction of the behaviour of complex socio-economic and physical systems, and using extensions of established economic, social and environmental indicators is unlikely to be adequate (Latawiec and Agol, 2015). Composite indicators have commonly replaced single-dimension indicators; e.g., the environmental sustainability index (ESI) (https://ec.europa.eu/eurostat/web/sdi/indicators – accessed 06/12//21). Hák et al. (2007) listed and assessed a number of SD indicators.

The UN Sustainable Development Goals (SDGs) are a collection of 17 interlinked global goals. A set of 102 indices has been structured with reference to the 17 SDGs. Each SDG goal has six indicators primarily attributed to it and 37 of the 102 indicators are multipurpose (used to monitor more than one SDG) (da Cal Seixas and de Moraes Hoefel, 2021; SDG indicators website: https://unstats.un.org/sdgs/ – accessed 07/12/21).

Standards

A standard is a widely accepted or approved example of something against which others may be measured. Standards allow meaningful evaluation, exchange and comparison of data, improve objectivity of judgement, aid recognition of crucial thresholds and limits, and support negotiation, law making and comparison (between sites, between countries and between years). Environmental standards may be divided into three broad groups: those concerned with ensuring human health and safety; those concerned with maintaining environmental quality; and those concerned with the quality of consumer items. Monitoring, modelling, auditing, standards, benchmarks and systems (including EMSs standards) help ensure ongoing objectives are set and met, check progress and warn of problems and opportunities (Heras-Saizarbitoria, 2018). Standards enable the establishment of benchmarks, and checking and stocktaking using those benchmarks. ISO14001 is a descriptor for a set of standards developed in response to concern for the environment, these represent a consensus by various standards bodies about procedures needed for an EMS (https://app.croneri.co.uk/topics/ems-standards/indepth – accessed 05/12/21).

Standards, units and benchmarks overlap a little; the latter are approved and provide standardised descriptions of best practice, procedures, and perhaps intents and goals (Hitchens et al., 2012). In the past national standards collected in, say, a French colony would have to be converted to units used in the USA, etc. Sometimes indicators are not easily comparable or the means of gathering data are more or less unique, making even rough comparison difficult. What is a useful standard in a temperate country may be meaningless when applied in the tropics, mountains or polar regions. Without worldwide appropriate standards and units it is difficult to research the structure and function of the environment and to monitor conditions. Before the 1950s international unions agreed standards for some fields, such as telegraphy and radio, but for the environmental sciences many improvements came only after the International Geophysical Year (1957–1958) (Olson, 2019). There is a risk someone will mistake one unit or standard for another. A NASA probe was lost at enormous cost because one of the teams involved used metric and another imperial; a command in one was interpreted by the probe in the other. It is not uncommon to see 'man hours', there is a huge difference between a well-fed worker and a malnourished one, or those in a chill environment and those in the tropics, and the hours might be those of women. Bundles still appear as a unit in studies of firewood; research papers talk of straw return and miss that there are various crop straws involved; biochar is frequently of unknown feedstock and mode of preparation, etc.

As research into environmental issues progresses and new technology appears, new standards are needed. The process involves various national and international institutes and standards organisations. Advances in medicine, ecology, toxicology, etc., sometimes force the revision of established standards. New standards are being developed which distinguish between groups of people; one example is pollution standards that may have to take into account the greater vulnerability of children to some compounds.

There are a number of ways of developing a standard; for example, a standard for checking that fruit does not exceed 'safe' levels of a pesticide might be based on a simple maximum residue level, or a sort of lump sum, or an acceptable daily intake, all of which assumes that consumers eat a given amount per day. It is consequently important that EM knows the characteristics of a standard as well as the levels measured by it, and the reliability of the measurements. The methods of data collection as well as the agreed units must be standardised. Taking the same meteorological measurements in the lee of a house and in open countryside or at various times of day probably gives quite different results.

Standards are of little use if they are not effectively enforced (Faure et al., 2015; Farmer, 2016). Another difficulty is that standards may sometimes be relaxed for economic/profit, political or strategic reasons. The expression REGNEG (renegotiating of regulations) has been applied where a developer succeeds in persuading authorities to relax or modify regulations (including standards), making it easier to meet standards or avoid assessments.

Eco-labelling is a form of standard used on many products to indicate how much they impact upon the environment. The consumer can judge one product against another and, hopefully, buy the greenest. But the focus is on product or service impact and says little about manufacturing or recycling impacts.

Benchmarks

Benchmarks provide reference points by which to measure the performance of something or set minimum targets, and are a means for sharing and promoting good practices (Staplehurst, 2009; Hitchens et al., 2012). Benchmarking also assists the comparison of one situation with another. For some a benchmark is just a level that can be aimed for, or it can be a waymarker against which to judge standards, compliance or progress.

Various tools can be used in the task of benchmarking; one is trend analysis which consists of time-series tabulations of data which enable the pattern of change to be assessed and possibly some future forecasting. Trend analysis can also be useful in performance appraisal. Many bodies provide benchmarking documents and guides; for example, educational curricula or pollution control. Some benchmarking relies on peer review.

Setting goals and objectives and getting an overall view

Without a clear goal EM is likely to be haphazard and unfocused. Goals are aims or purposes; target, aspiration and objective may be used as equivalent terms. Some university EM courses work with business management and law departments and encourage students to acquire tools like goals setting, producing a business plan and some understanding of law and environment.

Setting goals

At the earliest opportunity there should be simple brainstorming to exchange ideas which may be used to scope and establish goals. Brainstorming is likely to consist of workshops with stakeholder representatives or experts, or focus groups may be consulted. A focus group is a relatively informal meeting with stakeholders where the observer prompts discussion in a limited way but essentially listens (Shamdasani and Stewart, 2014). A more powerful tool for brainstorming and future scenario prediction is the Delphi technique. It is a workshop of experts and other stakeholders and is more orchestrated than focus groups, using controlled feedback, to get a pooling of opinions (Mukherjee et al., 2015). This tool is useful when there is less than optimal data and may be done via email or tele-conferencing. The Delphi technique minimises squabbles and deference to powerful peers. When brainstorming, environmental managers should be aware of the limitations of the data available. The search for goals and objectives may be initiated by a known threat or opportunity or by international or national agreements or guidelines; for example, SDGs.

There are other tools for setting goals and objectives. Some were developed by strategic planning, military strategists, policy research, public relations, business management, etc. The problem is that goals are often set before EM is consulted. When that is the case adequate solutions may be difficult. Politicians, citizens, businesses and bodies may change their goals before the original goal(s) are reached.

Guesswork is a last resort; funding bodies and those monitoring progress want a reasoned (ideally transparent) assessment and goals backed by data, modelling and theoretical argument. The sorts of tools which may be used to support goal-setting and decisions include: LCA, environmental risk assessment, hazard assessment, impact assessment, CBA and multi-criteria analysis (Ahraman and Sari, 2017).

Usually decision making demands a guiding framework, tools to identify options and to select from the options identified. There are also scenario-prediction or role-playing tools which may be used. Role playing and gaming are ways to act out possible future scenarios and decide goals. Decisions can be guided and supported by drawing up a decision tree (dendrogram) to show the relationships between a set of objects; the tree shows relationships and hierarchical clustering and can indicate chains of causation; computer programs are available for their construction. A commonly used technique is random forest, a machine learning classification algorithm which combines multiple decision trees into one result. This can be more useful than regression analysis if there are large numbers of variables (https://builtin.com/data-science/random-forest-algorithm – accessed 06/03/23).

There may be public involvement: the use of focus groups, questionnaire surveys, or even a referendum. These are likely to suggest the goals wanted by groups, and are not necessarily rational.

In addition to setting goals it is important to get an overall view before starting EM. Strategic environmental assessment (SEA) seeks to extend the focus from local/project to one which is broader: programme or policy or sector, even global (Pentreath, 2000; Dalal-Clayton and Sadler, 2005; Fischer, 2007; Marsden, 2008; Therivel, 2010; OECD, 2012; Sadler et al., 2015; Cutaia, 2016; Fischer and González, 2021). The EU SEA Directive came into effect in 2004 and requires Member States to implement SEA measures (Dusik and Sadler, 2016; Scotford, 2017).

Once an initial set of goals and objectives has been determined it can be tested and refined, using simple tools such as SWOT analysis or cost-benefit analysis (CBA). SWOT analysis is a simple tabulation of strength – S, weakness – W, opportunity – O, and threat – T, associated with a given choice. SWOT analysis is cheap, easy and fast. It offers a crude overview of a situation or proposal. SWOT was developed by management studies and provides a structure to describe a project or programme, and test the logic of the planning action in terms of means and ends. It focuses on how objectives will be achieved and what the implications of action will be (Nikolaou and Evangelinos, 2010; Ioannau, 2012).

Used to help set goals and for project or programme evaluation, the logical framework evaluation/analysis (LogFrame) approach can help users to think about plans (Green Logical Framework & Indicators: https://europa.eu/capacity4dev/public-environment-climate/book/29310/print; accessed 10/12/21; *The Logical Framework: a list of useful documents*: https://mande.co.uk/2008/lists/the-logical-framework-a-list-of-useful-documents/ – accessed 10/12/21).

A commonly used goal-setting (and progress appraisal) tool is CBA. This tool is applied to plans, projects, programmes and policies to try and calculate positive and negative impacts, in some cases in advance of a proposed development. It originally valued impacts in economic terms, which can mean problems for assessing environmental and social issues. Efforts to remedy this are usually indirect, using techniques such as opportunity costs, shadow pricing, contingency valuation and property values (Pearce et al., 2006; Barbier and Hanley, 2009; Moosa and Ramiah, 2014; OECD, 2018). CBA can help developers select from a set of identified development alternatives but it is not very objective.

CBA is less useful in developing countries because some people there are more likely to operate outside any formal market setting, they consume what they produce, and there are societies which value social or cultural things above material. Cost-effectiveness analysis seeks to select development alternatives on the basis of lowest monetary costs (i.e. best value for money). A goal is set, say an improved environmental standard, and assessors seek the least-cost way to achieve it.

Scoping

Once goals are set there is another useful preparatory approach: scoping. Collecting data is often expensive; it is therefore important to avoid poorly focused or excessive data collection. It is a good idea to scope first to assess what should be measured and how.

Pilot study

Once EM has selected a set of goals and objectives and scoped it makes sense to undertake a pilot study. A pilot study is a small-scale application and forerunner of the main effort; it can identify problems, develop tools and train personnel and offer some chance for adaptability and determine feasibility. But often it is argued that pilot studies raise costs, delay, and that small-scale results may be difficult to scale up. For a smaller exercise it may not be needed.

Life-cycle assessment

Often EM deals with processes: manufacturing, land use, carbon emissions, etc. Successive stages are likely to present different challenges and opportunities. Life-cycle assessment (LCA, or life-cycle analysis) seeks to identify impacts and demands at each stage ('cradle-to-grave' or 'cradle-to-cradle', when there is recycling) of manufacturing, service provision, etc. Impacts do not cease when goods leave a factory; there may be pollution associated with their use and ultimate recycling/disposal. It is used by manufacturers, mainly since legislation to require it in Europe and the USA. LCA can help identify stages where environmental measures are needed and are most effective (Lewis and Demmers, 2013). LCA is used to help evaluate the impacts and best practice at each stage in the provision of services, in manufacturing or consumption. The UK has been decommissioning nuclear power stations built when little thought was given to the dismantling and disposal; had it been, much could have been engineered in to help later.

The spread of LCA has encouraged practices such as design-for-environment, eco/environmental labelling, and other applications which seek to integrate manufacturing, design and service provision with environmental concern. Social LCA (S-LCA) has been used to assess social aspects of development, production, etc. (Crawford, 2011; Muthu, 2014). There is overlap with risk and impact assessment and LCA has been applied to efforts to attain SD (Jingzheng Ren and Toniolo, 2019; Muthu, 2021a). The LCA literature reflects diverse use (Curran, 2012, 2017; Klöpffer and Grahl, 2014; Schenck and White, 2014; Simonen, 2014; Vogtlnder, 2014; Hauschild and Huijbregts, 2015; Sellers, 2015; Jolliet et al., 2016; Hauschild et al., 2018; Passarini and Ciacci, 2021).

Participatory assessment

A range of participatory tools are used for goal setting and evaluating and monitoring people's needs, livelihood strategies, social capital, attitudes, traditional knowledge, vulnerability, etc. (Kindon et al., 2007; Rees, 2008; Dickensen and Bonney, 2012; Rodenbiker, 2021). In the past a failure to consult people commonly led to negative social impacts and valuable local knowledge and skills were missed.

Data are gathered with participatory rural appraisal (PRA); rapid rural appraisal (RRA); and participatory learning and action. Commonly the data collection is linked with research, learning and development activity (Reason and Bradbury, 2008; Narayanasamy, 2009; Mukh, 2010; Freudenberger, 2011; Kanwat and Kumar, 2011).

Monitoring

Monitoring aims to establish a system of observation, measurement and evaluation for defined purposes, the observation constant or regularly repeated. But most assessment techniques give a spatially and temporally restricted view which may not be representative. Monitoring may use such techniques repeatedly in order to build up a sequential set of observations. This is valuable because things differ during the life cycle of a development. There may also be changes in wider economic, social, political, ecological and environmental conditions. Without monitoring, it can be difficult or impossible to establish how things are performing. Monitoring is the process of keeping the health of the environment or ecology (and with social monitoring, of society) in view. Monitoring should be operated to agreed schedules with established and comparable methods. The focus may be on biology, chemical or other pollution; indeed, any aspect of the environment, ecology or society (Inyang and Daniels, 2009; Lindenmeyer and Likens, 2010; Ekundayo, 2011; Gitzen et al., 2014; Lindenmeyer et al., 2014; Goodenough and Hart, 2017; Wiersma, 2017; Brar et al., 2019; Higgins, 2019; Gemitzi et al., 2020; Johnson, 2020; *Environmental Monitoring and Assessment* – ISSN 0167-6369).

Monitoring is usually undertaken for a specific reason(s), for the systematic measurement of selected variables to:

- improve understanding of environmental, social or economic processes;
- provide early warning;
- help optimise use of the environment and resources;
- assist in regulating environmental and resources usage (e.g. it may provide information for law courts);
- assess conditions;

- establish baseline data, trends and cumulative effects;
- check that required standards are being met, or see whether something of interest has changed;
- document state of chemical sinks (e.g. carbon sequestration in soil), activity of sources and so on;
- test models, verify hypotheses or research;
- determine the effectiveness of measures or regulations;
- provide information for decision making;
- advise the public.

There has been increasing interest, spurred by transboundary problems, in developing international monitoring systems, especially for carbon emissions, ozone depletion, ocean acidification, nitrates, natural disaster risk, etc. An independent international research unit was founded in 1975 to assist organisations with monitoring – the Monitoring and Assessment Research Centre (MARC). This concentrates on biological and ecological monitoring, particularly pollution. The World Conservation Monitoring Centre was established in 1980 to monitor endangered plant and animal species. The UNEP has established the Global Environmental Monitoring System (GEMS), which is a co-ordinated programme for gathering data for use in EM and for early warning of disasters. The US Food and Drugs Administration monitors pharmaceuticals and foods. The WHO monitors disease challenges. In most countries, doctors, vets and other professionals report observed effects to national monitoring bodies.

Monitoring may aid understanding of environment and biota structure and function. Monitoring, surveillance and screening (the checking of a specific thing, e.g. a particular disease) are valuable but they can generate problems over who should administer, enforce and pay for them.

Ultimate environmental threshold assessment

This is a tool used to warn of the development of a critical situation that may at best be difficult to recover from. Derived from threshold analysis, this watches for a point at which it is known problems will start to develop catastrophically. The thresholds (or boundaries) may be global or local, environmental, social, economic, etc. The threshold has to be established by previous research (and ideally is a recognised standard). Early development of the tool was by national park managers looking for a precautionary planning aid; it is no good learning of a conservation problem after biodiversity has been exterminated. It has been used by those seeking SD (Guntenspergen, 2014; Kelley et al., 2015), and for the nine planetary boundaries proposed by the Stockholm Resilience Centre.

Remote sensing, GIS and GPS

There are fast, easy ways to assess and monitor or fix positions precisely.

Remote sensing Before the late 1960s remote sensing was restricted to aerial photography in visible, infrared and ultraviolet wavelengths, radar and sonar. To these have been added side-scan radar, light detection and ranging (LiDAR), interferometric synthetic aperture radar (InSAR), terrestrial laser survey (TLS) and other laser-based methods and a range of electromagnetic sensors. For aquatic applications, side-scan sonar and other techniques have been developed (Rees, 2008; Jones and Vaughan, 2010; Campbell et al., 2011; Shunlin

Liang and Jindi Wang, 2012; Gibson and Power, 2013; Fawaz and Long, 2014; Njoku, 2014). For ground surveys there are seismic, geomagnetic, gravity survey and motion detection. Gases and pollutants can also be sensed remotely.

Remote sensing data (physical and socio-economic, terrestrial, marine and atmospheric and extra terrestrial) can be gathered by satellite, aircraft, drones (in air or water), remote cameras/sensors, chemical sensors, seismic sensors, tide gages, temperature sensors, etc. (Nguyen, 2010; Khorram et al., 2012; Alcantara, 2013; Awange and Kyalo Kiema, 2013; Lavender and Lavender, 2015; Xian, 2015; Reddy, 2016; Wegmann et al., 2016; Xuan Zhu, 2016; Wiersma, 2017; Lindenmeyer and Likens, 2018; Byron, 2019; Ferrari and Rae, 2019; Petorelli, 2019; Brimicombe, 2020; Seenipandi et al., 2020; Elachi and van Zyl, 2021; Kanga et al., 2022; Mishra et al., 2022). Data acquired from a range of sources can often be displayed using a geographical information system (GIS) in real time (i.e. constantly updated information – the current situation) or frequently updated; the data may be retrieved and displayed or processed in a diversity of ways (Acevedo, 2015). So, if an EM wants a map of snow cover in the UK in March correlated with atmospheric pollution levels, it is possible provided the data has been stored.

There are several satellite sources available to non-military users: the LANDSAT (NASA/USGS) have provided data for decades, and others include SPOT, SENTINEL, ASTER, MODIS and GF (Ramachandran et al., 2011; Ünsalan and Boyer, 2011; McHale, 2012; Welti, 2012; Emery and Camps, 2017; Chuvieco, 2020). These are widely used for land use change assessment, resources monitoring, drought and climate change monitoring, deforestation and land degradation monitoring, biodiversity conservation, archaeology, pollution assessment, etc.

GIS GIS creates, manages, stores, displays, analyses (typically maps) different types of data (spatial and attribute). The GIS (software) connects digital data to a map, integrating location with descriptive information so patterns can be seen and things monitored. GIS has become a very important tool for EM (Heywood et al., 2011; Longley et al., 2015; De Lima, 2016; Xuan Zhu, 2016; Rustamov and Samadova, 2017; Bai Tian, 2019; Brimicombe, 2020; Mitchell, 2020). Costs can be reduced by using open source software (Neteler and Mitasova, 2008; Wegmann et al., 2016).

GPS A global positioning system (GPS) uses geospatial technology (satellites, receivers and software) to provide precise positioning, navigation, and timing data to users. GPSs have become a useful component in daily life and enable EM to easily, quickly and accurately fix a position (Hofmann-Wellenhof et al., 2010; Matejicek, 2010; DePriest, 2013; Earnest, 2015; Bhatta, 2021; Petropoulos and Srivastava, 2021).

Business and project evaluation monitoring tools

Business management and project evaluation have developed a range of monitoring and evaluation techniques, some used by EM. These gather data; for example, from local people or different groups in a culture. There are tools to help simplify complex situations. Open-access software can usually be found (Wegmann et al., 2016). Some off-the-shelf tools can help establish best practice or set goals (e.g. LogFrame) (World Bank, 1997; Nuguti, 2015). Project evaluation is also useful for deciding strategy and is often linked with impact assessment and provides an idea of how a development is progressing and what may have gone well or wrong.

Surveillance

Monitoring and surveillance merge with remote sensing. Surveillance is close repetitive measurement of selected variables over a period of time, but with a less clearly defined purpose than monitoring. It is more exploratory and can be undertaken to determine trends, calibrate or validate models, make short-term forecasts, ensure optimal development, and warn of the unexpected. Surveillance, like monitoring, can focus on the environment, people or an economy, and may:

- check whether regulations are complied with (without surveillance the setting of standards and rules is of little value);
- provide information for systems control or management;
- assess environmental quality to see whether it remains satisfactory;
- detect unexpected changes.

Where surveillance/monitoring seeks to establish the ongoing picture, it may be important to examine past conditions and establish trends to understand the present and permit extrapolation of possible future scenarios.

Environmental, social, health and economic surveillance/monitoring may focus at local, regional, national or global level or study 'pathways' (e.g. for pollution) (Lombardo and Buckridge, 2007; Morain and Budge, 2012). Surveillance and monitoring may be done at source (where something is being generated), at selected sample points, at random, along transects, or by sampling some suitable material or organism. For example, pollution may be checked by observing a smoke-stack, by a network of instruments, or by surveying lichen species diversity, occurrence patterns and growth. Regulatory monitoring checks its findings against set, in-house national or international standards or stated objectives.

Data may be gathered by orbiting or geostationary satellites, UAVs, reconnaissance aircraft, USVs and static terrestrial land or marine data-gathering stations linked to the data collector by radio or phone link. Satellite remote sensing data and GPS positioning/timing is often open source (or available at low cost and with little restriction) from platforms like LANDSAT, MODIS and the global navigation satellite system (GNSS) (Shuanggen Jin et al., 2014; Awange, 2017).

Rather than aircraft, balloons, UAVs, microlights or kites can be used at low cost (Toro and Tsourdos, 2018; Wich and Lian Pin Koh, 2018). Sensing and transmitting equipment is now small and inexpensive enough to attach to wildlife, even insects. Remote sensing can give information where access is difficult, sensitive or dangerous; not surprisingly, intelligence-gathering bodies initially developed much of the hardware and software. Civil liberties and privacy can be an issue with surveillance, especially when it comes to health, social and economic data (Karampelas and Bourlai, 2018).

Modelling

There is a huge diversity of modelling methods; all seek to clarify without misinterpreting the process under study, some allow forecasting, and some accept the input of alternative sets of variables to explore different scenario outcomes. So modelling can be a research tool, a decision support or means to see into the future (Jakeman et al., 2008; Gray and Gray, 2017; Jørgensen, 2017). The quality of data is crucial: 'garbage in garbage out'; even when data input is good and the model has been well tested, results should be interpreted with caution and users need to understand model limitations (Emetere, 2019).

Models are a caricature or simplification of reality (that may be their only role – to present a simple, clear picture). For non-analogue, mathematical models a set of equations are used to predict the behaviour of a variable or variables (Smith and Smith, 2007; Wainwright and Mulligan, 2013). Models developed for research may need modification for EM. Some predictions can be imperfect, but good modelling should cope with change and inadequate data and give useful results. There are: computer models, analogue models, conceptual models, participatory models, role-play exercises, and many others (Barnsley, 2007; Paegelow and Olmedo, 2008; Ford, 2009; Jørgensen, 2009; Mihailovic and Lalic, 2010; Fath and Jørgensen, 2011; Jørgensen, 2011a, 2011b; Holzbecher, 2012; Gray et al., 2017; Schnoor and McAvoy, 2019; Falconer, 2020; Fort, 2020; Sang, 2020; Clark et al., 2021; Kanga et al., 2022; *Environmental Modelling* – ISSN: 0304-3800; *Developments in Environmental Modelling* (book series), vol. 26 (2014), vol. 27 (2015), vol. 28 (2016), vol. 29 (2016), vol. 30 (2017), vol. 31 (2019), Elsevier, Amsterdam, NL). Conceptual models are used to see what needs study, to help formulate and check hypotheses and to organise ideas. It is important not to lose sight of goals and to become over-engrossed in the modelling/computing process. Modelling seldom gives all the answers or wholly accurate results so EM must allow for this (Beven, 2018).

Simulation or predictive models can provide EIA with an indication of what may happen in the future, and can help EM see how something is proceeding. Hydrologists set up mathematical hydrological models or catchment models, or scale physical (analogue) models of river channels or estuaries and release flows of water to study tides, currents, flooding, scour, etc. (Bolgov et al., 2002). Climatologists are developing general circulation models, and using powerful computers to try to establish likely future climate change. Input–output and other models have been used by regional planners and for integrated EM and strategic EM (Milford et al., 2005). Ecosystem simulation modelling is applied to specific ecosystems; social scientists use social modelling to predict socio-economic impact, and economists use economic models to try to establish micro- or macro-economic trends and to test ideas for manipulating an economy (Maurya et al., 2022).

Environmental auditing/assessment, eco-auditing, environmental accounting, SD auditing and environmental compliance auditing

There is some confusion over the terms auditing, accounting and assessment. Assessment is the gathering of, and effort to understand, data; audit is an official inspection of an organisation's accounts, ideally by an independent body; accounting is the recording and summarising of actions, achievements and transactions, and verifying and reporting on them. Analysis is detailed objective examination of something to understand its nature, function and structure. Evaluation is a final interpretation of value, trends, etc.

Auditing, appraisal, assessment and evaluation are used in planning (pre-development), to judge progress (during development or implementation) and when implementation is finished. The results do not just indicate development status; there may be clarification of what has happened which helps others in the future. There is a reluctance to undertake post-development or post-impact assessment or appraisals; the reason(s) may be that:

- money was made available only for implementation, and recurrent funds are scarce;
- expertise may have moved on and there is nobody to undertake it;
- interest has shifted to something new;
- nobody is keen to look for past problems.

Environmental auditing

Environmental auditing overlaps eco-stocktaking, eco-review, eco-survey, eco-audit, eco-evaluation, environmental assessment, state-of-the-environment assessment, the production of green charters and the checking of impact assessments to determine their effectiveness (Shrivastava, 2013; VanGuilder, 2014; Cahill, 2015; Pathak, 2015; Blokdyk, 2018). There are also close links to health and safety auditing and SD auditing (Pain, 2018; Prasad, 2018).

Eco-auditing

Eco-audit (corporate environmental auditing or EMS auditing) is a systematic multidisciplinary methodology used periodically and objectively to assess the environmental performance of a company, public body or, in some instances, a region. Eco-audit usage expanded as companies were held more responsible for the damage they caused and realised the need for a green image/corporate social responsibility. Eco-audits can be repeated at intervals to see how something changes or evolves.

There are two broad categories of eco-audit: 1) industrial, private sector corporate eco-audits; 2) local authority or higher level government eco-audits (sometimes called 'green charters'); these are more standardised than industrial (private sector) corporate eco-audits, and are commissioned by local authorities to show local environmental quality. Some countries now audit the environmental quality of new buildings to ensure that they use eco-friendly construction and do not waste energy.

Eco-audit is commonly an internal review of the activities and plans of a business or other body; or is undertaken by potential buyers of a company and sometimes environmental enforcement bodies. The end-product of an eco-audit is an audit report, sometimes released to the public, with an undertaking for ongoing repetition.

The decision to eco-audit has been mainly voluntary, but in relation to finance and company matters, auditing usually implies involuntary. There have been cases where shareholders have asked for eco-audits; aid agencies and banks often commission them before granting funding; NGOs may press for them; and insurance companies may require them before accepting a client or grant reduced premiums if there is a satisfactory audit. In the future governments may pass legislation requiring eco-auditing or adoption of EMSs. Eco-auditing is part of a growing shift from mere compliance with regulations to developing forward-looking EM strategies.

First-time audits (and EMSs) are usually more complex than follow-up audits. Voluntary adoption of eco-auditing, especially if it is handled in-house, poses risks from institutional politics (e.g. ministries may compete; companies may be rivals; internal squabbles). The cost of eco-auditing varies, depending on the complexity, novelty, thoroughness and local circumstances. Poorer institutions and small businesses may need aid to be able to afford auditing. Box 8.2 presents some eco-audit–EMS standards.

Box 8.2   Eco-audit–EM system (EMS) standards

BS7750

In early 1992 the world's first eco-audit standard was published – British Standards Institute's BS7750 Specification for Environmental Management Systems (British Standards Institution, 1994). A number of countries adopted it, and it was revised in 1993 and 1994 to make it more compatible with the recently introduced Eco-Management and Audit Scheme (EMAS) (Wenk, 2005). BS7750 is a means by which an organisation can establish an EMS. By 2006, BS7750 was being superseded by the ISO14001 series.

EMAS

The Eco-Management and Audit Scheme (EMAS) was launched in 1993 (Welford, 1992). EMAS goes beyond eco-audit to require an approved EMS and the production of an independently verified public statement. EMAS seeks to encourage industries in EU states to adopt a site-specific, proactive approach to environmental management and improve their performance. EMAS registration is voluntary. Participants write and adopt an environmental policy which includes commitments to: meeting all legislative requirements and ensuring continued improvement of performance; implementation of an environmental programme with objectives and targets derived from a comprehensive review process; establishing a management system (which includes future environmental audits) to deliver these objectives and targets; and issue public environmental statements (EMAS does not insist on full publication of audits). If these terms are broken, the organisation may be suspended from EMAS, and lose its right to a logo, which means loss of publicity advantage, and possibly increased insurance premiums or supplier, investment, or sales-outlet boycott.

Criticisms of EMAS include the charge that its auditing criteria are vague, that it disrupts the activities of an organisation, that it may reveal trade secrets, and perhaps cause public or workforce hostility.

ISO14000/14001 (now effectively the 'world standard')

The International Standards Organisation (ISO) has been developing a series of standards, some advisory, some contractual, broadly compatible with EMAS and BS7750. ISO[DIS]14001 was introduced in 1996, and the series also incorporates ISO14004. ISO14001 provides information on the requirements for an EMS, and ISO14004 has the elements needed and guidance on implementation of an EMS. The ISO14000 series are roughly equivalent to BS7750 and EMAS, but more user-friendly and easier to understand. These ISO standards are related to the ISO9000 series which are widely used by business worldwide and deal with quality systems registration. These standards, which deal with EMSs, have evolved from total quality management and are quality auditing systems. It is also desirable that they integrate EM quality standards with commercial quality management (product/service quality) standards and occupational health and safety quality management standards.

Environmental accounting

Environmental accounting may be used to support eco-efficiency. This is an approach geared towards ensuring competitively priced goods or services that also satisfy environmental goals (Aronsson and Löfgren, 2010; Bebbington et al., 2010; Baldarelli et al., 2017; Debnath, 2019; Bebbington et al., 2021; Do Carmo Azevedo et al., 2022).

The approach seeks to establish the *current* status of an ecosystem; stocktaking methods such as EIA focus more on the *future* effects. Ecological evaluation seeks to establish what is of value. Environmental audits can be conducted at company, institution, state, national or global levels, and may include: 1) a stocktaking or inventory-focus approach to the environment which seeks to review conditions and evaluate impacts of development (e.g. new systems of national accounts); 2) studies aimed at avoiding or reducing environmental damage; 3) means by which a body systematically monitors the quality of the environment which it interacts with or is responsible for.

SD assessment/audit and state of the environment accounts

SD assessment (sustainability assessment) may be used to police, identify opportunities and analyse situations, not just track progress towards SD (Ali, 2013; Morrison-Saunders et al., 2013, 2015; Gibson, 2016; Gokten and Gokten, 2018; Ren and Toniolo, 2019; Spangenberg, 2019; Bastante-Ceca et al., 2020; Jingzheng Ren, 2021b). It is often an aspect of eco-audit.

State-of-the-environment accounts set out a region's or a nation's environmental, social and economic trends and assets. The Worldwatch Institute (Washington, DC) publishes a *State of the World* guide annually, a form of state-of-the-environment account focused on progress towards SD (Cho et al., 2018; Baldarelli and Del Baldo, 2020).

Environmental assessment/appraisal

The term is applied to pre-development stocktaking. In the USA an environmental assessment means a concise public document which should provide enough evidence for a decision whether or not to proceed to a full EIA. Environmental assessment has also been applied to surveillance or screening such as checking drugs or industrial activities, and it is used for studies which seek to establish the state of an environment with less focus on impacts than EIA.

Supply chain auditing

Overlapping environmental auditing and eco-auditing is (green) supply chain audit and management (Emmett and Sood, 2010; McKinnon et al., 2015; Achillas et al., 2019). A company or organisation conducts a comprehensive environmental check on the products, materials and other inputs it purchases to manufacture or provide a service. The aim is to reduce environmental and other impacts throughout the life cycle of products or services. Investigators may check that a supplier deals with waste properly, does not put the environment, workers or the public at risk, pays adequately, etc. Supplier Ethical Data Exchange (SEDEX) is a global online platform that allows members to provide customers and partners with detailed information about their environmental, social and ethical performance to seek greater transparency across the entire supply chain.

Supply chain auditing puts companies and organisations under pressure to improve their environmental and human resources management.

Environmental compliance auditing

Environmental compliance audit (or assessment) seeks to determine whether the provisions of laws and rules are being complied with. When compliance is deemed to have been satisfactory an authority or professional body may issue a certificate which may enable the body to continue to practice and reassure the public, insurers and financiers. Certification should be the start of the next cycle of continuous ongoing improvement checked by repeated auditing/assessment (Logo, 2010; Rogers et al., 2016; Tricker, 2016).

Eco-footprint and carbon footprint

Eco-footprinting seeks to measure ecological performance of a target – an individual, group, a company, organisation, sector, transport network, supply of a commodity, or a region, city, island, valley, etc. It tracks the impact of the target and compares it with what the environment can provide. It does this by calculating how much biologically productive land and water (as area or area per capita) is needed with current demand and technology to provide inputs and dispose of outputs (wastes) safely. The result is a footprint in something like km^2 for the target, which offers a visual image that is easy for people to grasp and which can be compared with other situations (Vale and Vale, 2013; Collins and Flynn, 2015; Beyers and Wackernagel, 2019; Wackernagel and Beyers, 2019; Thornbush, 2021; Jingzhen Ren, 2022).

A slightly different eco-footprint which has spread with concern for global warming is the carbon footprint (overlapping the energy footprint), which assesses the amount of CO_2 emitted each year by a household, company, etc. (Berners-Lee, 2010; Franchetti and Apul, 2013; Muthu, 2016, 2021b; Fernández et al., 2017).

Integrated environmental assessment

Integrated environmental assessment (and integrated ecological assessment) are interdisciplinary processes which seek to collect, interpret and communicate the likely consequences of implementing a proposal (Rossini, 2020; Gupta et al., 2021; Jensen and Bourgeron, 2021). It has been applied to global warming issues, acid deposition, etc., and is similar to the Delphi technique, in that it draws on informed group opinion, usually through focus groups interviewed in a hands-off manner having been asked a set of questions. There is overlap with modelling and EIA and strategic impact assessment.

Cumulative impact assessment

Cumulative impact assessment is similar to cumulative effect assessment and seeks to identify impacts that happen when chains of causation intersect and impacts interact and/or accumulate to cause potentially unexpected issues. Identifying and predicting cumulative impacts can be difficult (Iqbal et al., 2011; World Bank, 2013b; Canter, 2015; Jones, 2016; Blakley and Franks, 2021).

Summary

- Even if goals are assessed, data collected and recommendations made, decision makers may fail to adequately consult the findings, understand them, or just 'shelve' them (i.e. ignore).

- Each situation demands the selection of a method, approach and set of tools (which often need refinement). The environmental manager(s) may be able to do this through simple desk research; however, new challenges often demand considerable research.
- Laws, socio-economic conditions and the environment may change, and new tools and approaches appear; often EM is more of a co-ordinating role and specialists will be gathered or hired to apply methods and tools.
- There is a huge diversity of tools, some unique to EM, but most borrowed from sciences, social sciences and business; these may need fine-tuning.
- Environmental managers seldom find tools that are adequately reliable and precise; important decisions are ideally based on the application of more than one tool and a number of sets of data. Some tools can generate a sense of false security (e.g. EIA), some take time and some produce results which lack transparency.
- Objective decision making is seldom unhindered; developments are often piecemeal, frequently hurried and constrained by funding and politics.

Further reading

Books

Barnsley, M.J. (2007) *Environmental Modeling: a practical introduction*. CRC Press, Boca Raton, FL., USA.

Bryman, A. (2004) *Social Research Methods* 2nd edn. Oxford University Press, Oxford, UK.

Burke, G., Singh, B.R. and Theodore, L. (2005) *Handbook for Environmental Management and Technology* 2nd edn. Wiley, New York, NY, USA.

Demers, M.N. (2009) *GIS for Dummies*. Wiley, Chichester, UK.

Gibson, P. and Power, C.H. (2013) *Introductory Remote Sensing: principles and concepts*. Routledge, Abingdon, UK.

Save the Children (2004) *Toolkits: a practical guide to assessment, review and evaluation monitoring* revised and expanded edn (Save the Children Development Manual No. 5 – prepared by L. Gosling and M. Edwards). Save the Children Fund, London, UK.

Therevel, R. (2010) *Strategic Environmental Assessment in Action*. Routledge (Earthscan), London, UK.

Journals

- *Environmental Modelling & Assessment* (Springer). ISSN print: 1420-2026; ISSN online: 1573-2967. www.springer.com/journal/10666 (accessed 20/12/22).
- *Environmental Modelling & Software* (Elsevier). ISSN print: 1364-8152; ISSN online: 1873-6726. www.sciencedirect.com/journal/environmental-modelling-and-software (accessed 20/12/22).
- *Geographical and Environmental Modelling* (Taylor & Francis). ISSN print: 1361-5939; ISSN online: 1469-8323. www.tandfonline.com/action/journalInformation?journalCode=cgem20 (accessed 20/12/22).

www sources

- Environmental data, UNSD: https://unstats.un.org/unsd/envstats/interlinks.cshtml (accessed 20/12/22).

Practice

- ❏ Environmental data, OECD: www.oecd.org/environment/indicators-modelling-outlooks/data-and-indicators.htm (accessed 20/12/22).
- ❏ Climate data sources, GIS: https://gisgeography.com/free-world-climate-data-sources/ (accessed 20/12/22).
- ❏ EM standards, ISO 14001: www.iso.org/iso-14001-environmental-management.html (accessed 21/12/22).
- ❏ List of EM standards and frameworks: https://advisera.com/14001academy/knowledge base/list-of-environmental-management-standards-and-frameworks/ (accessed 20/12/22).
- ❏ EM standards: https://app.croneri.co.uk/topics/ems-standards/indepth (accessed 21/12/22).
- ❏ Environmental modelling (EPA): www.epa.gov/measurements-modeling/environmental-modeling (accessed 21/12/22).

Chapter 9

Proactive assessment, prediction and forecasting

Chapter overview

- Futures studies
- Hazard assessment and risk assessment
- Environmental impact assessment
- Social impact assessment
- Other tools for assessing the potential for development and impacts of development
- Livelihoods assessment
- Vulnerability studies
- Technology assessment
- Health risk assessment and health impact assessment
- Computers and expert systems
- Adaptive environmental assessment and management
- Integrated, comprehensive and regional impact assessment, integrated and strategic EM
- Summary
- Further reading
- www sources

Futures studies

Futures studies (futures, futuristics, strategic foresight, futurology) has a long history. Futures studies are the systematic identification and study of alternative futures with the aim of recognising the best way forward. The approach is at best subjective so findings must be treated with caution.

In the 1970s computer modelling was applied (*Limits to Growth*), initially dismissed as lacking in rigour or even as pseudoscience. The approach of the Millennium (AD 2000) prompted further studies focusing on population, food, military threats, technology innovation, economics and business (Copenhagen Institute for Futures Studies, 2020). Science fiction explored futures long before the 1970s: many authors including Jules Verne and H.G. Wells showed remarkable foresight (Gore, 2013; Tetlock and Gardner, 2015; Harari, 2017).

Futures studies can prompt thought about development, alternative options, and encourage questioning of assumptions; often it presents more than one view of what might happen and this can help develop EM strategies. It may even suggest win-win policies and looks further ahead than other methods. Methods like the Delphi technique are futures studies, as is trend analysis (Hines and Bishop, 2015; Bell, 2017; Gidley, 2017; Helke, 2019; Copenhagen Institute for Futures Studies, 2020). Some futures studies adopt a dystopian stance and some are too optimistic, both unwise.

Predicting future scenarios

The following seek to be less speculative forms of futures studies/futurology.

Forecasting Forecasting is a part of planning, programme and policy formulation. Forms of forecasting have been traditionally used to decide, where and what to hunt, where to settle, etc., and may be good; but some approaches rely on magic and superstition. Since the mid-eighteenth century forecasting has drawn on rational observation, projection of trends (time-series) and hindsight (Hyndman and Athanasopoulos, 2021). Businesses use forecasting (Wade, 2012; Tetlock and Gardner, 2015; Börner, 2021).

Futures modelling and future scenario prediction Banking, investment and insurance developed forecasting and risk assessment methods, and military tacticians started trying to predict future scenarios during the Second World War. Time-series data are widely used as a basis for projections, and key indicators may be monitored for warnings or modelled to predict (Alcamo, 2009). Models include physical models (analogue) (e.g. laboratory tests, scale models of estuaries or catchments, etc.), statistical models (e.g. principal components analysis), computer models and systems models. The further ahead one attempts to make predictions, the less accurate they are likely to be.

The Delphi technique seeks to obtain a consensus about future developments from multidisciplinary panels of experts. The Delphi technique is used in modelling on ecology, biological conservation, healthcare policy and innovations, resources allocation, technology innovation, etc. (Mukherjee et al., 2015). It is useful for looking up to 15 years into the future.

Hazard assessment and risk assessment

A hazard is something identified and perceived to pose problems or danger. An environmental hazard can be a substance, state or event (natural or anthropogenic) which has the potential to threaten the surrounding natural environment and/or adversely affect people. A risk is the probability of an unwanted issue happening. A disaster is the realisation of a hazard; a catastrophe is a particularly serious disaster.

Environmental risk interrelates with social and economic risk (Measham and Lockie, 2012). Risk = probability × consequence. Thus there is a hazard/threat of drowning involved in crossing the Atlantic, but the risk of suffering it by using a row-boat is higher than going by jet or liner. Risk assessment (appraisal or analysis) is a loose term, it considers hazard and vulnerability: how people react to risk and their pattern of exposure. Risk assessment is the process of assigning magnitudes and probabilities to adverse effects of technical innovation or natural catastrophes.

The initial step is to identify hazards (hazard or threat assessment), then understand how these impact and if there is more than one how they might interact, then gather data and finally calculate risk/probability (risk assessment) and perhaps also determine the seriousness and timing (Smith, 2013; Dalezios, 2017; Rougier et al., 2018). These assessments deal with a diversity of threats, some are multidimensional risks, possibly indirect and involving interrelated physical and social impacts, and some are cumulative (National Academy of Sciences, 2009; Kapustka and Landis, 2010; Theodore and Dupont, 2012; Wren, 2012; Simon, 2014; Brennan, 2016; Greenberg, 2017; Bell et al., 2020).

Risk assessments often prompt disagreements (e.g. Lomborg, 2001, 2020) so EM must exercise cautious judgement when using them. It is desirable to identify opportunities and positive impacts, but risk assessment generally flags negative impacts.

One difficulty faced by risk assessment, and EIA, is assessing what is acceptable. Various groups, even within one society, may perceive and assess hazards and risks differently and vary in their vulnerability. The perception of risk is seldom based on rational judgements: people have 'gut reactions' or dread of certain things and little fear of other, perhaps more real, threats (Bryant, 2014). There are likely to be different risk perceptions from class to class, age group to age group, and for different religions and sexes: much depends on previous exposure or awareness through the media (Renn, 2010; Slovic, 2010). Perception varies from individual to individual, and for any given group through time. In general, people are more concerned about the short term rather than the long term, and by concentrated hazards: an air crash that kills 300 rather than the same number of fatalities from household accidents dispersed over a short time generates more concern. Some risks generate more fear even if they are less common (e.g. radiation hazards compared with traffic accidents). If people think they are in control, as car drivers, for example, they are less worried than as passengers on a train; yet car accidents are much the greater risk. Perception can be greatly affected by media and myth.

The 2004 Indian Ocean tsunami and recent wildfires, hurricanes, etc., have prompted politicians and citizens to consider some risks. There is a need to widen awareness of volcanic eruption, natural climate shifts and asteroid strikes and there is little concern for land degradation, the world food situation, population and atomic weapons. Some hazards only become evident through research without which the problem would be missed or only seen when well-developed; e.g., stratospheric CFCs' damage to the ozone.

Risk assessment may help identify coping strategies, threats to the environment or biota and humans. Risk assessment can focus on effects, pathways or factors involved and can involve weighing dangers against benefits (e.g. the threat of asbestos-related illness versus its value in protection against fire).

Risk assessment roots in the actuarial, investment and insurance professions, and spread to engineering, development of new materials, economics, healthcare and criminology. Drug companies use trials on groups of people. Insurance companies and bankers need to know risks before providing cover, mortgages or loans. Administrators may use risk assessment to reduce the likelihood that they could be accused of negligence if something goes wrong and for contingency planning. Unlike EIA, risk assessment tends not to address development alternatives.

Hazard and risk assessment usually use a template to help order the process to generate a statistical estimate of probability of occurrence of a certain level of impact (not a forecast but a statistical recurrence, e.g. a one in hundred-year chance of a serious flood). The probability estimate data may be used to produce a zoned map which can be used to determine land-use or building regulations, insurance premiums, select transport routes or site infrastructure, prepare contingency or emergency procedures.

Environmental impact assessment

Environmental impact assessment (EIA) seeks to blend administration, planning, analysis and public involvement in pre-decision assessment (Figure 9.1). Impact may be defined as the difference between a forecast of the future with a development occurring and a forecast without the development. To do that requires the first step of a baseline survey.

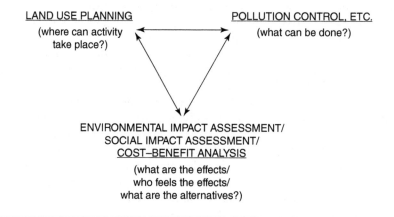

Figure 9.1 How impact assessment fits into planning.

Impact assessment is still imperfect and is widely misapplied or misused. The field has been dominated by EIA but there are approaches running parallel, with similar goals, exchanging information, techniques and methods. These include social impact assessment, hazard assessment, risk assessment, technology impact assessment, eco-auditing and CBA. These approaches are forward-looking, systematic, focused, interdisciplinary, comprehensive and generally iterative.

Box 9.1 An overview of environmental impact assessment

The following observations describe EIA:

- It is a proactive assessment, and should be initiated pre-project/programme/policy, before development decisions are made. In-project/programme/policy and post-project/programme/policy assessments are common. While these may not allow much problem avoidance, they can advise on problem mitigation, gather data, feed into future impact assessment, improve damage control and the exploitation of unexpected benefits.

- It is a systematic evaluation of all significant environmental consequences that an action is likely to have upon the environment.

- It is a process leading to a statement to guide decision-makers.
- It is a structured, systematic, comprehensive approach.
- It is a learning process and means to find the optimum development path.
- It is a process by which information is collected and assessed to determine whether it is wise to proceed with a proposed development.
- It is an activity designed to identify and predict the impacts of an action on the biogeophysical environment and on human health and wellbeing, and to interpret and communicate information about such impacts.
- It is a process which should force developers to reconsider proposals.
- It is a process which has the potential to increase developers' accountability to the public.

- It usually involves initial screening and scoping (to determine what is to be subjected to EIA, and to decide what form the assessment should take).
- It should be subject to an independent, objective review of results.
- It should publish a clear statement of identified impacts with an indication of their significance (especially if any are irreversible).
- It should include a declaration of possible alternative development options, including nil-development, and their likely impacts.
- Ideally there should be public participation in EIA.
- There should be effective integration of EIA into the planning/legal process.

Box 9.2 Typical step-wise environmental impact assessment process

Screening (phase 0) is concerned with deciding which developments require an EIA. This should prevent unnecessary assessment yet ensure that there is no escape when assessment is needed (in practice difficult). Screening may not be mandatory in some countries. Note that the term 'environmental assessment' is used for screening in the USA, but in the UK has been applied to EIA. In the USA if environmental assessment/screening (also called initial environmental evaluation) indicates no need to proceed to a full EIA, a statement of Finding of No Significant Impact is issued publicly, allowing time for objection/appeal before a final decision is arrived at.

Scoping (phase 1) overlaps phase 0 and should help determine the terms of reference for an EIA, the approach, timetable, limits of study, tactics, staffing and so on. By this stage the EIA should consider alternative developments. In practice, a decision as to how to proceed may already have been made by a developer.

Identification, measurement and evaluation of impacts (phase 2) may proceed with or without public review(s). A variety of techniques may be used to determine possible impacts: as human judgement is involved, this is an art rather than a wholly objective scientific process, regardless of the statistics used.

The difficulty of identifying indirect and cumulative impacts makes this a tricky and often only partially satisfactory process. This phase is much assisted if an adequate set of baseline data is available; often it is not, and extensive desk and field research is needed.

Checking findings (phase 3) may follow a public review and/or may involve an independent third party to ensure objectivity. A statement, report, chart or presentation is usually released, effectively the product of an EIA; this is termed the environmental impact statement (EIS) and is what the decision makers, environmental managers (and perhaps the public) have to interpret (Eccleston, 2011, 2014).

Decision on proposal (phase 4): in practice, where a development has already been decided on or is even under way, corrective measures can be perfected. It is a way of passing on hindsight knowledge to planners in the future. The EIS may not be clear or easy to use: some countries require irreversible, dangerous and costly impacts to be clearly shown. It also useful if alternatives and potential benefits are indicated. The environmental manager must be able to read the EIS and identify gaps, weaknesses, limitations. An EIA must not be allowed to give a false sense of security.

Implementation (phase 5): this is where an environmental manager is especially active. Unexpected problems may arise.

Monitoring and audit (phase 6): in practice this is often omitted or is poorly done. If planning and management are to improve, efforts should be made to assess whether the EIA worked well. It is also important to continue monitoring to catch unexpected developments.

Efforts to assess EIA are generally termed post-EIA audits. An EIA is a snapshot view (i.e. representative of just a moment in time), and ongoing monitoring or repeat EIA can help counter that.

Note: These are the idealised steps or phases: 0 to 6.

EIA can be a policy instrument, a planning tool, a means of public involvement and part of a pre-emptive framework crucial to EM. Attitudes towards EIA vary from the view that it is just a required rubber-stamping activity, or that it determines optimal development, to the idea that it has a vital role to play in improving EM and planning to achieve SD.

Using EIA to consider goals, realities and available alternatives should make it possible to identify the *best options* rather than simply acceptable proposals. EIA has tended to flag negative impacts but can also ensure that opportunities are not missed. It is important to stress that EIA should consider all options, including no development/no change (Perdicoúlis et al., 2012). In practice, much is retrospective, and EIA may take place after development is underway. This is still of value because it can help clarify problems and add to hindsight knowledge. Nevertheless, if EIA is done after decisions it is unlikely to be able to shift developers to less damaging options. At worst it may be cosmetic: done to reduce opposition, with the results ignored, side-stepped and not enforced. Added to which is the possibility that the EIA process fails to identify a problem or opportunity. EIA, like other EM tools, must not be allowed to give a false sense of security; those commissioning it must also know its strengths and weaknesses (Hanna, 2005, 2022; Raman et al., 2014; Momtaz and Kabir, 2018; Morrison-Saunders, 2018; Therivel and Wood, 2018; Carroll et al., 2019; Glasson and Therivel, 2019; Alex, 2020; Noble, 2020; Rosales, 2020; Hundloe, 2021; Manyuchi et al., 2021; Rathi, 2021; Fonseca, 2022; Islam and Nomani, 2022).

EIA has mainly been applied at project, to a lesser extent programme, and more rarely policy levels. There have been efforts to extend assessment to a higher level: multi-project, programme and policy applications at national, transboundary and even global levels. In 1991 UNECE launched the Convention on Environmental Impact Assessment in a Transboundary Context (Espoo, FI). This Convention was the first multilateral treaty on transboundary rights relating to proposed activities and provided for the notification of all affected parties likely to suffer an adverse transboundary impact from a proposed development.

Dealing with indirect and cumulative impacts

EIA seldom effectively considers indirect and cumulative impacts (Canter, 2015; Blakley and Franks, 2021). Direct impacts (first-order) are relatively easy to assess but indirect impacts (second-order and higher) lie along a chain of causation distant in time and/or space and may not be readily apparent. Second-, third- or higher-order indirect impacts can be explored with some techniques, but it is likely to be expensive and slower than identifying first order.

A cumulative impact is the consequence of more than one direct or indirect impact acting together (Jones, 2016). The component impacts may seem to pose no problem. Cumulative effects assessment (CEA) has mostly focused on negative cumulative impacts. However, it can assess positive impacts as well. Chemical and biological 'time bombs' are forms of cumulative impact: a compound accumulates, a biological process continues, without causing a problem, hidden until a threshold is suddenly exceeded, either through continued accumulation, or because some environmental or socio-economic change(s) triggers it. For example, pesticide gradually accumulating in the soil may suddenly be flushed out when acid deposition brings soil chemistry to a threshold. In practice CEA progress has been slow and results variable.

Environmental changes accumulate through many different processes or pathways:

- incremental (additive) processes (repeated additions of a similar nature $a + a + a + a$ ...);
- interactive processes ($a + b + c + n$ ...);
- sequential effects;
- complex causation;
- synergistic impacts;
- impacts which occurs when a threshold is passed as a consequence of some trigger effect (e.g. physical, chemical or biological 'time bomb');
- irregular surprise effects;
- impacts triggered by a feedback process (antagonistic – positive feedback which reinforces a trend, as opposed to ameliorative – negative feedback which counters a trend).

There are specific CEA methods which are partially effective, e.g. the component interaction matrix and the minimum link matrix.

Social impact assessment

Social impact assessment (SIA) seeks to assess whether a proposed development alters quality of life and sense of wellbeing, and how well individuals, groups and communities adapt to change caused by development/change. SIA should also identify how people can contribute and show what they need (Vanclay and Esteves, 2011; Burdge, 2015; Mathur, 2018; Therivel and Wood, 2018; Brusset, 2020; World Bank, 2022). EIA and SIA deal with opposite ends of the same spectrum and blend together. There is also overlap with cultural impact assessment. Some of the issues SIA deals with are difficult to quantify; e.g., sense of belonging, community cohesion (maintenance of functional and effective ties between group members), lifestyle, feelings of security, local pride, willingness to innovate, perception of threats and opportunities.

Social impacts include alterations in the ways in which people live, work, play, relate to one another, organise, and generally cope as members of a society (and involve lifestyle, community cohesion, mental health, etc.); while cultural impacts involve changes to the norms, values and beliefs of individuals that guide and rationalise their cognition of themselves and their society.

Early SIA assessed the impact of the Trans-Alaska Pipeline on the Inuit People. In general SIA has remained underfunded and neglected compared with EIA. The Three Mile Island incident (a US nuclear facility which suffered a near-meltdown necessitating evacuation of householders in 1983) used SIA to assess threats and public fears before restarting the reactor.

The diversity, complexity of SIA and the relative lack of funding mean it is less standardised than EIA and it has spread more slowly (Arce-Gomez et al., 2015; Wong and Ho Wing-Chung, 2015; Rawhouser et al., 2019; Kah and Akeneroye, 2020; Vanclay, 2020). Critics of SIA argue that it is too theoretical; too descriptive; weak at prediction; ad hoc; mainly applied at the local scale; and likely to delay development. Also, few of the theories it uses are tightly defined so it is difficult to make comparisons between successive studies.

SIA can help ensure that projects, programmes and policies generate fewer socio-economic problems (Vanclay and Esteves, 2011; Vanclay, 2014, 2020; Vanclay et al., 2015; Kvam, 2018; Woodcock et al., 2021). As with EIA, different socio-economic or socio-cultural impacts may be generated at various stages in a policy, programme or project cycle and SIA and EIA give a snapshot view. SIA has been applied more at project level.

Tools used by SIA include: social surveys; questionnaires; interviews; use of available statistics such as census data; public hearings; systems analysis; social cost–benefit analysis; the Delphi technique; marketing and consumer methods; field research, etc. Behavioural psychologists may be involved in SIA to ascertain whether stress has been or will be suffered, what constitutes a sense of wellbeing and so on. The SIA equivalent of an EIA baseline study is the preparation of a social profile to establish what might be changed and what would probably happen if no development took place. Field research techniques can be divided into direct and indirect. Direct observation of human behaviour may be open or discreet, conducted during normal times or times of stress. Indirect observation includes study of: changes in social indicators, patterns of trampling, telephone enquiries directed at selected members of the public, historical records, property prices, suicide rates, etc.

Communities can be monitored for changes using demographic, employment and human wellbeing data, so SIA often adopts a community focus. Alternatively, especially when aid donors commission an SIA, the focus is on target groups, typically the people(s) investment is supposed to help. An issues-oriented approach is another possibility, or a regional approach, or it is possible to make use of RRA and PRA. Much less progress has been made with assessing cumulative impacts than is the case with EIA. SIA should be conducted by competent social scientists familiar with the local situation. SIA has been used when dealing with gender issues, migrants, relocation, etc. (Nijkamp et al., 2012; UNDP, 2020a). In a few countries SIA is required before gambling or drinking licences are granted.

Other tools for assessing the potential for development and impacts of development

EM is interested in identifying irreversible impacts and thresholds as well as challenges and opportunities.

Ecological impact assessment

This considers how organisms, rather than people, will be affected by a challenge or development. Ecological impact assessment is concerned with establishing the state of the environment, whereas EIA focuses on predicted and actual effects of change. Ecological impact assessment has been applied to biodiversity loss. Ecological impact assessment may rely on selected ecosystem components as indicators or on ecosystem modelling (Brouwer and Ittersum, 2010; Ashton Drew et al., 2011; Ninan, 2020).

Habitat evaluation

This seeks to assess the suitability of an ecosystem for a species or the impact of development on a habitat. There may be more than one habitat affected by a challenge or development, in which case each is dealt with separately. This approach has been used by the US Fish and Wildlife Service in assessments of the impacts of US federal water resource development projects, by the US Army Corps of Engineers and by rewilders.

Land-use planning

This is a process which may operate at local, regional or national scale; land capability assessment, land appraisal, land evaluation, land suitability assessment and terrain evaluation feed into that process. A land-use survey indicates the situation at the time of study, and is not the same as a capability classification, which looks to the future (Silberstein and Maser, 2019).

Land capability classification, evaluation and appraisal There are various approaches for land capability classification (e.g. the ecological series classification or the Holdridge life zones system). Often the land-use planning approach adopted depends on a country's politics. It is widely felt that land-use planning is a valuable ingredient of EIA and in the quest for sustainable development, and that EIA can feed into land-use planning. In practice the two are often poorly integrated. There has been some renewal of use of the Holdridge life zone system by those assessing climate change impacts and soil carbon sequestration (Khatun et al., 2013; Jungkunst et al., 2021).

Land capability assessment, land evaluation soil capability assessment and land appraisal generally follow a proactive approach similar to that of EIA (scoping, data collection, evaluation, presentation of decision). The end-product is a description of landscape units in terms of inherent capacity to produce a combination of plants, animals and so on; it is also likely to reflect government development goals, market opportunities, labour availability and public demands (e.g. terraced agriculture may be possible but the labour needed is not available). There has been use of LiDAR, sidescan radar and other remote sensing to evaluate land capability, either using UAVs, aircraft or satellites, but thorough ground truthing (checking sensed data against real conditions on the ground) is needed in the process.

Land suitability assessment Land suitability assessment, analysis or evaluation is a rating of landscape units showing what development they might best support, for example what are the best crops or to select areas for irrigation. Land suitability assessment may depend on overlay maps of various landscape or development attributes or direct field observation of clues (something local people may traditionally do) (e.g. seek distinctive plants indicative of good soil). GIS and remote sensing (as for land evaluation) are widely applied to land suitability and capability assessment and ground truthing is usually part of the process (Constantini, 2017).

The universal soil loss equation and revised universal soil loss equations The universal soil loss equation (USLE) is a predictive tool which uses data on a wide range of parameters to estimate and predict average annual soil loss. Developed in the 1930s by the US Soil Conservation Service, it has been repeatedly modified to make it more suitable for various environments. There are now numerous revised (RUSLE) versions less influenced by land use, with better slope factor and other calculations, and some are available as computer programs. RUSLE is widely used to check on existing and likely future soil loss, and to select appropriate agricultural practices and crops to sustain production (Blanco-Canqui and Lal, 2010; Morgan and Nearing, 2011).

USLE/RUSLE should be used with caution: problems arise when data are imprecise or unavailable, and it is best applied in situations where water rather than wind erosion occurs (although there are versions intended to cope with wind erosion). A typical form of the USLE is:

$$A = (0.224)\,RKLSCP$$

Where:

A = soil loss;
R = rainfall erosivity factor (degree to which rainfall can erode soil);
K = soil erodability factor (soil vulnerability to erosion);
L = slope length factor;
S = slope gradient factor;
C = cropping management factor (what is grown and how);
P = erosion control practice factor.

Agroecosystem zones Agroecosystem zones provide a framework for considering a range of parameters over a limited planning term with the aim of promoting SD. An agroecosystem is an ecological system modified by humans to produce food or commodities, which generally means a reduction in diversity of wildlife. Agroecosystem assessment (or analysis) attempts rapid multidisciplinary diagnosis that includes ecological and socio-economic concepts and parameters. It considers not only the farming system but also household characteristics and regional, national, even global factors likely to affect the local community (Anderson et al., 2021; Krishna, 2021). The area under consideration is zoned, making use of a land-use survey or land capability assessment. Agroecosystem assessment needs to be approached with caution because it can lead to over-simple results.

Farming systems research

Farming systems research (FSR) is an open-ended, iterative, multidisciplinary, holistic, continuous, farmer-centred, dynamic process applied to agricultural research and development (it considers biophysical, social and economic factors, and seeks to integrate their study) (Darnhofer et al., 2012; Shaner et al., 2021). All approaches share five basic steps:

1 Classification – the identification of homogeneous groups ('target groups') of farmers.
2 Diagnosis – identification of limiting factors, opportunities, threats and so on for the target group.
3 Generation of recommendations – which may require field experiments, pilot studies and/or research station work.
4 Implementation – usually working with an agricultural extension service.
5 Evaluation – which may lead to revision of what is being done.

FSR is promoted as a way of increasing farmer participation in development, and of generating improved and appropriate approaches and technology. FSR includes study of factors which may be beyond the control of the farming community – world trade issues, global warming, etc. Unless some off-the-shelf approach is available, FSR can take time – sometimes years.

Participatory assessment approaches

Participatory assessment and monitoring is qualitative research or survey work, which seeks an in-depth understanding of a community or situation. There is overlap between agroecosystem assessment, FSR and participatory rural appraisal (PRA) and rapid rural appraisal (RRA). These approaches can assess the current situation *and* assist in the prediction of future opportunities and difficulties (Kanwat and Kumar, 2011). RRA has evolved rapidly and there is no single favoured methodology.

There has been a tendency to emphasise the strengths of RRA and PRA and to understate the problems (Freudenberger, 2011). Speedy data collection is also needed for urban environments, and RRA-type assessment has been developed to provide it.

Livelihoods assessment

The assessment of livelihoods is often a key element in improving environmental care, fighting poverty and reducing people's vulnerability. The focus has mostly been on rural livelihoods, although there is growing interest in urban situations. Livelihood assessment seeks to uncover how groups make a living, ways in which they can be disrupted, and explores the potential for improving livelihoods and making them more secure and resilient. EM is interested in assessing how people regard and interact with their environment in the past, now and in the future. Livelihoods assessment is also applied in disaster situations to determine the impact on people (FAO and ILO, 2009).

Vulnerability studies

There is interest in predicting change in people's vulnerability to physical and socio-economic changes. Some of this has been generated by concern about climate change and by recent eruptions, earthquakes, hurricanes, etc. Urban growth and population increase mean that people are more concentrated and less able to find food or water if normal conditions are disrupted. A number of key products are manufactured by a handful of factories which supply the world, and if these are disrupted the impact can be severe.

Environmental vulnerability assessments seek to identify the key factors affecting adaptive capacity and assess the ability of communities and ecosystems to cope with and respond to the combined effects of local stressors and climate change and variability (Shrâter et al, 2004). There are numerous applications to global climate change, biodiversity and human health (e.g. UN-Habitat, 2018).

Technology assessment

Technology assessment explores the impacts of technology and technological innovation to establish things will work in practice and what effect they may have in future. It follows a parallel path to EIA, and may involve evaluation of indirect and cumulative impacts.

An International Society for Technology Assessment operated from the USA in the mid-1970s, developing into the International Association for Impact Assessment (IAIA), which now promotes EIA, SIA, technology assessment, hazard assessment, risk assessment and related activities (Rip, 2015). The National Science Foundation in the USA also supports technology assessment, and Europe, Japan, Canada and Australia have established bodies.

Technology impacts can be through technology function or malfunction; operator failure; poor maintenance/ageing; poor design; faulty installation; terrorism; natural or human accident; or adaptations prompted by the innovation. Assessment practitioners are usually engineers, so socio-economic issues may not be well covered. The tendency has been to concentrate on morbidity and mortality but there is now interest in civil liberties and social aspects of technology innovation. It is possible that a major challenge to humans will come from AI or nanotechnology, with human creations turning on humans; technology assessment may prove vital for avoiding future catastrophe.

Technology risk may be posed by a current activity and by new, untested technology (Grunwald, 2019). There is a tendency for technological hazard to be exported to countries where laws, monitoring and enforcement may be less stringent.

Health risk assessment and health impact assessment

Health impact assessment seeks to predict impacts, physical or mental/social. This may allow the prediction of healthcare needs or help those seeking to improve health and safety measures. There is also interest in establishing the effect of disease on society, employment, taxation and agriculture; for example, establishing the impacts of HIV/AIDS on labour. A rapidly expanding field is health risk assessment often closely linked to safety assessment; it may involve assessing the risk posed by waste disposal, workplace environment, employment activities, noise, stress, etc. Alternatively, assessment may be more health focused, and conducted by medical and related staff; with damage claims soaring there is an incentive for pre-emptive assessment (Birley, 2011; Kemm, 2013). Global warming fears have prompted attempts to assess future disease patterns, difficult because many factors are involved in transmission and in determining human or animal vulnerability.

In 2005 fears of avian flu caused governments and agencies such as the WHO to meet and exchange information and to prepare contingency plans in advance of any pandemic. An outbreak of severe acute respiratory syndrome (SARS) 2002 to 2004 was contained quite effectively; however, in 2019 a similar coronavirus, Covid-19, proved far more difficult to control (and/or the response was poor) and caused a pandemic. Studies made since 2019 might help preparations for future outbreaks, although some governments seem slow to learn lessons or heed advice.

Computers and expert systems

There have been attempts to computerise impact identification monitoring and assessment. Computer techniques have also been used for interpreting impacts. Accidents like Chernobyl have prompted development of joint rapid impact assessment and data exchange systems. These are vital for coping with rapidly developing transboundary problems (e.g. airborne pollution). The EU has gone partway to developing such a system for radioactive fallout by establishing the EC Urgent Radiological Information Exchange (ECURIE) in 1987.

Expert systems (knowledge-based systems) can be valuable as an aid (not replacement) for skilled assessors. However, they may take a lot of research and time to develop. They are also used for environmental planning, eco-audit and EM, and to apply EIA to regional planning. These systems draw on heuristic (rule-of-thumb) reasoning to advise, provide support for decision making, or to aid data management (Rodriguez-Bachiller and Glasson, 2004).

Adaptive environmental assessment and management

Impact assessment generally adopts a snapshot approach (limited in time and space) as causal relationships are often not constant (e.g. monetary units may be devalued, the environment can alter, decision-making objectives change, etc.) such an approach can be ineffective. There is also a risk that a one-off impact assessment could discourage planners from adequate monitoring. The ideal is to ensure that assessment is continuous or repeated regularly and has a wide spectrum (Holling, 1978). Two approaches have evolved: adaptive environmental assessment (AEA) and adaptive environmental assessment and management (AEAM). These are broader than mainstream EIA, and have a bias towards coping with uncertainty. In addition, AEAM seeks to integrate environmental, social and economic assessment with management.

AEAM can be useful where baseline data are poor. It also encourages and facilitates multidisciplinary assessment. However, it can be demanding in terms of research expertise and may be slow.

Integrated, comprehensive and regional impact assessment, integrated and strategic EM

Impact assessment should be better integrated into policy making, planning and administration. The following approaches seek to cover more than just a restricted range of impacts, to do so over more than a snapshot of time, and at wider scales, and up through all project, programme and policy levels. Some of the approaches seek to cope better with indirect and cumulative impacts than mainstream impact assessment.

Integrated and comprehensive impact assessment

Integrated impact assessment is a generic term for the study of the full range of ecological and socio-economic consequences of an action. It seeks to promote closer integration of impact assessment into planning, policy making and management, adopting a tiered approach (Scow et al., 2018; Blakley and Franks, 2021).

Economists use econometrics and input–output analysis to explore economics and environmental linkages at regional level: for example, the impacts of irrigation associated with Malaysia's Muda Scheme (Bell et al., 1982).

Integrated regional environmental assessment

This is similar to the approach discussed above, having the following objectives:

- To provide a broad, integrated perspective of a region about to undergo or undergoing developments.
- To identify cumulative impacts from multiple developments in the region.
- To help establish priorities for environmental protection.
- To assess policy options.
- To identify information gaps and research needs.

There is no single methodology for doing this (UNEP: Integrated Environmental Assessment www.unep.org/global-environment-outlook/integrated-environmental-assessment – accessed 23/12/22). A solution might be to subdivide regions into smaller units for assessment, perhaps territories, ecosystems or river basins, although there may be situations where administrative regions offer better possibilities (Medeiros, 2020).

While most EIA and SIA are applied at project level, it is also desirable to assess at programme and policy level. The greatest promise probably lies with tiered assessments. These adopt a sequential approach with broad assessment at policy level (tier 1), e.g. impact assessment of national road policy; followed by more specific assessment at the programme level (tier 2), e.g. regional road programmes; and even more specific assessment of individual (road) project(s) (tier 3), i.e. local road construction. Efforts are made to cross-reference broad and specific assessments. Prior or parallel events in higher tiers condition events in tier 3, so it is unsatisfactory to look at a lower tier without also considering higher ones or vice versa. Tiered impact assessment can also adopt a multisectoral approach (horizontal tiers), if sectors were considered in isolation cumulative impacts might be missed (or a sector might get missed).

Strategic environmental assessment

SEA differs from mainstream project-focused assessments, in that it considers other projects, programmes and policies, cultural and other forces that effect on what is being assessed (projects can usually be studied in relative isolation). To cope with these challenges strategic environmental assessment (SEA) (or programmatic EIA) has developed (Dalal-Clayton and Sadler, 2005; Fischer, 2007; Sadler et al., 2011; Fischer and González, 2021). SEA should be undertaken early in the development process, like EIA or SIA, at the start of draft planning and programme development. SEA offers a means of viewing and co-ordinating development from policy and programme levels down to project level through a tiered approach (Sánchez-Triana and Ahmed, 2008; EU: Strategic Environmental Assessment https://webmail.talktalk.co.uk/cp/ps/Main/Layout#checkNewMail – accessed 23/12/22).

In 2001 the EU issued Directive 2001/42/EC which provided an SEA framework (Scotford, 2017). In 2003, at Kiev, a Protocol on Strategic Environmental Assessment (the 'SEA Protocol') was signed and there has been increasing adoption of SEA (Marsden, 2008; Cutaia, 2016; Dusik and Sadler, 2016; Campeol, 2020).

The SEA approach may cope better with cumulative impacts, assessment of alternatives and mitigation measures than standard EIA. SEA may be useful for implementing SD, because it allows the principle of sustainability to be carried down from policies and programmes to individual projects (Therivel, 2010).

Summary

- EM seeks to be proactive. Tools and techniques are consequently required to place more emphasis on prediction and assessment of future scenarios, selection of optimum strategies, etc.
- There is a move away from piecemeal (ad hoc) approaches to more standardised and accredited measures, and from individual project and local focus to one that is more strategic and integrated (even holistic).
- Impact assessment and the forecasting of future scenarios is seldom accurate and gap-free; EM must be cautious and seek multiple lines of evidence.
- Predictive tools can be problematic; EIA, SIA, hazard and risk assessments may give a false sense of security. This is because they mainly give a snapshot view. It is also difficult to reliably identify indirect and cumulative impacts.
- The unexpected must be expected.

Further reading

Becker, H.A. and Vanclay, F. (Eds) (2006) *The International Handbook of Social Impact Assessment: conceptual and methodological advances.* Edward Elgar, Cheltenham, UK.

Glasson, J. and Therivel, R. (2019) *Introduction to Environmental Impact Assessment* 5th edn. Routledge, Abingdon, UK.

Perdicoúlis, A., Durning, B. and Palframan, L. (Eds) (2012) *Furthering Environmental Impact Assessment: towards a seamless connection between EIA and EMS.* Edward Elgar, Cheltenham, UK.

Sadler, B., Aschemann, R., Dusik, J., Fischer, T.S., Partidário, M.A. and Verheem, R. (Eds) (2011) *Handbook of Strategic Environmental Assessment.* Routledge (Earthscan), Abingdon, UK.

Theodore, L. and Dupont, R.R. (2012) *Environmental Health and Hazard Risk Assessment: principles and calculations.* CRC Press, Boca Raton, FL, USA.

There are a number of futures journals: *Journal of Futures Studies*, ISSN 1027-6084; *European Journal of Futures Research*, eISSN: 2195-2248; *Futures & Foresight Science*, eISSN: 2573-5152; *Futures*, ISSN: 0016-3287.

www sources

❏ EIA Introduction, Centre for Science and Environment (India): www.cseindia.org/understanding-eia-383 (accessed 21/12/22).
❏ EIA Index of websites, IAIA: www.iaia.org/eia-index-of-websites.php (accessed 01/02/22).
❏ FAO Livelihood assessment toolkit: www.fao.org/fileadmin/templates/tc/tce/pdf/LAT_Brochure_LoRes.pdf (accessed 22/12/22).
❏ Strategic environmental assessment (SEA), OECD: www.oecd.org/dac/environment-development/strategicenvironmentalassessment.htm (accessed 01/02/22).

Part III

Global challenges and opportunities

Chapter 10

Resources
Character, opportunities and challenges

Chapter overview

- Resources characteristics and management issues
- Water
- Air
- Land and soil
- Energy
- Food and commodities
- Biodiversity
- Timber, wood fuel and charcoal
- Minerals
- Marine natural resources
- Indigenous peoples and natural resources
- Summary
- Further reading
- www sources

Resources characteristics and management issues

Key natural resources include food, water, energy, unpolluted atmosphere and room to live a satisfying life. Resources are also used for warfare and to satisfy fashions and consumerism. There are challenges of unsustainable consumption and problematic waste (Heinberg, 2010). Around AD 1800 the world shifted from mainly using renewable resources to non-renewable and the trend has become pronounced (Ashby, 2013). Much associated with resources is politicised (Ting and Vasel-Be-Hagh, 2020).

A resource is something humans realise has actual or potential value. Many things influence this (Young and Esau, 2016; Tan and Faúndez, 2017; Devlin, 2019). Resources may be

DOI: 10.4324/9781003189985-13

Figure 10.1 Tao te Ching – 'Taijitu' – 'ying/yang' symbol representing duality: good/bad or challenge/opportunity or resources/wastes. Source: Shutterstock / Svitlana Amelina (www.shutterstock.com/image-vector/ying-yang-symbol-harmony-balance-58014958)

subdivided into natural, human, technological, financial, organisational, cultural and aesthetic. Alternatively:

1 finite no matter what;
2 renewable if well managed;
3 renewable if reasonably managed;
4 renewable no matter what.

Or:

i.   easy to damage and swift to recover;
ii. easy to damage and slow to recover;
iii. difficult to damage and slow to recover;
iv. difficult to damage and swift to recover.

Past civilisations have collapsed through resource issues, often water. Water supply problems probably damaged the Khmer, and Easter Island is cited as an example of poor resource management (or perhaps rat introduction), although without clear proof of what destroyed forest and soil (Diamond, 2021). There may be little to show long term from exploiting resources. Norway, learning from others, invested oil/gas resources earnings in a Sovereign Wealth Fund to support future generations; other Europeans have been more profligate (Røste, 2021).

The value of a resource can depend on fashion. Environmental and economic changes and technology also create or reduce demands for resources. Human resources are important and can be grown through education and research or devalued by neglect and poor management (George and Schillebeeckx, 2018; Nunan, 2019; Cubbage, 2022). Identifying key resource(s) may be difficult; ecologists have used Von Liebig's law which determines what is crucial to the survival of a species in a given ecosystem.

Figure 10.2 Ruins of Angkor (Cambodia). The city/temple complex (9th–14th centuries AD) by the Khmer. For c. 500 years people maintained a culture and then perhaps water supplies failed.   Source: Photo by author, 2009.

Resource use depends on human skills, knowledge, attitudes, funding, etc. (Dodds and Bartram, 2016). There are natural resource-rich countries which suffer poverty, and natural resource-poor countries with a good standard of living. Often human resources are crucial.

Businesses and cartels wield power: most important resources are distributed and sold by a handful of companies/states. Human resources developed with effort can be stolen (spying, biopiracy, theft of traditional knowledge) and used elsewhere. Security and profit are vital; if individuals, companies or a state feel insecure they will not invest labour or money in developing, sustaining or improving resources.

Many resources need to be transported to consumers. There have been huge advances in air, road, rail and sea transport, cables and pipelines. Also there have been improvements in resource conveyance: refrigerated shipping from the 1890s; airfreight especially after the 1970s; and the cargo container. Road networks have improved. The Belt and Road Initiative is a mega-programme, a land and sea transport system spinning-off development across a vast section of the globe. The resulting environmental, economic, social and cultural changes should be carefully assessed and managed (Box 10.1).

Resource management and EM can adopt a sector, ecosystem or bioregion as a planning and management unit/focus, or apply political ecology and human ecology (Conroy and Perterson, 2013; IGI Global, 2016).

Global challenges and opportunities

Box 10.1   China's Belt and Road Initiative

Perhaps the largest investment in infrastructure in history, it will reshape the economies of a vast area. It has had little Western media attention and most citizens are uninformed. It will connect the PRC to the rest of Asia, Europe, Africa, the Middle East, Europe and indirectly (by shipping) to Central and South America, Sri Lanka, Africa, and through the Arctic, etc. (see Figure 10.3, page 165). Two major infrastructure sub-programmes – the Silk Road Economic Belt and the 21st Century Maritime Silk Road – will support a 'new system of global economic governance' (Williams et al., 2020: 128). Supporters argue it will help global economic development; opponents claim it will lead to 'debt traps', human rights problems and environmental impacts and is neo-colonialism/inclusive globalisation. Or perhaps. it might be compared with the Marshall Plan by which the USA helped Europe recover after the Second World War.

It is difficult to pin down what is involved or agreed, what is political, military or truly international. It might allow a range of nations to become involved in international lawmaking and environmental care, growth and development. Much depends on how effective co-ordination is and whether environmental protection measures are adequate and enforced.

For such a huge programme/initiative, EIA and SIA of component projects or the overall Initiative are sparse, mainly limited retrospective studies (Ascensão et al., 2018; Hughes, 2019; Lechner et al., 2019; Xiang Cao et al., 2021; Apostolopoulou, 2022; Chiu, 2022; www.theguardian.com/cities/ng-interactive/2018/jul/30/what-china-belt-road-initiative-silk-road-explainer – accessed 17/01/23; www.worldbank.org/en/topic/regional-integration/brief/belt-and-road-initiative – accessed 18/01/23; www.eesi.org/articles/view/exploring-the-environmental-repercussions-of-chinas-belt-and-road-initiativ – accessed 17/01/23).

Water

Water is needed for domestic use, irrigation and processing of food, industrial products, hydropower generation, navigation, biodiversity conservation, and disposal of waste (Molden, 2007). Water is seldom available in the quantity and/or quality that are desirable and management is often poor. The situation will get worse as population and per capita demand grow, land is degraded and climate changes; better management must be developed. Concern grows that there is, or will be, a water crisis (Pearce, 2019). Water can sometimes be a sustainable resource *if* adequately managed, although some supplies are finite. A number of countries have been buying or leasing land in poor countries because it is cheaper than developing irrigation in their own and unexpected impacts are of less concern: some have called this 'virtual water'.

Land drainage, soil compaction and urbanisation cause precipitation to run off quickly; the result can be erosion, flash flooding, reduced groundwater recharge and wildly fluctuating stream flow (Sene, 2013). Manure and fertiliser use leads to nutrient over-enrichment and sediment-contaminated water, which causes eutrophication, damages fish, shellfish and aquatic plants, silts up channels, lakes and reservoirs and impacts coastal waters. The sediment is a lost resource for the areas eroded. Sediment/dust transport is often episodic, generated during storms, especially after tillage and before crops provide ground cover. Soil and

164

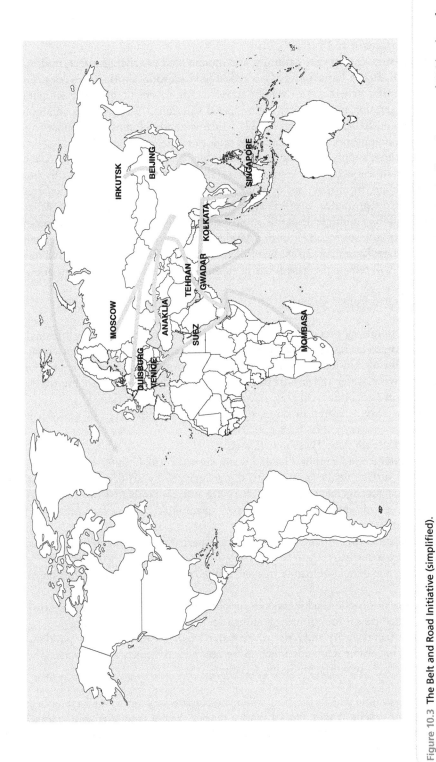

Figure 10.3 The Belt and Road Initiative (simplified).

Note: Blue = marine links: '21st Century Maritime Silk Road' (seaways and port developments); red = land routes (road and rail) portion west of PRC borders often termed the 'Silk Road Economic Belt'. Not shown are the proposed road links into E. Africa and W. Africa, pipelines, electricity lines, factories, hydro dams, etc. Source: www.theguardian.com/cities/ng-interactive/2018/jul/30/what-china-belt-road-initiative-silk-road-explainer (accessed 17/01/23). By L. Kuo and N. Kommenda & Brake, M. (Ed) (2020) Infrastructureatlas, daten und facten über öffentliche Räume und netze. Drukhaus Kaufman, Lahr, GR. ISBN 978-3-8692-220-6 (map drawn by Appen-zeller, L., Hecher, S and Sack)

water conservation approaches which reduce run-off and retain soil and moisture can help sustain agriculture, recharge groundwater, and should help cut flooding and silting. Groundwater, surface water and soil management should be comprehensive and integrated (Hipel et al., 2015; Jørgensen and Fath, 2020).

Precipitation may vary year to year naturally, and human land use change affects infiltration and runoff (Funk, 2021). Large parts of the world have seasonal shortages so reservoir storage or water transfer schemes are usually adopted. Improvement of water supplies through land and vegetation management nature-based solutions (N-BSs) is less common (Eslamian and Eslamian, 2017; Cassin et al., 2022). Much water development is engineering, politicised, inflexible, inappropriate, unsustainable, seldom resilient or adaptable.

Water will be in shorter supply in the future and more likely to be competed for (Alfredson and Engeland, 2020; Kulenbekov and Asanov, 2021). Demand will have to be met by less wasteful usage, recycling, from new less optimal sources and desalination. Water crosses borders above and below ground so conflicts are a possibility (Prud' Homme, 2011). Water management has to address multiple demands, multistate sharing, recycling and reduction of wastage and seek alternative sources (Lenton and Muller 2012; Tortajada, 2015; Management Association, Information Resources, 2018; Bogardi et al., 2021). Shrinking glaciers and less highland snow cover will reduce the base flows in summer of many rivers supplying populated lowlands (Pollak, 2010).

Freshwater is obtained from:

- Short-lived flows following precipitation (overland flow, runoff or temporary streams). These may be trapped in tanks (seasonal reservoirs), cisterns or gravel behind check-dams. Improvement of rain-fed cultivation is hugely important: it is accessible to poor farmers, does not waste water and is more likely to be sustained than irrigation. Improving rain-fed and run-off agriculture will play a key part in feeding the world.
- River flows. Flows may be stored by dams and water can be transferred by aqueducts/canals and pipes.
- Stable springs and stream flow. These may fluctuate.
- Groundwater (underground supplies) held in water-storing rocks (aquifers); renewed by precipitation and surface water seeping into the ground, or by artificial recharge and possibly vegetation management (assuming they are not contaminated or the aquifer damaged). With good stewardship some are sustainable, some are finite and vulnerable.
- Desalination. There are hopes that it may provide an alternative supply, currently limited because technology is costly.
- Trapping mist, fog and dew. Usable for crops and limited supply of drinking water; small-farmer suitable.
- Rainfall collection. Roofs and catchments can supply water. Some islands encourage this. Some highways shed runoff for cultivation alongside; urban areas could do more.
- Reuse. A few settlements purify and reuse waste water. Mainly used on amenity parkland or forestry. Drinking water demands much more complex treatment. Stormwater and sewage are seldom kept separate, hindering recycling.

Some groundwaters accumulated in the past when the climate was wetter or have collect slowly, so heavy use now is unsustainable. Groundwater may be contaminated with natural pollutants, agrochemicals, contaminants from damaged storage/spills, pipelines, sewers or salts from saline groundwater, irrigation wastewater or seawater. Capillary rise/evaporation can make sources salty. Decontamination of polluted groundwater may be difficult, slow or impossible. Overused sustainable groundwater may be permanently destroyed if an aquifer collapses or saltwater enters.

A growing trend is for water privatisation (Barlow, and Clarke, 2003). Privatised water may be as profitable in the future as oil has become. Business might be better able to fund development and conservation.

Recent estimates suggest by 2025 around two billion people will live in areas where water is scarce. The human demand, whether for domestic supply or irrigation, means less natural water for river, wetland and lake ecosystems. Human demands and nature have to be carefully integrated. Integrated watershed or river basin management offers ways forward to co-ordinate the management of water, environment and human activities.

Floods

Floods have been increasing because of: rising sea levels; poor soil and water management; land drainage; urbanisation (which speeds/increases runoff); and river channelisation which removes the moderation offered by meanders and floodlands (Pender and Faulkner, 2011). Flood damage costs are rising because people build with materials that are vulnerable and have easily damaged possessions. Authorities may fail to consider risks when installing infrastructure. There is a possibility storm patterns and associated flooding are altering through climate change.

Reliance on costly flood defences like New Orleans (USA) or Dhaka (Bangladesh) may be unwise. The former suffered in 2005 and the latter in 2022 in spite of mitigation measures. Alternative flood mitigation strategies include: resettlement from flood areas and replacement with appropriate land use (linear riverside parkland, conservation areas, etc.); improved whole catchment management; early warning and flood refuges. Recently a number of nations have re-established flood regimes: opening dam spillways to create artificial floods to flush away silt, keep reservoirs from being overwhelmed, and benefit riverine and floodland biodiversity. But there is sometimes reluctance to do so because managers want to retain water for power generation or irrigation. Flood protection levees seldom give adequate protection, can shift flooding to less protected areas, and prevent flood silt from fertilising riverside lands. Also, if breached, floodwater is retained and people may get a false sense of security.

Drought and reduced river flows

Climate change may cause water shortage and erratic or extreme weather including floods (Wilhite and Pulwarty, 2017). Drought is a complex hazard affecting rich and poor nations and may have roots in meteorological, land use, rising water demand, etc. It needs a multi-disciplinary approach to management (Wardlow et al., 2012; Cook, 2019; Iglesias et al., 2019; Mapedza et al., 2019; Ondrasek, 2020). Some regions may be subject to repeated drought, like parts of the USA, Northeast Brazil, etc. (Botterill and Cockfield, 2013; De Nys, et al., 2016; Perdew, 2018). Drought prone areas may have increasing human and livestock populations.

Recently there have been proposals for a Water Poverty Index to give an integrated assessment of scarcity. Societies respond to insufficient water in different ways: some innovate and reduce demand or find alternative supplies; others may abandon settlements and relocate. Drought management can be inappropriate mal-development: e.g. installing boreholes leads to livestock concentration and overgrazing.

Authorities may not to want to admit problems and sometimes ignore a 'backward' drought region's plight. The need for drought forecasting and management increases as populations in vulnerable areas grow. Response to drought includes: de-stocking grazing areas, improving water supplies, food aid, etc. Some measures must be applied with caution to avoid dependency or causing people to relocate into marginal environments where they further suffer/degrade the land.

Water resources management

Water resources management environmental awareness has been improving (Horne et al., 2017; Theodore and Dupont, 2019). Water resources management has to deal with expanding population, greater per capita water demands, rapid global change and new pollutants (Mysiak et al., 2010; Booth and Charlesworth, 2014; Teegavarapu et al., 2020). Greater emphasis has been placed on seeking SD of water resources (Ojha, 2017; Giridhar, 2019; Gude et al., 2020; Brears, 2021a; Tanil, 2021). Large irrigation systems are costly to install and maintain, so are mainly used for commercial crops. A fraction of the world's irrigation relies on efficient low waste techniques, so there is potential for water savings. There are opportunities for alternatives: runoff farming and soil and water conservation techniques such as mulching, stone-lines, terraces and microcatchments, particularly if these are based on low-cost, locally available materials and labour. Water management has to deal with sewage, livestock waste, storm drainage from urban areas (grey water), agrochemical runoff, and effluent from mining, industry, oil palm, rubber, aquaculture, etc.

Some countries responded to water supply problems (or were pressured) by importing cheap wheat and maize. This affects global food reserves and more people will depend on fewer producers, increasing vulnerability. In the grain-importing countries rural employment might deteriorate, reinforcing the drift of people to cities and overseas.

Developing rivers

Rivers are valuable ecosystems damaged worldwide by altered flow and pollution (White, 2019). The range and quantity of pollutants entering rivers and groundwater has expanded. Rivers damaged by pollution, poor flow, dredging and channelisation might in some cases be rehabilitated (Brierley and Fryirs, 2008; Darby and Sear, 2008; Boon and Raven, 2012).

Small reservoirs and tanks Large dams store years of river flow; small reservoirs (tanks) hold one rainy season's runoff for dry season use (or night flow for day use). Small tanks are cheap, pose little threat of failure or overtopping and cause few impacts. Ephemeral streams can be dammed with small check-dams to improve dry season water availability, help recharge groundwater, trap silt and control gullying and stream erosion. In South Asia tanks serve(d) a dual purpose: storage of water to improve and extend cropping and as a source of fertile silt for soil enrichment: the tanks are cleared using a bullock-cart at the end of the dry season and silt is spread on the land. Unfortunately, if labour costs rise or boreholes are installed, tanks tend to be abandoned.

Barrages Barrages divert water and moderate flow but impound no large reservoir, avoiding most large dam problems; however they do not store enough to meet demand in extended dry periods, and support limited electricity generation.

Large dams Large dams appeared in the 1920s for water storage (for domestic use, industry and irrigation), hydroelectricity and flood control; between 1950 and 2004 over 35,000 were constructed, more than 19,000 in China (Nüsser, 2014). Some modern dams are huge, e.g. the Three Gorges Dam (China). Globally, about 400,000 km^2 have been inundated by reservoirs, and between 40 and 80 million people have been displaced by dams since 1960. There are plans for huge schemes on the Congo, Nile, Mekong and many other rivers.

There are benefits from large dams: hydropower, regulation of floods, water storage for dry seasons or drought years, etc. However, dams and associated reservoirs may fail to yield benefits and can cause considerable environmental and social impact (Tortajada et al., 2012; Scudder, 2012, 2019; E. Shah et al., 2019; Taylor Klein, 2022). Experience should by now have helped reduce problems; however, it is striking how slowly engineering and dam management has improved. River development is overshadowed by inflexible, insensitive solutions, which for political, economic and other reasons continue to occur. Some problems result from a point-problem outlook; i.e. the developers do not look upstream or downstream at the whole issue.

The World Commission on Large Dams 1998–2005 was established to review large dam developments, set guidelines and assess alternatives; this role has been taken over by the UNEP Dams and Development Project.

Run of the river/stream hydro generation is underdeveloped as are smaller stream developments (micro- and mini-hydro). There is a long history of using streams for mills – so the reliability of flow is known in many countries. In the USA, UK, EU and elsewhere many small generation plants could be installed with little or no environmental impact, supplementing solar and wind power, post oil. Few hinder the passage of migratory fish, and fish-ladders can be easily installed.

Interbasin transfers There are opportunities to take water from rivers with periodic or seasonal excess and transfer it to other rivers, irrigation schemes or cities which have periodic or seasonal shortfalls. Numerous interbasin transfers have been proposed and some implemented. Interest is growing (Pittock et al., 2009; Boddu et al., 2011; Ghassemi and White, 2012; Liang Zhang et al., 2015; Hanasz, 2018). Transfers have been completed or are planned for Latin America, USA, India, Africa, Spain, UK, Turkey, Brazil and China (de Andrade, 2011). China part-completed a huge (over 1200 km) South–North Water Transfer Project in 2014, one of three selected routes S–N using the existing Grand Canal. Completion is planned for 2050 and it will channel flows (about one-tenth of that of the Mississippi) from the Yangtze River to Beijing. India has proposed the Indira Gandhi Canal and a National Water Grid (roughly 12,500 km) linking 14 northern Indian rivers to redistribute flows: the impacts could be huge.

Before transfer is embarked on, it should be asked whether a better approach would be to control water wastage and/or discourage development in the water-scarce areas. There are issues with impacts, sustaining transfers and potential for inter-state disputes (Howe, 2011).

Shared rivers Many of the world's larger rivers are shared by more than one country. If tension grows over the management there are two possible ways forward: 1) consultation, agreement and co-operation; 2) power politics ('hydropolitics') and possibly conflict. In the last 20 years there has been development of shared river governance, agreements and legislation (Megdal et al., 2013, 2020; Garrick et al., 2014; Boer et al., 2015; Dinar and Tsur, 2017; Jia Shaofeng and Lei Xie, 2017; Morre, 2018; Bréthaut and Pflieger, 2020; Kittikhoun and Schmeier, 2021). But there is potential for 'water wars', although as yet, inter-state disputes have seldom led to conflict (Ho, 2021). But competition for water increases the risk. Agreements on sharing rivers may be possible while, say, 80% of total flows run to the sea, but if demands rise, agreements may not hold and new ones would be less likely.

Most agreements have focused on quantity of flows and less on quality (Mirumachi, 2017; Antunes, 2021). A Convention on the Protection of Transboundary Watercourses and International Lakes was signed in 1996, but was only a start towards resolving problems. Laws relating to the sharing of river flows and groundwater are still developing (Schmier, 2013; Hollo, 2017; O'Bryan, 2018; O'Donnel, 2019). The UN Convention on the Law of the Non-Navigational Use of International Waters (1997) provides a framework for negotiations between countries. This defines the obligation not to cause harm to another and the right to reasonable and equitable use by riparian nations, but it is not law. Spain and Portugal were in 2005 beset by drought and in spite of long-standing agreements fell into dispute over sharing the Tagus and Douro Rivers.

Lakes and ponds

Lakes and ponds have suffered worldwide from: pollution, infilling, acid deposition, silting-up, the introduction of alien species, disruption of inflows, agrochemicals, plastic and other pollution, climate change, etc. (Freedman and Neuzil, 2017; Wun Jern Ng et al., 2018; Tundisi et al., 2019; Melack et al., 2021). The loss of biodiversity is enormous. Small ponds should be encouraged for landscaping and conservation.

Some lakes are of great age and contain rare endemic species now under threat. Aquaculture can cause impacts through collection of fish or other creatures for feed, discharge of effluent, deliberate or accidental release of cultured species that harm indigenous biota, introduced diseases, and use of chemicals. African lakes in 2023 are reportedly shrinking or rising for undetermined reasons.

Lake management needs a higher profile and better overall co-ordination, which integrates all the relevant fields and stakeholders, and which is preferably administered through a single lake development authority with adequate enforcement powers.

Irrigation, runoff collection and rain-fed agriculture

Huge efforts and expenditure have been directed at adapting land to fit crops; far less has been spent adapting crops and technology to fit the environment. The usual strategy has been large-scale commercial canal irrigation which is wasteful of water, difficult to sustain and liable to raise groundwater levels and contaminate land, rivers and aquifers with agrochemicals and salt. However, there are alternatives which demand much less water: drip-irrigation and hydroponics or runoff cultivation and improved rain-fed agriculture. The distinction between irrigated, runoff and rain-fed agriculture is not always clear-cut. Irrigation may be used to boost or protect rain-fed agriculture in dry periods. Integrated watershed management is one way forward (Wani et al., 2011; Nanwal, 2019).

There are many ways of improving rain-fed agriculture: crops that have shorter growing seasons, the introduction of tractor ploughing to allow infiltration and earlier planting, drought-resistant crops, fallowing, return of crop waste (like straw), careful use of fertiliser and green manure or biochar, stall-feeding rather than grazing livestock (reducing overgrazing and aiding infiltration and snow retention in colder climes), etc. (Rockström et al., 2009b; Tow et al., 2011; Nanwal and Rajanna, 2017). Improved rain-fed farming should improve security of harvest, boost yields and through soil conservation allow sustainable production. Some crops can make use of brackish water or salty soil, reducing freshwater demands.

Larger canal-fed irrigation schemes are increasingly difficult and costly to implement because the suitable sites have often already been developed (Hossein, 2011, 2013; Brebbia and Bjornlund, 2014). That is no bad thing because it is not unusual for these projects to fail to repay investment costs before becoming salinised or falling into disrepair and development moves on to repeat the process. It would be better to improve and support drip-irrigation using small-bore pipes for field distribution or even water and fertiliser conserving hydroponics.

The area of irrigated land per person (worldwide) has shifted from increases during the 1960s–1990s to a present decline (Postel, 1999). Large-scale irrigation currently feeds many people, but it uses much of investment in agriculture and is faltering. Roughly two-thirds of the world's irrigation needs repair. The problems include: inflexibility, vulnerability, narrow focus, insensitive engineering, wishful thinking by outsiders, reluctance to work with local people, and a failure to learn lessons in spite of considerable hindsight experience.

Archaeology prompted some interest in developing runoff agriculture; studies in the Negev Desert, North Africa, the Andes, southwestern USA and lowland Latin America have yielded potential strategies. In the past some cultures used runoff techniques and flourished in environments modern agriculture is unsuitable for. Some of these strategies are appropriate for modern small farmers; being inexpensive, sustainable and using local materials. Techniques can perhaps be improved by modern materials. Run-off collection microcatchments water smaller area plots, magnifying rainfall, and may be used in poor rainfall regions and/or on steep gradients or the level if soil conditions are suitable (Snobar et al., 2011; Han and Nguyen, 2018). Microcatchments can fill cisterns for humans, livestock and perhaps limited dry season homegarden watering. When assessing past strategies it is important not just to look at the technology but also costs, labour inputs and organisation.

Air

Air is a global commons resource, vulnerable and cannot be adequately monitored and protected without international agreement. Europe and the eastern USA experienced smog and acid deposition problems in the nineteenth century, and London fogs killed thousands up to 1952. The air pollution problems today are carbon emissions, sulphur emissions, acid deposition, tropospheric ozone pollution, stratospheric ozone scavenging, nuclear fallout and hazardous chemicals. Agriculture emits greenhouse gases; greenhouse gases are generated by ploughing, livestock and burning vegetation. Damaging pesticides pollution is common and nitrous oxide may be liberated by fertiliser use. Feeding people (agriculture) pollutes as much as industry.

Atmospheric pollution with CO_2 receives a lot of attention; other important atmospheric pollutants (methane, etc.) have had less. There has been progress in controlling partially burnt hydrocarbons from vehicles. Acid deposition has led to transboundary disputes for years. With a number of countries industrialising and with growing use of fertilisers, acid deposition will remain a problem.

Land and soil

Knowledge about current fertility, degradation status, C-sequestration and soil vulnerability to various threats such as global warming or acid deposition is often inadequate. UNESCO and the FAO published a *Soil Map of the World* in 1974. The UNEP, International Soil Reference and Information Centre, FAO, and other bodies conducted a country-by-country world soil degradation assessment, published in 1991 as the *Global Assessment of the Status of Human-Induced Soil Degradation* (GLASOD). In 2003 the European Soil Forum was established to explore threats to soil. The Soil Conservation and Protection Strategies for Europe (SCAPE) were established in 2005. For Central and Eastern Europe, the EU conducted a Pan-European Soil Erosion Risk Assessment in 2005.

Land degradation is the loss of utility or potential utility through the loss of or damage to physical, social, cultural or economic features and/or reduction of ecosystem diversity. Soil degradation occurs worldwide in rich and poor nations (Saljnikov et al., 2022). It may manifest as decline in fertility, pollution, erosion, desertification, vegetation change, compaction, salinisation, etc. There may be a single or a complex mix of causes, natural and/or human, sometimes indirect, cumulative and difficult to identify. Land degradation is likely to release CO_2.

Unless an adequate portion of profits or sufficient labour is reinvested in land husbandry, crop yields will fall or vegetation cover will be damaged and soil will degrade. Failure to take care of the soil and invest in it may result from poverty, population growth, insecurity, greed, ignorance, etc. In many regions land reform is as important as agricultural innovation. Researchers and agricultural extension services have developed guidelines for soil and water conservation. Unfortunately, soil conservation services in many countries are underfunded.

Well managed soil can be a sustainable (even improvable) resource and well managed land can sequester C; when misused the risk is that it will be degraded or even wholly lost and C is released and desertification spreads. It can be difficult, costly and even impossible to rehabilitate degraded soil. Soil degradation may be rapid and obvious – gullying, severe sheet erosion, crusts or hard layers – or gradual and insidious. The latter is worrying because it can go unnoticed until a threshold is reached: e.g. the minimum depth to sustain crops or natural vegetation is passed followed by disaster. Soils are removed by flowing water, oxidation, wind or both. Wind erosion commonly follows droughts, ploughing, wildfires or other ground disturbance; the sediment can wash or blow into streams, or it can also travel great distances as airborne dust. Wind-blown dust settles, some being fertile and some harmful. Volcanic eruptions are a natural source of sometimes fertile ash and dust (sometimes very abrasive and acidic).

Soil degradation can alter water availability, changing run off, soil moisture and snow storage, and by releasing soil organic carbon into the atmosphere speed global warming. There should be awareness following the 1930s USA Dust Bowl, and more recent problems in other parts of the world. Nevertheless, soil management has not had anywhere near adequate attention or funding. Perhaps 45% of the world's land surface is significantly affected by soil degradation. Even slight alteration of vegetation cover or drainage can trigger soil changes, possibly leading to permanent damage.

Desertification is a severe land degradation process leading to desert conditions (some term it 'browning'), which may be difficult to reverse. Although widely used, the term is imprecise and emotive. Desertification is usually seen as the product of mismanagement of vulnerable environments, manifest as damaged plant cover followed by soil degradation. Common causes are: overgrazing, woodfuel collection, bushfires, salinisation, pollution, the introduction of new species, etc. When faced by drought or desertification, the first questions should be: 'Is it natural?' 'Is it exaggerated by humans?' If nature is to blame, attempts to control the problem will probably waste resources; however, if humans are partly or wholly the cause, control might be possible.

There is a tendency for land use to expand in semi-arid areas during a good rainfall period and fail when precipitation declines, possibly leading to desertification. Desertification is more likely in drylands, uplands and harsh environments where vegetation is naturally under stress and easily damaged. The claim is often made that deserts 'spread', desertification occurs *in situ*. It can occur in humid environments where soils drain freely, even in Amazonia and Iceland. Desertification can happen where human populations are very low but frequently results from rising livestock and human numbers or increased and unwise exploitation or spread of pest animals and plants.

The UN Conference on Desertification in 1977 drew up a Plan of Action to Combat Desertification with poor results. The UN Sustainable Development Goals (SDGs) adopted in 2015 gave land degradation limited consideration. SDG 15.3 did have land degradation neutrality (LDN) as an aim. LDN is a concept adopted by the UN Convention to Combat Desertification – to assess, plan and monitor to achieve no further net loss of healthy land. The UN Decade on Ecosystem Restoration (2021–2030) launched in 2021 at last seems to show there is some concern.

Various estimates suggest c. 25% of the world's land area is desertified, affecting over one billion people. A number of researchers have questioned that some regions have serious environmental degradation and desertification (Thomas and Middleton, 1994; Fairhead and Leach, 1996; Leach and Mearns, 1996). Land degradation study and control has gained from developments in monitoring, modelling and social studies (Zdruli et al., 2010; Imeson, 2012; Kaswamila, 2016; Stringer and Reed, 2016; Zambon et al., 2019; Bhatnagar, 2022).

There are 2022 reports that conditions in the Sahel have improved compared with the 1970s–1980s, this greening (in the vegetation sense) is apparently due to more precipitation and because some farmers have been adopting soil and water conservation. However, some attempts to improve vegetation cover to aid soil, water and biodiversity conservation have had poor results; one strategy, plantations of eucalyptus species, has been tried in many countries but may transpire more moisture than the woodland or scrub replaced. Eucalyptus also intercepts precipitation and then sheds it as large, erosive drops from as much as 20 metres above ground. Beneath eucalyptus little wildlife survives and few jobs are generated by the plantations – choice of plants for rehabilitation needs care. Efforts have also been wasted on insensitive schemes to impose terrace or check-dam construction. The farmers may build measures so long as there is a grant or food aid but, if not convinced of the value or that maintenance efforts are worthwhile, they will abandon them.

The keys to countering land degradation and desertification are broadly to improve appropriate vegetation cover and soil fertility, encourage soil and water conservation, counter speculative settlement/land use and perhaps establish alternative livelihoods. The ideal is to seek long-term rather than short-term improvements, and to stimulate land users to initiate changes using their own labour and building on established (traditional) methods. Soil and water conservation efforts may not seem especially attractive to herders, ranchers and farmers: the return may not be obvious, slow, or it may appear to benefit others, not those working on it or paying for it.

Wetlands

Wetlands are diverse; some are permanently waterlogged and others only during flooding. Some are tropical and others temperate or polar, they can be near sea level or at any height up to alpine. Some floodlands get annual deposition of often fertile silt which is ideal for agriculture. Wetland nutrient regimes vary from nutrient rich to poor and from acidic to alkaline. Wetlands can develop thick coverings of peat and these can sometimes be unsuitable for farming. Some of the world's most productive and stable ecosystems are wetlands, and a number of sustainable agriculture strategies are agro-wetlands (e.g. rice paddies, Mexican *chinapas*, flood recession agriculture, etc.).

Swamps, peatlands in cooler and tropical environments, floodlands, coastal marshlands, mangrove forests and bogs are being lost at a worrying rate and with them biodiversity, coastal protection and carbon and methane sequestration (Proux, 2022). Coastal mangroves are important breeding grounds for marine organisms, and provide storm protection and other ecosystem services. The removal of mangroves exposes coral to silt, which might otherwise have been trapped. Seagrass beds (there are over 60 species of marine seagrasses) have suffered in many countries and play a key role in biodiversity conservation and carbon sequestration. Peat has been destroyed worldwide by agriculture, fire during dry periods, and for use as horticultural compost or fuel. As peatlands are destroyed or dry out, the carbon and methane they locked up is released to the atmosphere, accelerating global warming. Peat combustion can emit further carbon. Wetlands act as vital stopping places for migratory species. The main initiative to protect wetlands is the Ramsar Convention (1971) but there has been relatively little published on their EM.

Wetlands can be drained for agriculture, aquaculture, docks, malaria control, real estate development, etc. Larger wetlands such as the *Sudd* in the Sudan and the marshes of Iraq have suffered from agricultural drainage and warfare. In South America the huge *pantanal* wetlands could be damaged by navigation, ranching and development projects. In Amazonia areas seasonally flooded by rivers carrying silt (*várzeas*) are being damaged by dams or developed for rice and ranching.

Energy

Given adequate energy supplies, almost anything can be done: seawater desalinated, fertiliser produced, etc. By 2021 there were serious moves to end reliance on non-renewable sources and make renewable energy more competitive. Wind power and solar photovoltaic (PV) generation have developed rapidly but need energy storage and/or alternative power sources for low generation periods. Replacements for oil and coal are not established, although it seems likely that the shift will be to wind, solar and possibly hydrogen, biofuels, geothermal and improved fission generation. In 2022 there were tests of deep drilling using millimetre-microwave beam and plasma cutting which could unlock geothermal power – a more realistic dream than some proposals; however, wind and solar look more likely in the next 50 years. In the UK for example, PV expansion is being delayed by weak government support and a need to improve the grid.

In 2023 world electricity generation looked set to become carbon neutral. There are developments underway in long-distance transmission which could result in solar power transfers from sunny to colder countries and easier access to marine wind turbine generation.

Nuclear power relies on expensive and dangerous uranium fission reactors, which some countries have abandoned. Thorium/molten salt fission reactors might be safer but are undeveloped (Martin, 2012; Kamei, 2012). Fusion reactors would be an ultimate energy source but at best are 30 years off. Small modular (uranium) reactors look to be the BPEO stopgap to backup wind and solar generation. Methanol and ethanol have been adopted in a few countries. In 1976 Brazil substituted sustainable alcohol for gasoline/petrol nationally: all pumps deliver ethanol +25% petrol mix, the alcohol derived mainly from sugarcane waste.

Hydrogen, especially 'green' hydrogen, may play a significant role in future energy strategies. Hydrogen does not pollute when burnt and is less heavy to transport than batteries but easily leaks and may emit CO_2 when produced. 'Blue' hydrogen is produced mainly from natural gas and CO_2 produced in the process is captured and stored. There are other types/colours: 'grey' uses coal or oil land processing and lacks CO_2 capture and storage; 'brown' is produced from coal (or lignite) with much more CO_2 emitted. Nuclear power is used to make 'pink' hydrogen with no C emission. There are other less common types/colours: green hydrogen uses C-clean methods or is a natural gas and is currently about 4% of total global production. Nature-based solutions might one day make green hydrogen on a large scale.

There is much noise about post-petroleum energy and the Fossil Fuel Non-Proliferation Treaty is being promoted as a way to prevent all new oil wells. A few nations have signed by 2023 but the transition to non-petroleum low carbon energy looks like it will be a slow process.

Electric cars may not be especially green: batteries rely on rare metals, are heavy and difficult to recycle, have a poor range, limited life, and charging is likely to be from a non-green grid; their manufacture also yields a lot of emissions. Transport policies should consider alternatives such as ammonia, alcohol or hydrogen, although production of ethanol and hydrogen can involve land that could be used for food production. Waste and sewage could be used to produce biofuels. Current batteries rely on metals like lithium mined by a few nations and their production and disposal is dirty. Sodium batteries might prove a better option than lithium.

Very few countries have non-polluting, sustainable, affordable energy generation. Eco-Luddites who call for a return to pre-industry forget that most food today is produced with inputs of petroleum, and they fail to offer alternatives. A number of oil companies are researching post-petroleum energy policies. Biofuels are being adopted (biodiesel and alcohols from algae aquaculture, plant sources, rapeseed oil, soya, palm oil, sugar waste, maize, sewage, etc.). Biomass energy might compensate for their carbon emissions. But feedstocks, especially cereals and vegetable oils, encourage a shift to fuel cropping and may help cause food shortages. Biomass can be produced from sewage treatment beds of reeds, willow, algae, water hyacinth, etc. – waste to energy (Cross et al., 2021). Elephant grass on unfarmed land (e.g. *Miscanthus* spp) is another productive feedstock. Use of waste cooking oils for biofuel is likely to be on a very limited scale.

Wind power has progressed but turbines take space and blight landscapes. A partial solution is to site turbines at sea (strategically vulnerable) or on brownfield sites. Wind and solar facilities demand minerals that have to be mined and use energy in their production, and turbine blades are difficult to recycle. Energy conservation is needed alongside development of alternative energy.

There is a need to back up wind and PV with batteries or some other form of accumulator (pump-storage, etc.). Nuclear fission may be the BPEO but current large reactors are perhaps not the best way forward. Small modular reactors may be favoured but are not as efficient as large reactors but easier to build, repair and decommission, and probably safer. Chernobyl and Fukushima *should* have shown that secure power backup and cooling are as vital as reactor containment. Also, the problem of safe repositories for fission wastes has progressed slowly; Finland began construction of a deep repository in 2022.

Geothermal energy is under-exploited and might soon expand and offer a clean source for many countries with drilling advances and practical experience from Iceland and some other users. Fusion reactors are a hope – as ever '30 years away'. Ammonia, hydrogen or batteries may fuel short–medium range aircraft within ten years. In 2023 researchers in Cambridge (UK) announced a biotech prototype yielding carbon monoxide and hydrogen (syngas) when floated on water in sunlight that locks-up carbon – syngas fuel and C-sequestration. As with C-capture, energy substitution should focus on practical quickly achievable and not hoped-for solutions.

Energy conservation through better insulation and more efficient usage should play a large part in energy strategies. Presently energy use is often wasteful and there is a need to develop better technology and retro-fit inefficient usage. Discouragement of energy wastage through taxation, regulation and incentives is important.

Food and commodities

Hunger is due to too little adequate food being *available*; famine is the catastrophic impact of hunger. Malnutrition is caused by diet that is inadequate in some way: insufficient quantity, or poor quality, or an excess of some foods (obesity affects roughly 1.5 billion globally). World food reserves to cover one or more widespread failed harvests are inadequate. Self-sufficiency should not be confused with food security, a nation can have food security without being self-sufficient – *it may pay for imports if there is food to import*. If the poor weather of 1815–1816 (or drought rather than rain and chill) happened today there would probably be hunger globally – 1815–1816 was caused by a volcanic eruption, and another similar or more powerful could happen at any moment. Too much food comes from too few areas and too few and vulnerable types of crops. It would be wise to encourage diversification of crops to make production more resilient and less vulnerable and to generate and store greater reserves. World trade and economics do not support such goals. Probably a major difficulty with world food supplies will need to occur to provoke change. In 1972 and 1975 the USSR had crop failures and bought grain, sending world food prices soaring. A similar situation in future may be worse as some farmers have shifted from grain to growing inedible fuel/industrial crops.

The causes of famine include one or more of: poor organisation, inclement weather, crop or livestock or human disease, economic policies and labour issues, war (Cohen, 2021). In 1946 the FAO estimated at least one-third of the world population were starving. There has been progress, but the improvement has come from irrigation and Green Revolution approaches dependant on mechanisation, irrigation, oil-based agrochemicals and hybrid seeds. In 2023 a calorie of cereal in the USA took about 9.7 calories of inputs mostly provided by petroleum. Irrigation and rain-fed cereal production should aim to reduce pollution, stop land degradation, cut C-emissions, increase security and improve adaptability. Fertiliser supply is insecure with phosphates already in shortage. The hope is for technological and crop-breeding breakthroughs which reduce the need for water, fertiliser and agrochemicals. But improvements may not come, let alone meet rising demands.

There might now be a hidden crisis, whereby innovations have maintained crop yields, yet failed to halt soil degradation, leading to risk of a sudden decline of output. Also, worldwide urban growth is destroying considerable areas of farmland, some of it the best. The GLASOD programme suggested that between 1945 and the early 1990s, about 23% of the world's productive land had degraded to the point of uselessness. Hope is offered by productive and less vulnerable: vertical farming (highly productive, space-saving, indoor horticulture/hydroponics); aquaponics (growing aquatic plants and animals); hydroponics (growing plants without soil using water to circulate nutrients); mariculture (plant and animal marine aquaculture); simple drip-irrigation (precise delivery of water often with added nutrients); biochar (to increase sustainability, reduce polluted runoff, sequester C and raise yields); new and improved crops.

Globally sea fishing and mariculture provided about 155 million tonnes (*c*. half wild-caught) in 2022 and land-reared meat around 320 million tonnes. FAO estimates in 2021 suggested roughly two-thirds of the total seafish catch is 'sustainable' and that roughly one-third of stock monitored by the FAO is overfished. The situation is one of decline with roughly one-third of the world's 'meat' coming from the sea. Strategies at present place the emphasis on yield increases (catch), and much less on sustainability, security of harvest, reduction of environmental damage (to reefs, seafloors and non-catch species), equitable access and employment. A cut in fishing subsidies granted by a few nations would help conserve ocean fisheries.

The promotion of environmentally supportive and sustainable strategies to increase crops has been seen as a 2nd phase of the Green Revolution (the 'Doubly Green Revolution' phase) beginning after about 1980. However, much of today's agricultural research and investment supports commercial producers in favourable locations using unsustainable strategies and motivated by profit seeking. There is currently limited interest in SD, resilience, adaptability, pollution reduction or the millions of small farmers and herders living in harsh and remote environments (Mupambwa et al., 2022). In many regions smallholders are suffering a breakdown in traditional livelihood patterns as a consequence of development pressures, and the results are poverty, land degradation and relocation.

Organic farming is popular but not as useful as many assume; it is markedly less productive than typical 'high tech' farming, so more land has to be farmed (raising greenhouse gas emissions). It might be more sustainable and reduce need for chemical inputs and cut consumer exposure to pollution but crops cost more. Intensive outdoor organic production might supply some fruit, vegetables and livestock but organic staple crops like cereals or potatoes typically give around half that of conventional high input farming. There are new crops that could be developed and old ones that might be improved, one being the breadfruit (*Atrocarpus altilis*). Hydroponics and vertical farming and intensive small livestock production might become organic and cost effective. Some crops, grasses and plantation trees might be bred to better remove C from the atmosphere.

Globally, there *should* be enough to feed all today. World food production, if divided on a per capita basis, gives everyone around 2,700 calories per day, adequate for most needs. Yet, in early 2001, food emergency arose in 33 countries. In Africa, 18 million needed food aid 2001–2002. Recent estimates suggest that, worldwide, around 821 million lack adequate access to food while others become obese (FAO et al., 2021; Galanakis, 2021a, 2021b). About 210 million of these are in sub-Saharan Africa, 258 million in East Asia and 254 million in South Asia. People are hungry because food is unavailable in their location or they are too poor to afford it (even in the UK and USA some people resort to food banks).

Farming (and herding) is probably the most destructive human activity (Monbiot, 2022). Archaeology shows some pre-agricultural societies had more abundant and better balanced diets than agricultural successors for less labour input and, because they could move around, with less vulnerability. Agriculture, when it functions well, feeds many more than hunter-gathering or shifting cultivation, but it has serious drawbacks; the Anthropocene needs new forms of industry and agriculture/fisheries. A worrying aspect of modern food is that 60% of human calories globally depend on just four crops: wheat, rice, maize and soya. Reducing the vulnerability of such a situation to disease, climate and unrest should be a priority.

Those whose livelihood is subsistence agriculture have little saved against failed harvests and struggle to invest to sustain and improve production. Such agriculturists grow low value crops so agricultural industries tend not to invest in improving things. Consequently, commercial agriculture may be flourishing, while close by subsistence agriculture supporting considerable numbers of people is weak.

Monocropping is widespread, often with genetically identical crops which can be more vulnerable than traditional diverse crops to environmental threats. Mechanised production means less employment, agrochemical pollution and compacted soil leading to land degradation – a 'high price for cheap food'. Technology and trade give agroindustry control over seeds, inputs and marketing. This may drive many small farmers from the land with adverse environmental impacts. Safe and accessible GM crops need to be developed with an eye to improving food security, reducing land degradation and avoiding expensive inputs. The advances of the 1st and 2nd phases of the Green Revolutions (1960s to near present) rely on natural mutations for new crops/livestock which are relatively rare; GM speeds things up and opens a greater range of opportunities (Toensmeier, 2016).

In many countries livestock and cropping are increasingly separated: livestock and poultry manure have become problems rather than resources and cropping shifts to chemical fertilisers. Some recognize a 3rd phase of the Green Revolution marked by awareness of the need to move away from oil-based inputs and scarce fertilisers and to adopt GM seeking resilience and sustainability.

An alternative to intensification is extending food and commodity production into new areas; sometimes these areas are vulnerable to land degradation. In 1990 about 11% of the Earth's land was cropped for food or commodities (excluding grazing land), most lying within 35° of the Equator. Easily developed land is now largely occupied; further expansion will require breakthroughs that will overcome lack of moisture, poor soils, pests, etc. It will also probably require social and community developments (Jain, 2010; Kilby, 2019).

Food security

Food security is a mechanism which tries to ensure people will not starve (Ashley, 2016; Ferranti et al., 2018; Zhang-Yue Zhou, 2020; Mazo and López, 2021). That means adequate surplus stored in more than one place to prevent total loss in a disaster. Over 60% of cereals produced by developing countries in 2002 were consumed by livestock, most of which were located in developed countries; reducing meat consumption could feed more people. Agricultural output slowed to around 1.0% by 2001 with cereal demand outpacing production between 1999 and 2001. Sub-Saharan Africa is where food production has lagged farthest behind; between the 1960s and 1999 per capita food production fell by about 20% and food imports increased and things have got no better since (Miralles-Wilhelm, 2021).

Agriculture also produces 'non-food' products: beverages (tea, coffee, etc.), raw materials (rubber, feedstocks for plastics, paper, etc.), energy crops (biofuels), fibres, spices, vegetable oils (e.g. canola, sunflower, palm oil, etc.). While grain prices tend to be rising, other food or commodity prices may be behaving differently; e.g., coffee fell markedly in 2000 to about one-third of what it was in 1993 but rose in the 2020s. Where agriculturists produce a commodity that falls in price they may have to sell more to obtain vital farming inputs and buy their food. In such circumstances farmers and herders tend to neglect land management and migrate to find employment, or undertake exploitative production of livestock, narcotics, etc., damaging the environment. Similar impacts may be caused by withdrawal of subsidies, grants and low-cost loans. WTO agreements have affected how countries can subsidise inputs and apply tariffs and other controls to support domestic production and discourage foreign competition; hopes for free trade and 'open' markets may have unwanted impacts on agriculturists and their environment and on food security.

Pests like locusts and livestock diseases can suddenly cause disaster. Effective pesticides and vaccination have helped, but recent warfare, austerity and pesticide restrictions have hindered control of locusts. The outbreaks of foot-and-mouth and BSE in the UK since the 1990s show sudden threats can still happen.

Rubber, oil palm, timber or wood-pulp plantations reduce biodiversity and drive people to clear forest to farm elsewhere. There is relatively limited experience of these plantations' sustainability; it is uncertain how often trees can be renewed without degrading land. These crops may expand as petroleum use declines. Soya production has been expanding; in 2022 Brazil was the world's largest producer, followed by the USA, Argentina and China. The spread of soya has been aided by market demand and the development of varieties which can withstand humid climate and unfavourable soils.

Faced with a problem of migration richer nations may see the value of supporting farmers in areas people relocate from, paying more to counter land degradation, improve security and profits.

Better storage of food

Food is lost between harvest and consumption to rot and pests; one estimate suggested about one-third (Nicastro and Carillo, 2021). Those who store food on a large scale are developed-country or multinational commercial organisations seeking profit. It would be wise to encourage more local or regional food reserves. And it is important at all scales to reduce rot/pest spoilage (Blakeney, 2019; Galanakis, 2021b). Concern over the safety of fungicides and pesticides has made it more difficult to reduce losses. Better, low-cost means for safely drying, fumigating and rodent control and safe insect control are needed.

Biodiversity

Biodiversity is the diversity of different species together with genetic variation within each species. Biodiversity is a resource *and* a responsibility (Grillo and Venora, 2011; Lovejoy and Hannah, 2019). Biodiversity loss is a serious global problem (Kolbert, 2014; Crist, 2019; Haring, 2020; Sepkoski, 2020; Hobohm, 2021). The UN Decade on Biodiversity 2011–2020 made poor progress; some feel nature/biodiversity should be granted legal status/rights and a few states have begun to do so.

Improved agriculture means areas dominated by a few crop species. Agro-industrial corporations support fewer crop varieties and these are bred or genetically engineered and protected by legal controls so there is less chance of farmers saving viable seeds. Growers become dependent on business, reducing food security. For most of the time humans have farmed, crops have grown in fields with ancestor and related species, so that there has been crossbreeding which generates new varieties and maintains crop genetic diversity. Monoculture, commercial seeds, and biodiversity loss has reduced crossbreeding.

Companies may collect biodiversity and restrict access (probably collecting with little or no payment to source areas). The progenitors of domesticated plants, livestock, fish, and raw material for biotechnology and pharmaceutical use have restricted distributions in the wild, and can become extinct as land is disturbed, climate changes, there is acidification, etc.

Some field crops and wild organisms have been contaminated by DNA from domesticated varieties and genetically modified organisms (GMOs) (Ferry and Gatehouse, 2009; Newton, 2014; Shetterley, 2016). The Cartagena Protocol on Biosafety 2003 governs transboundary movement of living GMOs. Many countries have developed their own legislation and the UN SDG 15 provides broad encouragement (Gillespie, 2011; Watson and Preedy, 2016; Leonelli, 2021). The Convention on Biological Diversity (1993) accepts that countries would better conserve their biodiversity if they could make money from it. As well as encouraging conservation, the REDD+ initiative supports sustainable use of biodiversity and sharing benefits.

Ex-situe conservation, like seed banks and collections of frozen material (cryo preservation) such as bacteria, stem-cells, fungi, bryophytes and algae, offer backup to *in-situe* conservation. Duplication of collections reduces risk of catastrophic or incremental loss through global warming, land degradation, disasters, mismanagement or warfare. Kew maintains an expanding seed collection and the Norwegian Government has a seed vault underground in Svalbard, where the permafrost secures storage (Fowler, 2016). Marine and freshwater aquatic biodiversity can also be conserved *in-situe* and *ex-situe*.

Figure 10.4 *Rafflesia* sp. (Malaysia), the flower of the plant, which is parasitic on the roots of the *Tetrastigma* sp. vine, is *c.* 1 m across. The surrounding forest to within a kilometre was disturbed by farmers resulting in biodiversity loss. Less spectacular flora, fauna and microorganisms are also under threat here and may have potential value.   Source: Photo by author 2008.

Biodiversity stocktaking is an important first step in conservation. An assessment should be made to determine what is vulnerable. Next, efforts should be made to identify the causes of loss; these can be indirect, cumulative and complex. It may also be desirable to assess any present utility of species to be conserved (Lameed, 2012). Once all this has been done it should be possible to develop a biodiversity management strategy (Hawksworth, 2009; Underkoffler and Adams, 2021). There are NGOs which focus on specific ecosystems or species and those with more general biodiversity conservation interests. Funding biodiversity conservation remains a challenge (Gasparatos and Willis, 2015; Drechsler, 2020).

Recent studies suggest about one-third of the world's *protected* areas are under serious threat from humans and natural changes (Jones et al., 2018); declaring a reserve is one thing, good management and enforcement another (Perrings, 2014). The Aichi Biodiversity Targets 2010–2020 failed; only 6 of the 20 UN biodiversity protection targets were achieved. In 2022 the '30 by 30' proposals to set aside 30% of global land and sea by 2030 to conserve biodiversity looked like they might stand some chance of success. But the Convention on Trade in Endangered Species needs improvement.

Conservation and livelihoods can conflict: efforts to protect forests may be seen by locals involved in the timber industry as a threat; attempts to restrict fishing methods or access to fisheries can result in opposition. There is a need to involve people, gain their support and, wherever possible, to offer them livelihood opportunities (Sandbrook et al., 2013; Martin, 2017). If not involved, locals may poach or resist conservation in other ways. Locals are increasingly involved in policing, administering and servicing conservation areas, and can bring considerable traditional and local knowledge (De Boef et al., 2013; Baldauf, 2020; Berkes, 2021). An example is the Communal Areas Management Programme for Indigenous

Reserves (CAMPFIRE – Zimbabwe). However, in a number of places, participatory approaches have not been successful. Conservation should be to protect biodiversity, not a bolt-on to socio-economic development that under achieves.

Conservation can be linked to ecosystem services/benefits, a forest reserve may lock up carbon, protect a slope or catchment, support tourism, etc. The Great Barrier Reef World Heritage Area Strategic Development Project links conservation with over 60 organisations in reef-related food production, job creation, recreation, cultural heritage protection, etc. Developed countries benefit from funding biodiversity in developing countries and citizens should be made aware. Treaties and controls can help reduce the trade in endangered species. Tourism taxes, airport taxes and levies and debt-for-nature swaps can help fund conservation.

Conservation areas should be linked by corridors for biodiversity to spread from reserve to reserve although surrounded by developed land. The corridors may even be highway or rail verges or narrow fringes of streams. The practice draws upon island biogeography theories (Hilty et al., 2006). Conservation areas suffer human incursions and penetration of pollution, noise, livestock or domestic animals, and so have a much less effective outer zone vulnerable to edge effects. These buffer zones might be used for: collecting forest products, hunting or tourism to support locals and help pay for conservation (Rai, 2012). Even if a reserve core is large enough to maintain biodiversity and has resilience against climate change, disasters can strike: wildfires, the arrival of pest species, pollution, warfare, encroachment by squatters, etc. In New Zealand recently an eco-terrorist introduced damaging opossums to successful bird reserves.

Given the threat of climate change, conservation areas should ideally span as great an altitude range as possible.

Timber, wood fuel and charcoal

Thousands of years ago cultures were destroying forests for shipbuilding, smelting and wood fuel (Radkau, 2012). Logging, wildfires, charcoal production, land clearance, grazing, tree diseases and acid deposition continue to take a toll. By the eighteenth century a few countries were attempting forest conservation and replanting and between the 1870s and 1914 assessments of world timber resources and proposals for improvements were published.

The response to dwindling timber supplies has been afforestation or reforestation, generally with single species exotic plantations, often conifers or eucalyptus. Unfortunately, these strategies seldom support as much biodiversity as natural forest. The demand for timber, woodchips (for paper and plastic production) and veneer also causes much deforestation of lowland tropical, subtropical and temperate forests (Dauvergne and Lister, 2011). This yields a one-off profit compared to potentially sustainable selective harvest. Selective logging is better than clear cutting but opens the forest up to further damage if it is not policed, and may fail to leave enough quality specimen trees, seed dispersers and pollinators for good reproduction. Acid deposition, climatic change, etc., can impact even remote uncut/undisturbed forests.

Non-wood resources (forest products) and other ecosystem services are extracted from forests and woodlands: waxes, fibres, nuts, oils, etc. (Shmulsky and Jones, 2019). Woodlands and forests are sources of pollinating organisms important for surrounding regions.

The causes of forest loss are commonly complex and controls may be ineffective and can mean regional climate changes and carbon emissions, prompting the REDD+ initiative which involves giving value to rainforests by pricing the carbon they store (O'Sills et al., 2014; Angelsen et al., 2018). Deforestation countermeasures may help cut C-emissions (Williams, 2010; Runyan and D'Odorico, 2016; Horning, 2018; Nikolakis and Innes, 2020; Pearce, 2021).

Many rely on wood fuel and charcoal (Foley, 1986; Leach and Mearns, 2009). In drier woodlands growth is slow so removal of timber can have considerable impact. The poor may be unable to afford to shift from wood/charcoal. Charcoal use for smelting ores declined in most countries after the 1800s. However, there is still demand because it is preferred for cooking and it provides activated charcoal for industry. Charcoal producers often strip large areas of tree cover to supply their kilns and can be difficult to control.

Pelletised wood may be made of wood/waste from lumber or veneer production, peat, perennial grasses, fast growing willow, etc. (Dahiya, 2015; Pandy et al., 2021). Biomass emits carbon on combustion which *should be* compensated for by new planting (net carbon neutral/green) (de Jong and van Ommen, 2015; van Swaaij et al., 2015). It may be dovetailed to sewage or waste treatment.

Some tree planting is intended to counter global warming but this carbon-offsetting should be on brownfield sites or degraded land not replace natural tree cover with fast growth species. Planting by a developed-country business often takes place in poorer regions or countries and can have socio-economic and environmental impacts. Checks on what is actually planted and survives are weak so carbon-offset can be greenwashing.

Minerals

Minerals are mined, extracted from boreholes or from evaporated seawater/saline lakes. Ocean floors may be exploited soon. Impacts include noise, dust, pollution of streams and sometimes hazardous waste slurry (Jain et al., 2016). Mining may be open-cast (excavations) or underground (tunnels) or dredging of streambed materials; often it generates waste spoil that presents challenges to restore/rehabilitate. Small-scale mining recovers gold, tin, silver, diamonds, etc. Miners conflict with locals, silt streams and pollute the environment with mercury or cyanide (Hilson, 2005; Lahiri-Dutt, 2018).

Open-cast mines should store topsoil for rehabilitation and bank funds for remedial work, pollution control and compensation. Post-mining land can usually be restored to something like its former state or pits and quarries may be allowed to flood or return to nature. But some waste is prone to spontaneous combustion or is toxic and leaks problem pollutants. Mining impacts the immediate vicinity, workers and bystanders; those further afield are affected by transportation of ore, processing, hydroelectric generation for refining, dust, noise and workers' settlements. Stream and groundwater may carry mine pollution great distances, and some is very harmful and persists for a long time, even after mining has ceased. Most mining is temporary, although salt evaporation and recovery from geothermal emissions can be managed indefinitely. Mining companies may go out of business before there is rehabilitation or problems manifest. In some countries mining is one of the few employers and workers migrate to it, causing labour shortages in the areas they desert, which can lead to environmental degradation.

Deep ocean mining, dredging cobalt, phosphates or manganese nodules, has only recently begun to develop, but damages sea floor biota (Sharma, 2019, 2022). Petroleum and natural gas exploitation developed after the 1970s on a number of continental shelves. It has resulted in a few serious environmental incidents, like the Gulf of Mexico *Deepwater Horizon* spill (2010). Gravel, tin, gold and diamond dredging are already established in some shallow seas.

There is a threat of mineral exploitation in vulnerable environments such as the Arctic and Antarctica. Activity in the Arctic looks set to expand. Diamond (2005) noted some mining companies were as good at EM as anyone could reasonably expect.

Marine natural resources

In the past whaling and sealing decimated stocks, but was finally controlled and virtually ended in the 1960s, and the stocks are recovering. Hydrothermal vents have yielded useful material for biotechnology development but as yet have not been exploited for minerals and should be mapped and protected. Marine crustaceans (krill) are increasingly exploited and that may have serious impacts on wildlife. Ocean fisheries have been discussed.

Jellyfish are a problem in some seas due to nutrients pollution, posing a threat to fishing, nuclear power and shipping (they clog nets and cooling water intakes). Wind turbines at sea seem to have little negative impact. As yet marine currents and wave energy have been little exploited, and few tidal barrages have been constructed. For many years several nations disposed of hazardous waste in ocean deeps; hopefully this has now been prevented. Control of marine pollution and better management of fish, shellfish, etc. are priority needs. There has been a serious decline in sharks as a consequence of demand for their fins and marine mammals and seabirds suffer from pollution, noise generated by shipping, fishing nets and lines.

Acidification of oceans deserves more attention as global warming may cause seas to become acid enough to alter plankton populations with catastrophic impacts.

Coral reefs

Reefs protect shores, conserve biodiversity, act as nursery areas for many species, support tourism, etc. Once-flourishing coral reefs are now dead or dying, due to one or a combination of causes: climate change/rising sea temperature, acidification, crown-of-thorns starfish and triggerfish, silt from land degradation, sewage, industrial pollutants and agrochemicals, raised UV levels and disease. Coral bleaching is widespread and may be due to be bacterial disease, perhaps following weakening by other causes, and temperature rise (as little as 1°C). Sea temperature increases and sea level changes seem a probable consequence of global warming, and are likely to inflict coral damage.

When causes of coral loss are understood perhaps there will be opportunities to establish new reefs, seeding robust corals at shallower depths to compensate for sea level rises, using cultured polyps and substrate such as rubble or old tyres. EM might thus assist natural slower adaptation.

Indigenous peoples and natural resources

Indigenous peoples around the world have been establishing and exercising their rights over natural resources. The idea that indigenous peoples are always sympathetic to their environment, and will not exploit it, is wishful thinking. Indigenous rights can sometimes allow environmental degradation, and central governments find it difficult to intervene after granting them. Some indigenous peoples have sought profit or have been exploited by non-indigenous entrepreneurs seeking to evade environmental controls. They have also resisted large mining companies and other potentially exploitative developers and helped develop SIA. A recent development: the extractive reserve combines conservation and sustainable exploitation with local livelihood provision.

Summary

- Modern livelihoods require a diversity of natural resources; shortage of just one could cause serious problems.
- Water supplies must be better managed and valued more highly.
- Water resources have mainly been developed by managers concerned with economic and engineering goals. EM can better integrate the needs of all stakeholders.
- Hunger and malnutrition are problems which plague poorer countries and threaten developed countries. Food production and food security need more attention and investment.
- Agricultural improvements since the 1960s have been based on improved seeds, fertilisers and irrigation, but may be unsustainable and degrade the land.
- Agricultural efforts since the 1960s have possibly masked worsening soil degradation and yields might quickly fall. In future agricultural improvement will need to boost yields, sustain production, yet cause much less environmental damage, be adaptable and find alternatives to oil.
- Biodiversity conservation should duplicate reserves, establish *ex-situ* genetic collections and prepare for global warming. Enforcement of conservation needs improvement.

Further reading

Adams, W.M. (2003) *Future Nature: a vision of conservation* (revised edn). Earthscan, London, UK.

Conway, G. (1997) *The Doubly Green Revolution: food for all in the 21st century*. Penguin, London, UK.

Dauvergne, P.P. and Lister, J. (2011) *Timber*. Polity Press, Cambridge, UK.

Galanakis, C.M. (Ed) (2021) *Food Security and Nutrition*. Elsevier (Academic Press), London, UK.

Hunt, C.E. (2004) *Thirsty Planet: strategies for sustainable water management*. Zed Books, London, UK.

Imeson, A. (2012) *Desertification, Land Degradation and Sustainability*. Wiley-Blackwell, Oxford, UK.

Molden, D. (Ed) (2007) *Water for Food Water for Life: a comprehensive account of water management in agriculture*. Routledge (Earthscan), Abingdon, UK.

Runyan, C. and D'Odorico, P. (2016) *Global Deforestation*. Cambridge University Press, New York, NY, USA.

www sources

- Consultative Group on International Agriculture: supports 15 centres around the world which promote the Green Revolution and work to improve agriculture. www.cgiar.org/ (accessed 12/04/22).

- Food and Agriculture Organisation. www.fao.org (accessed 12/04/22).
- International Soils Reference and Information Centre: soils and sustainable land use information. www.isric.org (accessed 13/04/22).
- International Water Management Institute: improving water and land resources management for food, livelihoods and nature. www.iwmi.cgiar.org (accessed 13/04/22).
- Soil and Water Conservation Society (USA). www.swcs.org (accessed 14/04/22).
- UNEP World Conservation Monitoring Center: has a database of the world's protected areas, budgets, information sheets, and much more. www.unep-wcmc.org/en (accessed 14/04/22).
- International Union for Conservation of Nature and Natural Resources. www.iucn.org (accessed 14/04/22).
- UN Decade on Ecosystem Restoration (2021–2030). www.decadeonrestoration.org/ (accessed 21/04/22).

Chapter 11

Population increase, global warming, pollution, biodiversity loss and diseases

Chapter overview

- Identifying key challenges and opportunities
- Population increase
- Global warming
- Pollution and wastes
- Diseases
- Summary
- Further reading
- www sources

Identifying key challenges and opportunities

Between 1984 and 2015 the world was post-Cold War in a period of relative peace, but now hopes for peace are uncertain and arms spending is rising. Pollution, land degradation, population increase, renewed nuclear weapons proliferation and biodiversity loss are all self-inflicted challenges added to natural threats.

Interest in challenges has developed (Harris and Roach, 2013; Shapiro, 2015) but they (and opportunities) are mainly addressed in an ad hoc manner after they become obvious. There is a need to widen focus from carbon emissions control to other things and to be pro-active (Hulme, 2009; Benjaminsen and Svarstad, 2021). EM tools and strategies are improving and there is more integration between environmental science, planning, social studies, economics and political studies.

DOI: 10.4324/9781003189985-14

Global climate change is real, but if enough causation is natural, control may be difficult. Adaptation, resilience improvement and vulnerability reduction are valuable whatever the causes and if emissions controls are slow or fail.

Global challenges which deserve attention are:

- *Global warming* – widely accepted. Nations are negotiating. The focus is on reducing emissions and not enough on reduction of vulnerability and improving resilience and adaptation.
- *Pollution and waste* – huge challenges; more polluters have become aware and some seek reduction.
- *Serious pandemic* – Covid-19 has proven fears valid, but have lessons been learnt for the future? Plenty of other diseases threaten, but in general too little is being done. The WHO must be better supported to monitor and advise.
- *Population change* – receiving less attention nowadays than in the 1970s–1980s. Some nations have declining/ageing populations, others are growing fast. Gross world population must be reduced to help adaptation and resilience, and cut resource use and pollution.
- *Food shortage* – too little response to this threat. Improved agriculture since the 1960s has given a false sense of security. Soil, water, weather, volcanic and warfare problems can suddenly contribute to food shortage; land degradation and population growth add to the longer term challenge.
- *Food vulnerability* – dependence on a few crops, grown in a few countries and too little stored.
- *Soil degradation* – environmental scientists recognise the challenge, but citizens and governments do not give it enough attention.
- *Inadequate knowledge of the state of carbon and methane sinks* (oceanic, terrestrial vegetation, and soil) – sudden releases could drastically speed up global warming.
- *Water resources* – generally people undervalue water.
- *Biodiversity loss* – serious damage, terrestrial and marine. Some citizens are aware and governments provide some support. Too little is being done and enforcement is weak.
- *Loss/reduction of the Gulf Stream* – to cause chill weather conditions in Europe and eastern USA. Little or no contingency planning.
- *Volcanic eruption* – a threat which is inadequately perceived. Potential to suddenly cripple world food supplies. Air transport across the Atlantic and Europe was hit by a small Icelandic eruption in 2010.
- *Asteroid/comet strike* – has attracted some attention. Monitoring for early warning has started. More should be spent but some risk will remain because a few bodies will approach suddenly or evade sensors.
- *Tsunami* – normal tsunami risk awareness has been intensified since 2004; there has been some expenditure on early warning and coast defences, but not for the Atlantic or North Sea. Risk of mega-tsunami (possibly > 100 m waves) has been flagged but there has been little serious preparation.
- *Ozone layer damage* – threats accepted and some success in threat reduction. Must be better enforcement of controls, there is still misuse of CFCs.
- *Seismic threats* – citizens are often well briefed, there are mitigation measures and contingency planning and some retro-engineering and improved building regulations. Some peoples fail to change habits to reduce risk and some building may be less sound than required.

- *Refugees/migrants* – there are predictions of perhaps huge numbers if global warming, unemployment, soil degradation, warfare and sea level rise are not dealt with. Target nations might reduce their attractiveness. International law is weak. Source areas could be assisted. Displaced people might cause damage and trigger conflict.

EM must guard against maladaption: e.g. it was feared the 01/01/2000 Y2K bug would cause chaos (potentially it could have). Vast sums were spent preparing for this challenge that did not materialise. Governments, companies and institutions mainly plan for less than five years; long-term development is neglected. Attention is directed towards what can be done and seen to be done in a term of office. EM often demands planning and management that look a long way ahead. In a democracy people usually support measures which give short-term gain. Research increasingly concentrates on what can be seen to yield (or 'spin-off') quick, useful/applied results ('useful' may be interpreted with reference to fashion, political correctness and what non-scientist managers believe). This is important but could discourage original, non-applied (blue sky) study, which in the past has given much to human development. Research funding and institutes of learning are geared towards established fields and current academics are less likely to have tenure allowing them to conduct non-applied research without career risk, plus there is an establishment resistance to new ideas. The ozone hole was found in 1987 because of academic, not applied, research.

Transboundary challenges

Problems cross borders making it difficult to monitor, hold perpetrators accountable and make them pay for solutions. Management of transboundary issues is addressed by treaties and international legislation but progress is slow (Evans, 2012; Biermann, 2014; Brukmeir, 2019; Park and Kramarz, 2019; Orsini and Morin, 2021).

Global challenges

Global challenges can be split into two part-overlapping categories: human-caused and physical. Many demand international co-ordinated efforts for avoidance, adaptation, mitigation, establishment of early warning, monitoring, etc. Human-caused (anthropogenic) challenges may involve politics and international competition, yet demand impartial monitoring, policing and legal negotiation. Global challenges can involve current expenditure or abstinence to ensure that something is passed on to the future. Altruism, which benefits people in the future or at a distance, is quite new; most governance, legal systems and human behaviour are not geared towards EM needs.

Even when people are aware of a challenge they may be slow to take action. Some threats may be rare but cause terrible damage. Studies by environmental historians show past environmental changes and abrupt disasters have impacted people, that similar things might recur, and offer hints on vulnerability reduction and adaptability improvement (Fagan, 2008; Diamond, 2012, 2020, 2021; Svenson, 2012; Hobson et al., 2014; Malyan and Duhan, 2019; van Bavel et al., 2020; MacKinnon, 2021).

Sustainable development SD strategies should be workable, adaptable, robust and resilient. Ideally, initiatives should be diverse, duplicated and widely spaced. Then, if one element/strategy is faulty, damaged or destroyed there should be inputs and skills somewhere that can be used to recover. Effective co-ordination is vital (Vigilance and Roberts, 2011; Schreckenberg et al., 2018).

Population increase

In the 1970s neo-Malthusians saw population growth as the cause of key problems. But linkages are not simple; in some regions environmental degradation results from population increase but there are situations where problems are in no way related to it. Nevertheless, the world is finite and high global human numbers mean increasing demands, making it less easy to respond to challenges and sustain needs. In 1960 the global population was *c.* 3 billion, today it is over 7.5 billion and the 2002 UN projection for 2050 was 8.9 billion (medium projection; some optimists suggest 6 billion naturally if birth rates fall) (Lutz et al., 2014). The population of China seems to be falling. Improving education is a key factor for slowing population growth. Migration and ageing complicate population growth impacts. Some fear population decline (Angus and Butler, 2011; Emmott, 2013: Bricker and Ibbitson, 2019). But human survival long term is unlikely unless a global population of 3 billion, ideally less, can be reached within *c.* 100 years. The problem is agreeing what population share each country has of the total.

Global warming

Humans have always been affected by natural climate fluctuations. The last interglacial period saw global temperatures at least 3°C above the present and probably ice-cap free poles; the sudden Little Ice Age (AD 1300–1800) was much colder than now (Fagan, 2000; Frankopan, 2023). Global warming generates huge discussion (Gore, 2006, 2013; Philander, 2008; Giddens, 2009; Monbiot, 2009; Collings, 2014; Moran, 2015; Dessler and Parson, 2016; Mann, 2021). There has been a rise in atmospheric CO_2 from *c.* 280 to 420 ppm since the 1750s and a warming trend since the 1850s. But there was warming and rising sea levels before humans had any effect (Fagan, 2008, 2013, 2019; Fagan and Durrani, 2021). How much human activity has changed natural trends is not that clear.

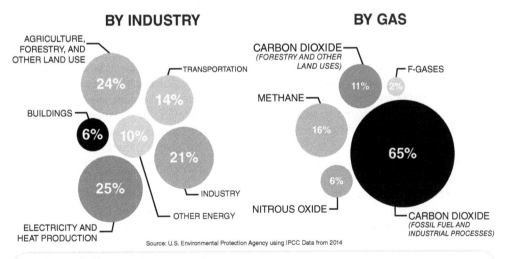

Figure 11.1 Global greenhouse gases emissions 2020: left by industry/sector; right by gas. Source: IPCC (2014) *Climate Change 2014: Synthesis Report. Contribution of Working Groups I, II and III to the Fifth Assessment Report of the Intergovernmental Panel on Climate Change.* IPCC, Geneva. F-gases mean fluorinated gases (includes refrigerants)

Global warming arguments can be bitter (Lomborg, 2004, 2009, 2010, 2020; Shellenberger, 2020). In 2023 focus was on carbon *emission controls* (mitigation), and little concern was given to vulnerability reduction and increasing adaptability and resilience. Resilience and adaptability should receive far more attention. Also, workable and affordable carbon sequestration approaches should be developed (Biermann et al., 2010; Kokotsis and Kirton, 2015; Gates, 2021; Mann, 2021). There should be more attention to methane, nitrogen, nitrogen compounds and activities that improve C-sequestration in soil, vegetation and oceans.

Rich countries account for 12% of the global population today but are said to be responsible for around 50% of all the greenhouse gases released from fossil fuels and industry over the past 170 years. Food production emits C so large populations contribute. The 1997 Kyoto Protocol is an international and legally binding agreement to reduce greenhouse gas emissions; by 2022 over 190 nations had signed – signatory developed nations have mandatory targets, developing countries have voluntary. The Kyoto Protocol required developed country signatories to cut greenhouse gas emissions back to a target by 2010. And it allows the trading of emissions quotas between countries which have surplus and those likely to exceed allowances.

Unfortunately, hopes that the Protocol would return greenhouse gas emissions to something resembling 1990 levels by 2010 were misplaced (Zedilo, 2008; Aldy and Stavins, 2010; Koh, 2010). The EU has established a Carbon Emissions Treaty Scheme (EU-ETS) to combat climate change by reducing greenhouse gas emissions cost-effectively. It is the world's first major carbon market and is thought to have cut EU emissions 35% between 2005 and 2019 (https://ec.europa.eu/clima/eu-action/eu-emissions-trading-system-eu-ets_en – accessed 08/05/22). This allows participants to import CO_2 credits to add to national allocations (a tradable emissions quotas type scheme). Some fear the scheme will increase EU energy costs without cutting emissions enough.

Between 1990 and 2000 emissions from industrialised countries fell by about 3%. By 2022 at least a third of the world's largest nations plus many cities, states and public bodies had made net-zero carbon pledges.

Education and fashion can prompt changes; in the West meat consumption is falling and might help a little to reduce C-emissions. Agriculture causes C-emissions through use of chemical fertilisers, irrigation and soil disturbance but this might be reduced by using approaches like biochar. Efforts to lower methane emissions need attention. Simply ploughing-in straw, not burning it, would considerably reduce C-emissions.

In 2005 the USA, Australia, Japan, South Korea, India and China consumed around 45% of the world's energy and caused approximately 50% of global carbon emissions. These countries agreed to try and develop cleaner and more efficient technology as a way to reduce emissions without damaging economic growth: the Asia–Pacific Partnership on Clean Development and Climate. Supporters of this Partnership claim it is easier to develop technology and promote its use than to alter economics and people's energy use habits. Can technological solutions such as practical, economical sequestration of CO_2 be developed fast enough?

Use of chemical fertilisers to feed growing populations and sudden natural disasters may also raise or lower global temperatures. So can fashion; e.g. rather than less polluting compact cars the 2020s has seen increasing ownership of SUVs. China and India now manufacture things for the West – effectively transfer of emissions.

In 2022 the Intergovernmental Panel on Climate Change (IPCC) published their *6th Report* on impacts, adaptation and vulnerability to global warming/environmental change (www.ipcc.ch/report/ar6/wg2/ – accessed 08/05/22). This report argues that achieving *limiting warming* of around 2°C (3.6°F) still requires global greenhouse gas emissions to peak before 2025, and be reduced 25% by 2030. Things in 2024 do not look promising. The report warned that more than 1.5°C above pre-industrial levels will have serious impacts and that current plans are not likely to avoid that. In 1992 at the Rio Earth Summit 154 nations agreed a diplomatic framework for addressing climate change – this Conference of Parties (COP) meets annually with COP28 in 2023.

Global warming might be causing more erratic and extreme events (heatwaves, storms, floods, etc.). Raised levels of CO_2 are likely to affect plant photosynthesis and crop productivity and cause ocean acidification. Acidic oceans might have caused at least one prehistoric mass extinction. Much depends on speed of change; humans, flora and fauna need time to adapt. Greenhouse gas emissions and warming can trigger indirect and cumulative impacts which may not be obvious and runaway changes are possible.

Control of sulphur, dust and smoke pollution – positive developments – could increase warming. There are global warming impact forecasts, focusing on sea level rise, health, agriculture, severe weather events, etc. (Schlesinger et al., 2007; Keskitalo, 2008; Navarra and Tubiana, 2013; Quante and Colijn, 2016; US Global Change Research Program, 2019; Sanjay et al., 2020; Lyubchich et al., 2021). Knowing in advance what the impacts might be gives nations and businesses opportunities for profiting as well as preparation to mitigate, adapt or avoid difficulties (Nakashima et al., 2018).

Large projects, infrastructure, crops, agricultural practices, etc. should be designed to deal with uncertainty and be as flexible as possible (Salih, 2009). That is often not the case.

Sea level rise

As Earth warms, less water is locked in ice sheets and glaciers and warming expands oceans resulting in sea level rise. There has been a rise globally of *c.* 24 cm since AD 1880. Archaeology suggests seas were rising long before industrialisation, though in some areas siltation or tectonic movements have raised land (Church et al., 2010; Fagan, 2013). A rise of 1.1 m by 2100 was predicted by the IPCC in 2019. A 0.5 m rise could displace over six million people from Bangladesh alone. And storm surges and tides can magnify the impact of sea level rise.

Around 30% of the global population live close to the current sea level, 8 of the 10 largest cities are close to the coast. A lot of agricultural land is also near the current sea level, as is vulnerable infrastructure like nuclear reactors. Given time and money defences might be built, new land reclaimed, drainage measures improved and building practices developed (Watson and Adams, 2011; Galen et al., 2022). Some islands and coastlands are already impacted and may not have opportunities or resources to prepare (Ward, 2010; Pilkey and Pilkey, 2019; Englander, 2021). Sea level rises have more impact if mangroves or reefs are damaged and if construction edges nearer the coast. Tectonic or isostatic recovery can elevate or depress land faster than sea level rises and in some areas subsidence or erosion increases risk. Land drainage to the sea (overwhelmed during storms and high tides) is a widespread problem. Severe flooding in the USA resulted from hurricanes Katrina (2006) and Irma (2017).

Adaptation to rising seas includes: abandonment of coastal land to recreational or conservation use, resettlement and floating buildings. The Netherlands and Venice have so far coped. If regions and islands are submerged, will they still exist as states, retain rights and be entitled to relocation or compensation? Law as well as engineering must respond (Gerrard and Wannier, 2013).

Pollution and wastes

There must be management of *outputs* (waste and pollution) as well as *inputs* (food, water, living space, etc.). The ultimate goal of those seeking to control pollution is zero waste management (ZWM). ZWM is achieved by minimising waste; maximising recycling; reducing consumption; reuse of products; and encouraging sustainable products (Rathoure, 2020). Historically, humans spent more effort attending to inputs than to outputs. Growing, sedentary populations and the invention of potent compounds have made it vital to manage outputs. Pollution is the introduction, deliberately or inadvertently (by humans or nature), of substances or energy (heat, radiation, noise) into the environment, resulting in deleterious effects. Contamination is the presence of elevated concentrations of substances in the environment, food, etc., not necessarily harmful or a nuisance. Pollution involves contamination, but contamination need not constitute pollution. Pollution and waste offer opportunities as well as threats; some can become useful resources. Nature generates toxic or nuisance compounds (e.g. volcanic ash). Most pollutants are difficult to deal with once released into the environment; some become a global problem as they spread through the ocean or atmosphere (Chandra, 2016; Muller, 2016; Jingling Liu et al., 2017; Spellman, 2021).

Waste is movable material that is perceived, sometimes erroneously, to be of no further value. Once discarded, it may be a nuisance or a hazard, but some may turn out to be a resource. As waste may give rise to pollution, it is necessary to view both together. Pollution and waste management can focus on: 1) prevention and avoidance; 2) collection and disposal; or 3) reclamation/treatment/mitigation (sometimes difficult and costly or impossible). Prevention involves avoiding/catching waste/pollution before release.

Open-cast and underground mines generate a lot of often problematic spoil which can pose a risk of spontaneous combustion and/or can leak and contaminate streams and groundwaters. Mine waste often prompts attempts at remediation and vegetation cover restoration; some open-cast pits provide sites for refuse landfill or become lakes. Toxic, acidic, radioactive and unstable mine wastes can pose particular threats and may need costly remediation/bioremediation that should be borne by the mines, not left for others in the future. Wherever possible, waste and pollution disposal should be integrated with employment generation, appropriate industrial development, etc. For example, incinerators can provide district heating and ash or biochar as raw material.

The poor accumulate higher levels of pesticides because they are exposed through work and poor housing. In the UK poorer areas of cities are often downwind (east) of pollution. Children and women may be differently exposed and physically more vulnerable than the males. Pollution can be primary, having an effect immediately, or secondary (indirect), the product of interaction after release with moisture, other pollutants and/or sunlight. The effects may be direct, indirect or cumulative, felt intermittently or constantly, immediately or after delay; affecting the atmosphere, soil, waterbodies, groundwater, or be restricted to certain organisms, produce or localities. The effects of pollution may be short term or longer term; pose a hazard or a nuisance; be toxic or non-toxic; take the form of a chemical, radiation, heat, light, noise, dust or smell problem. The environment may render pollution and waste harmless (an ecosystem service) until a threshold (absorptive capacity) is exceeded, after which there will be gradual or sudden problems, perhaps permanently destroying the capacity to treat the emissions.

Standards and monitoring techniques need constant improvement. What was considered safe ten years ago may no longer be accepted; also new compounds are being produced. A 'safe' background level of a pollutant may become dangerous to organisms near the top of a food web as food organisms concentrate it by feeding or absorption (bioaccumulation/

biomagnification). Some pollutants become concentrated in certain tissues of higher organisms (e.g. DDT and polychlorinated biphenyls (PCBs) in the fat; radioisotopes like strontium-90 in the bone; and radioiodine in the thyroid). Wastes and pollutants can also be concentrated by tidal action, sudden rain-out by storms, seasonal snow melt, chemical bonding to certain soil compounds, localised interception of contaminated rainfall or dust, etc., to form hotspots.

Pollutants or wastes initially discharged into water or the atmosphere may change systems; dust may settle on water and sink, or polluted water may form aerosols. Pollution sources may be: point (e.g. an explosion), linear (e.g. a road) or extensive (e.g. dust from a desert). Releases can be continuous, single, repeated, brief, extended, random or periodic. The distance pollution and waste disperses depends on its qualities, where and how it is released. Gases or dust are affected by the height of release, their temperature relative to the air, weather conditions (especially windspeed), and their density or particle size, presence of inversion layers, whether any obstacle is encountered and the texture of that obstacle, etc. Dispersal in water is affected by an equally diverse set of factors. Dispersal is enhanced if material is projected into a jet stream, storm or ocean current. It is not uncommon for a temperature inversion in the atmosphere or a waterbody to effectively put a lid over an area, trapping pollution. Stratified in that manner only limited levels mix and disperse compounds, conditions become stagnant and pollutants get concentrated.

Pollution and waste background

Greenland ice cores contain soot and heavy metals (especially lead) deposited from Roman times onward. Heavy metals, pesticides, radioactivity, PCBs and CFCs are toxic, carcinogenic, mutagenic or harmful in a variety of ways, even at very low concentrations, and some persist for a long time. Background contamination may reduce organisms' immunity to infections. From the 1840s pollution spread and increased with industrial activity.

Before the 1980s it was common for industry and government bodies to ignore or hide harmful on-site or off-site impacts and it was difficult for workers, consumers or bystanders to seek damages. When action was taken it was usually cleaning up rather than preventive. The burden was/is not borne by those who polluted, and sometimes people far removed, perhaps unaware, suffered/suffer. Cleanup costs may be incurred generations after a business had ceased. A post-1970s shift towards making the polluter pay and encouraging prevention is far from complete (Yung-Tse Hung et al., 2012). Many sites where pollution and waste have accumulated have not been recorded and pose a hazard. Dangerous material discarded in containers could escape when seals deteriorate.

Too strict a control, and business struggles and may relocate or go bankrupt; too lax, and people and environment suffer.

Pollution and waste associated with urbanisation and industry

Cities and built-up land have heat storage characteristics and roughness that differ from non-urban. These can cause a heat-island effect: city areas warmer than their surroundings, causing airflows that may trap and recirculate pollution. Most cities have dereliction, brownfield sites, which can be used to reduce demand for new land although to decontaminate and rehabilitate it for settlement use can be costly.

Cities in Europe and North America began to install waterborne sewage systems after the 1850s. The technology spread and many of the world's sewerage systems are overtaxed and consume large amounts of water. Cities may discharge sewage contaminated by small-scale industrial activity, making treatment more challenging. Modern sewerage design can reduce blockage/silting up, for example by installing stepped or ovoid cross-section pipes. Waterborne sewage systems pose problems:

- the cost of installing, extending and maintaining sewerage;
- failure to separate storm water, sewage and industrial waste, making treatment and re-use more difficult;
- treatment of sewage before discharge is commonly unsatisfactory;
- blockage often due to fat, wet-wipes, plastics.

Where long sea-outfalls were once satisfactory they are now inadequate. Treatment usually generates sludge contaminated with pathogens, toxic heavy metals, phosphates, pharmaceuticals, nitrates, etc. In the past sludge was sent to landfill sites, spread on agricultural land, or dumped at sea. Those strategies are now widely banned (Chen et al., 2017). De-watering and high-temperature incineration is an alternative. If biochar proves as useful as hoped de-watered sludge may become one pyrolysis feedstock, offering a land improvement resource. There are water-conserving sewage disposal systems: low volume water flush to septic tank, earth-fill latrines, composting toilets, aquatic plant treatment beds, etc. Most of these systems generate sludge requiring regular removal and treatment, and in some cases this must deal with heavy metals if the material is to be used.

Agricultural pollution

Intensification of agriculture generally leads to pollution from agrochemicals, livestock manure, tillage, irrigation, plastics and crop processing. If not carefully managed, fertilisers and tillage cause microorganisms to emit greenhouse gases. Better soil management reduces pollution, and may lock up carbon. The behaviour of pollutants may alter with global climate change (Capri and Alix, 2018; Pereira, 2019).

Chemical fertilisers Between 1952 and 1972 UK agricultural output rose about 60%, thanks largely to chemical fertilisers, although changes in crops grown make it difficult to assess. On a world scale, fertilisers have played a major part in increasing crops. In 1950 the world used about 14 million tonnes of N-fertiliser; by 1985 this had risen to about 125 million tonnes. In the late 1970s on average developing countries used 28 kg ha^{-1} and the developed countries 107 kg ha^{-1}, most for large-scale grain and export crop production. Fertiliser use is likely to increase but there are uncertainties about the long-term impacts and thus sustainability of cultivation using agrochemicals; one problem can be a net loss of organic matter from the soil (reducing C-sequestration), and in some areas acidification and zinc or sulphur deficiency. However, chemical fertilisers offer advantages over organic fertilisers:

- They can be easier to store, handle, apply and transport than most natural fertilisers.
- There is less smell and lower risk of pathogenic contamination, although well-composted organic material is virtually pasteurised.

Arable and livestock farming are often no longer integrated; the former must rely on chemicals and the latter presents a waste disposal problem. Ideally the waste should be returned together with domestic refuse. But to do so will require sophisticated bioremediation or incineration facilities because of heavy metals. Incinerated wastes can provide electricity and/or district heating.

Chemical fertilisers can also cause eutrophication of waterbodies and streams and increased harmful nitrates in groundwater. Phosphates have been accumulating in soils, river and lake sediments as a consequence of fertilisers and other pollutants. Studies in Europe suggest that even if the application of phosphates is reduced, the problems may grow for years due to leaching of existing material from soils and more rapid mobilisation (through soil acidification and warming); it could lead to a six- to ten-fold increase in river and groundwater contamination within decades. Such levels raise problems for domestic water supply, and for the ecology of rivers, lakes and shallow seas. In the UK, borehole studies suggest conversion of pasture to arable using N-fertilisers causes high groundwater nitrate levels with periods when water supply bodies have had to issue bottled water because of the health threat. Improvements will come slowly because nitrates and other chemicals may continue to be released for 50 years or more after controls.

Phosphates are mined in only a few countries and supplies are dwindling (they are also vital to many industrial processes); hopes of mining phosphate nodules from the ocean floor have not yet been realised and to do so may have biodiversity impacts. It would make sense to cut back on use of phosphate (and other) fertilisers. There are ways of controlling fertiliser use: reduction of price supports for crops; regulation of crops grown; quotas or permits which seek to limit expansion; set-aside (withdrawal of land from production); taxation of fertilisers. Reduction of fertiliser use will reduce crop yields unless effective green manuring or crops able to fix nitrogen can be developed/adopted.

Costly slow-release liming has been used to reduce leaching of heavy metals, phosphates and nitrates through warming/acidification. In wheat lands, planting winter wheat in rotation with white clover might help to cut nitrate leaching, lower chemical fertiliser inputs and discourage some pests. Biochar addition to soil might really help hold nutrients, carbon, heavy metals and other pollutants reducing need for inputs and cutting pollution problems, *if it works*.

Agricultural waste and pollution This includes livestock excreta, silage effluent, soil greenhouse gas emissions, salt and soda build-up, crop residue, nutrient and agrochemical runoff, eroded sediment and plastic waste. These can be countered:

- excreta – treatment and incineration;
- quotas – limits on quantities a farm may produce;
- use as a raw material – e.g. straw for strawboard/cardboard;
- set-aside – withdrawal of land from production, provided remaining agricultural land is not more intensively used;
- return straw – plough-in straw not burn it; sequesters carbon and improves soil organic matter content and cuts C emissions from burning;
- biochar – sequesters carbon and holds nutrients and pollutants in the soil. The benefits need research; might be a vital technology.

Manure is now a problem: a UK farm of 40 ha with 50 cows and 50 pigs presents waste disposal problems equivalent to a town of nearly 1,000 people. UK livestock in 1993 produced 2.5 times the country's total human sewage. The Netherlands in 2021 had a *c.* 40 million tonne 'manure mountain', more than twice what could then be disposed of. The Netherlands and Denmark are establishing incineration plants supporting district/horticultural greenhouse heating and some manure is composted. Stored in pits or lagoons, livestock waste generates methane, ammonia and hydrogen sulphide, which cause nuisance smells, damage vegetation and are strong greenhouse gases. If such slurry escapes it causes serious stream, lake or groundwater pollution through chemical oxygen demand, biological oxygen demand, harmful bacteria and parasites, excreted antibiotics or growth-promoting hormones, steroids and sometimes heavy metals (especially copper and zinc added as growth accelerators to feed). Disposal on to farmland is illegal/unwise, de-watering and bioremediation or incineration is now needed.

Livestock digestion generates methane, reduction can be through dietary supplements, breeding low emission stock, etc. Clearing land for agriculture is often done with fire, generating soot and greenhouse gases. Fires in Indonesia, Amazonia, Venezuela, Mexico and elsewhere pose a transboundary health risk and contribute to global warming (Scott, 2018).

Plastics waste

Plastics waste is a serious problem globally, a component in landfill disposal and as litter in streams and oceans. For decades the wastes have blocked streams, been eaten by marine and terrestrial wildlife with ill effects, and disfigured the landscape (Wani et al., 2019). Global annual plastics waste in 2022 was roughly 355 million tonnes, of which around 8% was recycled. Many countries have enforced a change to biodegradable plastic bags or paper bags since 2000. Substitution is unlikely to be 100% adopted. Plastics in landfill and the environment may degrade to form hazardous compounds and decomposition can be very slow: most plastic bags take 20 years and some drinks bottles last 450 years (the latter in seawater).

Figure 11.2 Floating plastics waste (a small portion of the total which includes waterlogged and microplastics) on a Southeast Asian river.　Source: Photo by author, 2011.

On average each person in the UK in 2022 threw away 100 kg of plastics. The solutions are 'reduce, re-use, refill'. Reduction by substitution is already possible for plastic film: degradable film made from soya or pea proteins or milk. Plastic containers can be substituted with waxed paper cartons, reusable glass bottles, etc. Reuse demands sorting refuse, cleaning and treatment. In some cases a plastic can be used indefinitely but most can only be reused very few times and for lower grade products. Ultimately much ends up as landfill or goes to incineration. In 2023 the EU and UK tightened laws restricting single-use plastics which should help cut waste.

Where plastics waste can be collected, cleaned and sorted economically, reuse/recycling is viable. Some countries require householders to sort plastics (Letcher, 2020). Where plastics waste is degraded or difficult to sort or clean it may be burnt for electricity generation or shredded, treated and used for: substitution for road surfacing, as feedstock for biofuel production, as blast-furnace fuel, etc. (Subramanian, 2019; Cirino, 2021). It should be possible to design plastics that can be remanufactured easily and more plastics are now biobased or can be substituted using soya, maize, cellulose, fungii mycelium, etc.

Plastic waste might be treated in bioreactors using: mealworms or beetles, synthesised enzymes, fungi and suitable bacteria. Polystyrene can be digested by beetle larvae (e.g. *Plesiophalmus davidis* and *Zophobas morio*); polyethelene by waxworms (various species) and the fungus *Pestalotiopsis microspora* and bacteria *Ideorella sakciensis* also digest polystyrene and tetraphthalate. Cattle stomach bacteria may have potential for bioreactors. Commercially viable bioreactors are being developed.

Plastics degrade slowly and disperse, harming wildlife and equipment such as pumps. Phantom fishing by lost or discarded nets and long-lines (which may remain un-degraded for centuries), does tremendous damage. Problems might be reduced if biodegradability could be built in, but fishermen do not want equipment to deteriorate before it gives service. In December 2022 the first global treaty on plastics pollution was convened.

Microplastics Microplastics are affecting the whole globe, even remote places: deep oceans, Antarctica, etc. (Defu He and Yongming Luo, 2021). The minute particles (< 5 mm size) of various plastics are ingested or breathed in, enter food webs, air and soil of terrestrial and marine/aquatic environments and contaminate organisms (Muthu, 2021c; CIEL, 2022; Rocha-Santos et al., 2022; World Bank, 2022). Microplastics concentrate in localities or organisms and form biofilms. The sources include eroded larger plastic debris, fragments from clothing (much released by washing machines), tyres, and breakdown and abrasion of coatings. All are human-caused, date from the 1950s on and are difficult or impossible to treat/remediate once dispersed into the environment. The impacts are not well researched and many of the plastics involved are persistent. The fragments may also carry chemical and microbiological contaminants. Reduction of sources is currently the only option.

Plastics (macro and micro) waste in oceans Plastics reach the sea, washing down rivers or being dumped from shipping. It is not practical to fit strainers to the worlds' estuaries but they can be added to discharge pipes and channels. There are serious problems (Bergmann et al., 2015; Ahamad et al., 2022; Bonanno and Orlando-Bonaca, 2022). Both micro- and macro-plastics debris can concentrate on beaches or in eddies (gyres), some far from shore. Perhaps concentrations can be skimmed and dealt with. There are five known large accumulations offshore, one being the 1.6 million km^2 'Great Pacific Garbage Patch' between California and Hawaii. Roughly 8 million metric tons of plastics pollute the seas annually. About 60% of shipping *should* be covered by a 1989 treaty requiring no dumping of plastics at sea (Annex V of the International Convention for the Prevention of Pollution from Ships).

The WTO was trying to establish a treaty to cut plastics pollution in oceans in 2023.

Some plastic waste can be collected but only a fraction of the total released (Niaounakis, 2017); one approach is to use vessels or booms that catch and skim-off floating plastic. This only treats a small area, is costly, and microplastics and plastics that have sunk are missed.

Heavy metals

These include debilitating or toxic compounds depending on degree of exposure (italics indicate a serious risk to humans and wildlife): *lead, mercury, cadmium*, occasionally *chromium*. Less commonly: iron, zinc, aluminium, barium, beryllium, bismuth, copper, cobalt and manganese (also other metalloids). *Arsenic* may be considered a heavy metal (Wang et al., 2009; Marfe and Di Stefano, 2020). They pose a threat on land, and in oceans, food and water (Furness and Rainbow, 2018) and management must be based on careful measurements and study. They may be waterborne, or airborne as aerosols or dust.

Lead pollution has taken place locally for millennia; other compounds have become more of a problem and more widespread since the 1900s. Lead pollution of water and food peaked in the nineteenth century, falling after lead pipes and soldered food cans were discontinued. Paints, industrial and domestic, still pose some threat of lead contamination; redecoration may release dust from material applied decades previously. Increasingly paints and finishes are made free of harmful materials. About 90% of the atmospheric lead today probably derives from leaded petrol, long subject to controls. Ice cores from Greenland clearly show the pollution rising after the 1750s and accelerating from 1925. Lead reduces birth weights and children are vulnerable as they accumulate the metal and mental development may be negatively affected, especially if exposed in the early years of life. The poor are likely to suffer greater exposure. Many countries, starting in 1972, insist on the use of non-leaded petrol. Atmospheric lead in the UK has fallen from a peak in 1974.

Air pollution

Nature can cause air pollution such as volcanic ash, aerosols and gases; dust, salts and soda can blow from drylands, algae cast on beaches or lakesides can decompose and emit toxic gases, toxic red-blooms of plankton poison lakes, rivers and sea. Airborne pollen and spores and clouds of insects can also be a serious nuisance. Between the 1860s and 1950s many UK, USA and European cities had sulphur dioxide-rich winter smog (fog contaminated with smoke and/or other contaminants) caused by industry, domestic coal fires and vehicles. That problem has been much reduced (although set back by the popularity of log-burner domestic fires since about 2000 and partly shifted to India and China where goods are increasingly manufactured).

Worldwide there has been an increase in nitrogen dioxide-rich (photochemical) smogs, mainly caused by combustion engine vehicles, where there are still, sunny weather conditions. Traffic or industrial pollution cause smogs at high altitudes because there is less shielding from UV (Mexico City is an example). Other emissions caused by vehicles include partially burnt hydrocarbons including the dangerous volatile organic compounds (VOCs) (which include toluene, ethylene, propylene and benzines), toxic dust, heavy metals and noise. VOCs are formed mainly from diesel exhausts, cause respiratory diseases and may be carcinogenic; they also play a part in tropospheric ozone production and acid deposition.

A problem is tropospheric ozone formed from partially burnt hydrocarbons from vehicle exhaust or power station flue gases and sunlight. The WHO considers 60 ppbv of ozone to be dangerous to humans (the UN suggests 25 ppbv as a safe limit); in 1992 in Mexico City levels sometimes exceeded 398 ppmv. Other cities exceed safe ozone levels in hot, still weather. Tropospheric ozone can depress crop yields by 5% or more. In the UK from 2018 estate agents (realtors) published local pollution levels for houses on sale (Gurjar et al., 2010).

Air pollution from combustion-engine vehicles has been reduced by exhaust catalysers, cleaner fuel (reduced sulphur and lead content), lean-burn and direct-injection engines, capture of evaporated fuel, non-polluting vehicles (driven by electricity, natural gas, hydrogen or fuel cells), restrictions on use of polluting private cars, such as road tolls or high parking charges and other vehicle use restrictions, car sharing and improved public transport (exhaust catalysers do not reduce CO_2 emissions). But numbers of vehicles and distance travelled by car have increased globally and fashion has countered improved technology by encouraging larger SUVs. The UK proposes to ban sales of hydrocarbon vehicles within a decade and hydrocarbon domestic boilers by 2025, prompting the use of air- or ground-source heat-recovery pumps.

Pollution from thermal power stations (coal, natural gas or oil burning) and industry can affect the environment at considerable distances off-site, one impact being acid deposition. In the coming decades China and India face air pollution from coal combustion and have expanding industry and road traffic (Delang, 2016; Gardiner, 2018).

Acid deposition Acid deposition results from contaminated precipitation: snow, rain, mist, fog (wet deposition) or dust particles, aerosols and gases – especially SO_2 (dry deposition). Unpolluted rain is pH 5.6–7.0 (slightly acidic), sometimes 8.5 (slightly alkaline) (lemon juice is about 2.0 – note pH 4.0 is 10,000 times more acid than pH 8.0). Acid deposition is commonly 5.0 but can fall as low as 1.0.

Wet deposition is caused when atmospheric pollutants react with moisture to form droplets of nitric or sulphuric acids. Ecosystems on non-calcareous bedrocks (having little capacity for buffering – e.g. granite) are vulnerable to acid deposition. Deposition may be episodic or ongoing and the impact varies: some localities may be exposed to prevailing winds; others get storms of acid rain and/or receive sudden snowmelt carrying the accumulated deposition of a whole winter which would have caused less damage delivered over several months. Sulphur compounds from coal or peat burning form SO_2 and in moist air sulphuric acid; burning fossil fuels (vehicles, electrical generation or industry) releases several forms of nitrogen and may form nitrogen dioxide in the air which reacts with moisture to form nitric acid (Brimblecombe et al., 2007; Visgilio and Whitelaw, 2007).

Lakes, rivers and forests have suffered from various forms of air pollution, including acid deposition (Smith, 2012; Rothschild, 2019). Seas and oceans are at risk of acidification as CO_2 levels rise and are being impacted by acid deposition. Some soils or waterbodies withstand acidification better than others (and may even become more fertile) by buffering the pollution through alkaline material within or reaching them from underlying basic (alkaline) rocks. In temperate and colder environments soils over slow-weathering, non-alkaline bedrock are more likely to become acidic; in warmer climates already naturally acidic, aluminium-rich soils are vulnerable. Soils treated with ammonium-rich fertiliser may suffer acidification whether or not there is acid deposition. Volcanic eruptions, sea spray, weathering of gypsum and gas emissions from forests, grasslands and marine plankton can lead to natural acid deposition. In the Pennine Uplands (UK), peat moorlands have suffered acidification since the 1750s and this has damaged plants vital for continued peat formation.

During the 1960s damaging acidification of Scandinavian waterbodies was linked to air pollution from Europe and the UK; controls to reduce emissions took decades to adopt. By 1977 Western Europe, parts of North America and several other countries were clearly suffering damage to acid-sensitive plants and animals, and increased maintenance costs for infrastructure. Acidification may cause aquatic systems to suffer mercury methylation (release of harmful levels of mercury from sediment or bedrock) and soils may liberate aluminium compounds toxic to plants. Weakened biota may suffer diseases as a consequence.

Most years volcanoes vent less SO_2 than the UK's power stations did in 1987, but occasional eruptions release vast amounts, which can cool global temperatures for a few to several years. Human SO_2 emissions have more significance in terms of acid deposition than climatic cooling. Probably about 70% of acid deposition globally is due to SO_2 (largely produced by combustion of coal), and about 30% due to nitrogen compounds – nitrogen dioxide (NO_2) and nitric oxide (N_2O). Greenland ice cores show a two- to three-fold increase in sulphate and nitrate deposition during the last *c.* 100 years, mainly attributable to acid deposition.

Acidification is spreading in Russia, China, South America, India and South Africa (ApSimon et al., 2013). Northern polar regions receive acid deposition mainly in the spring (visible as atmospheric 'Arctic haze' and as soot particles in the snow), along with aerosols, dust, pesticides, heavy metals and radioactivity; the sources are EurAsia, Europe and North America. There is concern that this haze will trap solar radiation and warm the Arctic causing more melt. Also, slow-growing tundra lichen and mosses may accumulate pollution and die or grazing animals may suffer heavy doses of pollutants.

It is possible to map areas of vulnerable soil, vegetation and waterbodies, and to superimpose forecasts of future acid deposition. Upland cloud forests which intercept precipitation are vulnerable to acidification, as are epiphytic plants and already acidic tropical rivers. By the time there are obvious signs of acid deposition there will have been serious damage to sensitive ecosystems. How acid deposition damages soil and vegetation can be difficult to unravel and impacts vary, even from plant to plant of the same species. Plants may not be damaged directly: it may be that symbiotic bacteria or fungi and other soil microorganisms are affected and a plant then has less support in its quest for nutrients or resistance to disease and pests. Vulnerability of plants is affected by position, altitude, soil, moisture availability, and various other factors, so it is difficult to forecast. Plants that grow in exposed positions can trap pollution and, as they are in a harsh environment, they are already under stress and vulnerable. Broad-leaved trees appear less susceptible, although in Europe and North America they are increasingly showing dieback. The process can be slow, taking up to 40 years, so is probably underway in many areas without manifesting. By the mid-1980s over half of Germany's coniferous forests were showing signs of dieback, some devastated.

Ozone Ozone (O_3) poses two pollution challenges: in the stratosphere and troposphere. High in the stratosphere the photochemical processes maintaining O_3 are disrupted by pollutants, notably CFCs. The result is a *reduction* in O_3 and its solar radiation shielding (ozone holes and ozone thinning). At lower altitude – a few km (troposphere) – O_3 can *increase* and accumulate through sunlight acting on air pollutants (photochemical reactions involving traffic pollution, domestic fires smoke, emissions from fertilisers, etc). O_3 is a strong greenhouse gas and can be corrosive (Dameris and Fabian, 2014). O_3 changes threaten to impact human health, damage wildlife, markedly reduce crop productivity and degrade construction materials (textiles, paint, rubber, plastic, etc.). O_3 varies temporally and spatially, influenced by environmental factors like winds and sunlight, and may reach harmful levels at considerable distances from the sources of pollutants involved in its genesis.

Stratospheric ozone O_3 depletion in the stratosphere (*c.* 10–50 km) has been addressed with some success (Müller, 2012; Fabian and Demeris, 2014; Abasi and Abasi, 2017; Bower, 2017; Bower and Ward, 2018; Brasseur, 2020). But there are global trends and seasonal, periodic, temporal and regional variations in O_3 levels. Ozone-scavenging compounds already released can remain active for hundreds of years. Some CFC-11 is still produced and illegally used. So controlling emissions is hindered.

Polar ozone holes grow after the long, dark and cold winter when sunlight triggers photochemical reactions.

Tropospheric ozone The problem is not depletion but increased O_3 levels which can be enough to damage natural vegetation, cut crop productivity and degrade construction materials (Tiwari and Agrawal, 2018; Agrawal et al., 2021). O_3 is also the third most potent greenhouse gas. Some areas have O_3 concentrating periodically or seasonally, for example in valleys or basins, in mountains (locations where solar radiation reacts with pollutants trapped by the topography) and where wind and temperature inversions trap pollutants and prevent smog dispersing. O_3 affects even remote rural areas (Cracknell and Varotsos, 2012; Singh, 2022). O_3-related tree dieback is one effect. Natural emissions, wildfires and biomass burning contribute to O_3 production.

Pesticides

Pesticides are compounds used to kill, deter or disable pests for the following purposes:

- to maximise crop or livestock yields;
- to reduce post-harvest losses;
- to improve the appearance of crops/livestock;
- for disease vector control (infested animals or humans), combating locust and other insects;
- for preservation and maintenance of buildings, clothing and boats;
- to control animals which hinder transport and access, slow shipping or block pipes and channels;
- for aesthetic, pet care or leisure reasons.

Some natural pesticides are available (organics), but are not necessarily harmless (pyrethrum or nicotine). Some synthetics are very toxic or carcinogenic. DDT was one of the first synthetic organochlorines, used during the Second World War. There has been a trend to replace organochlorines with 'safer' organophosphate and pyrethroid insecticides and fungicides. Organophosphates can be more toxic than organochlorines but are less persistent (Rathore and Nollet, 2012; Singh, Singh et al., 2022). Some insect control has been achieved by breeding and releasing control species (nematodes, wasps, etc.), sterile males or spreading pest-specific fungi, viruses and bacteria. Baiting with poison, pheromone or sticky traps can also be used.

Pest control commonly reduces field and storage crop losses by 20% or more (Tait and Napompeth, 2018). With restrictions on pesticides, crop losses may rise. Pests flourish if they acquire immunity and/or predators are poisoned by pesticides. Pesticides and fungicides are widely used because consumers demand blemish-free produce. Pesticide use today tends to be clumsy and poorly planned.

Global challenges and opportunities

In 1972 the use of DDT was banned in the USA (but not its manufacture and export). The problems associated with pesticide use are:

● poor selectivity of many pesticides (i.e. not specific in terms of what is killed or injured);
● overuse;
● toxicity and slow breakdown;
● tendency to be concentrated by the food web;
● the effects of long-term use of pesticides on soil fertility is little known;
● pesticides may degrade, be mixed or misused, and marketing may mislead;
● cumulative effects on the global environment are poorly known.

A pesticide ideally kills, disables or deters a specific pest and affects nothing else; many fail at that. There are other impacts: on-farm contamination (of workers, livestock, crops, soil, wildlife, etc.); off-farm drift (to nearby woods, housing, streams, etc.). Tracing impacts and proving liability can be difficult. Much pesticide is used pre-emptively and is probably not necessary. About 50% of all pesticides are applied to wheat, maize, cotton, rice and soya; much of the rest goes on to plantation crops such as cocoa. About half of known pesticide poisonings and at least 80% of fatalities have occurred in developing countries, yet these use only 15% to 20% of the world's pesticides so it is a management issue.

Pesticides are costly to develop and there could be reluctance to withdraw a compound if there is some fault. Manufacturers may resist developing focused pesticides because it restricts sales. Side-effects may only become apparent after extensive use. Pesticide problems can be reduced by:

● banning dangerous compounds;
● developing alternatives such as biological control or integrated pest management;
● restricting trade of pesticide-contaminated produce;
● controlling pesticide use by monitoring, inspection and licensing to ensure sensible procedures;
● developing less dangerous pesticides;
● controlling prices of pesticides to discourage excessive use;
● education to discourage unsound strategies;
● rotation of crops to upset pest breeding and access to food;
● hand or non-chemical weeding;
● encouraging agencies to cut funds for pesticides;
● use of biochar which might bind pesticides in soil reducing contaminated runoff;
● treating drinking water to remove pesticides.

Most countries have established departments for reviewing pesticide use, but there are still problems in disseminating information about pesticides and with their control. The FAO issued an International Code of Conduct on the Distribution and Use of Pesticides and in 1990 the UNEP, WHO, OECD, ILO, the Pesticides Action Network (PAN) and other bodies and NGOs tried to improve pesticide use and controls. Databases and networks are established to assist with monitoring and control. The FAO and WHO set up the Codex Alimentarius Commission (Codex System) to establish food standards; one of its tasks is to check on pesticide residues in produce (and each year to publish information). Under GATT agreements the Codex has increased influence over the way countries set their food and agriculture standards.

Integrated pest management (IPM) could reduce use of pesticides and make pest control more focused. IPM involves study of the best mix of crop and pest control techniques to use.

It must be co-ordinated with conservation, land and water management, social and economic development, etc., and uses pesticides only as a last resort in a judicious manner (Coll and Wajnberg, 2017).

As with chemical pesticides, there is a need for caution over biological controls. History has taught control organisms may become a problem. Genetic engineering may be a double-edged sword. It offers alternatives to pesticides but also threatens problems if dangerous traits were to be passed to other species, or a modified organism runs out of control.

Pesticides are used to control mosquitoes that carry malaria, yellow fever, dengue, etc. There are ways to reduce such pesticide applications: release of sterile mosquitoes, nets, application of kerosene to standing water, stocking waterbodies with mosquito-eating fish, and enforcing laws that discourage pest-breeding sites.

Herbicides, fungicides, etc. In addition to insecticides there are:

- herbicides – weedkillers (poisons or hormones);
- molluscicides – to control snails, mussels, barnacles;
- fungicides – to control mould and fungi; limited development of fungicides to control human disease which means vulnerability;
- growth 'hormones' – to stimulate plant/animal growth, milk yield, crop flowering, ripening, and so on;
- rodent poisons;
- algicides;
- anti-helminths – to destroy worms in livestock and humans;
- anti-fouling compounds – for ship hulls, pipes, etc.

Anti-fouling compounds cause serious damage to wildlife in sea and freshwater aquatic environments. Wartime use of defoliants causes collateral damage for decades. Glyphosphate herbicides were in wider use before carcinogenic notoriety.

As an alternative to herbicides, weeds may be weeded-out, burnt, hoed or discouraged with cover crops, plastic sheet or mulch. There are also opportunities for using weed-specific grazing insects, fish, bacteria, fungi and viruses. Crops may be bred with weed resistance or deterrence. Global change might favour weeds more than crops. There is too little known about pollution and global warming impacts on potentially vital soil microorganisms and fungi.

Antibiotics pollution

Antibiotics are used to treat bacterial infections. There is a risk especially if used as a prophylactic or if courses are not completed that disease organisms survive and pass on genetic information; the antibiotic then faces organisms that are immune (Keen and Monforts, 2012; Hashmi, 2020; Manaia et al., 2020). Human movements and gatherings are also a factor in spreading the problem. Many diseases now resist some or most antibiotics. Little is known about the effect of antibiotic contamination in soil and other ecosystems.

Hormones and similar materials

Hormones have been widely used for decades. There are problems with environmental contamination from contraceptive pills. In many countries wildlife has been affected: fish and reptiles suffer sex change and altered rates of tumours. Some chemicals leached from herbicides, plastics, PCBs and other materials act like hormones impacting health and repro-duction – in some cases skewing the sex of organisms (gender benders), something global warming is also affecting.

Noise and light pollution

Noise pollution has become a nuisance to humans and a threat to wildlife on land and at sea. Whale and dolphin strandings have been linked to noise pollution. Traffic noise may pose a threat to human health and sonic booms from aircraft can disturb wildlife and damage buildings (Kurra, 2021). The noise from wind turbines seems to little disturb humans and wildlife. Very low frequency sound impacts remain poorly researched. Workplace noise reg-ulations have been introduced in most countries.

Light pollution can hinder astronomical research and disrupt wildlife behaviour (song-birds, migrating birds, turtles, etc.). It is also indicative of wasted energy. LED streetlights and restricted operating hours may help.

Solid waste (trash/refuse/garbage/landfill)

A lot of 'less hazardous' waste is sent to landfill, the tips often poorly sited and managed (Cossu and Stegmann, 2018; Pariatamby et al., 2019). Suitable landfill sites may be difficult to find and their selection should be careful (Wong et al., 2012; Ulas, 2016; Whittaker, 2018). Less hazardous waste may develop into something threatening after burial, as things like packaging, inks and so on decompose, mix, react or simply leak to groundwater or streams.

Ideally households and manufacturers should be aiming at zero waste: reducing disposal and recycling. Household waste is usually wet and mixed, making recycling problematic. If materials are economically recoverable they become a resource, saving on mining and much processing. Refuse disposal is often done by urban authorities but can be based on self-help, which cuts the cost and may benefit poor people by offering artisan recovery/recycling oppor-tunities. Street corner skips/dumpsters can be collected and taken to the tip or recycling plant or collection is from households. An alternative is to offer an incentive to people to collect waste and bring it to a recycling plant, composter or incinerator (Mariciano et al., 2021).

Cities generate large quantities of refuse: in developed countries this can be 500 to 800 tonnes per day per million people. This may be landfilled (compressed or as collected), incin-erated, dumped at sea, recycled or composted. Domestic waste in the UK is typically (ap-proximately): 7% plastics; 8% metals; 10% fines (dust); 1% miscellaneous textiles; 20% waste food and other easily decomposed material; and 33% paper products. Poor countries tend to produce waste less rich in plastic and paper packaging; however, the trend is towards a plastics increase. There has been a world decline in the use of reusable glass bottles since the 1960s (and calls for reintroduction in the 2020s). Tin, aluminium and steel can be easily recovered from refuse but plastics are a challenge due to the sheer diversity and the problem of identifying and cleaning them.

Over 90% of UK refuse was landfilled in the 1990s and other nations also rely on it. An adequate layer of clay and a membrane should be put in place before tipping, and after com-pletion used to cap and seal. Many redundant tips and some of those currently in use meet none of those standards and present a hazard. Sometimes domestic refuse is mixed with in-dustrial waste (co-disposal), nuclear waste, power station fly-ash or sewage sludge. There is also a need to minimise damage from escaping methane, heat or fires, and to ensure that vermin are not a problem. Flies and rodents cause difficulties some distance from landfill sites and scavenging birds pose a threat to airports within several miles.

Provided that decomposition of organic matter is vigorous, a refuse tip should generate enough heat to kill most harmful organisms. Chemical contaminants are a different matter. Garbage is often rich in heavy metals and tip archaeology in the USA suggests that paper products are a source of various contaminants due to the printing inks, waxes and sealants. Methane, spontaneous combustion, toxic compounds and subsidence limit the future land use of many tip sites. Gas can sometimes be collected and used for power generation or heating. Management should be an ongoing process for decades after burial.

Pollutants are often disposed of in the sea; in shallow estuaries, enclosed seas and continental shelf shallows, mixing and dispersal may not be enough. Waste sealed into containers and dumped in ocean deeps may leak. Once dumped into the deep ocean, inspection and remedial action is difficult. Waste tyres are a fire and pollution risk and are best treated by

Figure 11.3 A revegetated refuse tip, now a country park (Clyne Valley, Swansea, UK). Before the tip the area had coal mines, an arsenic factory, limestone quarrying, non-ferrous metals production and brick kilns.   Source: Photo by author, 2023.

remoulding and reuse (only possible for less-worn tyres); combustion in district combined heat/power plants and cement kilns; or they might be used for reef building. Tyre 'reefs' reportedly attract fish, become encrusted with marine organisms and do not leak pollutants. Another possibility is to shred waste tyres to make road-surfacing material or construction materials.

Contaminated land has sometimes been built upon; infamous cases include Love Canal (USA) and Lekkerkirk (The Netherlands). The solution is adequate record keeping, careful rehabilitation and land-use restrictions.

Illegal waste disposal (solids, liquids and gases): fly-tipping poses health threats, and damages landscape and wildlife. It is one of the most widespread means of side-stepping the polluter-pays principle. Fly-tipping may be by householders, traders, manufacturers or dishonest contractors. The solution is surveillance and checking fly-tipped waste for clues to its origin, then enforcement of severe penalties. As pollution controls are tightened there is a temptation to export nuisance or hazardous substances to where regulations and public resistance are relaxed. There are two ways of doing this: 1) a factory can be relocated in a developing country; 2) waste or pollutants can be shipped for disposal. If hazards are transferred employees and local people may not appreciate risks, or may be forced by circumstances to accept them in return for employment. Companies may make inadequate declarations about the materials they are using. There is a need for inspection of sites and carriers, so that all involved know what is present (hazard labelling has been developed).

Illegal or careless disposal of waste to streams and groundwater is a problem in rich and poor nations. Tracing the point of disposal can be difficult. Sometimes old sewerage systems get overwhelmed during storms and cause pollution or accidents may breach containment.

Radioactive waste and pollution

Some radioactive emissions are natural: e.g. radon from igneous rocks, uranium ores, etc. For c. 80 years production of uranium has affected miners, enrichment plant workers and the global environment. Atomic weapons pollution, military and civil nuclear power plant accidents, and contamination from industrial and medical isotope sources have also appeared in this period. Between 1945 and 2005 there were about 2,000 nuclear test explosions (c. 130 aboveground in the atmosphere, the rest mainly underground). The 1963 Limited (or Partial) Test Ban Treaty sought to end tests in the atmosphere, under water and in space. The 1967 Nuclear Weapons Test Ban ('Test Ban Treaty') cut aboveground testing although non-signatory nations might have continued. Inspection of nuclear facilities by the International Atomic Energy Commission has been refused or hindered by a number of nations.

Some underground tests fail to contain radioactivity. There are weapons test contaminated areas: in southern USA, in the Russian Arctic around Nova Zemlya, in what was Soviet Central Asia, the Gobi Desert (PRC), near Muroroa Atoll (French Pacific), Montebello Island (Indian Ocean), Maralinga (South Australia) and in Pakistan, India and North Korea. Accidents have led to the loss of several nuclear submarine reactors at sea, and there have been at least 54 re-entries of nuclear isotope-powered satellites, some of which scattered radioactive debris upon the Earth. Many nuclear weapons have been lost at sea and on land. Civil nuclear power stations and military processing facilities have too often suffered major accidents e.g. Kyshtym, USSR (1957); Winscale, UK (1957); Three-Mile Island, USA (1979); Chernobyl, USSR/Ukraine (1986); Fukushima Daiichi, Japan (2011). The world needs electricity and even clean, tested and safe hydroelectricity can kill and pollute on par with the worst of nuclear accidents (Mahaffey, 2014). More radiation (particles) is emitted from coal power stations in fly-ash than from nuclear power stations (excluding accidents).

There is a growing problem of high-level nuclear waste disposal (even with reprocessing of spent fuel). This waste generates heat for decades, can be corrosive and unstable. Without careful long-term management to contain and cool waste an explosion and scattering of dangerous material is likely and warfare raises the risks. Management may need to continue for thousands of years. The waste may be attractive to terrorists so there is a security challenge. Developing repositories which are safe is costly and people seldom want such a site anywhere near them (Sanders and Sanders, 2020). A few countries have started to develop deep geological repositories and some of these function as international facilities (charging for disposal).

In the 2020s the energy/carbon transition will force countries to build fission reactors. Fission could be safer and cheaper. Small modular reactors are less efficient and produce more waste than large reactors but they could be the BPEO to fill an energy gap. Making them modular allows easier repair/updating/decommissioning. SMRs can be built in factories quickly, rather than *in situ*, and buried for extra shielding. It is easier to provide strong containment for smaller reactors. It might be possible to develop cleaner and safer fission reactors – e.g., using thorium fuel/molten sodium cooling – but little progress has been made so far (Martin, 2012; Clynes, 2015; Roy, 2021).

Low-level waste is stored in shallow ponds, discharged into rivers or the sea, pumped down deep boreholes, dumped in containers in deep ocean, landfilled or reprocessed. These options are inadvisable or illegal.

There is still a lot to be learned about safe levels of exposure to gamma, beta and alpha radiation. Concern has been raised about alpha-particle-emitting tritium (released from nuclear installations in the belief it poses little hazard). Many obsolete nuclear power stations, weapons production plants and military reactors require disassembly using robot equipment and material will need to be put in deep repositories. Decommissioning the UK's obsolete nuclear stations may cost £30 billion, to be added to disposal costs of ten nuclear submarines.

In 1989 the world had at least 356 nuclear power-generation reactors operating or under construction, 239 units had been shut down and 100 were being decommissioned. Sweden discontinued atomic power, no easy decision given that 52% of Swedish electricity came from nuclear generation in 1992. Other countries will continue to depend a great deal upon it for decades: in 1989 the former USSR received about 14% of its electricity from nuclear reactors, France 73%, Japan 27%, Belgium 59%, the UK 23%, Germany 28%, Switzerland 40% and Spain 40%. Roughly 1% of the world's electricity is generated by nuclear reactors.

There are dangers in nuclear generation and until there are better alternatives it may be the BPEO to back solar and wind power. However, the need for generation in calm and dull conditions (avoiding C-emissions) means fission reactors are needed in some countries unless workable means to store wind-generated or PV electricity can be developed.

Electromagnetic emissions Electromagnetic radiation (non-ionising) electromagnetic force (EMF) emissions are produced by microwave ovens, radar transmitters, power cables, transformers, radio and TV broadcasting, telecommunications equipment (including mobile phones), computers and high-voltage transmission lines. They can cause difficulties with radio and TV broadcasting, research activities, control systems in cars, aircraft, weapons, etc., and measures are taken to shield against them and to control sources. Epidemiological studies in the USA and by the Swedish National Board for Industrial and Technological Development hint high-voltage power cables might cause cancer; worries about mobile telephone receiver and transmitter emissions are also unproved. So far, there is no proof that

EMF of less than 100,000 hertz is dangerous to humans. However, until proven safe, EMF should be treated seriously. It may prove necessary to shield equipment much more carefully and to keep transmission lines and housing apart.

Electromagnetic emissions also come naturally from the sun and space and are potentially a serious challenge.

Treating pollutants and waste

Some wastes can be easily treated provided they are not dispersed or contaminated; others are a challenge (Tiefenbacher, 2022). Treatments seek to neutralise a material chemically/biologically, bind it to something (e.g. vitrification), or destroy it by heat. To avoid emission of dangerous fumes or dust, incinerators must achieve complete combustion at high temperature under controlled conditions. To treat PCBs effectively requires over 1,200°C for at least 60 seconds. Things can go wrong, so it is wise to site hazardous waste incinerators in remote areas or on ships that can move to a suitable place. However, there has been criticism that shipboard waste incineration may be difficult to oversee and an accident means widespread and possibly untreatable ocean contamination; in EU and North Sea waters a moratorium on their use is in force. America has companies which offer mobile (trailer-mounted) incinerators, which can be taken to where decontamination is needed. In the future particularly dangerous compounds may be treated in incinerators at over 9,000°C using solar power or plasma-centrifugal furnaces. Treatments can be expensive: PCBs incineration or bioremediation (treating with microorganisms) costs US$2,000–9,000 per tonne of soil/waste treated in 1995. At present there is no cheap, effective way to decontaminate fissured rocks or clays that have been deeply infiltrated by materials such as PCBs or dioxins.

Bioremediation has developed rapidly (Bharagava, 2017; Bharagava, 2020). Topsoil can be formed into ridges, treated with bacteria and left for bioremediation or material can be transported for treatment in a bioreactor. Fermentation and oxidation in bioreactors may be sufficient to treat many pollutants. Bacteria and yeasts are being developed to hopefully neutralise hazardous compounds (including chlorinated hydrocarbons and waste oil). For low-level wastes, vermiculture/vermicomposting (use of earthworms to break down waste), composting or fermentation can be suitable, yielding compost and methane (Unnisa and Rav, 2012; Edwards et al., 2021). There is interest in some of the bacteria found deep underground or around deep ocean hydrothermal vents, as they may be effective in converting heavy metal pollution into recoverable sulphates. But escape of something like a plastic destroying microorganism might not be good.

Chemical treatment ranges from simple disinfection (e.g. maceration and chlorination or ozone treatment) to complex detoxification plants that chemically convert hazardous materials. Asbestos, widely used for construction, insulation, fireproofing and in vehicle brake and clutch linings, poses health problems during manufacture and through dust it liberates when disturbed. Blue and white asbestos present the greatest threat; brown asbestos less. Inhalation or ingestion, particularly of white or blue asbestos, causes asbestosis, a chronic, debilitating, often fatal respiratory disease that can manifest itself decades after exposure. The dust can be carried on the wind and workers using the material may contaminate people downwind, and their families and friends through dust on clothes. In developed countries controls have been tightened but in many developing countries they are still inadequate.

The Basle Convention, which came into force in 1993 (amended 1995), is intended to regulate international trade in hazardous waste and especially to ensure that hazard is not exported to developing countries. Unfortunately, it has gaps, and a number of nations did not ratify the Convention.

Recycling and reuse of waste

Some waste can be reused easily; e.g. glass, scrap metals, wood. Others require sorting, washing, chemical, heat or other processing. A country might recover 80% of its waste paper or plastic but recycle little, instead using it for district heating or send it abroad for disposal; another may recover 10% but recycle/reuse most of it. Recycling tends to be 'downcycling', producing lower value material that often cannot be reprocessed another time. Complex goods mean complex recycling which could be part alleviated by modular construction, also allowing replacement parts to extend life and reduce waste.

Sorting for recycling can be done by the state, companies, householders, individual scavengers, scrap dealers or 'pickers', or the waste producers (Doron and Jeffrey, 2018; Landsberger, 2019). Reverse vending has been tried: a waste recovery company or citizens are paid by a manufacturer for return of cans, bottles, etc. (El-Gendy, 2021). Estimates suggest about 5% of Japanese, 15% of European and 10% of US plastics were recycled in the late 1990s. Even if plastics or metals can be identified, pieces may be attached to other materials, or have a coating or contamination that is difficult to remove. Some plastics degrade or absorb chemicals, reducing their value for recycling. Crude sorting is sufficient if the aim is to recover a limited range of materials such as aluminium, glass, low-grade plastics, iron and combustible material for fuel. Recovered material is often bulky for a given weight, making transport and storage costly, and the end product may be of low value.

Recycling may not be as environmentally desirable as it first seems; some studies suggest waste paper processing generates more pollution than burning it for electricity generation and district heating. Ensuring materials are labelled might reduce the cost of recycling. Glass can be recycled indefinitely and each time saves on energy compared with production of new material (in 2022 Europe recycled about 49% of its glass). Reuse of soft drink or milk bottles requires a decentralised network of refurbishing/manufacture and centralised supermarket retailing is unlikely to encourage a return to heavy reusable bottles which cost more to transport. Some reusable bottles get damaged or are not returned raising costs.

If a firm arranges to recycle its products, it may be able to restrict sales of salvaged second-hand parts and thus profit. Steel and aluminium recovery is worthwhile: the latter saves *c.* 95% of the electricity used in making fresh aluminium. Paper can be recycled up to four times before the fibres are damaged too much. The increasing use of disposable nappies (diapers) poses recycling and health hazards, which might be countered by establishing laundry, delivery and collection services, but consumers need to be assured of very high standards of hygiene.

Rural and urban agriculture could use composted refuse and wastewater if these are free of heavy metals and problem materials. Organic wastes like used cooking oil, human urine and faeces, food waste and some plastics can be used as feedstock for biodiesel, alcohols or methane. In 2022 much of the biodiesel in the EU was made from palm oil and in the USA *c.* 33% of the maize crop was used to make ethanol which is blended into gasoline. Both those feedstocks would be better used for food.

Substitution Materials that become waste or pollute might be substituted. Plastic bottles replaced waxed card cartons in the 1980s and there could be a return (there are tested manufacturing methods). Bioplastics using natural feedstock are being developed, as are biodegradable plastics. Less polluting hydrocarbon fuels are available; low-pollution detergents, VOC-free paints, etc. Manufacturing infrastructure, costs and public preferences play key roles; for example, it is possible to return to using glass bottles but these are heavy so many companies would struggle to make a profit if selling and recovering them.

Electrical and electronic equipment waste The amount of electrical and electronic hardware has increased, and the complexity of items challenges recycling. There are profitable opportunities processing circuit boards, phones, etc. (Hester and Harrison, 2009; Prasad and Vithanage, 2019; Hussain, 2021).

Diseases

There are now far better methods to understand and counter disease than in the past. However, people live in denser populations which are less isolated, with more travel, shorter transit times, plus the appearance of bioterrorism. So the risk of disease spread has risen. Environmental change, pollution and land development destabilises existing and potential disease vectors/disease reservoirs and human involvement can make people more vulnerable to illness (King, 2017; Ingegnoli et al., 2022).

Human disease challenges

Since the 1950s there have been huge successes with vaccination (smallpox control, TB, polio, etc.); amazingly there are citizens who question vaccination putting others at risk. Some diseases spread fast and have catastrophic effect; proactive research, monitoring and preparations are vital. Vaccine development is important and did speed up during Covid-19 (but some facilities have since been shut). EM can assist vector control, monitoring, developing responses and identifying opportunities (Sala et al., 2009; Atlas and Maloy, 2014; Maha, 2017; Hurst, 2018; Friis, 2019; Ingegnoli et al., 2022). Diseases mutate and new ones appear – challenges do not diminish.

The plague bacteria *Yersinia pestis* is still present around the world. Plague vaccines and antibiotics are available, but the disease might become antibiotic resistant or authorities might fail to react fast enough to help victims, or it could be deliberately weaponised to hinder treatment (Harper, 2021). Between 1334 and 1353 plague killed 75–200 million around the world. Covid-19 by 2023 has killed roughly 5–17 million. In 2023 health researchers were trying to assess Covid (Hassanien et al., 2021); the impacts include experiments in home working, disease monitoring/control, supply of protective equipment, testing and vaccine development/manufacture. Whether the pandemic frightened governments into better preparations for future disease challenges or lessons were learnt by people and healthcare is debatable (Andrews et al., 2021; Gates, 2022).

After AD 1492 Europeans carried smallpox (*Variola* virus) and influenza to the Americas; perhaps 25–55 million died in outbreaks recurring until recently – more died in the Old World. Some introduction was deliberate to help colonial expansion: some unsettled areas today were once populous. Victims of flu, anthrax, smallpox, etc., have been buried in permafrost and with global warming these and other human and livestock diseases might reappear. Influenza killed 50–100 million worldwide in 1918 and it remains a serious threat. Mosquito control and coordinated vaccination have helped control yellow fever. The movement of migrants pose an ongoing risk. Human history has also been shaped by malaria, probably humankind's greatest killer; it remains a threat, although cheap vaccines have recently been developed (Winegard, 2020).

While the impact of HIV/AIDS has been global Africa has been especially hard hit. Globally it has caused 27–48 million deaths. Advances in treatment are often too expensive or socio-political factors hinder control (King, 2017).

Crop disease challenges

Most crops are threatened by disease (Schumann and D'Arcy, 2012; Chand and Kumar, 2016; Kumar and Droby, 2021). Global reliance on just four cereals is unwise; they are vulnerable to fungal (rust or blast) diseases which spread readily as airborne spores. One of these: wheat stem rust (Ug99 – *Puccinia graminis* f. sp. tritici), presently common in Africa and the Middle East, could potentially hit world wheat production.

Without biodiversity it is more difficult to respond to new disease threats or environmental changes. Maize, potatoes and most cereals have wild/traditional stocks that are lost or under threat. There have been efforts to collect and securely store material: e.g. Kew seed collections and the Svarlbard seed vault (which some brand neo-colonial exploitation). Bananas are especially vulnerable because widely grown Cavendish varieties which replaced disease vulnerable cultivars are descended from rare or extinct wild material, do not set seeds and now easily suffer from fungal *Fusarium* wilt. Breeding new crops to keep up with disease challenges can be difficult. One positive step is the 2019 International Treaty on Plant Genetic Resources for Food and Agriculture.

Crop disease countermeasures include: pesticides/fungicides, quarantine/import inspection/import bans, attractant pheromones/traps, release of insect parasites or predators, protective barriers (mesh greenhouses), insect pest deterrent compounds and support for wild control organisms (Chaube, 2018). The spread of diseases has become easier: where once travel took weeks by sea (effectively quarantine) it can now be done in hours. Potatoes are threatened by fungal, bacterial and viral diseases and by soil nematodes and insects like the beetle (*Leptinotarsa decemlineata*). The loss of potatoes to disease in Ireland (Eire) in the late 1840s caused heavy mortality and mass migration to the USA. European vineyards were decimated in the 1860s by the aphid Phylloxera which originated in North America (possibly spread by faster steam shipping); recovery was slow.

Livestock disease challenges

Many livestock diseases (and human diseases) originate from wildlife and risks can change due to global warming, land clearance and growing contact between livestock/humans/wildlife (Vicente et al., 2021). Diseases may 'jump' species and gene exchanges can alter transmission and virulence. In Asia and Southeast Asia aquaculture, rice farming, pig, duck and chicken raising brings wild species, livestock, humans and mosquitoes into contact, furthering the spread of new diseases or problem strains of existing microorganisms.

Livestock diseases management can be through monitoring, inspection, vaccination, import controls, vector control, drugs, culling wildlife and/or infected livestock, quarantine, etc. The UK has suffered serious foot-and-mouth outbreaks since the 1920s, with those of 1967 and 2001 being severe. An outbreak in 2007 started through a laboratory accident. The 2001 UK outbreak was probably the country's worst, and hit cattle, pigs, sheep and goats. The financial and farmer morale impacts were huge and nearly 4 million livestock were slaughtered and burnt. African (classical) swine fever virus poses an ongoing threat worldwide (there were EU outbreaks in 2019). Wild boar provide a reservoir for swine fever which can infect domestic pigs. Once wild boar were not a problem but they are now common in the EU and in the UK as a result of farm 'liberations', escapes and ill advised rewilding.

The rabbit in Australia, Chile, the EU and UK has been hit anew by myxomatosis caused by poxvirus originating from South, Central and North American mammals. Old World and Australian rabbits were decimated following deliberate virus introductions in the 1950s.

In Australia rabbits have developed immunity and made a recovery but there were reports in Europe in 2022 of re-emergence of a virulent viral strain that could threaten rabbit extinction.

Wild animal biodiversity and disease

Biodiversity is especially threatened when confined to one or a few islands, reserves or forest fragments; disease can mean extinction. In 2023 widespread loss of wild birds occurred in Western Europe, UK, Greece and Israel with outbreaks even on remote islands; H5N1 avian influenza was the cause and it might jump to the poultry/egg production industry. The origin is thought to be chicken production in Asia.

In the UK the introduction of grey squirrels in the nineteenth century has led to extensive forest damage and near extinction of native red squirrels due to a virus carried by the greys. American red signal crayfish have transmitted disease to native UK crayfish, decimating them. The 2020s fashion for rewilding needs to be monitored, given the potential for serious impacts, even through the return of once-native species. Distributions of bats, insects and rodents that may harbour human and crop/livestock diseases might change with global warming and should be studied. Worldwide amphibian populations have been suffering fungal infections (*Chytridomycosis*) since the 1990s. Many frog and toad species are extinct or threatened as a result.

Chemical pollution, heavy metals, radioactivity (including altered solar radiation), climate change and noise can all weaken organisms or alter their behaviour (or gender) and render them vulnerable to disease. Europe has seen a marked decline of pollinator species, like bees, blamed on pesticides, parasites, habitat loss and diseases. Loss of pollinators impacts crop production and biodiversity survival. Other insects in the EU are declining, obvious to any driver who remembers more windscreen strikes in the past. EM can integrate human, animal and environmental health monitoring (Atlas and Maloy, 2014).

Some tree disease challenges

Trees are weakened by pollution, waterlogging, climate change, frost or drought and become more susceptible to bacterial, fungal and viral infections. New disease vectors can appear or existing ones change their behaviour/range. Transport and trade developments alter transmission of diseases/pests. In the UK the incidence of tree diseases and pests seems to be rising (Urquhart et al., 2018). The following are some examples of temperate species currently challenged; around the world there are far more.

Dutch elm disease Dutch elm disease affects *Ulmus* species and is carried by wood-borer/bark beetles which fly considerable distances. It appeared in northwest Europe about 1910 and might have been the cause of elm declines in prehistoric times. Between 1920 and the 1930s 10% to 40% of European and UK elms were lost. Worse was to come, especially for the UK: in the late 1960s a more aggressive species introduced on imported plants had a huge impact (> 90% loss) on one of the country's most common trees (*U. procera*); by the 1970s the landscape was drastically changed. The USA, Canada and New Zealand also had problems but American elm species possess more resistance (US National Parks Service, 2018).

Oak dieback Acute oak dieback has happened periodically in the UK and Europe for centuries but in Germany since roughly 1990 and the UK since roughly 2000 it has become pronounced. France had an outbreak in the 1970s. In the UK mature trees (*Quercus robor* and *Q. petraea* over 50 years old) seem more prone. Changes in environmental conditions and/or insect defoliation seem to be the root cause and affected trees die in around 2 years. The USA has problems with insect attacks on oak species.

Ash dieback disease Samson (2018) explored ash crown dieback (Chalara dieback disease – *Hymenoscyphus fraxineus*), in the UK. It has been suggested it will kill 80% of UK ash (www.woodlandtrust.org.uk/trees-woods-and-wildlife/tree-pests-and-diseases/key-tree-pests-and-diseases/ash-dieback/ – accessed 06/12/23). It is caused by a fungus that spreads by windblown spores. It originated in Asia where it causes little damage to local ash species; however, after arriving in Poland around 1992 it has devastated European ash (*Fraxinus excelsior*). Attempts to manage the disease have poor results and are costly (Anon, 2012).

Conifer diseases Pine tree species, but not other conifers, in the UK have been affected by needle blight fungus (*Dothistroma septosporum*) since about 1990. The problem also affects North America and Europe. North America also has problems with pine wilt disease which is caused by pinewood nematodes carried by beetles commonly called sawyers. Insect pests, especially caterpillars and wood-borer pine bark beetles, have also damaged North American pines since the 1990s.

Summary

- Recognising important global challenges does not mean that they can be reliably forecast.
- Modern peoples are more vulnerable to natural and human-caused disasters.
- Humans cause many of the challenges they now face.
- EM, whether in non-urban or urban environments, must seek effective handling of outputs as well as adequate inputs.
- Identifying a threat is no guarantee people or governments will prepare for it, especially if preparation is costly. Maladaption is possible as Y2K showed.
- Pollution management demands the establishment of ethics as well as effective monitoring, regulations and enforcement.
- Many pollution and waste management issues are transboundary, even global.
- The polluter-pays principle has become much more widely adopted since 1970.
- Much effort and money are being spent on carbon emission controls. There must be better adaptation to warming, and efforts to reduce vulnerability and improve resilience.
- Human population increase is far down on the agenda but should be given more serious consideration.
- Covid-19 might have raised awareness of disease threats.

Further reading

Angus, I. and Butler, S. (2011) *Too Many People?: population, immigration, and the environmental crisis*. Haymarket Books, Chicago, IL, USA.

Cartwright, F.F. and Biddis, M. (2001) *Diseases and History: the influence of disease in studying the great events of history*. Sutton, London, UK.

Chasek, P.S. (ed.) (2005) *The Global Environment in the Twenty-First Century: prospects for international cooperation*. United Nations University Press, Tokyo, JP.

Global challenges and opportunities

Houghton, J. (2004) *Global Warming: the complete briefing* 3rd edn. Cambridge University Press, Cambridge, UK.

Liu, Jingling, Lulu Zhang, Zhijie Liu (2017) *Environmental Pollution Control*. Walter de Gruyter, Berlin, GR.

Oldfield, F. (2005) *Environmental Change: key issues and alternative approaches*. Cambridge University Press, Cambridge, UK.

Spellman, F.R. (2021) *The Science of Environmental Pollution* 4th edn. CRC Press, Boca Raton, FL, USA.

www sources

- Global warming: Intergovernmental Panel on Climate Change: www.ipcc.ch (accessed 06/06/22).
- World population growth: https://ourworldindata.org/world-population-growth (accessed 06/06/22).
- Journal: *Global Change Biology* (Wiley), eISSN: 1365-2486. https://onlinelibrary.wiley.com/journal/13652486?utm_source=google&utm_medium=paidsearch&utm_campaign=R3MR425&utm_content=LifeSciences (accessed 11/10/22).
- Biodiversity loss (2020 statistics): https://earth.org/data_visualization/biodiversity-loss-in-numbers-the-2020-wwf-report/ (accessed 06/06/22).
- Covid-19 impacts: www.undp.org/coronavirus/socio-economic-impact-covid-19 (accessed 06/06/22).

214

Chapter 12

Human and natural causes

Chapter overview

- Challenges caused or exacerbated by human activity
- Urban environments
- Migration
- Genetic modification (GM)
- Warfare and terrorism
- Tourism
- Natural challenges – terrestrial
- Natural challenges – extraterrestrial
- Summary
- Further reading
- www sources

Challenges caused or exacerbated by human activity

Humans cause many challenges. One is population, which has grown from roughly 1 million in 1000 BC to around 8 billion now. Today many individuals are consuming much more than they would have in the past. There are other human challenges: poorly managed technology, consumerism, weapons, pollution, etc. Those too poor to enjoy high consumption also degrade the environment by being unable to adopt ways offering less damaging sustainable livelihoods. Human negligence, greed, fashions, misfortune and aggression degrade nature and threaten civilisation and perhaps survival.

Poverty

Poverty is widely identified as a cause of environmental and social challenges; the linkages are complex. There are countries with low incomes which have progressed with social development and environmental care, and there are rich nations with serious environmental degradation and deprived lower income groups. Before the 1990s poverty reduction had priority over the luxury of EM. Poverty reduction and EM are now more commonly seen to be equally important and interwoven.

Other human challenges

In the 1970s technology was blamed by some for environmental problems; also the Judaeo-Christian worldview which, some argued, led to mal-development. The cure proposed was to reject economic growth and technology; but, with the world so altered and a huge population, return to an imagined 'natural' pre-modern idyll would condemn millions to starvation and not prevent further damaged ecosystems. Some environmentalists currently adopt that stance. But a human-altered globe must in future be steered by sensitive and informed EM. The Anthropocene Epoch has started and is defined by human impacts but hopefully it will not end with human disaster. Technology has been criticised but used wisely it is valuable to EM.

Market forces cause problems and have not been controlled as economists might have hoped. Economic reform will hopefully play a part in resolving challenges and in exploiting opportunities. A significant amount of the negotiations on environmental challenges focuses on how to pay and who pays. After 1945 the establishment of international bodies such as the UN, FAO, UNEP, international courts and laws have helped address global challenges but need to evolve fast.

Citizens, businesses and countries are more interdependent than in the past. An earthquake, volcanic eruption or war could wreak havoc worldwide. International linkages bind trade, livelihoods, politics, cultures and much more. Globalisation is a growing influence on the environment and humans (Kroll and Robbins, 2009; Newell, 2012; Christoff, and Eckersley, 2013; Newall and Timmons Roberts, 2017).

No society has avoided environmental and socio-economic problems, whether it is consumerist, socialist, tribal, etc. Some have been better at managing them. Explanations for problems include:

- *Malthusian and neo-Malthusian* – population exceeds environmental (and possibly social) limits.
- *Poverty and marginalisation* – people are forced to misuse resources.
- *Ignorance* – the implications of an activity are unknown, because knowledge about the structure and function of the environment is poor.
- *Greed* – selfishness, and more recently consumerism, advertising and fashion, prompt exploitation.
- *Warfare* – damage is deliberate or incurred as a side-effect (fires, displacement of people, fallout/nuclear winter, etc.).
- *Common resource ownership* – individuals seek to maximise their use where there is no effective control.
- *Expropriation* – colonial power, occupying force, powerful company or state take control of resources.

- *Dependency* – weak groups are compelled and encouraged to cause damage by more powerful/controlling groups.
- *Faulty ethics* – humans damage nature because religion/worldview encourages it.
- *Poor valuation* – the value of something is not understood.
- *Policing problems* – citizens and businesses are not monitored or controlled.

In negotiations on environmental issues some nations and businesses are willing to allow others to pay to resolve things. While a few may be altruistic, many seek profit and advantage. Most wait to be convinced and forced or encouraged by laws and treaties. More are likely to spend if assured that costs will be reasonably borne by all, and that benefits justify expenditure. In negotiations there is a need for reliable and unbiased data and an informed and honest adjudicator with 'teeth' who is respected by all. UN bodies currently fall far short of this.

Certain countries have or will have a falling and elderly population, others are young and growing, so seeking agreements that impose a per capita payments solution has very different impacts. Some countries are run by elite groups who do not see general welfare or world conditions as priorities. There are countries with a harsh environment and poor resources endowment and others with better conditions – the foundations from which to address global challenges/opportunities are not simple, fair or level.

Urban environments

For most humans the future will be city living. By 2013 cities consumed *c.* 75% of global energy, produced over 50% of global waste and covered *c.* 3% of global land surface (UNEP, 2013). In some developing countries urban areas are very different from those in richer nations, and within each country cities vary, sometimes markedly (Spence et al., 2008; Beall et al., 2010; Birch and Wachter, 2011; Atkinson et al., 2020). There is usually difference between districts of any city. Some cities are planned; many grew up in a haphazard way or reflect obsolete economic, defensive or colonial needs. There are cities that have been established for centuries and some built in the past few decades. As well as general urban growth, mega-cities (> 10 million), far larger than anything in the past, have appeared. Servicing and governing such large settlements is challenging and urbanites out-vote and out-influence their hinterland, attracting development away from elsewhere.

Cities in developed countries tend to have greater population density, more commuters and cars, consume more energy and water and generate more waste. Cities differ in degree of industrialisation, and whether it is dispersed among dwellings or contained in zones separate from or close to housing. Some cities are modern (Rowe and Limin Hee, 2019), and some have scattered, old, low-rise buildings with a great deal of urban and peri-urban wasteland or agriculture. Rapid, and poorly planned, urban growth poses problems for EM (Benna and Garba, 2016). But there are cities which have made EM progress, e.g. Singapore. China and India are rapidly urbanising and provide valuable lessons on urban EM (World Bank, 2014a).

Urban ecology has developed to consider city ecosystems (Douglas and James, 2015; Barbose, 2020; Douglas et al., 2021). One development is the SMART city that uses information and communication technologies to collect data to manage resources, and improve the quality of services and citizen welfare (Townsend, 2013; Houbing Song et al., 2017; Elias Bibri, 2018; Anthopoulos, 2019; Cathelat, 2019; Bobek, 2020; Vinod Kumar, 2020).

But SMART data harvesting could endanger human rights and its application does not necessarily ensure efficiency (Cugurullo, 2021; Minaei, 2022).

The eco-city is a settlement modelled on the self-sustaining resilient structure and function of natural ecosystems (Caprotti, 2015; Tai-Chee Wong and Yuen, 2015; Jingyuan Li and Tongjin Yang, 2016; Mostafavi and Doherty, 2016; Liu et al., 2017; Zhifeng Yang, 2017; Huapu Lu, 2020). China has explored development of eco-cities and the sustainable city-region (Yanarella and Levine, 2020).

Cities modify run-off, making it more erratic, usually polluted and with greater flows. Few cities collect rainwater or reuse wastewater. Cities pollute the air, and cause a 'heat island effect' with a warmer cool season and hotter warm season within and downwind (Douglas et al., 2011, 2021; Zubelzu and Fernández, 2016).

Squatter settlements

Cities may have areas of overcrowding, poor housing, lack of infrastructure, etc. Until recently, squatter settlement caused authorities to evict people, bulldoze shacks and drive settlers elsewhere. Since the 1980s the numbers involved, plus the growth of civil rights and media interest, have reduced such treatment. Growing numbers of authorities and NGOs have started to try to improve water supplies, housing, waste disposal, public transport, etc. Improvements are often hindered by unplanned urban growth and the poverty of settlers. Squatter settlers are frequently a large proportion of total city populations; for example, in Recife (Brazil) they account for over half.

Squatter settlements commonly make use of steep or swampy land rejected by others so are at risk whenever rainfall is intense. EM often has to work with those with poor housing and limited options.

Workshops, small and medium enterprises (SMEs)

In some cities the home is also the workplace, which means exposure of the family and neighbours to noise, fire risk and harmful materials. Scattered small workshops and householders undertaking contract work are difficult to supervise and protect (Figure 12.1).

Urban sprawl

Often the best agricultural land gets built upon. Disturbance and pollution associated with urban growth impacts large areas. Urban and periurban horticulture could help lock up carbon and feed people, and denser, high-rise building could help cut sprawl.

Pollution and waste associated with urban areas

Cities may rely for drinking water on wells or low-pressure pipes; both can become contaminated from leaking drains, chemicals and oil escapes. Managing waste is therefore crucial (Kuhad and Singh, 2013; Rajaram et al., 2016; Muthuraman and Ramaswamy, 2018; El-Din and Saleh, 2019; Marciano et al., 2021; Gille and Lepawsky, 2022). Most cities have traffic pollution, although a few have limited vehicle ownership, or 'leap-frogging' (using advances made elsewhere so avoiding development costs and delays) has led to the construction of improved public transport.

Figure 12.1 Southeast Asian shop-house/workshops. This family make furnishings. Note child's cot, TV, and cans of paints and solvents – a threat to children and fire risk.   Source: Photo by author 2006.

Sustainable cities/urban SD

The concept of cities functioning as sustainable ecosystems is relatively new (Newman and Jennings, 2008; Richter and Weiland, 2012; The Worldwatch Institute, 2016). Sustainable urban development, governance, management and design are more rhetoric than success (Rydin, 2012; Zetter and Watson, 2016; Cohen and Guo Dong, 2021; Metzger and Lindblad, 2021; Cheshmehzangi et al., 2022). Some SD measures are obvious: recycle as much water as possible; plant trees and shrubs; build to reduce need for heating/aircon; counter noise; discourage pollution; use trash/refuse and sewage productively; establish energy-efficient public transport systems; collect rainwater; keep sewage and grey water separate; generate solar power; support urban and periurban agriculture and vertical farming (Zetter and Watson, 2016). UN SDG-11 calls for urban SD (Al-Zu'bi and Radovic, 2019). Japan is one of the most urbanised countries, has made progress with urban SD and city dwelling has left extensive natural areas. Urban strategies must be adaptable to face challenges (Cheshmehzangi and Dawodu, 2019; Hamin Infield et al., 2019). Cities interact with huge areas and urban SD has not been attained if the wider environment is damaged trying to achieve it.

Urban water supply and wastewater disposal

Finding adequate water, conveying it safely to consumers and disposing of waste water are key tasks (Goonetilleke et al., 2014; Srivastav et al., 2022). The private sector has taken over many city water supplies (Mambretti and Brebbia, 2012; Bell, 2018; Bolognesi, 2018; Köster et al., 2019; Watson, 2019). Sometimes this is simply sale from tanker vehicles, but piped water supply systems and sewerage improvements may be paid for by business. Another route is community self-help; typically an NGO or authority provides pipes and other hardware, and citizens labour to install and run things. If storm water (grey water) is kept separate from that polluted with sewage or industry it can be applied to amenity vegetation, like golf courses and landscaping, or be treated more easily for recycling (Eslamian, 2016). However, the usual practice is to allow storm water and sewage to mix so when there is a rainstorm the treatment plants are be overwhelmed and release untreated effluent (with plastics) into rivers. Water supply and sewage disposal are fields where SMART technology is being applied (Di Nardo et al., 2021).

Improved urban waste disposal

Landfill is widely practiced and can be improved (Chandrappa and Das, 2012; Cossu and Stegmann, 2018). Waste sorting is one means and in poorer countries is usually undertaken by the informal sector (UN-HABITAT, 2010). Developed countries with organised refuse collection generally require households to sort materials. Some nations have waste incineration (with or without sorting), providing electricity generation and/or district heating, or they use biogas/composting plants (Kumar, 2016; Meissener and Lindner, 2016; Godfrey, 2021). Refuse and sewage are sometimes contaminated with heavy metals, pharmaceutical, etc., limiting opportunities for use as compost or irrigation (unless biotechnology can resolve this). City sewage worldwide is generally water-carried and often discharged after only rudimentary treatment. Screening storm and sewer discharges at least reduces the amount of plastic reaching the sea and should have higher priority.

Whether water-based or waterless sewage disposal is adopted, sludge residues have to be safely and cheaply disposed of, ideally utilised (heavy metals permitting) (Taşeli, 2020). Forestry and amenity land can use some heavy metals contaminated waste or it may be reduced to ash and used in construction. Urban inputs and outputs (including pollution) should not degrade non-urban areas and ideally be a resource for them – EM should seek such urban–non-urban integration.

Urban and periurban agriculture

Urban and periurban agriculture (the former within, the latter around cities) are rapidly developing in richer and poorer nations. Allotment gardens have a long history and can be highly productive. Havana and other Cuban cities have decades of experience with intensive urban allotment horticulture as has the UK and France (van Veenhuizen, 2006; Philips, 2013; de Zeeuw and Drechsel, 2015; Roggema, 2016; WinklerPrins, 2017; Diehl and Kauer, 2021; Pickard, 2022). There are also unofficial undertakings (guerrilla gardening) (Hardman and Larkham, 2014) using neglected brownfield sites for amenity, crops or species conservation.

In the 1990s, Shanghai (China) was supplied with over 80% of its vegetables from urban and periurban farmers using sewage compost. Around Calcutta sewage is added to fishponds and used to grow water hyacinth (*Pontederia crassipes*) as fodder for livestock.

Pontederia has good potential for sewage treatment and use as feedstock for charcoal, bio-char, biogas and compost (heavy metals and disease or pharmaceuticals contamination of wastes permitting).

Urban and periurban agriculture can be very productive, reduce transport costs and generate livelihoods. Poultry or pig rearing can too, but it poses risks: avian flu, SARS and other diseases (perhaps Covid-19) have originated from interaction between fowl/livestock, mosquitoes and humans.

Urban intensive production Hydroponics, aquaculture, apiculture, and intensive small livestock and poultry production will spread as the technology drops in cost, becomes accessible and reliable and as entrepreneurs realise the potential (Lohrberg et al., 2016; Wiskerke, 2020; Caputo, 2022). In parts of rural South Asia expensive tube-wells and pumps prompted farmers to seek alternatives; they have grouped together to buy and mount motor-pumps on bullock carts (shared use) and substituted bamboo tubes and coir strainers for expensive drilling parts: bamboo tube-wells. There will hopefully be similar cost-cutting and improvement initiatives in future.

Vertical farming and hydroponics Some new developments appear to offer huge potential and EM benefits. One is vertical farming that uses cheap-to-run LED lighting which generates little heat and can be 'colour-tuned' to maximise plant growth. Many racks of plants can be stacked with small 'footprints', indoors, even underground, often in cheap disused facilities, where weather, season, day-length and pests/diseases are not a problem. The approach uses only a little water and fertiliser so is efficient and highly productive. Schedules are more reliable and quality better than field crops, and the market is on the doorstep cutting transport (Despommier, 2010).

Another multiple intensive cropping technique is hydroponics which in simple forms can be relatively cheap. It offers large yields with limited environmental damage and vastly reduces water and fertiliser needs while using relatively little land. A number of initiatives in the 2020s have proved profitable, suggesting a future for these techniques.

By offering intensive alternatives to extensive cropping these approaches may cut land clearance for farming or help return some land to nature. These approaches are also less vulnerable. They provide vegetables and soft fruit; cereals and potatoes may still have to come from extensive fields unless new approaches can be developed.

Urban rehabilitation and improvement

Degraded areas (brownfield sites) include former factories and filling (gasoline) stations. Redeveloping these for housing may mean very expensive decontamination/remediation to protect against lead and other contaminants and those responsible for the pollution may be long gone. Non-housing usage offers a way forward, e.g. solar or wind generation.

Urban transport improvement Road congestion, air pollution and noise are associated with urban transport. One solution is to reduce private car commuting, another to encourage home working, which Covid-19 might have done in some nations. Some cities are poorly served with public transport, having grown up with the car (e.g. west coast USA). Many cities rely on bus and taxi transport, but thanks to leapfrogging (adoption of technology developed elsewhere) some have excellent rapid transit systems (Vasconcellos, 2013; Yedla, 2013). In South and Southeast Asia motorcycles and three-wheeled two-stroke auto-taxis are common (Figure 12.3). These offer cheap, effective transport, consume little fuel and cause

Figure 12.2 Salad crops for markets of Malaysia and Singapore grown under sheet-plastic and mesh, smallholdings with cheap low-tech hydroponics. Less intensive than vertical farming, it still offers low water demands, is much less polluting, less vulnerable and more sustainable than open-air horticulture. It provides multiple crops under relatively controllable conditions.   Source: Photo by author, 2011.

less congestion than full-size vehicles (and should be easy to convert to battery power). Some cities have car entry restrictions or taxation to reduce traffic in inner areas or encourage car sharing. Electric bicycles and scooters are fashionable but unsuited to cold wet weather and conflict with cars and pedestrians. Electric vehicles rely on costly, heavy, fire-prone, inefficient batteries and due to their weight may pose a threat to other vehicles (something given little forethought). Electric bicycles, scooters and three-wheelers may have potential in crowded developing country cities and are spreading in richer nations. Heavy transport is unlikely to adopt batteries (ammonia or compressed hydrogen seems more likely).

Improving urban energy supplies Transition is hopefully underway to renewable/sustainable, non C-emitting electricity, biogas, etc. (Nelson, 2011; Michaelides, 2018). In poorer cities many use fuelwood, charcoal, dung, kerosene or bottled gas, even if electricity is available. Fuel takes a significant proportion of poorer people's incomes. Household photovoltaic (PV) cells provide limited amounts of electricity. These PV systems have been falling in cost and improving in efficiency and robustness, and are now being adopted in poor countries to power a TV and limited lighting, allowing extended hours of working and evening study for children. In richer cities PV arrays generate enough for household use, car charging and to

Figure 12.3 Bicycles and motorcycles outnumber cars in some towns and cities. Source: Photo by author, 2007.

sell to the grid. More could be done to site and encourage PV along routeways, on roofs or on brownfield sites, rather than set up arrays on agricultural land. Household heating and electricity generation units (gas or diesel, some using waste incineration heat) are also used. Cities worldwide could make much more use of building design to trap solar radiation for heating and/or reduce the need for air-conditioning. Biogas adoption has been mainly at household scale in poorer nations. Sunlight can heat water for washing in warm climates using simple heat exchangers on roofs or walls.

A few countries have shallow geothermal sources easily tapped for district heating and even electricity generation or horticulture/aquaculture: e.g. Iceland. New drilling approaches should open up deeper geothermal energy in more countries. If earth- and air-recovery heat-pumps can be improved in output and reduced in cost they may prove useful, but retro-fitting these to existing systems is sometimes problematic. There should be efforts to discourage inefficient air-conditioning units, which are widespread.

Sustainable fuelwood plantations can help with fuel supply; however, plantations are often eucalyptus, typically on common land, which means a range of unwanted environmental and socio-economic impacts. Perhaps better are plantations of other fast-growing plants: willows, reeds or aquatic plants, possibly fertilised with sewage as a treatment system. The biomass can be converted to charcoal briquettes, woodchips or biochar (if the effluent is low in heavy metals).

Urban–rural linkages

Urban wages, usually being higher than rural, can cause rural poverty and rural–urban migration, robbing the countryside of labour and triggering social problems and land degradation. Urban areas usually have key facilities/services and dominate national policies. These factors drive or attract rural folk into cities, stressing urban services. Governments may hold grain prices down for the city poor, depressing rural agricultural profits. Importation of cheap foreign grain to feed city people (there has been a shift to wheat as a staple in some countries) can have a similar effect. Falling national food production leads to more dependence on a few world food surplus areas and increased vulnerability.

Cities and environmental change

Environmental change can disrupt food and water supplies to cities (Bulkeley, 2013; Seto et al., 2016; Rosenzweig et al., 2018). More urban areas will suffer from rising sea levels and some have installed protection bunds and floodgates on rivers; these are not an adequate long-term solution, are costly and give citizens a false sense of security. The alternatives are to discourage vulnerable land use, provide flood and storm refuges and early warning.

Numerous cities depend on rivers fed by distant rains and snowfields and warming may reduce summer flows. Storm drainage and water supply systems are built to tight budgets, and once constructed changes are difficult to make and there is little flexibility for adaptation. Environmental change may have urban health impacts, but forecasting is difficult because many factors are involved. There are predictions warming will cause the spread of malaria; however, in Europe back to the 1450s conditions were cooler than now, yet malaria was common. Land drainage, pollution, quinine and window glass probably had more influence than climate on transmission.

It is unwise to expect that there will be relatively gradual and predictable global warming (Cartwright et al., 2012; Wamsler, 2014). Preparations should focus on reduction of vulnerability and improving adaptability and resilience (Hamin Infield et al., 2019; Fields and Renne, 2021). It is also wise to 'design with nature' and adopt nature-based solutions (Roaf et al., 2009; Kabisch et al., 2017).

Migration

Migration may be relocation within or from region to region and/or rural to urban areas or involve crossing borders (World Bank, 2014b). Refugees are displaced by warfare, unrest, natural disaster or persecution. Relocatees (migrants and refugees), temporary, seasonal or indefinite stay, are displaced by drought, land degradation, socio-economic changes, natural disasters, lack of opportunity, desire for self-advancement or a wish to evade the law. Some are tempted to relocate by media and hearsay. 'Economic migrant' is sometimes applied to those moving by choice. Asylum seeker is a migrant wanting refugee status/sanctuary. Refugees have usually left suddenly and were forced out. The distinctions between refugee and migrant/forced migrant are blurred and very politicised and estimates and predictions of numbers vary wildly and are unreliable (Mayer, 2016; McLeman and Gemenne, 2018; De Haas, 2023). Movement of people can cause unrest (Jones, 2016). Numbers may grow as global warming takes effect, and/or because of land degradation, land grabbing, economic pressures and media (Piguet et al., 2011; McLeman, 2013; Vinke, 2019). Some claim there were 32.6 million relocated worldwide in 2022 but many will soon return and estimates are difficult and unreliable. It is easy to predict future warming will cause more relocation, but a challenge to prove.

There is a large eco-refugee literature (Bardsley and Hugo, 2010; McLeman and Gemenne, 2018; Krieger et al., 2020a, 2020b; Palinkas, 2020; Ginty, 2021). Vince (2022) suggested there might be 3.5 billion 'climate refugees' if global temperatures rise 3°C by 2050. The legal issues involved in forced migration are far from resolved (Behrman and Kent, 2018). A portion of eco-refugees may be from low-lying states flooded by sea level rise and find they are stateless; the world has yet to decide how to deal with this.

One route to reducing relocation is to counter land degradation and lack of opportunity in source nations. Movement is also triggered by the attraction of destinations; the media plays a part, presenting attractive lifestyles, access to welfare, etc. – perhaps inadvertently by rebroadcasting rich nation's soap operas. Counter-propaganda may be one route to discouraging some movement. The affluent and some businesses also migrate, to tax havens or where there are more relaxed laws or they feel secure.

Migrants can play a useful role in some economies; some are highly skilled. Earnings remitted home can aid families and help poor economies. As some host populations age, refuse to do menial jobs or lack skills, migrants may take on the tasks.

Religious migrants (pilgrims) and tourists generally make only short stays.

Genetic modification (GM)

GM is part of biotechnology and there are risks associated with it, necessitating very strict monitoring and control (Bodiguel and Cardwell, 2010; Lynas, 2018). Escape of genetic material is a danger, one of many routes being pollinating insects, or a laboratory accident. GM offers potential solutions to global challenges and promises new opportunities. GM potential is apparent for C-sequestration, pollution treatment, food production, waste treatment, disease response, etc. (Popp et al., 2012; Bury and Grunewald, 2015; Watson and Preedy, 2016; Newton, 2021). But hope of a 'biotechnology revolution' is being met with citizen suspicion, fuelled by media and social media (Regis, 2019).

The terms genetically modified organism (GMO), genetic modification (GM), transgenic and genetic engineering (GE) are used with lack of care. GM can include normally bred organisms involving no lab work (most crops and livestock): a modern chicken is GM, very different from a skinny wild progenitor, but achieved through nothing more than selective breeding. The transfer of genes happens naturally all the time, the mutations and changes more or less random. Modern GM can turn selected genes on and off as required. GE and GMO can enable fast gene changes that would be unlikely or impossible naturally and transfer of material between species. GE and GMO problems include unexpected impacts, intellectual property issues, problems with access to seeds, risk of national or corporate controls or threat of terrorism.

With growing populations, problems sustaining production of food, land degradation, environmental change and a need to cut pollution and fertiliser inputs, GM offers solutions and threats (Ferry and Gatehouse, 2009; Halford, 2012; Kinchy, 2012; Falck-Zepeda et al., 2013; Tharp, 2014; Scott, 2018; Chaurasia et al., 2020: Tzotzos et al., 2020).

Warfare and terrorism

Neither warfare nor terrorism are new but the scale, diversity and ability to disrupt have grown. Technology delivers nuclear, chemical and biological weapons and much improved conventional means of maiming and destroying. Undeclared war (state terrorism) has become an issue with a few nations causing disruption by proxy but denying blame. Terrorism has not yet accessed all destructive advancements but has become a curse that is difficult and expensive to monitor and control. Both war and terrorism result in environmental damage in conflict areas and indirectly (Brauer, 2009; Closmann, 2009; Smith, 2017). The passage of military vehicles may cause serious long-lasting soil compaction and erosion. Some countries restrict agriculture and other land uses decades after conflict due to contamination and unexploded ordinance/mines. Vietnam is still suffering collateral damage from unexploded ordinance and defoliants over 50 years after the war. Munitions manufacture diverts spending from EM and unrest hinders environmental monitoring and conservation.

Nuclear war and global warming may be *the* two major challenges to humans. Nuclear weapon numbers have increased, after reductions and signs of controls between the 1970s and 2000. Defence increasingly uses IT and AI that is vulnerable to disruption or malfunctions. EM may help by warning of potentially misunderstood disruptive solar flares or meteorite strikes. But war can now develop rapidly through error or aggression and there are around 13,000 nuclear warheads.

There has been concern about 'environmental warfare': conspiracy theorists suggest drought or downpours can be engineered, but proof is lacking. Certainly rivers can be diverted and herbicide can destroy vegetation easily. The 2nd Gulf War saw burning oil fields. Crop and livestock and perhaps human pandemic diseases (e.g. cereal blight or foot-and-mouth) are easily spread and outbreaks are deniable.

Opportunities for EM have been generated by improved satellite and UAV-based remote sensing originally for the military. Military activity and discarded munitions can discourage access, enforcing conservation.

Tourism

Green tourism (ecotourism) seeks to reduce environmental impacts, educate citizens on environmental issues and improve the environment; it has the potential to offer sustainable livelihoods and bring together various stakeholders who otherwise would probably not cooperate (Reddy and Wilkes, 2015; Honey and Frenkiel, 2021). Undesirable tourism (mass tourism or overtourism) still degrades environments and host cultures.

There has long been environment/nature-based tourism; i.e. aspects of nature attract visitors: walking, climbing, skiing, bird watching, viewing scenery or wildlife, sport fishing, etc. Twenty years ago, tourism planning would have come from economists; today some planning will be multidisciplinary including EM. There has also been a shift from 'develop-now-cope-with-problems-as-they-appear' towards an environmentally and socially aware approach.

Green/ecotourism

Tourism provided one in fifteen of all jobs worldwide by 2002 and is expanding. Green tourism may be little more than opportunism and hype (Ballantyne and Packer, 2013;

Schweinsberg and Wearing, 2019). Sometimes the only basis for claiming to be green is that the company donates a fraction of its profits to environmental charities and writes that off against tax. The following shows the diversity of green tourism:

- *Eco-active tourism* – those willing to pay quite large sums to *participate* in conservation, environmental monitoring, citizen research or visit particular environments.
- *Hard-core ecotourism* – those who make minimum negative impact/offer benefits to environment and host population. Tourist interest is genuine and tourists learn and have their attitudes affected. These tourists are willing to endure indifferent accommodation and catering to enjoy pristine sites.
- *Dedicated tourists* – limited negative impacts and reasonable benefits for the local economy and some contributions towards environmental care. These tourists do learn something and they are not just present because they are bored by mass tourism or keen to return with tales and photos to impress others. They choose bird watching, diving, golf, fishing, photography, hill walking and climbing rather than general environmental or cultural interests. Standards of accommodation and availability of alternative activities and attractions are important.
- *Marginal* – little benefit to the locality and some negative environmental and socio-economic impacts. These probably learn a little. Possibly they have become bored with mass tourism and are seeking 'an experience' (perhaps to impress others). Comfort is important and attractions have to be enhanced (photographing (and disturbing) penguins, 'swimming with dolphins', mountain and cross-country biking, four-wheel-drive trips, etc.). They are more tolerant of crowding, unlikely to accept non-air-conditioned accommodation or basic food and beverages. Includes many cruise ships.
- *Casual* – on the whole a negative impact. They have little interest beyond an entertainment visit, make a minimal contribution to the local economy and environment and learn little. Many stay within the boundaries of hotels or beach resorts, which often have a poor eco-footprint.

Planners should research the customers, and try to environmentally propagandise them before arrival (Seraphin and Nolan, 2019). It is possible to reduce the socio-economic and environmental impacts of all categories of tourists through appropriate building, energy and water conservation, and land use zoning. Better ski, diving, mountaineering and walking equipment allow access to situations that could only be reached by specialists 30 years ago. Cheap air travel means people can travel to more remote sites. Mountains, islands and isolated coral reefs are increasingly visited. Trampling, wildlife disturbance and oil spills from cruise ships are a concern in regions like Antarctica/sub-Antarctic and the Arctic/sub-Arctic, and some of the highest peaks of the Himalayas are littered with discarded gear, refuse and corpses.

Tourism may be established without sufficiently involving local people, they then feel alienated and exploited and may have lost access to traditional resources (Zeppel, 2006). Tourist behaviour may offend, and frequently employment and supply needs are not met locally.

Tourism can be seasonal and easily damaged by shifts in fashion, bad publicity, disasters and unrest. Sole reliance on tourism is not wise, and authorities should seek a mix of mutually beneficial dovetailed activities: agriculture, fishing, resource extraction, conservation and craftwork. Some countries seek to control tourism impacts by restricting numbers of tourists and charging individuals who are admitted large fees (e.g. Bhutan increased charges

Figure 12.4 Crocodile farm Tonle Sap (Cambodia). A tourist attraction also yielding skins. Source: Photo by author 2007.

from £54 to £165 per day in 2022; in Indonesia the charge for landing to see Komodo dragons rose from £11 to £208 in 2022).

Even before the 1920s tourism was seen as a way of paying for conservation areas (e.g. the Yellowstone Park in the USA), and as a means of educating tourists to better themselves through contact with nature. Green and economic pressures have led to cruise ships and airlines seeking less polluting engines.

The First Asia Ecotourism Conference was held in 1995, and 2002 was the UN International Year of Ecotourism. Ecotourism should be a symbiotic relationship, whereby the environment attracts tourists, and tourists pay a significant amount for EM; hopefully it is sustainable and educates tourists (Wood, 2017; Fennell, 2020). Some ecotourism is like citizen science: paying to work on conservation, land rehabilitation, environmental research, etc. Assessing green tourism performance demands indicators and benchmarks (Sungsoo Pyo, 2012).

One strategy is to zone areas according to their sensitivity to give protection to pristine and vulnerable localities; buffer areas around these help to protect them. Less sensitive outer zones could be used for mass tourism, and intermediate zones could support smaller scale ecotourism and occasional day trips by those based in mass tourism zones. Green tourism can use redundant mansions, old plantation buildings, etc. for accommodation.

Ecotourism can be established with reasonable investment and limited socio-economic change (Honey, 2008).

(a)

(b)

Figure 12.5 Yucatán (Mexico): a) Mayan pyramid – Chichen-Itzá. Beyond the sites there is little tourism impact and some profits can be spent on EM. b) Some hotels in the region are low-rise, adobe-walled and thatched. Traditional construction cuts the need for imported materials, and cement manufacture. The thatch and open-plan designs reduce or eliminate the need for air-conditioning. Some revenue is generated for EM by tourist taxation.   Source: Photo by author 2004.

Sustainable tourism

Sustainable tourism has expanded (Cater et al., 2015; Epler Wood, 2017; Legrand et al., 2017; Raga, 2017; Coghlan, 2019; Edgell, 2020; Sharma, 2020; Slocum et al., 2020; Spenceley, 2021). Fragile environments pose especial challenges (McCool and Bosak, 2016). Golf courses, trekking, angling and bird watching can usually be made sustainable and green.

In the UK, Kew and other botanic gardens have evolved a secondary recreational role that pays for some of the upkeep and research. In Cornwall (UK), the Eden Project was opened in 2001 in a former china clay pit (quarry). It is a development of large enclosed environments filled with exotic plants which provides an attraction for tourists, conserves endangered species, supports research and seeks to educate the public (Smit, 2001). The Project has also created employment and funds for environmental improvement in a relatively poor region.

EM and tourism

Tourism authorities can distribute green informative brochures, offer in-flight videos, establish visitor centres, run publicity on TV, place articles in specialist magazines and newspapers, and websites can be established to attract tourists (Evans, 2015).

Studies of livelihoods–environment interactions were conducted in a mountain valley in Austria in the 1970s; one of the findings was that, as tourism developed, locals invested money and time in hotels so dairy farming was less careful and sought to save time. Because cattle were stall fed and not grazed on pastures as before, the grasses grew and offered a poor snow anchorage so that avalanches were more of a threat and ski pistes held a thinner layer. Reduction of grazing also damaged the display of spring and summer wild flowers: negative impacts upon both environment and tourism. SMART tourism has been explored as a route to SD (Katsoni and Segarra-Oña, 2019). Codes, guidelines and ethical standards have been published by the Ecotourism Society of Australia and in the USA by the (US) Ecotourism Society.

Natural challenges – terrestrial

Natural environmental challenges have always affected humans but they are less able today to move and recover (Leroux, 2005; Emanuel, 2007; Hulme, 2009; Booker, 2010; Bunker, 2018; Cranganu, 2021). Environmental threats can have a long recurrence interval, so there is a temptation to view them as too rare to be bothered with. Such events might have a vast impact on wellbeing and survival, so ignoring them is very unwise. Much of the time change and evolution is gradual and slow, but it is punctuated with sudden intense challenges.

Box 12.1 Challenges – natural terrestrial

- *Climate change* – Periodic and quasi-periodic (ENSO, Gulf Stream, and others), random and unexpected, or major global climate trends that may be recognised as they develop in time to allow useful responses. Human impacts have altered natural patterns, making the future less certain.

- *Weather events* – Storms, tornados, floods, frosts, hail, dust storms, drought.

- *Pests, wildfire* – Much influenced by weather events.

- *Stratospheric ozone damage* – Caused by volcanic eruption or other natural events and human pollution.
- *Major volcanic eruption or fissure out-pouring* – This might cause climate change; also ash deposition over large areas; possibly major changes to the atmosphere, enough to damage plant production and affect organisms. One threat, as yet unproved, is the vern-shot – a huge blast of hot CO_2, other gases and molten rock projected into the upper atmosphere in greater quantities and for longer than normal volcanoes. The cause is a hot plume rising in the Earth's magma, which breaks through the crust. If this were to happen little could be done to miti-gate or avoid the impacts; perhaps a cause of some past mass extinctions.
- *Lesser volcanic eruption* – More fre-quent than the preceding threat; nev-ertheless could cause cooling and a loss of crops globally for a few to sev-eral years if an ash/aerosol cloud is generated. Also, could damage the stratospheric ozone layer with impacts associated with this. Air transport is easily disrupted.

- *Evolution of epidemic diseases* – The threats from influenza and other potential problems are still attracting inadequate attention.
- *Tsunami – normal and mega* – The for-mer can be tracked to offer warnings; the latter are difficult to mitigate.
- *Ocean/atmospheric circulation shifts* – Some are concerned about the threat of a weakened Gulf Stream (perhaps disrupted suddenly by global warm-ing) leading to a colder Western Europe. Other ocean circulation phe-nomena may also be prone to change.
- *Poorly perceived threats* – 1) ocean outgassing of methane or sulphides due to global warming, leading to sudden global warming; 2) major solar flare or cosmic ray bombardment; 3) ocean turnover problems like sud-den upwelling of deoxygenated and nutrient-poor water which damages plankton and other organisms; 4) acid-ification of oceans which damages plankton productivity; 5) alteration of global cloud cover; 6) geomagnetic variation (weakening and then field reversal) during which more radiation penetrates to the Earth's surface.

It is impossible that all challenges or opportunities will be foreseen; so it is important to try to ensure human vulnerability is reduced, adaptability is enhanced and key things are se-curely cached for helping in disaster recovery. Recognising opportunities deserves more at-tention; the focus tends to be on threat recognition.

Natural climate change

There are datasets from ice cores, ocean sediment, tree-rings, etc., which reveal temperature, sea level, floral and faunal changes, volcanic eruptions and geomagnetic shifts over many thousands of years. The message is that climate, sea level, etc., are not naturally steady and change is complex and not understood enough (Flannery, 2007). Global warming looks to be a challenge with an increase of 2°C or more by 2050; 4 or 6 °C would be a disaster (Lynas, 2007). Humans have made change more uncertain and possibly worse (Parkinson, 2010; Cooper and Sheets, 2012). If change is natural, mitigation is difficult.

Past changes have sometimes been sudden. Examples include: the Younger Dryas (*c.* 12,900 to 11,700 BP) shift to cold conditions; the Little Ice Age (roughly AD 1350 to 1850) a 500 year-long cold snap in which thousands of European peasants starved (Fagan, 2019, 2000). Today sudden eruptions or asteroid strikes could hit world harvests for a few to several years.

Recurrent weather events

Climate is the average of weather. Weather events range from days to perhaps a five year duration and may be periodic, random or quasi-periodic with local, regional or global impact. Monsoon seasons vary in strength, duration and extent and various ocean-atmosphere oscillations influence Africa, Asia, the Americas and other parts of the globe. Medium duration quasi-periodic events include the El Niño–El Niña Southern Oscillation (ENSO) and the North Atlantic Oscillation (NAO). El Niño events are a warming of the seas off the Pacific coast of South America normally cooled by the Humboldt Current. The result is a chain of impacts on wildlife and agriculture, storm damage and severe weather events felt as far away as Australasia. El Niña is a cooling from the norm, the other extreme from El Niño, and also causes impacts. These quasi-periodic events recur, early signs are worth watching for and likely progression is known. But with global warming, ENSO and similar events may be getting stronger and less predictable (McPhaden et al., 2002).

The NAO is a less understood ocean–atmosphere interaction than ENSO. A high-pressure cell over the Azores and a low-pressure cell over Iceland (high NAO) 'flip' when westerlies weaken, a low develops over the Azores and a high over Iceland. The pre-flip situation keeps European winters warmer (if stormier), after the flip the (low NAO) warming westerlies decline allowing cool air to penetrate south, chilling Europe. The process can be sudden, catching governments unawares.

Short duration extreme weather events can happen almost anywhere; some claim their severity and frequency are increasing and patterns/extent are changing due to global warming (Lusted, 2018; Fares, 2021). Extreme weather events may be brief but can have severe consequences, as Hurricane Katrina demonstrated in 2005, with c. 1800 deaths and damage of at least US$125 billion in the USA. Severe winds, heavy rain, cold or snowy conditions, sudden frosts (especially in areas seldom affected), hailstorms, heatwaves, duststorms, tornados and floods generally come as a surprise (with at best a few days' warning), even if recurring often. Forecasting drawing on satellite remote sensing has much improved, providing hours or even days of early warning, but citizen response and post-storm relief may be poor. In storm-prone areas building regulations can help ensure less structural damage and people may invest in shelters or have support from civil defence bodies with sheltering or evacuation programmes. Building research seeks to develop more storm/wildfire/earthquake resistant housing. There are developing countries (e.g. Cuba) that have made better recoveries from hurricanes and storms than US states in recent decades, probably due to citizen attitudes.

Some regions are prone to tornados due to relief and wind systems, for example Midwestern USA. Weather radar and broadcast warnings may help; where these events are common it is wise to build storm shelters and ensure buildings have resistance. In forested and grassland areas lightning is a major cause of wildfires and loss of mature trees. Storms can trigger landslides and mudslides.

Dust/sand storms are a problem in many regions and their incidence will probably increase in extent, magnitude and frequency as a consequence of global warming and/or land degradation. The dust often includes fertile topsoil highlighting the loss of farmland and damage to soils. Dust/sand accumulation downwind of source areas can be a problem. Dust travels great distances; Europe has dust falls from the Sahara which can be a nuisance, and a little falls on Latin America and the Atlantic providing nutrients for plants, plankton and ocean food chains. Loess deposits many metres thick over huge areas testify to severe dust transport long before human activity. Dust storm areas include: Midwestern USA, West-Central Asia/China and parts of Africa and the Middle East. The EM response has been to try and hold dust/sand with planted vegetation, trapping devices or sprayed films (with or without seeds for revegetating) and the planting of wind-breaks, shelter belts, biofilms and protection of dry woodland from overgrazing. China, the Sahelian nations, Iran, Iraq and the USA have all invested in dust/sand storm mitigation and avoidance. The role played by dust in cloud, precipitation and climate change needs more research.

Frosts cause huge damage, impacting commodity crops. Florida, USA was hit in 1894–95, 1957–58, 1977, 1985, 1989 and 2022, cutting citrus production. Frosts hit Brazilian coffee production in 2021 and many (especially higher quality Arabica-type) bushes were killed; recovery will be slow and worldwide coffee prices have increased. Severe frosts at higher latitudes can kill flora and fauna which take years to recover. The impact may even be felt in coastal waters resulting in shellfish and fish kills. Frosts are not easy to predict but if forecast there can be hours to mitigate damage.

Ice storms are characterised by freezing rain (glaze events); moisture in the air and precipitation freezes on contact with vegetation, wires, roads, etc. Power transmission fails and roofs may collapse through the weight. Aircraft can be vulnerable and railways may halt if points (switches) freeze up. Ice storms are difficult to predict and mitigate but technology can help with preparation.

Fogs occur when air masses of different temperatures meet and moisture droplets are formed. Other than causing transport chaos, fogs have helped form smogs (smoke plus fog) in industrial areas, causing respiratory illness, deaths and infrastructure and vegetation damage. A problem in the UK when coal usage was common were smogs like that in 1952. Smog is still a problem in a number of countries, especially China and India. In sunny conditions photo-chemical smogs can cause serious damage, including some acid deposition: road vehicles and industry are the cause, and control of emissions reduces the incidence, but many cities are still blighted by it.

Floods Floods can be divided into floods and flash floods (the latter often combined with mudflows/debris flows). Floods often occur where stream flow is hindered. Flash floods happen suddenly often at a distance from the rain, snowmelt, or channel blockage by ice or debris that cause them. Warnings from upper basin sites may allow mitigation. Due to the amount of debris carried, flash floods can be very damaging and leave a lot of material to be cleared. Poor infiltration can be a flood cause as well as downpours or channel blockages.

There have been huge extensive floods in the past as evidenced from geomorphology and folk myths; for example the Channel Scablands of Washington State, USA result from ice melt or drainage of ice-impounded lakes during glacial-warm fluctuations (e.g. 18,000 to 14,000 BP). Tsunamis may also cause flooding.

The incidence and impacts of flooding are affected by human activity and have increased due to: installation of land drains, chanellisation of rivers and streams, compaction of agricultural land, land use changes like conversion of forest to grazing land, changing soil and vegetation to concrete or tarmac watersheds, dumping trash in streams, and change in crops grown in a region. Cheap land is often floodland and modern developers get tempted to build on it, resulting in flood damage; in the past people exercised caution and usually built on high ground. Modern constructions are more vulnerable with electrical wiring, soft furnishings and vulnerable finishes; a flooded traditional building can often be quickly renovated.

Awareness of flood risk does not mean people adapt: inappropriate housing is common and simple affordable solutions are often not adopted (Watson and Adams, 2011).

The Netherlands is a culture that has become more flood-adapted. Insurance companies map flood risk and trends. Crucially important services like nuclear power stations or waste repositories *should* be sited to avoid flooding. Yet the UK is updating one nuclear station quite near sea level where there is a pronounced tidal range, and where a tsunami reached *c.* 7 m in AD 1607 (Hall, 2013). Another facility, Sellafield (UK), includes high-level radioactive waste storage close to the coast and not far above sea level; if the waste is not transported away to deep underground storage, flooding may one day happen and could contaminate a huge area. Fukushima (Japan) showed nuclear stations need secure backup electricity; hopefully others have learnt the lesson.

Drought Humans have long been affected by drought and their vulnerability increased when hunter gatherers turned to agriculture. Civilisations have fallen through drought and it poses a threat now (Montgomery, 2007; Wood and Sheffield, 2012). Drought is often not a weather/climate event; many other factors can be involved. Some droughts last years and become the norm, which is the case in swathes of Australia. Peoples may lose some or all of the drought mitigation strategies they once had; for example: grazing on alternative pastures when times are hard may now be prevented by borders, common land has come under ownership and has been fenced, social unrest and terrorism discourage land use, and available groundwater can run short or overuse of wells causes them to become saline. Irrigation may counter drought threat, but sometimes it compounds the problem.

Drought is not confined to drylands; it can hit wetter environments, even rainforests in Amazonia or Australasia (during low rainfall periods especially after forest disturbance), the result can be permanent vegetation damage especially if there are wildfires/peat fires. In the 1980s the Panama Canal ran short of water (in a humid environment). Even the 'rainy' UK suffers droughts, e.g. in 1976 and 2022, exacerbated by wasteful consumption and infrastructure leaks. Some droughts seem linked to ocean-atmospheric (ENSO-type) events. Drought and land degradation can generate relocatees: Northeast Brazil and the *Maghreb* (a part of North Africa) have seen this.

Much has been written about increasing drought: North Africa once provided grain for the Romans and is now considered 'too dry' to farm. The question is has drought been natural or caused by loss of skills, labour and know-how, colonialism, or poor farming degrading the land (or a mix of some or all of these)? The Negev (Israel) had Nabatean peoples farming until around AD 650, but there is no cropping today; experiments show the micro-catchments used by Nabateans still work in present 'drought' conditions. The US Midwest is a region of recurrent climatic drought that becomes exacerbated by speculative and careless agriculture failing to invest in land care. Drought and land degradation are commonly interrelated. People are attracted or driven into areas prone to recurrent drought during a wet phase and suffer when rains fluctuate. The drivers of drought include population increase, outmigration, lack of land and poverty, climate change, unrest, herders keen to increase stock numbers, land/crop speculators, or governments encouraging cropping to generate foreign exchange (or a combination of causes) (Eslamian and Eslamian, 2017; Cook, 2019; Funk, 2021).

In a number of drought-affected regions human and animal populations have soared and 'drought has followed the plough'. Settlers include small farmers or herders, commercial agriculture or ranching. Research on human–drought linkages has generated schemes to counter the problem (Mapedza et al., 2019). One possible success is the huge Grain-for-Green Program launched by the Chinese Government in 1999 and implemented in 25 provinces by 2002 to mitigate the effects of soil erosion and drought and restore ecosystems by planting trees on former steep farmed areas/failing grazing (Delang and Zhen Yuan, 2015).

Planting tree belts in sub-Saharan Africa since the 1980s to counter the Sahelian 'drought' and land degradation may have had an effect, or perhaps the greening (revegetation) is because moister weather conditions have returned.

Dryland woodlands and grasslands are vulnerable if mismanaged (Botterill and Cockfield, 2013). Proactive governance/management is the best response to drought, drawing on science (Bressers et al., 2016; UNESCO, 2016; Wilhite and Pulwarty, 2017). Brazil, for example, has moved from response to proactive drought management (De Nys et al., 2016). Insurance schemes against drought may also prove an effective measure (Iglesias et al., 2019). Monitoring and recognition of drought and early warning are key measures for EM (Funk and Shukla, 2020; Ondrasek, 2020). Satellite remote sensing and UAVs have greatly improved these tasks.

Wildfires Fire is a natural occurrence in many environments, something that regularly (if not predictably) recurs and to which flora and fauna are adapted. Nature triggers forest and grassland fires with lightning and some birds have been recorded carrying and dropping embers to cause fires and drive prey. Humans cause fires directly or indirectly, intentionally or unintentionally (Pyne, 2021); they also settle fire-prone areas worldwide. Mediterranean environments and dry open forest and grassland have flammable sap ensuring a fire front that quickly passes sparing trees; fire-adapted plants have thick bark and seeds which germinate after fire. Rainforest can burn in dry spells or if disturbed, but are not really fire adapted and suffer badly. Land use changes can result in increased underbrush or grass accumulation and perhaps introduced plants, raising fire risk and making burns hotter. Monitoring dry matter accumulation offers some risk prediction, and fire breaks, fire watching, fire-resistant buildings, pre-emptive burns and brush clearance/control help control outbreaks.

Wildfires in 2018 in California (USA) alone did around US$16.5 billion damage. Australia has suffered recent serious wildfires, as have Portugal, Spain, Italy and even parts of rainy UK. Introduced plant species like eucalyptus, conifers and some grasses may support hotter wildfires. Rather than recognise unwise settlement, inappropriate land management or arsonists, media and citizens blame 'increased wildfires' on global warming. Land use management, building regulation and precautionary controlled burns are preferable to fires whether due to warming or humans (Paton, 2015; Manzello, 2020; Tedim et al., 2020).

Peat covers around 3% of the world's land surface and peat fires occur in tundra, temperate, subtropical and tropical environments; they can be difficult to extinguish, and release CO_2. There are also fires in coal seams (which can ignite naturally through oxidation) and where gases, oil or bitumen escapes from the ground, these too contribute to global warming and can be difficult to extinguish.

Controlled burning (*swaling*) is a traditional method for managing moorland in the UK, and grasslands elsewhere (e.g. US prairies). In the UK old ericaceous brush is fired, and the regrowth provides grazing for livestock and game, supports flora and fauna that would otherwise be shaded out and reduces the risk of major wildfire. The practice has come under scrutiny for perhaps causing peat damage and erosion. Some conservation NGOs oppose the practice and others advocate it. Burning-off vegetation is used by many shifting cultivators (*swidden*/slash-and-burn) to clear and release nutrients; burning is also favoured by ranchers; and those establishing farms. Burns frequently get out of hand and jump to biodiversity conservation areas, and in rainforest and moist subtropical forests that are not fire-adapted the impact is severe; with seeds destroyed and fauna killed or driven out recovery can be almost impossible – much of the biodiverse *Mata Atlantica* (Atlantic Forest) in Brazil has been lost.

Huge areas of Brazil and Indonesia, Australia and Midwestern USA are annually fired by farmers, herders and settlers. In 2007 Singapore and Malaysia had to close schools and airports and life outdoors became unhealthy because of smoke from fires hundreds of kilometres away in Indonesia. Wildfires impact on riverine, lacustrine and inshore environments as sudden bursts of ash and nutrients get flushed by rains causing aquatic impacts. Soils affected by fires can suffer long-term changes to hydrology and soil structure. On islands wildfires may cause havoc to endemic flora and fauna.

Large asteroid strikes (and large volcanic eruptions) can trigger vast wildfires; the K-T Boundary event is marked by charcoal particles globally followed by evidence of a decline in tree pollen and later a pteridophyte spore 'spike' suggesting pioneer fern regrowth. There are fears that a major nuclear war or moderate volcanic eruption would have similar effects: ash clouds and a several years long global 'nuclear winter'.

Landslides, mudslides and avalanches Landslides, mudslides and mudflows threaten dwellings and communications (Sassa et al., 2007, 2013; Davies and Rosser, 2022). In the past people avoided settling risk locations but nowadays such caution may have been lost or economic pressures prompt unwise development. Steep slopes, deeply weathered soils, lack of vegetation to anchor substrate, heavy seasonal rains, earthquakes, frosts, poorly planned road and rail cuttings, careless soil disturbance, etc., are triggers to these massflows. Overgrazing, wildfires, even acid deposition can reduce anchoring vegetation and increase risks. Some choose to build on slopes at risk attracted by position/views. Landslide impacts also can be indirect, damming streams or diverting stream flows that then causes flooding. The Cumbre Vieja Volcano on La Palma, Canary Islands poses a threat: a massive fault-related landslide could cause a mega-tsunami impacting the USA, Europe and Africa. Little can be done to prevent it and monitoring is unlikely to give much warning.

Undersea mudslides down continental shelves often cut telecom cables and generate tsunamis, the mudslides triggered by earthquakes or perhaps smaller tsunamis. The Storegga Event off Norway (*c.* 8150 BC) caused tsunamis leaving debris in Scotland at least 30 m above present sea level (the sea level at the time would have been lower than now so it was large). Tsunamis might cause the release of methane held in marine deposits causing climate impacts. Volcanic activity can cause ash slides into rivers and activity under ice sheets may melt enough to cause flash floods and slurry flows (*lahars*). Not all slides are rapid, some are slow but still challenge land users.

Mapping, remote sensing and terrestrial monitoring systems may offer early warning; buildings, roads and railways can be designed and sited to reduce the impact of massflows (Walker and Shiels, 2013). The UK Aberfan Disaster (1966) occurred when a coal mine spoil tip (sited on ill-drained slopes) slid through a junior school killing 109 pupils and 5 staff. Waste tips and tailings ponds on level or gently sloping sites also pose a threat: an example is the 2019 Brumadinho Dam Disaster, Brazil. A retaining bund (bank) broke (missed by bi-weekly checks) and a mudslide of toxic iron ore waste killed 270; a similar disaster but with less loss of life occurred quite close by in 2015. In 1963 the Vajont Dam Disaster, Italy was caused by a landslide triggered by earthquakes resulting from the weight of the reservoir acting on weak rock strata. The slide caused a tsunami that overtopped the dam, resulting in as many as 2500 deaths and much damage. These examples had problems flagged before they happened and should have had adequate monitoring. There is ample knowledge about slope stability (angle of stable repose for the substrate), slippage risk, need for monitoring, risk mapping, remote sensing and monitoring techniques, revegetation of slopes and regulations that can reduce landslide, mudslide and avalanche risks (see *Landslides*: ISSN 1612-510X; eISSN 1612-5118).

Some countries are extra vigilant for mass movements during ENSO events or in wet or melt seasons. In 2006 the International Programme on Landslides was established by the UN to improve knowledge and work to reduce disasters.

Avalanches are mainly snow and ice mass movements where slopes accumulate snow cover; they move swiftly and must be planned for in affected regions, necessitating snow-sheds, snow anchoring structures risk mapping, tree planting, development zoning rules, monitoring and control teams, pre-emptive triggering with explosives, and strengthened buildings and services.

Earthquakes

Earthquakes directly damage and trigger landslides, avalanches, tsunamis and perhaps volcanic eruptions and outgassing. Earthquakes kill thousands and damage costs huge sums. Most events take place near the edges of continental plates, others around fault lines and a few near large dams; rarer events can take place anywhere. In the active regions there may be monitoring perhaps giving some early warning but elsewhere events are largely unmonitored and unexpected. Risk and vulnerability mapping provides some help. There are hopes electromagnetic and luminous discharges or AI and new algorithms will enable better early warning.

People and developers may not heed risk mapping or learn from past quakes. Afghanistan suffered in 2022 in spite of long experience because people were either too poor and/or unwilling to change traditional construction and continued to live in vulnerable homes that collapsed. Simple building regulations and adjustments to construction methods/materials might help in future but people have to be able to afford it and comply. Enforcement of building standards is important as the quake in Turkey, 2023 has shown. Retro-engineering structures can be difficult. Mexico City has had different impacts in quakes for building from various period: pre-1970s and post-1990s buildings fare best because before the 1970s construction was sturdier, and after 1990 quake-resistant engineering had developed; 1970–1990 saw a building boom with poor construction. Nuclear waste repositories, biodiversity vaults, atomic power stations, dams, etc., should be sited and constructed to minimise quake damage. Civil preparedness is in part a matter of improving public awareness and encouraging changes. Japan has a tradition of light construction which is a good response to quakes, although much is flammable and shocks can cause fires. Prestigious buildings tend to adopt modern earthquake mitigation engineering; humble schools, etc., may not.

Tsunamis

Tsunamis are triggered by earthquakes, volcanic eruptions, marine mudslides, marine outgassing, large landslides or meteoroid strikes. They are common in quake-prone regions and shores facing them. Tsunami threat affects vast areas, including shores of large inland lakes (also vulnerable to *seiches* – standing waves on landlocked waterbodies and land-bordered seas). A tsunami has greater than normal height waves and a long wavelength so they wash in for longer and progress further. On arrival, shelving seafloor, bays and estuaries can magnify impact. The waves do huge damage: flooding and water pressure on arrival and damage as the trash-laden wave drains. There may be several waves and they can trigger mudslides causing further tsunamis.

Japan has suffered over 20,000 deaths since 1946 from relatively moderate tsunamis. Much Japanese development is close to the coast and vulnerable and there has been loss of mangroves and reefs which offered some protection. Earth bank or concrete defences help with small tsunamis, but as Fukushima showed, can be easily overwhelmed. The Fukushima (Tokoku) tsunami was up to 40 m high. Early warning is now possible but gives limited time if tsunamis are generated nearby. Satellites and ocean buoy systems can provide hours of warning when waves travel longer distances.

Tsunamis can be divided into moderate (< 40 m wave onshore) and super or mega- (> 40 m, perhaps more than 300 m). While regular tsunamis are quite frequent mega-tsunamis are less common. Palaeoecology and folk myths suggest huge waves do happen. It is possible to make meaningful responses to regular tsunamis, but mega-tsunamis are beyond management (Singh, 2009; Joseph, 2011). The 2004 Indian Ocean tsunami (51 m high in Sumatra, 3.5 to 10 m along African coasts) caused > 230,000 deaths (Engel et al., 2020), including up to 10,000 km away where there should have been eight hours of warning.

Volcanic eruptions

Over the last century volcanic activity has killed > 82,000 with roughly 17 eruptions accounting for most. The chance of an eruption cooling the climate enough to hit global harvests for several years is roughly 1 in 170. Eruptions can occur where there is no known previous activity but the threat is mainly confined to mapped regions. Active areas do shift over long 'geological' time and there have been periods when volcanicity was more marked; for example in the Miocene era (12~14 million BP). Perhaps seven mass extinction events might have been related to volcanic activity (Loughlin et al., 2015; Papale, 2015).

Worldwide there are roughly 550 known active volcanoes on land and at sea. Others lie fully or partially dormant, some having been recorded as active in historical records. Modern instruments have enabled some prediction by identifying ground temperature change, earth deformation, gas emissions, seismic activity, etc. (Trombley, 2006). But some eruptions develop before response can be organised. Most volcanoes threaten at a local or regional scale, but ash and sulphate aerosols can reach the stratosphere and spread globally, or tsunamis can be generated affecting distant shores.

Ash can halt air transport as the small Icelandic Eyjafjallajokull eruption showed in 2010 (jet engines seem especially vulnerable); and if enough fine particles reach the upper atmosphere global cooling and acid deposition occur. Eruptions could spread abrasive ash capable of killing flora and fauna for hundreds of kilometres, damaging infrastructure and farmland on a continental scale and seriously affecting the globe. In 2022 a moderate submarine eruption near Tonga cut intercontinental communications cables and ash prevented flights to the archipelago for days.

Some volcanoes are explosive: blast damage, pyroclastic ash flows, pyroclastic falls (incandescent lava fragments from the size of a pea to a boulder), tsunamis and hot mudslides. Ash-falls from Mt. Saint Helens, Washington State, USA, in 1980 devastated at least 520 km². Tambora (Indonesia) erupted in 1815 killing perhaps 100,000, cooling global temperatures *c.* 1°C and reducing harvests worldwide for a few years; Krakatoa (1883, Indonesia) hit populated areas, killing over 36,000. Mega-eruption fissure or caldera volcanoes are potentially globally catastrophic and might appear in Yellowstone, the Cascades and in Alaska (USA), the Vesuvius region (Italy) and Indonesia. Vesuvius had an eruption in 3780 BC, much larger than the AD 79 eruption; more eruptions are likely and the City of Naples lies close by.

Certain volcanoes are more likely to explode or emit ash than others; some have slow lava flows and some rapid. Most, if not all, eruptions in human memory are from moderate or small strato (mountain) volcanoes. Prehistoric fissure eruptions have been huge, forming the Deccan Traps of India around 66 million BP: lava and tephra in places over 2 km thick covering over 500,000 km². In the USA the Columbia River Basalts (40–60 million BP) cover over 200,000 km² and average 1 km thickness. These outpourings must have seriously challenged life on Earth and might be due to asteroid strikes or terrestrial causes.

Pyroclastic flows are incandescent (*nuée ardent*) surges of ash and gases which avalanche at 200 kmph or more, can ride upslope, cross several kilometres of sea or lake, and are

(a)

(b)

Figure 12.6 a) Pompeii: street excavated from under about 3 m of ash from the AD 79 Vesuvius eruption. b) Mt. Etna: A house engulfed by lava in the mid-2000s; the family were evacuated from the roof. Later the day this photo was taken an eruption destroyed new buildings a short way upslope.   Source: Photos by author 2006.

perhaps several hundred degrees Centigrade. Pompeii and Herculanium suffered in AD 79 and St. Pierre (Martinique) had 30,000 deaths in 1902. Other than evacuation before a flow there is little that can be done. For gentler eruptions simply warning people to clear ash from roofs would save both lives and costs.

Lahars are mudflows/flash floods associated with volcanoes; ash-covered slopes are mobilised by rainfall, melted ice and snow, flows from crater lakes or rivers dammed by lava.

Where eruptions happen under mountain ice caps or glaciers in more level terrain, glacier outburst are likely, leading to flash floods and mudflows sometimes on a wide front rather than in a channel (in Iceland termed a *jökulhlaup*). The floods and mud can travel to suddenly inundate valleys and lowlands. One that hit the Columbian city of Armero in 1984 killed 23,000. The routes *lahars* take can be predicted and some warning is usually possible.

Outgassing

Some eruptions emit gases, with ash and aerosols that disperse to cause acid damage sometimes hundreds of kilometres from the eruption (and climatic impacts on a global scale are possible). These acidic aerosols, gases like sulphur dioxide, sulphides and CO_2 can drop global temperatures. The Icelandic Laki eruption in AD 1783 sent gas clouds and ash across much of Europe. Not all outgassing is associated with eruptions; toxic emissions may take place in crater lakes, caves and valleys with little warning. Flora, fauna and humans are sometimes killed by these, one example being at Lake Nyos (Cameroon) where gas (possibly CO_2) was released from the crater lake and killed over 1,700 sleeping people and livestock in 1986. Some lakes and large dam reservoirs may accumulate gas from volcanic activity or vegetation decomposition and then release it under certain temperature conditions or when there is an earthquake. Methane may escape from deep below ground and from oil wells, coal mines, landfill and sewage, from the sea floor, frozen tundra soils/peat that thaws; it is also emitted by wetlands, biomass and termites (Reay et al., 2010). Apart from strongly contributing to global warming there is a risk a methane cloud (mixed with air) could spread and ignite. Methane emissions deserve more study (Khalil, 2000).

Methane accumulates in seabed sediments as gas hydrates or in solution at depth (Demirbas, 2010). Such deposits are deeper nearer the Equator; warming or disturbance/ pressure reduction can result in outgassing. Oil rigs in the North Sea and off Alaska have triggered a few outgassings. Earthquakes, sediment slides down continental slopes, or global warming might also cause outgassing. The risk is of accelerating global warming, perhaps causing a runaway.

Ozone changes

As discussed earlier humans have upset Earth's tropospheric and stratospheric ozone (O_3) (Dameris and Fabian, 2014; Abbasi and Abbasi, 2017). Various synthetic compounds have been released and some remain effective O_3 scavengers for centuries (e.g. CFCs) (Muller, 2016). Nature also causes ozone changes; sulphur dioxide emitted from volcanoes, geomagnetic fluctuations, varying solar activity and gamma-ray outbursts from stars may impact on stratospheric O_3. Marine plankton generate ozone-scavenging gas (dimethyl sulphide – DMS): the resulting increase in solar radiation damages plankton DNA reducing their activity, perhaps a sort of 'plankton-ozone thermostat'. Any stratospheric ozone loss is a concern for EM because it could upset CO_2 cycling or hit food chains causing loss of biota. Increased penetration of UV-radiation can damage flora and fauna on land and at sea and cause increased mutations.

Geomagnetic change

Changes in geomagnetism may have terrestrial causes or be prompted by increased solar radiation. The Earth's geomagnetism rises and falls in strength, field pattern and polarity (Glaßmeier et al., 2009; Merrill, 2010). Changes in field pattern result in 'wandering' of the

Magnetic North Pole. Polarity changes (magnetic field reversals) follow weakening of geomagnetic strength and appear to happen roughly every 330,000 years taking around 5000 years to complete, with the Magnetic Pole moving from one hemisphere to the other. Reversals happened throughout geological time and are used for dating and palaeolocation. During reversal protection from incoming radiation may decline to dangerous levels, perhaps for years. Between each reversal there seem to be around 20 fluctuations in geomagnetic intensity. When geomagnetism declines, more solar radiation penetrates to the Earth's surface and may impact flora, fauna, cloud cover and climate. The impact is greater at higher latitudes and altitudes. There is an assumption that there will be a full reversal in around 2000 years. Geomagnetic changes could affect crops, biota, navigation, telecommunications, pipelines, electricity grids and climate. If a fluctuation reducing shielding coincides with a solar storm, the storm's impact will be greater.

Altered ocean–atmosphere–land interactions

Temperature or salinity changes might alter oceanic circulation (Maser, 2015). There are signs that during the Postglacial melting the Gulf Stream weakened, reducing western European temperatures suddenly. The Antarctic Convergence (where cold polar water meets warmer more northern seas) has shifted position in the past impacting sub-Antarctic islands and southern landmasses. In the Northern Hemisphere salmon ascending rivers from the sea provide nutrients that benefit inland flora and fauna; reduced salmon runs may damage land ecology. Human pollutants reach the ocean and have growing impact on organisms and nutrient cycling.

Sea level change may alter depth over continental shelves and flow of currents impacting marine organisms. Mountain building may over very long periods alter rainfall patterns; the rise of the Himalayas and the Andes probably impacted weather patterns and rates of erosion on a huge scale. Altered snow cover, cloudiness and albedo through vegetation change due to introduction naturally or by humans of new plant species, deforestation, extensive building, etc., can change weather. Increased dust blown from land during dry periods, after land degradation or perhaps settling from space, may affect ocean plankton growth, especially if it is rich in iron oxides, which could reduce CO_2 emissions and perhaps cool global climate.

Macrofaunal challenges

Sometimes the causes of problems with larger organisms are natural and sometimes human.

Locust Locust and grasshoppers sometimes swarm; this may be seasonal or follow moist periods, is random or quasi-cyclic. In the Sahel and *Maghreb* (Africa), Middle East and Western Asia into South Asia and China locust swarms were a regular occurrence before the 1970s and can still appear. Pesticide reduced incidence but DDT and other insecticides are now discouraged prompting control by release of sterile male insects or use of pheromone traps. Control depends on willingness of countries/international bodies to spend enough on controls and on access to breeding areas. Recently control has faltered due to poor funding and in some breeding areas political unrest hindering access, resulting in reappearances (Everard, 2019). It is unclear whether global warming will alter rainfall patterns and favour locust breeding. The USA had serious locust problems up to the 1890s when the threat suddenly disappeared (why and whether the species is extinct is unclear); perhaps land use or climate changed but there was no human intervention.

Other insect pests Insects damage crops, infrastructure and act as human and livestock disease vectors. Mosquitoes carry many serious diseases: malaria, yellow fever, dengue, West Nile virus, zika virus, etc. Mosquitoes transmit H5N1 avian influenza virus among poultry and wildlife. In the past malaria and yellow fever had a huge impact on humans, but improved drugs, vaccines, release of sterile mosquitoes and environmental measures have reduced the problem. Dengue has increased, often due to garbage which holds rainwater and allows mosquitoes to breed; public health bodies can discourage this. The threat of crop insect pests is such that many nations monitor imports and travellers.

Ticks transmit human diseases: scrub typhus, Lyme disease, babesiosis, ehrlichiosis, etc. They also transmit livestock diseases. Ticks spread into new areas due to shifts in the populations of their hosts (deer, etc.) and their predators. In parts of the USA woodland regrowth and settlement/recreation have brought deer into closer contact with humans and may be to blame for an increase in Lyme disease. Bedbugs have spread after the Covid-19 pandemic but are not significant transmitters of disease. Reduviid bugs infest poor housing in Latin America and transmit *Trypanosoma cruzi*, the parasite that causes Chagas disease which has serious long-term effects.

Sandflies and blackflies transmit diseases to humans in Asia, Africa and Latin America. Sandflies spread a number of types of leismaniasis. Blackflies carry onchocerciasis, the second most frequent cause of blindness globally, mainly in Africa and to a lesser extent in Latin America. Tsetses are large biting flies found in tropical Africa, vectors of sleeping sickness (trypanosomiasis) in humans and cattle. Some dry woodland and savannah areas are unsettled due to the problem. Screw flies can damage cattle and ruin hides and sometimes infest humans. Termites are spreading to temperate environments and global warming is likely to see further expansion affecting buildings and insurance rates. In the USA species of fire ants are a pest, having been introduced from Central America, damaging property and injuring people. African honey bees escaped from a research facility in Brazil and have spread and interbred with more docile native bees through Latin America and the USA; they are aggressive. In Europe a number of wasp and hornet species are spreading, including species from Southeast Asia. They pose a threat to other insects and anyone who disturbs a nest.

Wood boring insects, mainly beetle species, damage buildings and furnishings worldwide. With global warming distributions of these and other nuisance insects like cockroaches and termites are likely to change. Worldwide bees and moths, important pollinators, have been dwindling.

Rodents Rat and mice are pests worldwide, spreading diseases, destroying food, wildlife, poultry and crops, and damaging infrastructure. Rodents have damaged indigenous wildlife on many islands. Successful eradication programmes to protect seabirds and other wildlife have been successfully mounted but have had to overcome opposition from some environmentalists.

Tree damage resulting from the introduction of the grey squirrel to the UK is endangering some woodlands, and they have caused the decline of native red squirrels. Burrowing rodents (e.g. *Myocastor coypus*) can be a problem, hindering land use and weakening banks and channels. In North America, Belgium, Russia and Chile beavers (*Castor* species) can cause drainage problems and damage forests, something rewilders might note. In several countries native and introduced porcupine species damage trees and infrastructure. However, some rodents play a vital role in dispersing and triggering germination of plant seeds.

Bird pests A number of birds are crop pests, e.g. quelea (*Quelea quelea*) in Africa, sparrows in Asia and starlings in Europe and the USA – also pigeon species. Aggressive seagulls have become a problem in parts of Europe and various species can endanger aircraft and spread diseases. Introduced green parakeets (*Psittacara holochlorus*) plague wildlife and fruit growers in southern UK and are an introduced pest in at least 34 other countries.

Bats Bats can damage fruit crops in the tropics and appear to be hosts for a range of potentially serious human and livestock diseases. However, they are valuable consumers of insects, pollinators and seed dispersers.

Large fauna European wild boar (*Sus scrofa*) damage forests and crops, attack people and cause road accidents in many countries. In parts of North America feral pigs are a similar problem. Monkeys (e.g. macaque species) are a nuisance in some parts of South Asia and Southeast Asia, as are some baboon species in South Africa. Wolves, large cats, venomous snakes, crocodiles, alligators and sharks can be a problem. Florida has growing populations of introduced snake and fish species competing with native plants and animals and posing a threat to humans (especially Burmese and Rock pythons and the snakehead fish).

Invasive species

Plant and animal species disperse naturally or are assisted (intentionally or accidentally) by humans. Some of the invaders (termed alien species by biologists) become a nuisance or a threat; occasionally they improve an ecosystem and help existing flora and fauna. What constitutes a nuisance or threat frequently depends on viewpoint; Australians see rabbits as a nuisance but sheep (both introduced) as an economic benefit; the latter probably cause far more land degradation (Wilcox and Turpin, 2009; Pearce, 2015). Himalayan balsam (*Impatiens glandulifera*) is a nuisance in parts of the Northern Hemisphere, crowding out native species and obstructing streams. Buddleja/buddleia (*Buddleja davidii*) has colonised Europe and sometimes damages buildings. Lantana (*Lantana camara*) is a widespread invasive pest in the tropics and is toxic to livestock, but it supports bees and birds. European hedgehogs (*Erinaceus europaeus*), growing rare in Western Europe, have become problem predators on some UK offshore islands and in New Zealand where they were introduced to conserve them. The list of nuisance flora is long.

Threat invaders Flora and fauna that present a threat to native species or humans are numerous. Some are introductions, thought benign or beneficial, which turn out to be a serious problem. Japanese knotweed (*Reynoutria japonica*) was introduced to the UK for landscaping in the late nineteenth century and now costs huge sums to control and can render real estate unsellable. Guam has problems with snakes that arrived as stowaways and now endanger native species and short-circuit power lines. Problem species around the world include: mosquito species, zebra mussels, Colorado beetles, cats, rats, snakehead fish, fire ants, termites that attack infrastructure, Chinese carp, certain earthworms, African giant snails and lionfish. Cane toads (*Rhinella marina*) introduced to Australia as a biological control for sugar cane infesting rodents have spread across the continent and are seriously endangering many native animal species.

There should be careful trials before any introductions and air and marine transport checks need to be improved to help prevent accidental introductions. One means by which marine species spread is in the ballast tanks of shipping; there should be effective tank sterilisation to prevent this. Hobby owners of exotic animals should be much more strictly regulated.

Useful aliens Some introductions have been beneficial: Ascension Island has forest on once barren mountains established with non-native species which have improved spring flows, and the vegetation now supports formerly struggling local species. Dung beetles have proven a benefit when introduced to some countries.

Rewilding

Rewilding seeks to return an ecosystem to a former state or to re-establish 'lost' species. Efforts range from allowing undisturbed natural dispersal and regrowth, to encouraging tree planting/seeding, re-introduction of extinct species, etc. There may be a need to protect re-wilding areas with exclosures (areas fenced to prevent damage or competition) or do it on islands. More than the welfare of the rewilding species must be considered. Some things present fewer challenges, e.g. rewilding butterflies, or wild flowers.

Sensible rewilding is valuable and aids conservation; however some rewilders refuse to assess and manage impacts or listen to reason (Rotherham and Lambert, 2017). Careless animal importers and owners allow escapes. Commercial farms have had escapes (boar, mink, etc.), or stock were released by eco-terrorists, and in the UK these now cause huge damage.

Re-introductions may need symbiotic organisms, root fungi, pollinators, seed dispersal species, etc., to survive. There have been successes: in Chad the oryx (*Oryx dammah*) was hunted to extinction around 1889; in 2016 survivors from zoos and a herd in Texas were released, now Chad has more than 400 (Monbiot, 2014; Pearce, 2015; Pereira and Navarro, 2015; Pettorelli et al., 2019; Kerr, 2022).

Natural challenges – extraterrestrial

Awareness of extraterrestrial threats has developed largely as a result of space programmes and geological research (Rampino, 2017).

Fluctuations in solar radiation

The Sun 'drives' Earth's climate, something those focused on carbon emissions should bear in mind (Haigh and Cargill, 2015). Solar radiation fluctuates over the longer term and there can be sudden and intense outbursts (coronal mass ejections or solar flares, geomagnetic storms or solar storms) or decreases (weak sun events – e.g. 5480 BC). Understanding of these fluctuations and monitoring is incomplete (Bothmer and Daglis, 2007; Strieber, 2012; Odenwald, 2015; Miyake et al., 2019). Assuming a mass ejection was seen immediately, warnings might enable some mitigation.

Sunspot activity has long been known to fluctuate and records go back to the 1500s. Records suggest there is sunspot periodicity/quasi-periodicity, the most favoured being a 22-year cycle and one of *c.* 11.2 years (the Hale cycle); others are suggested: 5 years, 80–90 years, 170–200 years. Cycles might affect weather, agriculture and wildlife reproduction. When sunspot activity increases stratospheric ozone damage is likely, allowing more radiation to strike the Earth's surface. Low sunspot activity shows some correlation with storms, lightning and cooler conditions. One such 'weak sun' period was the Dalton minimum (AD 1790–1815). The Little Ice Age *might* relate to low sunspot activity or asteroid strikes. Solar variation can suddenly counter or enhance global warming.

The world depends on GPS, telecommunications, undersea cables, data storage, satellites, electronic chips, pipelines and electricity grids, all of which are very vulnerable to solar storms (Solocova et al., 2021). There are too few spares available to rapidly repair damage by an event hindering recovery. There might also be disruption of Earth's radiation shielding allowing enough radiation to reach the surface to damage biota, crops and humans. High latitudes might be more vulnerable. Solar storms have been suggested as a cause of failed past civilisations (Schoch, 2021).

A strong solar storm in 1859, the Carrington Event, disabled telegraph systems globally and discharges injured telegraphers and caused fires. Tree ring and C_{14} evidence suggests much stronger past events, one, the Miyake (or Charlemagne) Event of AD 774–775, roughly 14 times stronger than the Carrington. Another happened in AD 994 believed to be about 60% as strong as the Miyake. Modern electronics are far more vulnerable than nineteenth century telegraph systems. In 1989 a gentle solar storm disrupted satellite use and knocked out the electricity grid in Quebec (Canada) for nine hours. New York and Europe experienced an event in 1921 which cut telegraphy and caused fires. A solar storm in 2012 might have caused serious damage to Europe if it had happened earlier when Earth's orientation would not have provided shielding. Most modern airliners are fly-by-wire (i.e. their controls are electronic rather than mechanical and thus possibly vulnerable to interference) and today's ships crews who seldom use traditional navigation techniques would be lost without GPS.

It is possible to fit circuit–breakers, surge control devices and shielding for vulnerable infrastructure. But protection is poor and patchy. Data storage underground makes sense as does keeping stockpiles of critical equipment like transformers and electronic chips. In 2022 many car manufacturers suspended production due to shortages of electronic chips; a storm would be far more disruptive.

Fluctuations in radiation and dust from beyond the Solar System

Little is known about extra-solar radiation; however, supernovae (various types of star explosions) happen. Astronomers record them; one was observed by the Chinese in 1054 BC, and estimates suggest there are around six per 1,000 years close enough to be a problem. Supernovae cause gamma-ray and neutrino outbursts that would, if within 200 light years, cause serious damage on Earth, possibly mass extinctions. There are claims that supernovae events and/or meteoroid strikes caused disasters 13,000, 16,000, 41,000 and 5 million BP. A supernova event might generate dust/micro-particles and meteoroids as well as radiation and, as these differ in speed, phases of damage might happen millennia apart (but warning could be centuries ahead – if seen). These might subject Earth to radiation, wildfires, atmospheric pollution, global cooling and algae blooms.

Events might leave dust clouds Earth passes through manifest as micrometeorites and reduced solar radiation, perhaps leading to ocean plankton damage. The Solar System might pass through dusty space and/or regions where supernovae are more common every *c.* 26 million years (perhaps causing mass extinctions 65 million and 440 million years BP).

If an event *did* hit around 13,000 BP this *might* explain loss of large Quaternary fauna. Hypotheses abound for this loss, probably too sudden to be only due to humans over-hunting (Macphee, 2018). In mainland UK the once common hedgehog became endangered in the last 30 years, in full view, yet no one is sure of the cause(s); unravelling Quaternary megafauna extinctions is understandably difficult.

Impact of bodies from space

Awareness of extraterrestrial impacts, beyond folk myth and popular writers, is recent. Geology, space research and observed strikes (especially the Shoemaker-Levy asteroid hit on Jupiter in 1993) made clear this is a threat to be taken very seriously. It is also something that probably can be mitigated at affordable cost. Technology is part-developed and most worthwhile, but is unlikely to give 100% coverage due to objects approaching from certain angles. Warning systems could enable most large bodies to be deflected and/or other preparations made. Already there is tracking, surveys and a few asteroids and comets have been sampled (some material returned) to try and understand composition and how to destroy or deflect them. In 2022 NASA's *DART* mission intercepted and 'diverted' a 176 m diameter asteroid. Space telescopes, ground radar stations and citizens linked by the internet may help with early warning. Better knowledge of orbits and where bodies originate should help too (Belton et al., 2004). In 2022 NASA has allocated over US$250 million to research on asteroid detection/deflection.

Small bodies constantly bombard Earth and most burn up; larger bodies are a threat (Nugent, 2017; Trigo-Rodríguez, 2022). Even a 5 m diameter meteoroid passing through the atmosphere and exploding could be misinterpreted by militaries and initiate a nuclear exchange, or might hit a city. A large strike would do fearsome damage: global wildfires, climate effects, tsunami and volcanic activity. Estimates suggest bodies the size of a football strike about once a month; 50 m diameter every 50 to 100 years; 500 m every *c*. 10,000 years; 1 km roughly every 100,000 years; 2 km every 500,000 years; 10 km every 50–100 million years. Assessments put the risk of a 1–10 km meteoroid strike in the next 1,000 years as similar odds to death in a car crash in the UK in 2001.

Myths and geomorphology suggest that around 13,000 BP and 2,300 BP a comet and/or a storm of small iron particles blasted much of North America and possibly other continents (Bobrowsky and Rickman, 2007). Small particle and dust clouds might give little optical or radar warning. It is not only asteroid size that matters; number, character/composition, angle, speed of approach and where there is a hit must be considered. Asteroids are probably solid and rocky, carbonaceous, metallic or ice. Loosely cemented collections of debris probably compose comets so they are probably more tenuous and dusty/gaseous and icy bodies but may have lumps capable of re-entering and exploding in the air or at surface, and anything big enough no matter how unconsolidated approaching at > 30,000 kmph will cause damage. Meteoroids are 'space rocks'; meteors are those entering our atmosphere and burning up as a shooting star; meteorites are those that survive passage through the atmosphere to strike or explode in the air. A bolide is a largish meteor or comet that explodes in the air. Strike location is important, not simply that it misses populated or vital areas – geology also matters. Had the K-T asteroid hit somewhere other than a shallow calcareous seabed, fallout damage may have been far less.

Recent strikes have been enough to devastate significant areas, e.g. Tunguska, Siberia AD 1908 flattened *c*. 2150 km² of almost unpopulated *taiga*. It is thought to have been an airburst of a small asteroid or comet generating about 12 megatons. Chelyabinsk 2013 was a bolide/meteorite or comet, the airburst about 40 km from the city still caused *c*. $US 33 million damage and many injuries (it was a *c*. 19 m body). In the last century or so there have been other quite large meteorite strikes which missed settlements. There are craters in spite of Earth's active weathering, some several km across and more have been obscured by sea, forest or erosion. Legends, texts, tree ring data and palaeoecology suggest comet, meteoroid or supernovae events affected humans at: 41,000 BP (supernova?); Australasia *c*. 34,000 BP (radiation and impacts?). Various authors suggest another disaster occurred at 11,500 BP.

What can be done to mitigate or avoid a strike? Early-warning monitoring and deflection technology is being developed (NAS, 2009; Lunan, 2014; Trigo-Rodríguez et al., 2017; Schmidt, 2019). The legal problems involved in using nuclear weapons as a means of deflection have been discussed (Marboe, 2021). Table 12.1 lists ways to reduce the threat.

Table 12.1 Possible ways to reduce asteroid impact threat

Avoidance
☐ Map and track known objects.
☐ Monitor for objects as far out from Earth as possible (to increase time for response). Some objects will evade detection by approaching at certain angles, because of their shape and/or composition escaping optical or radar systems. Optical detection is probably more effective and with longer range. Task undertaken by international and/or national bodies.
☐ Monitor using citizen science. Objects have often been seen by non-professional astronomers. The internet can be used to employ unused memory on home computers. Military (obsolete) radar might be modified and used to monitor.
☐ Intercept bodies with mass impactors or nuclear warheads (if there is time and laws allow). Technology available but response would need weeks, probably more. Post-2010 space launchers have improved.
☐ Improve understanding of composition of bodies – recent progress, including landings, mass impact tests and return of material. More research needed. The NASA *Deep Impact* probe (launched 2005) impacted the Tempel-1 Comet in 2013 and found it relatively solid.
☐ Develop lasers to move or disintegrate bodies – effective measures years off.
☐ Develop rocket motors/explosive charges to move a body – some years off.
☐ More research on supernova and comet threat – threat under perceived.
Mitigation
☐ Store data, tools, seeds, manuals, etc., to aid recovery in safe and duplicated locations (already done for some data). Store crucial seeds for agriculture and biodiversity conservation in duplicated secure bunkers – a start but much to be done.
☐ Store food reserves in secure and duplicated locations. Difficult with large global population – civil defence provisions less since the 1980s.
☐ Underground bunkers or sturdy buildings – useful against radiation or dust strikes, if there is enough warning and at altitude to avoid tsunamis.
☐ Prepare plans for response – crowd control, shelter and evacuation. Some regions and cities may have these – not very effective for hurricane response in USA in recent decades. In the USA the Federal Emergency Management Agency (FEMA) has made studies.
☐ Design infrastructure to be more resilient and less vulnerable – situation worse than before 1980 due to electronic chip dependence and lightweight building.
☐ Expect more than one problem – a strike might be multiple and each could trigger a tsunami that then causes more by setting off mudslides down continental shelves hours later. Climate change, perhaps for long periods.
☐ In the far future settlement on Mars, Europa, etc. could help avoid extinction (Smith and Davies, 2012).

Summary

- Humans have the capacity to reduce the threat of disasters. Awareness of problems has grown in the last 50 years.
- Population growth gets limited attention. There may be some countries with falling populations but overall global population looks set to grow.
- Twenty-first century humans are poor at controlling war and terrorism which challenge survival.
- Currently the focus is on carbon emissions control; other threats and opportunities deserve more attention.
- Whether in rural or urban environments, effective handling of outputs – waste and pollution – is vital.
- Refuse and sewage disposal are pressing problems.
- Urban environments are diverse. Urban populations have expanded to include more than half of all people. Much of that expansion has been in developing countries, where some cities are now huge.
- Energy production and pandemics have become obvious problems since 2020.
- Plastics pollution is a serious problem.
- Water supply is increasingly difficult.
- Environmental change and land degradation may increase migration.
- Gene modification has developed rapidly since 2000 and offers opportunities for EM but also threats, so must be monitored and regulated effectively.
- Infrequent terrestrial and extraterrestrial threats deserve more attention.

Further reading

Booker, C. (2010) *The Real Global Warming Disaster: is the obsession with 'climate change' turning out to be the most costly scientific blunder in history?* Continuum Publishing, London, UK.

Firestone, R., West, A. and Warwick-Smith, S. (2006) *The Cycle of Cosmic Catastrophes: flood, fire and famine in the history of civilization.* Bear & Co., Rochester, VT, USA.

McLeman, R.A. (2013) *Climate and Human Migration: past experiences, future challenges.* Cambridge University Press, New York, NY, USA.

Parkinson, C.L. (2010) *Coming Climate Crisis? Consider the past, beware the big fix.* Rowman & Littlefield, Lanham, MA, USA.

Pearce, F. (2015) *The New Wild: why invasive species will be nature's salvation.* Beacon Press, Boston, MA, USA.

Smith, G. (Ed) (2017) *The War and Environment Reader.* Just World Books, Chicago, IL, USA.

www sources

- Journal: *Environmental Challenges* (Elsevier). Print ISSN 2667-0100 – www.sciencedirect.com/journal/environmental-challenges (accessed 10/10/22).
- Global Development Research Center (GDRC) virtual library: environmental management – www.gdrc.org/uem/ (accessed 24/08/22).

- The Sustainable Tourism Gateway (GDRC): sustainable tourism – www.gdrc.org/uem/eco-tour/st-about.html (accessed 24/08/22).
- Institute for Global Environmental Strategies (Japan): Asia-Pacific EM – www.iges.or.jp/en (accessed 24/08/22).
- UNHCR Climate Change and Disaster Displacement: eco migrants – www.unhcr.org/uk/climate-change-and-disasters.html (accessed 24/08/22).
- New ITU Global Portal on Environment and SMART Sustainable Cities: UN urban EM – www.itu.int/hub/2020/10/new-itu-global-portal-on-environment-and-smart-sustainable-cities/ (accessed 24/08/22).
- The International Ecotourism Society: ecotourism guidelines/workshops/etc. – https://ecotourism.org/ (accessed 26/08/22).
- CNEOS (Sentry System) NASA asteroid watch – https://cneos.jpl.nasa.gov/sentry/ (accessed 24/08/22).
- NASA/JPL Asteroid Watch: assessment and watch for bodies – www.jpl.nasa.gov/asteroid-watch (accessed 24/08/22).

Part IV

Responses to global challenges and opportunities

Chapter 13

Mitigation, vulnerability, resilience and adaptation

Chapter overview

- Mitigation
- Vulnerability
- Adaptation
- Resilience
- Summary
- Further reading
- www sources

Mitigation

Mitigation means efforts to cure, soften or slow the negative impacts of change. Some challenges can be cured/avoided if caught early. Mitigation usually has to be started when a challenge is known, underway and it only softens or slows; it overlaps with adaptation. Mitigation can be helped by EIA/SIA, forecasting and modelling (Seiegel, 2016; Blokdyk, 2020a).

Mitigation has a history associated with disaster recovery and avoidance. Some soil degradation, subsidence, flood, landslide, earthquake and avalanche mitigation were practised before 1930. In the 1980s drought mitigation got attention; frost mitigation has a long history. The USA (California) and Japan have been developing earthquake, cyclone/hurricane and tsunami mitigation since the 1960s. Business has developed risk mitigation (Melianda, 2009; Jha, 2010; Magdalena, 2021). Climate change mitigation has developed since around 2010. Mitigation involves education, public relations and morale boosting/stress reduction/support as much as physical and governance measures. Sometimes, mitigation can backfire and exacerbate a problem (maladaptation).

Mitigation can be accidental through an unexpected impact of development; e.g. railways across India did more than improve communications, they enabled the transfer of grain, making it easier to respond to drought and famine.

Natural hazards and disasters mitigation

Mitigation measures have been widely applied to natural hazards and disasters to support alleviation, reduce impacts and improve relief and learning to reduce threats in future (Peters, 2016). Mitigation is sometimes easy; e.g., where radon gas seeps from the ground causing health problems a survey and installation of a cheap ventilation system in a building is a virtually 100% solution; the problem is one of problem-awareness, funding and motivation. Some challenges have impact zones so people can be evacuated and mitigation focused. Some problems are not seen coming, the scale is too big and/or there are no countermeasures making mitigation difficult.

Mitigation and recovery has generated a large literature on logistics, management, planning, etc., much based on aid agency experience (Islam and Ryan, 2016). Insurance is a means of mitigating post-problem economic hardship and of aiding recovery (Jerolleman and Kiefer, 2013; Alvarez, 2016). Flood mitigation has a long history (Knight and Shamseldin, 2006; Kabisch et al., 2017; Sen, 2018; Ferreira et al., 2022).

Anthropogenic impacts mitigation

Today many developments undertake EIA/SIA; these are required to forewarn of need for mitigation issues or provide information supporting mitigation (Manyuchi et al., 2021). In the 1970s oil/gas pipelines were being constructed in Alaska and northern Canada, more followed in Russia, the EU, China and elsewhere, and impact mitigation measures were applied (Aloqaily, 2018). Large dams have often adopted EIA and mitigation measures like relocation of the affected. Railway and road developments in China and Tibet have considered mitigation measures. Pollution and waste management mitigation has developed (Aravind et al., 2021; Schäli, 2022). An example of reactive mitigation is Bangladesh where tube-wells were provided to improve rural health (so avoiding shallow contaminated water) but authorities were unaware groundwater had an arsenic risk; mitigation was undertaken after people became ill and the cause was identified, but it was then easy (Akmam, 2017).

Nature-based solutions for mitigation

Before 2000 mitigation was based on engineering; aid; compensation, etc. Today nature-based solutions (N-BSs) are frequently adopted: floodable areas are rewilded so they hold floodwater and reduce the height of further floods, coastal lowlands are returned to nature and people encouraged to move, mangroves are planted to reduce storm damage, etc. (Brears, 2020; Dhyani et al., 2020; Ferreira et al., 2022). Climate change mitigation should, where possible, use a N-BS like farming to sequester carbon in soil (Trumper et al., 2009; Kabisch et al., 2017; Pawłowski et al., 2020). It is claimed N-BSs are resilient (environmentally and socially) and more likely to be sustainable (Dhyani et al., 2020). The UNEP called for N-BSs usage to double 2021–2030 and triple by 2050.

A number of nations have embraced forest planting to mitigate land degradation, erosion and drought. These efforts have had some excellent results but sometimes maladaption has resulted when the wrong species were selected or people displaced (maladaptation is poor or inadequate adaptation. It is a trait that is more harmful than helpful, in contrast with an adaptation, which is more helpful than harmful).

Climate change/environmental change mitigation

Climate/environmental change is happening, avoidance is unlikely and mitigation is wise but likely to have limited effect (Lawson, 2008; Wei-Yin Chen et al., 2012; Wollenberg, Tapio-Bistrom et al., 2012; Surampalli, 2013; Publications Office of the European Union, 2020; Štreimikienė and Mikalauskiene, 2021; Sumi et al., 2021). The Paris Agreement of 2015 requires countries to mitigate to keep global temperature rise this century to less than 2°C above pre-industrial levels, and to pursue efforts to limit the increase to 1.5°C. This looks unlikely and by AD 2050 a global rise of 3°C is possible. A middle of the road forecast is for average surface temperature to increase (from 2013 levels) by 0.3 to 4.8°C by the end of the century (Svensmark and Calder, 2007). Problems like global warming will continue to develop even when mitigation is effective – emissions cannot be rapidly removed from the atmosphere.

Mitigation may buy time to adapt to a challenge and it could help avoid runaway change and reduce damage. There is no shortage of proposals and strategies; mitigation technology and policies are being developed but many have a long way to go. Adaptation and vulnerability reduction are vital; mitigation may help reduce and/or slow the challenges and cut costs of adaptation and vulnerability reduction. Mitigation is presently focused on controlling carbon emissions, but things like methane and ammonia emissions, land degradation, biodiversity loss, etc., need attention as well (Yamaguchi, 2012). Some businesses and countries might be less than committed to mitigation.

Emissions control mitigation of climate change There are a number of ways to mitigate climate change: tax emissions; ban emissions; carbon trading; technology to control emissions; alternatives to emitting practices; enhancing carbon sequestration, etc. All these are developing (and focused overwhelmingly on C).

Whatever measures are taken, efforts should be made to ensure mitigation does not damage economic development (Martens and Chiung Ting Chang, 2010; Latin, 2012; Yamaguchi, 2012).

Low carbon energy Reduction of carbon emissions is an obvious global warming mitigation measure, but the problem might be slowed not cured and even with controls warming will continue for centuries. All energy should be sustainable, secure and not greenhouse gas emitting; technology is advancing and there might be breakthroughs but the focus should be on what is likely to have an effect soon. It is possible to go beyond reduction to negative emissions technology, agriculture and N-BSs to remove C but warming will still take time to slow and stop (Muthu, 2020; Shideler and Hetzel, 2021; Rackley et al., 2023).

Algae could be used to help treat waste, produce fuel, raw materials and capture C; it is possible there will be advances in mimicking photosynthesis to produce fuel from sunlight with negative C-emissions. Already proven are wind turbines, solar PV, hydroelectric and nuclear fission (Speight and Singh, 2014; Magalhães Pires and da Cunha Goncalves, 2019).

Carbon capture and storage and sequestration to mitigate climate change Carbon capture and storage (CCS) is the process of collecting and storing CO_2 before it enters the atmosphere from industry, combustion gases/emissions, etc. Carbon sequestration is the process of storing C in a natural pool/sink and/or enhancing that storage if possible (Sonwani and Saxena, 2022); CO_2 can be captured from the atmosphere through biological, geochemical and technological/physical processes. Many use CCS and sequestration as if synonymous or drawdown is used as a blanket term. CCS/sequestration/drawdown are developing rapidly (Hester and Harrison, 2010b; Smit et al., 2014; Rackley, 2017; Herzog, 2018; Magalhães Pires, 2019; Rahimpour et al., 2020; Goel et al., 2021; Pant et al., 2021; Shideler and Hetzel, 2021). The problems are: to deal with emissions cheaply; do it fast; avoid storage leaks; find uses for captured C not pumped to sequestration or held in pools/ sinks.

Some progress has been achieved in financing CCS (Ussiri and Lal, 2017; Youngseung Yun, 2017; Aresta and Dibenedetto, 2021). Strategies are being developed but there is a need for better understanding of the behaviour of C in ecosystems (Beran, 2013). Sequestration measures might cause environmental problems; for example, afforestation can lead to reduced groundwater, etc., so EM should assess impacts before mitigation. An ideal is to use captured CO_2 as a resource to help pay for capture or even profit from it (captured carbon and use – CCU); e.g. CO_2, water and sunlight might prove a means of making liquid fuel or captured CO_2 can be locked in gypsum or cement when making building materials. Agriculture, forestry, and other land use or marine activities can be managed to try and reduce carbon emissions or even capture carbon (Alongi, 2018; Singh, 2018). Seaweeds (especially kelp species) are good at CCS and may also help reduce ocean acidity and provide food or raw material for plastics or feedstock to make methanol. In 2023 mechanical/technological attempts at CCS were performing at a disappointing 40% of hoped-for efficiency. It should be remembered that efforts to develop CCS, DAC and CCU have only been underway for a few years. Various researchers suggest that to limit global warming to 1.5°C around 10 billion tonnes of CO_2 will need to be removed from the atmosphere each year by 2050.

There are a number of CCS/sequestration approaches and overlapping terms:

- *Direct air capture* (DAC) – compress or filter or absorb carbon from the air (perhaps claim payment or credits). Compression demands an energy input which should not cancel out efforts. At present DAC is not as developed as biological means of CCS/sequestration or as cost effective.
- *Pre-combustion and post-combustion* (CCS) – pre-combustion removes carbon and then burns fuel; post-combustion catches carbon in exhausts/emissions. This is converted to gypsum, cement, used enhance horticulture (pump to greenhouses), as industrial CO_2, dry ice, charcoal/biochar, etc. Then store captured CO_2 by geological storage (pump into salt caverns, disused mines, old oil wells; use to enhance gas/oil recovery). Some promise in these approaches.
- *Sequestration* – enhance natural pools in soil: soil organic carbon (SOC), in vegetation, wetlands, etc. (Halldorsson et al., 2015). 'Blue sequestration' consists of pumping captured gas into the sea to be locked-up by algae or plankton (may be risky). Or scatter compounds (such as iron particles or fertiliser) at sea to stimulate bio-capture by plankton. Convert vegetable matter, refuse or sewage to biochar and bury, hopefully also enhancing agriculture, reducing fertiliser loss and pollution, and reducing soil degradation. Agriculture might be made more SOC friendly; marine liming or fertiliser use may be risky – a runaway sequestration might occur; biochar is very promising. If plant breeders can insert genetic material to generate pyrenoids in crops, grasses and plantation trees safely it may be possible to improve growth rates by roughly 30% locking up a lot of C. This would involve taking genetic material from algae. Promising approaches but some might runaway.
- *Soil carbon storage* – soil improvement, forestry and farming can be managed to enhance carbon storage in natural carbon pools. It is important to have effective understanding,

measurement and monitoring of SOC and an understanding of soil microorganisms. These approaches are promising.

- *Bio-capture* – manage and/or expand natural vegetation, agricultural and forestry carbon pools. Including wetlands, kelp beds, peatlands, tundra, saltmarshes, seagrass beds, kelp, mangroves, marine plankton, algae, etc., ideally, enhance capture carbon and then sequester or use it. These approaches are promising and also as an adaptable food source.
- *Spread ground rock on soil* – may sequester C, but quarrying, grinding, transport and spreading emit C and it may be costly. Unlikely to be practical.
- *Carbon emitters are encouraged to fund compensatory (or better) tree planting or other CCS*. A governance approach with promise.

In those countries retaining or expanding coal-fuelled generation pre- and post-combustion capture should be improved (Stevenson, 2017). Launched in 2008 the (UN) Reducing Emissions from Deforestation and Forest Degradation Programme (or REDD) and the separate REDD+ (voluntary) Programme launched in 2013 seek to cut carbon emissions. REDD+ involves biodiversity conservation, sustainable management of forests and enhancement of forest carbon stocks, ideally improving livelihoods (Lyster et al., 2013).

Soil CCS/sequestration: agriculture and other land uses As knowledge of SOC behaviour improves land use, practices might be adopted to hold carbon in the soil rather than emit it and perhaps increase sequestration (Singh, 2018; Lal and Stewart, 2019). CO_2 is only one of the gases released from soils by agriculture and degradation to cause warming; others, like ammonia and methane, need to be monitored and controlled. Emissions result from use of fertilisers, tillage, irrigation and/or livestock rearing (Reay et al., 2010; Liebig et al., 2012; Gerber et al., 2013). CCS can be achieved by tree, grass or bamboo planting but there should be monitoring to avoid side-effects of growing, harvesting and burial (Kumar and Ramachandran Nair, 2011; Nath et al., 2020). By reducing need for fertilisers and improving agricultural sustainability and productivity, biochar may offer more benefits than C-sequestration and is safe.

Agriculture, commercial, small farmers, paddy cultivation, etc., can probably be managed to improve CCS (Liebig et al., 2012; Wollenberg Tapio-Bistrom, et al., 2012; Lorenz and Lal, 2018; Bhattacharyya et al., 2020; Pawłowski et al., 2020; Datta and Meena, 2021). Wetland management can be used for mitigation. One possibility is to link rural development and food security improvements with carbon storage by supporting farmers to undertake appropriate agricultural mitigation practices (Lipper, 2011; Oelbermann, 2014; Toensmeier, 2016).

Biochar Biochar (organic matter pyrolysed in an oxygen-free environment at less than 600°C) can be added to soil hopefully to improve and sustain agriculture and perform CCS. Or biochar/charcoal can be simply buried as CCS. Biochar in soil may reduce need for fertiliser, improve moisture retention, help control heavy metal pollution and cut agrochemical runoff. Studies and widespread usage since 2000 suggest agricultural benefits are real (Scholz et al., 2014; Ralebitso-Senior and Orr, 2016; Shakley et al., 2016; De la Rosa, 2020; Singh and Singh, 2020; Tagliaferro et al., 2020; Kapoor et al., 2021; Kapoor and Shah, 2022). But there is a need for further research to check benefits and if confirmed promote best ways to produce and use biochar (Bates, 2010; Taylor, 2010; Ladygina and Rineau, 2013; Ralebitso-Senior and Orr, 2016; Bates and Draper, 2019; Sarmah and Barceló, 2021; Tsang and Yong Sik Ok, 2022). The approach seems very promising and safe.

A risk is that poor biochar production and quality control and suboptimal application will deter usage. Biochar pyrolysis is sometimes allowed to produce char, lacking many beneficial effects – standards and quality controls are needed. Pyrolysis should probably be at *c.* 550–600°C (Yong Sik Ok et al., 2016, 2018; Manya and Gasco, 2021). Archaeology in Amazonia (exploring *terra preta* – black earth soil sites) and elsewhere where 'slash-and-char' were once used prompted research into biochar for sustainable agriculture (including difficult environments) (Taylor, 2010; Piccolo, 2012; Mingxin Guo et al., 2016). An example of the past suggesting ways forward (Lehmann and Joseph, 2015).

Bruges (2010) warned incorporation of biochar in carbon trading could have negative consequences. Use of rural/farm wastes as feedstock might result in less return of crop residue to the soil and/or reduced availability of construction materials (Bruckman et al., 2016). Potential feedstocks are: refuse, sewage sludge, aquatic plants or algae used to help decontaminate sewage or remove excess nutrients from waterbodies and streams (provided they are not too heavy-metals contaminated).

Earth cooling/geoengineering mitigation of climate change Geoengineering could be used for climate change mitigation; some of the techniques are safe; e.g. capture of CO_2 directly from the air (DAC) and enhancing weathering to lock up carbon. One safe form of geoengineering is human contraception; too large a population is behind most EM problems. Some geoengineering is quick and dirty, the risk is an approach may be difficult to control, perhaps dangerous and/or difficult to reverse, e.g. spreading reflective dust in orbit (MacCracken, 2012; Wei-Yin Chen et al., 2012; Lenton and Vaughn, 2013; Harrison and Hester, 2014; Morton, 2015; National Research Council of the National Academies, 2015; Santos, 2019).

Geoengineering mitigation techniques include:

- *Reduction of incoming solar radiation by spreading reflective or filtering aerosols or particles in the atmosphere or space.* Technology and costs probably viable but risks considerable. Cannot be undone rapidly, if ever, and a volcanic eruption could suddenly disastrously increase cooling. Possible legal issues.
- *Increase reflective cloud cover.* Technology not far off, costs probably manageable. One route is for ships to spray moisture into the atmosphere. Reversibility and impact on farming and biodiversity unknown.
- *Solar shields in orbit.* Technology some way off. Reversibility? Might not be reversible, legal issues and perhaps costly.
- *Biochar/char/black carbon burial.* Developed and being applied. Risks and costs seem acceptable/manageable.
- *Seed oceans with suitable material(s) to stimulate plankton* which capture and sink with carbon. Risks not established; some small-scale experiments conducted. Legal issues?
- *Cut CO_2, methane and ammonia emissions.* Technology not much of a challenge. Risks low.

Techniques need development and some may not be safe or cost-effective or fast enough. If geoengineering is used it will need strict management and legal measures (Morton, 2015; Gerraester and Hester, 2018; Letcher, 2019). Human activity has upset natural climate change and now requires EM to resolve things (Keith, 2013; Baskin, 2019). There are important EM questions to be resolved: How long have we got before there is a need for risky drastic geoengineering? What are the environmental and socio-economic risks of approaches? Who polices the geoengineers? How can laws hold them responsible for problems? What are the ethics of geoengineering? How best to manage and pay for it? There is no shortage of mitigation measures and proposals (most speculative) (Lomborg, 2010; Hawken, 2017, 2021; Gates, 2021).

Mitigation of pandemic (e.g. Covid-19)

Countries tried to limit Covid-19 by imposing travel restrictions, requiring masks, restricting meetings, etc., with limited success. As the disease took hold home working was often supported and grants or other forms of relief offered to soften the socio-economic impacts. Vaccines reduced fatalities and serious debilitation and cut pressure on health services, but were not universally available or welcomed (with opponents on social media). The economic, psychological, educational and social impacts are still being assessed in 2023 (Kennedy, 2023). Future pandemics may be more of transmissible and/or more virulent diseases – Covid-19 has offered lessons (it seems largely missed), shown vulnerability and need for resilience. It also showed weaknesses in the WHO and national disease response.

Vulnerability

Vulnerability is the characteristics of a person, group, organism, technology, governance or system that affect capacity to anticipate, cope with, resist and recover from the impact of a hazard. Sensitivity is the degree to which something is affected, adversely or beneficially. Sensitivity may vary in time and with other conditions – for example vegetation may be more likely to suffer if damaged when setting seed. Resistance is the ability to withstand a challenge and partly depends on degree of exposure. Exposure is not just a matter of quantity and time; it also depends on other things like whether it is slow or fast or a repeat process. Exposure to fast change is likely to pose more challenges. The linkages between vulnerability, resilience, resistance, exposure, sensitivity and adaptation are complex but broad vulnerability reduction should be a starting point for EM (Gheorghe, 2005; Pasteur, 2011). Exposure also has an element of luck; e.g. a challenge may vary in impact with location, time, etc.

A challenge may not be perceived and can be misunderstood or ignored (Heimann, 2020). Nowadays many citizens lack skills, opportunities and social capital to cope with challenges in the way their ancestors could, even in the 1940s. They hope technology and governance will protect them. Vulnerability has increased for many reasons often acting in concert:

- There is a large human population.
- People depend upon more complex and sensitive infrastructure, services and administration.
- Several countries have nuclear weapons; the threat of global war is real.
- In the past people had the option of migration if conditions became unfavourable; today most usable lands are densely settled and there may be borders or fenced boundaries so that is less straightforward.
- In the past famine foods (used in hard times) were a fall-back, but today these are either no longer available or people have lost knowledge to access them.
- Rapid modern transport disperses diseases and pests effectively and fast.
- Modern crops are mainly produced by agribusiness. Seeds are hybrids, and if there were a disaster, survivors could find seeds they saved were infertile or reverted to something undesirable.
- In many nations people can no longer draw on help from extended families and communities.
- Citizens have lost common skills their grandparents had.
- There may be weaker religious, social and moral supports.

- People are less used to hardship.
- Biodiversity loss reduces options for crop and pharmaceuticals production.
- Trade and agricultural development strategies discourage growing and storing sufficient food reserves.
- Globalisation and reliance on petroleum/natural gas mean that critical resource supplies can fail.
- Global environmental change means unknown and unexpected challenges, not just repeats of the past.
- Expanding settlement, crowded cities, misuse of antibiotics and the concentration of migrants and refugees who are poor, ill-housed and malnourished raise risks of disease.
- Housing and possessions are easily damaged: electrical systems, electronic chips, etc.
- People are dependent on mobile phones and cars; the latter shifted from analogue instruments and parts to electronic around 2015, which has increased vulnerability to sudden failures through solar storms or difficulties getting spare parts because of few electronic chip producers.
- In many countries, rules and measures are weakly enforced.

Vulnerability is a concept that links the relationship that people have with their environment to social forces, institutions and cultural values (Ten Have, 2016; Fuchs and Thaler, 2018). Care is needed to ascertain whether vulnerability is an inherent property of a group or system or is contingent upon a particular scenario of external and/or internal challenges and responses.

Most environments, organisms, humans and institutions are vulnerable to some degree. Vulnerability is a function of the character, magnitude, frequency and rate of environmental variation to which a system is exposed, its sensitivity and its adaptive capacity (ability to beneficially respond to change); for example a single challenge may not be a serious problem but if repeated it might become one (Zakour and Gillespie, 2013). The disturbance or innovation may be social, cultural, economic, technical or environmental (or a mix). Environmental and other changes can impact on already problematic conditions. There may be unexpected, perhaps indirect, cumulative effects, which are difficult to predict, study and model. Vulnerability for humans and some organisms is partly a product of history and learning and peoples are shaped by leadership and politics; so, experience and governance is likely to be crucial to their ability to withstand a challenge or grasp an opportunity. Technology may increase or decrease vulnerability.

Vulnerability can change in subtle ways, e.g. exchanging traditional building panels and gutters for plastic might make housing more vulnerable to wildfire and storm. There is a need to regularly assess innovations (technology, social change, cultural change, etc.) to see if they affect vulnerability. Vulnerability increases if people settle hazardous locations like floodplains or low-lying coastal sites. Global environmental change is adding to natural threats and making known threats less predictable. Those driving development may gain from ill-advised actions and leave a trail of increased vulnerability for others. The growing shift from mixed subsistence agriculture to less diversified cash-cropping may benefit farmers but mean more consumer vulnerability through market fluctuation, dependence on external inputs, and risks associated with intensive cropping.

Vulnerability and natural hazards

Hazards can be grouped into terrestrial and extraterrestrial, or by degree of threat, potential for recovery, on a cost basis, etc. Even very serious disaster can be prepared for by offering a few survival chances (Newitz, 2013). Some groups may have habits, money or skills that mean they will be able to cope with a hazard when others suffer (Birkmann et al., 2014).

Vulnerability and agriculture/food

Vulnerability assessment should play a key part in improving food security. Food, crucial for human survival, has been subject to little vulnerability assessment (Salinger et al., 2005).

Vulnerability: social aspects

The social sciences, disaster avoidance and relief, healthcare, psychology, urban planning, crime, poverty, economics, etc., have developed an interest in social vulnerability sometimes overlapping with environmental vulnerability (Phillips et al., 2010; Thomas et al., 2013; Bankoff and Hihorst, 2022).

Vulnerability and business

Business is becoming aware of vulnerability, especially supply chains, critical infrastructure and production line vulnerability (Anbumozhi et al., 2020; Blokdyk, 2020b). Companies increasingly espouse the 'triple bottom line'; that is, aiming to maximise economic, environmental and social performance (Linnenluecke and Griffiths 2015; Bals and Tate, 2017; Fiksel, 2022). Pursuit of the triple bottom line demands a multidimensional assessment of vulnerability and resilience:

- *Economic vulnerability/resilience*, which reflects the financial strength and stability of the company/enterprise, including the economic vitality and diversity of the communities in which it operates, the supply chain that it rests on, and the markets that it serves.
- *Social vulnerability/resilience*, which reflects the human capital of the company/enterprise, including the capability, teamwork and loyalty of its workforce and the strength of its relationships and alliances.
- *Environmental vulnerability/resilience*, which reflects the operational efficiency and effectiveness of the company/enterprise in terms of resource use and waste management and its ability to protect and nurture the natural ecosystems in which it operates.

IT/cyber and AI vulnerability

Reliance on IT brings vulnerability; hackers may spread malware and computer viruses, spy, blackmail, or misuse data; other criminals may steal or change data. Power surges caused by faulty generation and electromagnetic radiation could cause havoc with systems and data storage (Subrahmanian et al., 2015). Storage media may become corrupted and as equipment becomes obsolete records are difficult to read; paper/parchment may prove better!

Interference with online voting systems may affect environmental policy making. Citizens and environmental bodies may face excessive commercial or state monitoring (Magnussen, 2020). One way to reduce vulnerability is to educate IT users about risks and install antivirus and anti-spyware software. IT cables and satellites are vulnerable to solar storms, tsunamis, warfare/terrorism, even trawler nets. Backups like long endurance high-flying UAVs to transmit in an emergency and duplication of cable links would be wise. Replacement parts – chips, transformers, etc. – are made in a few factories and there are usually limited stockpiles; the world's electronics could be difficult to service/repair after a problem so spares should be kept.

Artificial intelligence (AI) is attracting considerable interest and generating some fear. The adoption of AI may have impacts on employment, privacy of data, business and governance effectiveness. There are also fears humans may 'lose control' of key processes and essentially become obsolete – certainly routine tasks may be done by AI. AI may enable faking of data, faking photos and movies, rigging of elections and broadcasting of convincing fake news. It is too soon to be sure of the threats but vulnerability studies should be started.

Vulnerability and economics

Unless living off-grid, citizens are vulnerable to economic shifts. Economics can be hit by a plethora of problems, from inept politicians and foolish voters to storms. In a globalised world disruption of a critical resource can impact seriously and suddenly (Cook et al., 2010). People living close to poverty or in poverty are generally most vulnerable; some types of employment are more vulnerable to challenges than others, as are those unable or unwilling to relocate. Challenges may hit certain social groups, genders or cultures and miss others.

Insurance, friendly societies, unions, food banks and government interventions may help mitigate economic vulnerability. Dependence and reduced motivation are risks and constitute ongoing vulnerability. Ecology and economics are essentially forms of housekeeping and wise housekeepers plan ahead, live within their means and seek backups.

Vulnerability of wildlife and environment

Conservation/biodiversity management and environmental protection have to assess vulnerability before developing legal, protection or management measures (Woolaston, 2022). Those setting up reserves may miss checking vulnerability. Wherever possible reserves should be established that are resilient, duplicated and backed by *ex-situe* collections. Wildfires, pollution, invasive organisms or people, inherent lack of sustainability, global warming, warfare and illegal encroachment commonly destroy reserves.

Vulnerability and SD

SD strategies must try and reduce vulnerability, improve resilience and should be adaptable. Initiatives should be diverse, duplicated and widely spaced. Then, if one element/strategy is damaged or destroyed there should be inputs and skills somewhere safe, which can be used to recover.

Poverty, environmental degradation and SD are commonly linked, so it makes sense to address them in a co-ordinated way. When people suffer poverty they are unlikely to give much attention to environmental issues. SD demands investment of some current resources into maintaining things for the future, and for poor people with little to spare this is a dilemma.

Vulnerability to environmental change

Adaptation takes time, so vulnerability to sudden and/or severe challenges has terminated many projects and cultures (Diamond, 2005; Erdkamp et al., 2021). Adaptation is less likely if a society is already under some stress (Adger et al., 2001).

Vulnerability and climate change Vulnerability to climate change has generated interest (Leary et al., 2008a). *The Vulnerability Sourcebook* (Federal Ministry for Economic Cooperation and Development, 2017) distinguishes four components that determine whether, and to what extent, a system is vulnerable to climate change: *exposure, sensitivity, potential impact* and *adaptive capacity*. Climate change vulnerability is intertwined with social issues, politics and policy making (Füssel, 2005; Kasperson and Berberian, 2011; Teebken, 2022). Climate change vulnerability assessment has been developing (Leary et al., 2008b; Pielke, 2013; Mason, 2019; Monirul Alam et al., 2021; Štreimikienė and Mikalauskiene, 2021).

Vulnerability to sea level rise, ocean warming and acidification Perhaps 25% of humanity are vulnerable to sea level rise; many cities have already spent huge sums countering it and a significant amount of the world's farmland and infrastructure are threatened (Siegel, 2020). Impact of sea level rise varies from locality to locality because of isostatic recovery, tectonic movements, prevailing winds, storms, tides, loss of reefs and mangroves, etc. Engineering solutions (barriers, embankments, river barrages) are now augmented by N-BSs, retreat and land use change (Pilkey et al., 2016).

The threats of ocean acidification and warming should prompt much more vulnerability assessment.

Vulnerability assessment and mapping

Vulnerability assessment is the process of identifying, quantifying, and if need be prioritising the vulnerabilities of an environment, organism, technology, individual, group or system (Morgan, 2011; Bankoff and Hihorst, 2022). Vulnerability assessment can also provide an indication of how sustainably humans are living within their environment. Vulnerability assessment may be undertaken to improve scientific understanding; to inform about specific targets; or develop adaptation strategies. It can focus upon: 1) the exposure of an individual, group, place or system to threat(s); 2) the sensitivity of an individual, group, place or system to threat(s); and 3) the adaptive capacity of the individual, group, place, or system to resist impacts, cope with challenges and regain functions. Unfortunately, efforts to reduce vulnerability sometimes fail to address (or even identify) causes and instead ineffectively react to symptoms.

There are various approaches to vulnerability assessment (Morgan, 2011), such as establishing indices and modelling. A vulnerability index is a measure of the exposure of an individual, group or system to a perceived hazard or range of hazards. Typically, the index is a composite of multiple quantitative indicators used to calculate a single numerical result. The environmental vulnerability index (EVI), was developed by the South Pacific Applied Geoscience Commission (SOPAC), UNEP, Small Island Developing States (SIDS), collaborating countries, institutions and experts. The EVI simultaneously examines levels of risk and conditions experienced now and with future events. It uses 50 indicators to give a composite picture of environmental vulnerability and resilience. Since 2004 there has been a published listing of over 230 countries, each assigned to one of five EVI classes ranging from extremely vulnerable to resilient.

Vulnerability assessments in the social sciences often use qualitative measures and may explore coping strategies. Vulnerability and resilience can be studied and addressed at different levels from a narrow issue to sectoral, country or even global scales.

The sustainable livelihoods approach can flag vulnerabilities. A livelihood can be seen as the capabilities, assets and activities required for a means of living; it may be sustainable if it copes with and can recover from stresses and shocks. The sustainable livelihoods approach or framework recognises capitals (natural, physical, human, social and financial) and argues that an understanding of livelihoods is vital if they are to be adapted and sustained in the face of challenges. Social capital can be the glue holding a group together. It may help people withstand physical, economic and social challenges (supports coping strategies) and might be a foundation for improvements. Understanding and monitoring social capital is important for any group or region wishing to assess or reduce vulnerability or sustain development.

Vulnerability mapping is easy with GIS; most insurance companies have maps of flood, wildfire and crime risk and can overlay at-risk groups. Avalanche and landslide risk/vulnerability maps are available for more affluent communities and their infrastructure. Geomorphologic factors determining some threats are also easily mapped (Bankoff et al., 2004).

Vulnerability reduction

Efforts to reduce vulnerability sometimes fail to address (or even identify) causes and so do little to prevent recurrence. Reduction of vulnerability and improvement of resilience demand a multidisciplinary and multisectoral approach. Creating or strengthening supportive institutions can be valuable. It is likely that measures to reduce vulnerability will be more cost effective than measures to try to protect against threats.

Engineering, architecture and psychology deal with vulnerability. Active fields are vulnerability reduction in new builds and through retro-engineering (Kidokoro et al., 2008; Masterson et al., 2014; Giorgi et al., 2022).

Adaptation

To adapt (adjust) is to become suited to challenges/conditions/opportunities. Adaptation is the process of achieving the changes in traits, technology, etc., needed to adapt. Adaptive capacity is the potential or ability to respond to challenges/change. Some adaptation is temporary because it is based on technology or organisations, which could fail or be overtaken by socio-economic or environmental change. There may be spontaneous adaptation (even unconscious) but much is steered and managed by government or international bodies (Taylor, 2015). Adaptation is not just response to external natural, social and economic challenges and opportunities; there are also internal social and economic challenges and opportunities.

Adaptation may be spontaneous, unconscious, learnt, encouraged or enforced. It may be temporary, permanent and irreversible, maintain wellbeing or involve a degradation of standards; it may improve resilience or reduce it.

Examples include: vaccination programmes, growing crops in new areas as climate changes, migration, conserving and developing alternative water and energy supplies, etc. Some of the adaptation and resilience literature assumes severe future disruption (dystopia) (Hill and Martinez-Diaz, 2020). Disaster might happen, doom need not (Newitz, 2013; Torp and Andersen, 2020); to assume a dystopian future is speculation; cautious planning is wiser. EM should check that adaptation is objective, not a response to assumptions of dystopia/catastrophe or based on over-optimism (Thompson and Bendik-Keymenr, 2012; Taylor, 2015; Kondrup et al., 2022). There may be a risk of runaway climate change but it is not inevitable (Koonin, 2021).

Adaptation is the core 'filter' of evolution, a failure to adapt behaviour or move (moving is adaptation) limits survival unless an organism is very resilient, specialised or lucky. One of the reasons for the success of humans has been past adaptability, innovative ability and mobility. A basic adaptive strategy is dispersal – spread an organism widely and hope some survive. Adaptation can be by individuals, groups, households, industry, state or

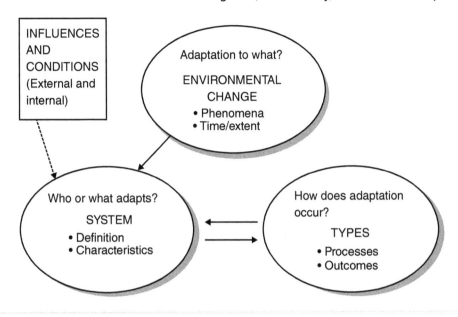

Figure 13.1 Adaptation.

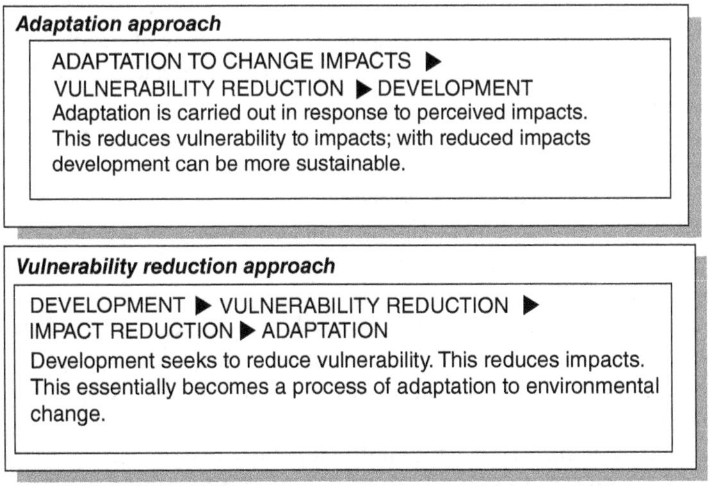

Adaptation approach

ADAPTATION TO CHANGE IMPACTS ▶
VULNERABILITY REDUCTION ▶ DEVELOPMENT
Adaptation is carried out in response to perceived impacts.
This reduces vulnerability to impacts; with reduced impacts
development can be more sustainable.

Vulnerability reduction approach

DEVELOPMENT ▶ VULNERABILITY REDUCTION ▶
IMPACT REDUCTION ▶ ADAPTATION
Development seeks to reduce vulnerability. This reduces impacts.
This essentially becomes a process of adaptation to environmental
change.

Figure 13.2 Adaptation approach compared with vulnerability reduction approach.

international bodies (Magee, 2013). Adaptation may be by all or some. Response is influenced by the outlook of advisors, decision makers and citizens, and is shaped by a complex of factors.

Adaptation is linked to vulnerability and can be anticipatory, concurrent or reactive, spontaneous or planned. Adaptation can be misjudged leading to greater vulnerability and/ or less resilience. In time an organism, organisation or technology may change and develop altered vulnerability (reduction or increase of vulnerability). The mechanism may involve change of habit or physiological or psychological adaptation. The timescale varies from short to long term. Institutions formally or informally shape expectations and behaviour of individuals and societies. In order to make adaptations, it may be necessary to strengthen existing or establish new institutions, or weaken those that are a hindrance.

Reasons for failure to adapt to challenges: a label (*italicised*) and definition for each reason followed by possible remedies (**in bold**):

(1) *Ignorance* – the threat is not perceived or well-enough known. (**Improve research**)
(2) *False analogy* – a problem is judged by observer(s) against their (perhaps inadequate) experience so crucial issues are missed. (**There should be a careful assessment of hindsight**)
(3) *Insufficient detail* – researchers, planners, managers have over generalised; and/or the problem is more widespread or developed faster than expected; and/or resilience was not as good as expected; and/or the response to environmental change took an unexpected route; and/or study was too general, or the coverage patchy, or there was difficulty in making measurements. (**Carefully check data**)
(4) *Observation over too short a time-span* – with slow processes there may be poor records, and faulty memories; and/or funding and the administrative situation may not permit adequate study. (**Seek better modelling and/or information from historians, palaeoecology, etc. Ideally use more than one line of evidence**)
(5) *Observer/manager detachment* – elites make decisions and are socially and spatially separate. As decision-making is increasingly urban, there may be insufficient awareness of non-urban situations. Failure to get local information or to inspire local activists. (**Improve accountability and try to empower a wider range of people**)
(6) *Evidence not questioned* – published material/desk research not adequately checked. (**Question established assumptions**)
(7) *Reactions are ill-timed* – decision makers wait for backing from voters, or for money to become available, or for the problem to develop so that it is evident that it is surely happening. (**Adopt precautionary principle**)
(8) *The problem is somebody else's* – decision makers judge they can delay while someone else acts (indeed, it might even weaken or distract a rival). (**Improve accountability/ public oversight**)
(9) *Inappropriate beliefs or ethical/political outlook* – for example, a land degradation problem may be wrongly attributed to climate or inept local peoples, but in reality causation may be say globalisation or national taxation. (**Seek dispassionate, objective assessment**)
(10) *Symptom mistaken for cause* – it is vital to identify causation and not react to misconceptions. (**Objective and thorough research and ideally multiple independent lines of evidence are needed**)

Deep adaptation

There are environmentalists who assume future chaos (Pogue, 2021). Deep adaptation is one such concept and supports a social movement that expects climate change will disrupt food, water, power, etc. The concept, established in 2018 by J. Bendell (UK), advocates measures to adapt to assumed collapse of livelihoods. The concept has been criticised for weak science but has spawned a movement and has been grasped by some environmentalists (Shellenberger, 2020; Bendell and Read, 2021).

Adaptation to climate change

Adaptation to climate change has generated advocacy and proposals (Garvey, 2008; Leary et al., 2008b; Schipper et al., 2009; Pelling, 2011; Seidl, 2011; Ash et al., 2013; Field et al., 2014; Leal Filho, 2016, 2019; Markandya et al., 2016; Alves et al., 2018; Dale, 2022; Pindyck, 2022; Pörtner et al., 2022; Rivera et al., 2022). There are also regional and sectoral studies (Ford and Barrang-Ford, 2011; Phillips and Barinaga, 2021; Sanderson et al., 2021).

Adaptation may benefit some or all, or may damage others. Need for adaptation varies from place to place, depending on the sensitivity and vulnerability; it usually incurs costs and will not prevent all damage. Adaptation may demand appropriate technology, governance, agricultural change, socio-economic change, etc. (Magnan et al., 2020). It involves reacting to challenges and grasping opportunities (Moser and Boykoff, 2013; Kahn, 2021). There are various response scenarios:

1) business-as-usual – no dramatic changes in the way problems are faced hardly encouraging adaptation;
2) technological and ethical advance – new technology and changed attitudes enable better adaptation and vulnerability reduction;
3) disaster, population growth, war, sudden extreme event, human inertia – any one hinders adaptation and vulnerability reduction;
4) powerful groups buy adaptation and reduced vulnerability – others probably will be abandoned.

Concern for climate adaptation often focuses on: Africa (due to poverty and challenging environment); low-lying Bangladesh; and Pacific islands (Ludwig et al., 2009; Oguge et al., 2021). The world's important food production areas deserve to be given more priority for adaptation.

Climate change may involve unpredictable and sudden extreme events, not necessarily warming (Fares, 2021). Some individuals, groups, companies and nations may suffer less or even seek to profit (Stern, 2007; Schellnhuber, 2009; Stern and Patel, 2009; WBGU, 2009).

Adaptation to sea level rise

There are a number of adaptations: move to higher ground; engineer barriers, raise buildings or float them; retreat and abandon land to rising water to serve for conservation, recreation or other compatible uses. It is unlikely sea level rise can be stopped or even slowed much.

Where there are large populations vulnerable to rising seas and with limited options for retreat there are practical and legal problems; international law needs to consider how refugees will be accommodated and solutions financed (Barnett and Campbell, 2009).

Adaptation of agriculture, food production and forestry

Agriculturalists have tried to adapt to change from earliest days but there were, and still are, frequent harvest failures and famines. Modern agricultural improvements have focused on increasing yield; adaptation, resilience, sustainability and land degradation have been much neglected. The *World Development Report 2008* explored how agriculture might improve yield, become resilient, adapt to and help mitigate climate change and how it could be made more environmentally sustainable (World Bank, 2007).

Agriculture is largely managed as agroecosystems which can be modelled to guide adaptation and resilience. Adaptation might be left to farmers or farm families, with or without state or agency help or promoted top down. Much adaptation will be crop by crop and locality by locality. It takes time to breed new adapted crops, especially with non-GM approaches (Lobell and Burke, 2010; Edden et al., 2011; Fuhrer and Gregory, 2014; Sejian et al., 2015; Bryant et al., 2016).

Climate change may impact food supplies by altering crop-weed competition, disease occurrence, changing photosynthesis and water supply, altering growing season length, etc. (Sarkar et al., 2019). There are three 'metabolic types' of photosynthesis: C_3, C_4 and CAM. Shift to hot and arid conditions would favour C_4 and CAM plants; C_3 plants generally function across a broader temperature range and include rice, wheat and potatoes. Most plants are C_3 with only about 4% C_4 (including some grasses, sorghum, sugarcane and maize). CAM photosynthesis occurs in plants adapted to dry environments. C_4 photosynthesis is more efficient than C_3 photosynthesis (by as much as 50%) in warmer climates. Increased CO_2 (up to a point) improves photosynthesis only slightly in most C_4 plants, but leads to higher biomass yield in C_3 plants (the yield might be mainly stem and leaf, not grain, etc.). Many perennial weeds are C_4 and annual weeds tend to be C_3. Weeds may move from one part of the Earth to another with climate change and could be favoured or discouraged by raised CO_2 and temperature. Crops, weeds and pests interact and make predictions about environmental change difficult. Factor in possible future water shortages, salinisation, and so on and deciding best adaptation becomes complex.

Forests and plantations of even fast-growing species take time to grow, so adaptation should be underway now, planting species suited to likely future conditions (McNulty, 2023). Intensive high-yield techniques like vertical farming, aquaculture, hydroponics and possibly small livestock are adaptable and may either be moved to new sites or are practiced in controlled environments so are more resilient.

Adaptation and SD

SD is a challenge under stable and predictable environmental and socio-economic conditions. To stand a chance of success, SD needs adaptable approaches, early warning of problems, dispersal of key resources and resilient institutions. There has not been much time to test how SD functions long term. Historians have noted that few societies operated for more than a couple of centuries before a challenge ended them. Even so, past societies may offer ideas for future adaptation (biochar and microcatchments being examples) (Ganpat and Isaac, 2017). Adaptation to environmental change could be undertaken in ways that support SD and vice versa, and this is termed 'integrated adaptation'. Under-represented in EM in general are the future generations; they have no say in decisions which will affect them.

Unattractive as it may be, SD might not yet be a viable way forward until survival problems like global warming, biodiversity loss and environmental degradation are resolved. Lovelock (2006) and others argued for sustainable retreat using nuclear power and GM to buy time until the environment can be better managed and human numbers fall to a level at which SD measures are worthwhile.

Adaptation of cities, buildings and infrastructure

Buildings, power grids, city infrastructure, transport, etc., can be adapted (at a cost). For example, buildings can be insulated, glass tinted, white paint applied and ventilation improved to reduce the need for air-conditioning (Bicknell et al., 2009; Roaf et al., 2009; Ayyub, 2018).

Adaptation – financing the changes

If change is gradual and perceived, the costs of impacts and adaptation can be spread, but rapid change and sudden disasters pose problems. The question is who pays for adaptation: individuals, groups, states or international financing (Bouwer and Aerts, 2006; OXFAM International, 2007; USAID, 2007; Agrawala and Fankhauser, 2008; Ruth, 2008; UNEP, 2021)?

Resilience

Resilience is an important concept with roots in ecology and ecosystem studies that has spread to EM and planning (Gunderson et al., 2010; Walker and Salt, 2012; Chandler and Coaffee, 2017; Grove, 2018). Resilience is the ability to withstand or recover from difficult conditions (Blewitt and Tilbury, 2014). Or the time required for an ecosystem to return to an equilibrium or steady state (stability) following a perturbation. Resilience is also used by: physics, engineering, healthcare, social studies, military studies, psychology, etc. (Biggs et al., 2015; Brown, 2016; Leal Filho, 2019; Mukherjee and Shaw, 2021).

Resilience thinking is vital for EM in an unpredictable Anthropocene (Grove, 2018; Coaffee, 2019). A resilience perspective is increasingly adopted for understanding and modelling environmental, social and economic systems (Walker and Salt, 2012; Ungar, 2021). Record keeping and learning from the past is part of improving resilience. However, a system might reorganise and recover in ways that differ from previous recoveries, thus the past/experience is not a wholly reliable indicator of the future.

Disturbances of sufficient magnitude or duration may force an individual, group or ecosystem to reach a critical threshold; if exceeded there will be a shift to a different scenario (perhaps irrecoverable). So, it is important to establish what critical thresholds are and then try and monitor them. An environment with rich species diversity or a human livelihood based on a range of things should be more resilient than simpler systems. When people rely on simple systems with little diversity change may more easily shift them into a more vulnerable condition.

There may be a need for short- and/or long-term resilience. Individuals, groups, nations, ecosystems and socio-economic systems differ in resilience; this has led to systems, holistic and interdisciplinary approaches (Fekete and Fiedrich, 2018; Jones, 2019). EM and resource management may adopt strategies like adaptive resource management and adaptive governance to try and improve resilience. Creating or strengthening supportive institutions can improve resilience (Aldrich, 2012).

Resilience assessment and modelling

Various disciplines and agencies have been exploring ways to assess resilience (Schipper, 2015; FAO, 2016; Sharifi, 2016; Serfilippi and Giovannucci, 2017; Jones, 2019). Thresholds occur in ecosystems and in social systems; in social systems they are usually referred to as tipping points. Johan Rockström and others (2009a) identified nine processes that regulate the stability and resilience of the Earth. This enabled the establishment of nine Planetary Boundaries (thresholds). Carey (2015) claimed four of these had already been passed. In 2022 another (pollution/plastics) had been passed (Rockström and Gaffney, 2021; Stockholm Resilience Centre, Stockholm University, SE: www.stockholmresilience.org/ – accessed 5/10/22).

Resilience in practice

The concept of resilience must be used with caution (Sapountzaki, 2022). Three broad steps can be recognised using a system approach: *describing* the system, *assessing* its resilience and *managing* its resilience (Walker and Salt, 2012). Graaf-van Dinther (2021) suggested five 'pillars of resilience' (which can be assessed or modelled):

- threshold capacity;
- coping capacity;
- recovery capacity;
- adaptive capacity;
- transformative capacity.

Resilience can be studied and addressed at different levels (and more than one threat can be explored at any time) from a narrow issue to sectoral, country or even global scales. It is wise to seek to improve resilience to as wide a range of threats as practicable and not to rely on adaptation or mitigation (National Academies, 2012; Brown, 2016; Tortajada, 2016; Fuchs and Thaler, 2018). Co-operation and collaboration are vital in any resilience improvement strategy (Fiksel, 2022; Goodman and Mesa, 2022). Resilience-building is an ongoing process with no real end.

Resilience to climate change There is some interest in resilience to climate change (Leal Filho et al., 2016; Ninan and Inoue, 2017; Zommers and Alverson, 2018; Leal Filho, 2019; Zolnikov, 2019; Hutter et al., 2021; Kaushik et al., 2021; Monirul Alam et al., 2021). However, it should be strengthened.

Resilience and sea level rise Sea level has been rising for centuries so humans should have resilience skills. There are many forecasts but a global rise of 110 to 770 mm between 1990 and 2100 is likely. Local conditions cause variation. Research and practical measures have been developing (Heidkamp and Morrissey, 2019).

Resilience: food, agriculture and forests Famine remains a threat. Much of the globe's food comes from monocultures with little diversity. Climate and crop disease resilient agriculture should be a priority (Reddy, 2015; Paloviita, and Järvelä, 2017; Lipper et al., 2018; FAO, 2022; Jatoi et al., 2022; Leal Filho et al., 2022). There is a need to consider resilience not just of the crops and soil but also of pollinators, seed dispersal organisms, and competition between weeds and crops (Gardner et al., 2019). Resilience can be improved by diversifying supplies, by improving and dispersing storage, by changing tastes and demand, innovative technology, better transport networks, etc. The threat of a moderate volcanic eruption (like Krakatoa) cooling global climate causing several years of reduced/no harvests is as real as global warming and should prompt food stockpiling (a vast challenge).

Resilience and social issues

Disaster relief, development planning, law, psychology, healthcare, policing, etc. have become increasingly aware of the need for resilient approaches (Balducci et al., 2020; Brears, 2021b). In particular assessing and improving community resilience and communication (between citizens, departments and subjects/fields) (Wilson, 2012; O'Donnell and Wodon, 2015; Sharifi, 2016; Howarth, 2019; Torres and Jacobi, 2021; Goodman and Mesa, 2022).

Citizen life experiences and support from society help determine resilience; a resilient people or group can become reluctant to change and weak if the wrong support and guidance is given – critics of a welfare state would flag this. Recently some EU countries have proposed reintroducing a form of national service to enhance resilience in citizens.

Resilience: urban areas and infrastructure

By 2030 roughly 70% of humanity will be living in urban areas, so urban resilience is very important (Coaffee and Lee, 2016; Garcia and Vale, 2017; Kershaw, 2017; Borsekova and Nijkamp, 2019; Bahadur and Tanner, 2022). Cities need food, water and energy – all should be made resilient (Shaw and Sharma, 2011; Burayidi et al., 2020; Graaf-van Dinther, 2021). Architects and city planners are exploring the resilience of urban communities (Caniglia et al., 2017).

Resilience and disease

The Covid-19 pandemic of the last few years has given many lessons on disease resilience (Leal Filho et al., 2016; Ramanathan et al., 2021). However, it does seem resilience to disease challenges is not certain, even when threats have been foreseen, if governance is weak.

Resilience and SD

There are similarities between resilience and SD (Blewitt and Tilbury, 2014; Achour et al., 2015; Benson and Craig, 2017; Dayton et al., 2018). Resilience and SD question the free market model within which global trade operates; a free market encourages specialisation which weakens resilience by allowing systems to become accustomed to and dependent upon their prevailing conditions and by reducing diversity.

Resilience and terrorism, war and unrest

Terrorism is difficult to predict and improved resilience and reduced vulnerability help reduce its impact, for example: organisational preparations and changes to infrastructure and policing (e.g. plastic litter bins or none to reduce bomb damage, more CCTV surveillance, etc.). Health services in a number of countries are more aware of nerve gas and other CBN (chemical, biological, nuclear) threats and have flexible contingency plans. Transport of hazardous materials or potential weapons has been made more secure. IT, telecom, data storage/ transfer and computing need better resilience; vulnerability can be reduced but there is also a need for alternatives like UAVs to replace lost satellite or cable links and diversified energy supply that improve resilience (Linkov and Palma-Oliveira, 2016). Resilience needs constant updating; for example 3D printers offer terrorists a new tool which can enable dissemination and construction of weapons or other potentially harmful devices, evading customs and security checks. New surveillance and controls must keep pace.

Resilience and business

Supply chains are far from resilient; business has been moving to stocking parts just in time from a few sources partly to reduce storage costs, but this is vulnerable when there is a strike, unrest, natural disaster, political pressure or pandemic (Anbumozhi et al., 2020; Kummer et al., 2022). Resilience would be improved by stockpiling or multiple lines of supply.

Resilience and economics

Economic resilience is important (Rose, 2018). The need is to build climate change and sudden event resistant local and global economies (Agrawala and Fankhauser, 2008; Ninan and Inoue, 2017). Until recently profit and returns to taxpayers, shareholders, etc., were the motivation; now there is awareness of a need for resilience and bodies like insurance companies and investors are seeking it.

Summary

- It is apparent that much modern development is unsustainable, vulnerable, not resilient and difficult (or impossible) to adapt.
- In the last 20 years realisation of weaknesses has started to prompt change and some argue a *resilience thinking* paradigm has developed.
- Some predictions, recommendations and theorising assume a dystrophic future, but with inadequate foundation for such assumptions, coupled with strident advocacy; this is dangerous.
- Trends, outcomes and early-warning thresholds (tipping points) are not adequately established so responses must be flexible.
- Adaptation is vital: 'mitigate we might, adapt we must'. Maladaptation is a risk.
- In a world characterised by change in ecological, social and economic systems, it is important to 'expect the unexpected' and manage to enhance resilience.

Further reading

Books

Ash, A.J., Stafford Smith, M., Parry, M., Waschka, M., Guitart, D., Palutikof, J.P. and Boulter, S.L. (Eds) (2013) *Climate Adaptation Futures*. Wiley-Blackwell, Chichester, UK.

Gates, B. (2021) *How to Avoid a Climate Disaster: the solutions we have and the breakthroughs we need*. Penguin, London, UK.

Lehmann, J. and Joseph, S. (Eds) (2015) *Biochar for Environmental Management: science, technology and implementation* 2nd edn. Routledge, Abingdon, UK.

Seiegel, F.R. (2016) *Mitigation of Dangers from Natural and Anthropogenic Hazards: prediction, prevention and preparedness*. Springer, Cham, CH.

Walker, B. and Salt, D. (2012) *Resilience Practice: building capacity to absorb disturbance and maintain function*. Island Press, Washington, DC, USA.

Zommers, Z. and Alverson, K. (Eds) (2018) *Resilience: the science of adaptation to climate change*. Elsevier, Amsterdam, NL.

Journals

Mitigation

- *Journal of Climate Change* (IOS Press). Print ISSN 2395-7611; online ISSN 2395-7697. www.iospress.com/catalog/journals/journal-of-climate-change (accessed 11/10/22).
- *Mitigation and Adaptation Strategies for Global Change* (Springer). Print ISSN 1381-2386; online ISSN 1573-1596. www.springer.com/journal/11027 (accessed 11/10/22).

Adaptation

- *The International Journal of Climate Change: impacts and responses* (AAP). Print ISSN: 1835-7156; online ISSN: 2833-4140. www.gov.uk/renew-adult-passport (accessed 10/10/22).
- *WIREs Climate Change* (Wiley). Online ISSN: 1757-7799. https://wires.onlinelibrary.wiley.com/journal/17577799 (accessed 11/10/22).
- *Change and Adaptation in Socio-Ecological Systems* (De Gruyter). ISSN 2300-3669. www.degruyter.com/journal/key/cass/html?lang=en (accessed 10/10/22).
- *Nature Climate Change* (Springer). Print ISSN 1758-678X; online ISSN 1758-6798. www.nature.com/nclimate/ (accessed 10/10/22).
- *Journal of Climate Change* (IOS Press). Print ISSN 2395-7611; online ISSN 2395-7697. www.iospress.com/catalog/journals/journal-of-climate-change (accessed 11/10/22).
- *Climate and Development* (Taylor & Francis). Print ISSN: 1756-5529; online ISSN: 1756-5537. www.tandfonline.com/journals/tcld20 (accessed 11/10/22).

Resilience

- *Adversity and Resilience Science* (Springer). Print ISSN 2662-2424; online ISSN 2662-2416. www.springer.com/journal/42844 (accessed 10/10/22).
- *Resilience* (Taylor & Francis). Print ISSN 2169-3293; online ISSN 2169-3307. www.tandfonline.com/loi/resi20 (accessed 11/10/22).
- *International Journal of Disaster Resilience in the Built Environment* (Emerald). ISSN: 1759-5908. www.emerald.com/insight/publication/issn/1759-5908 (accessed 11/10/22).

www sources

- IPCC – Intergovernmental Panel on Climate Change (UN). www.ipcc.ch (accessed 13/10/22).
- Resilience Alliance. www.resalliance.org (accessed 04/10/22).
- Centre for Climate and Environmental Resilience, Newcastle University, UK. www.ncl.ac.uk/environmental-climate-resilience/ (accessed 05/10/22).
- Stockholm Resilience Centre, Stockholm University, SE. www.stockholmresilience.org/ (accessed 5/10/22).

Chapter 14

Technology and social developments

Chapter overview

- Technology development
- Green attitudes, social media and other communications
- Fashion
- Politics
- Summary
- Further reading
- www sources

Technology development

Science is the accumulation of knowledge and understanding about the world and beyond; technology should be application of science for the benefit of humans and other organisms. Technology interacts with people's worldviews, it can change them and is shaped by them (Kaplan, 2017).

Stone tools and weapons, fire, grazing livestock and smelting were impacting the world millennia before cultivation and growing population (Golding, 2017; Pritchard and Zimring, 2020; Pacey and Bray, 2021). Probably some large animals were driven to extinction and vegetation was changed by prehistoric humans. In the last 10,000 years or so agriculture has largely changed from shifting cultivation to sedentary and has undergone a shift from hand-labour to animal-power and then mechanised (since the 1930s); the use of coal (after 1752) and then oil from the 1930s. Oil is now transitioning to less polluting technologies. Some are now calling for a transition to a Fourth, or even a post-Fourth, Industrial Revolution more suited to the challenges of the Anthropocene (Rifkin, 2011).

DOI: 10.4324/9781003189985-18

From Victorian times until the 1980s there was talk of controlling nature by building huge dams, even using atomic geoengineering, establishing large irrigation and drainage projects, etc. By the 1970s it was clear there were problems. It now falls upon EM and technology to repair the damage and find a way to work with nature. The right technology must be developed and supported. In the USA the Amish people judge and select technology to serve their cultural views, discarding what they see as superficial or harmful; however, many if not most humans seek technology/possessions to bolster pride with little consideration other than what it does for their status.

Anti-technology

Concern about technology is not new – the Prometheus myth is over 2500 years old. In Nottingham (UK) 1799–1816 a group led by Ned Ludd opposed textile machines fearing job losses; these 'Luddites' smashed machinery. Similar nineteenth century European worker activists spawned the word 'sabotage' by throwing wooden clogs (sabots) into machinery. Understandably nineteenth century polluting, coal-fired technology also generated opposition from a romantic movement.

In the 1970s neo-Luddites criticised technology (Farvar and Milton, 1972; Fluss and Frim, 2022) and prompted concern for limits and reassessments leading to some alternative, appropriate and green technology (Lomborg, 2001). The 'barefoot economist' Fritz Schumacher argued for human-scale local-focus, decentralised and appropriate ('Ghandian') technology, something attractive to many environmentalists but hardly productive enough given population growth. Much anti-technology literature is stronger on advocacy than workable alternatives; some is misleading (Heusemann and Heusemann, 2011; Shellenberger, 2020; Jensen et al., 2021). There is a need to be aware of Anthropocene challenges and the problems of governing and feeding a huge diverse population while coping with environmental change (Kaczynski, 2020). Technology must be carefully assessed to seek the BPEO and address key opportunities and challenges.

Opposition to nuclear power and biotechnology, especially GM, is understandable and has a basis of common sense (Carlson, 2010). But these technologies will probably be vital stopgaps in dealing with Anthropocene challenges. Citizens are often informed by weak journalism, misleading social media and conspiracy theorists, not objective research (Sachsman and Valenti, 2020). The risk is valuable tools will be delayed or discarded.

Anthropocene: technology working with nature not against

Between 2.5 million years and 11,700 BP the Pleistocene Epoch saw shifts between glacials and interglacials; from 11,700 BP to the present the Holocene Epoch has offered relatively stable and warming conditions with seas rising from approximately 130 m below today's levels. Although there were setbacks (*sudden* cold phases, droughts, famines, epidemics, etc.), humans increased their numbers, especially since the start of the Industrial Revolution: worldwide technology and civilisation have clearly benefitted humans and enabled multiplication, but that population increase, pollution and frequent warfare now pose serious environmental, social and economic challenges.

The Anthropocene Epoch (effectively underway although start date has not formally been established yet) is founded on the recognition of human impact. Its start marks a shift from a natural world to one overwhelmingly and irreversibly changed by humans. There is now wide acceptance Earth is finite and endangered but not that it must be managed by humans with great caution (Bonneuil and Fressoz, 2017; DellaSala and Goldstein, 2018; Ellis, 2018; Benner et al., 2021; Butler, 2022). Managed is not the best word; humans must co-evolve with Earth rather than blindly exploiting it. This poses many questions, including: Can human civilisation survive? Is co-evolution possible? How difficult/painful will the transition be? How many humans per nation?

The Anthropocene is not the first time life has conflicted with Earth and caused global ecosystems shift; there have been a number of such events. For example, *c.* 2.5 billion years ago some microorganisms started to emit O_2 – other life forms could not cope and died or survived only in relatively restricted anoxic environments. Technology will be an important part of any shift to co-evolution: rejection of technology would spell disaster for humanity and probably much of the remaining biota. Earth and some organisms will survive a failed transition but civilisation and humans may not. EM must ensure technology is not simply rejected but becomes focused. Many assume coming dystopia but it is better to see the challenge as 'not as a race against catastrophe but as an age of emerging coexistence with Earth' (Lynch and Veland, 2018: 69). Many environmentalists assume the Earth needs saving; but nature can simply discard humans. Humans and their civilisations need saving from collapse.

Earth has to be managed using thoughtful (not careless) technology (Moore, 2016; Lynch and Veland, 2018; Baskin, 2019; Biermann and Lövbrand, 2019; Simon, 2020). Humans have to sort out politics/governance for the Anthropocene soon. An Anthropocene 'gap' now exists: it is not clear how to shape EM, politics, society, economics and technology (Galaz, 2014; Hamilton et al., 2015; Chandler et al., 2020; Horn and Bergthaller, 2020; Merchant, 2020; Carilllo and Koch, 2021). Technology, society, economics and natural systems are intertwined and have to be managed in an integrated way. The shift to working with nature will not be easy and may not be painless (Hamilton, 2017; Hoffman and Devereaux Jennings, 2018; Symons, 2019; UNDP, 2020b; Abram et al., 2022). Climate problems are but part of the challenges. Key questions include: How to steer consumerism and technology? What does nature now need of humans? What population should we aim for?

Green technology / clean technology / biotechnology

There has been expansion in green technology / clean technology / biotechnology, and nature-based solutions (N-BSs) (Lofrano, 2012; Savarimuthu et al., 2022). Inspiration and innovation may come from study of natural systems or organisms (Huu Hao Ngo and Tian Cheng Zhang, 2016; Arya, 2017; Singh and Kumar, 2018; S. Shah et al., 2019; Sivasubramanian, 2020; Arora et al., 2022; Sadiku, 2022). This feeds into bioremediation and phytoremediation: the use of bio-tech to recover/improve degraded ecosystems such as mine spoil, heavy metals polluted wastes, salinised fields, etc.

In 2022 a modest city may have more people than the whole world 4000 years ago. With 80% of humanity expected to be urban dwellers by 2050, green technology must apply to cities (Ozge and Yalciner, 2012; Zhenjiang Shen et al., 2018; Fitzgerald, 2020; Tomar and Kaur, 2020; Vasenev et al., 2020; Chakraborty, 2021). Concentrating people in cities and cutting their environmental impact may enable rewilding of some non-urban land.

Biomimicry (biodesign or biomimetics) is the study of nature to yield inspiration to shape materials, processes, structures, systems, architecture, etc. (Ashby, 2013; Primrose, 2020). Examples include: artificial photosynthesis for energy or fuel (not yet perfected), and useful enzymes (Cohen and Reich, 2016). Nature has inspired circular economies that make use of waste products as resources which technology can draw upon for inspiration (Bragdon, 2021; Gonen, 2021).

Figure 14.1 Green technology is not all new or high-tech. The Lynton-Lynmouth (UK) cliff railway has given > 130 years of clean energy, free and safe operation, run solely by water and gravity. A small stream fills the tanks, providing weight to the uphill car that raises the downhill car; on reaching the bottom the water is released into the sea and the uphill car is filled.   Source: Photo by author 2003.

Technology and EM

EM can guide technology development to deal with challenges and opportunities and ensure interrelationships are considered and negative impacts are avoided (Kuhad and Singh, 2013; Huu Hao Ngo and Tian Cheng Zhang, 2016; Arya, 2017; Bharagava, 2020; Singh Hussain et al., 2022; Chowdhary et al., 2023). Technology has assisted EM, especially the development of GIS, CCS, remote sensing, monitoring, GPS, communications and data processing (Hoalst-Pullen and Patterson, 2010; Behera and Prasad, 2020; Rai et al., 2022).

Technology can appear from a perceived need, sometimes through chance or research but develops mainly for profit, fashion/consumerism, warfare and political power (Muralikrishna and Manickam, 2017; Rathoure, 2020; Rai et al., 2021). It is vital to build-in flexibility to adapt technology to the unforeseen and to have backups, two things often lacking today. Adopting technology can be unpredictable; e.g., AC versus DC transmission was largely determined through rivalry between two individuals: Edison and Tesla. Individuals with a mission and money are influential in technology development and in EM; space technology, electric vehicles and EM programmes have been funded by billionaires when governments have failed to do so. EM should seek formation of an international body to judge and shape technology for addressing Anthropocene challenges and opportunities.

Environmental technology management Environmental technology management seeks to reduce negative impacts and improve environment wherever possible (Pattberg and Zelli, 2016; Bolwell, 2019). Popular demand is not the best way to select technology: the BPEO might be unwelcome, involving citizen cost or inconvenience. Technology can be driven forward or held back by governments, commerce or individuals and institutions (Wong Shiu-Fai, 2006). Monitoring and forecasting is valuable for technology management to try and ensure a proactive, problem-free approach (Porter et al., 2011). Technology development can support SD (Leach et al., 2010; Bolay et al., 2014; Vos, 2015; Abraham, 2017; Pekmezovic et al., 2019; S. Shah et al., 2019; Behera and Prasad, 2020; Wenshan Guo et al., 2021).

Technology transformation and transfer The way a technology is used can be transformed: by innovations, reorganisation, changing the image it has, making it more efficient, green or profitable, etc. Often the key transformation is reducing costs: drip-irrigation was at first complex, prone to algae and sediment blockages and costly; light-proof pipes, algicides and cheap emitters led to much wider uptake. Computers were once expensive mainframes requiring skilled programmers but now there are cheap and user-friendly laptops.

Technology transfer is the process of conveying things from innovation and development to a wider usage, perhaps to a different country, group, level of skill, or from sector to sector. In some cases laws, intellectual property rights, patents and copyright may have to be relaxed or modified. Typically the technology will need associated training, support and supply industries, etc. The transfer can be formal, informal or even insidious, but is often driven by consumerism. Commercial espionage is one means of transfer, another 'piracy'; both give no reward to the technology researcher/developer and sap their sales. The 3D printer may aid such illicit activities by allowing easy reproduction of parts but used wisely with legal controls it makes supply of some spares and maintenance fast and easier.

The transfer can be from a person or organisation, top-down or more grass-roots, it can be through governments, NGOs, inspired by media, etc. (Marcus, 2015). There is a story of peasants who resisted innovation: an agency failing to get new crops adopted set-up 'trials'; the plots were fenced inadequately but enough to make farmers think something was being protected. Stolen seed was soon welcomed in the region. Generic drugs and vaccine production are areas of technology which pose transfer problems, ownership and quality control issues, but when agreement is reached to relax royalties and patents consumer costs are much reduced and usage expanded.

Metals and agriculture spread thousands of years ago; the steam engine was rapidly adopted from the 1830s (Frankel et al., 2013). Currently there is transfer with India and China acquiring from the USA and EU but soon India and China will drive and develop much more technology change (Preeg, 2008; Dahlman, 2012). It is vital that low carbon emissions technology, CSS and green technology, pollution control technology, food production technology and technology to improve adaptation and resilience are developed and transferred (Andersen et al., 2007).

Transfer of a technology should not lead to dependence on outside suppliers for spares, cause unemployment, etc. In reality such dependence is cultivated for profit or political ends. Technology transfer may be easy if no obsolete forerunner is present; the adopters can 'leapfrog' to state of the art. Old technology generally slowly mutates to new. Technology change may hardly be noticed; few in the 1980s foresaw the impact on the lives of millions by mobile phones. EM needs to monitor what is advertised and encourage 'useful technology' not 'vanity/profit technology'.

Industrial ecology

Industrial ecology, the quest for industrial systems that operate more like natural ecosystems, embodies a multidisciplinary and systems approach. Nature is also a source of ideas and guiding rules, for example seeking symbiosis: using waste from one activity as a resource in another (Klöpffer and Grahl, 2014; Deutz et al., 2015; Clift and Druckman, 2016; Jolliet et al., 2016; Bourg and Erkman, 2017; Hauschild et al., 2018; Spencer, 2018; Xiaohong Li, 2018).

Life cycle assessment and life cycle design

Two approaches support industrial ecology: life cycle assessment (LCA) and life cycle design (LCD). LCA is a methodology for assessing environmental impacts associated with all the stages of the life cycle of a commercial product, service or process. For a product it would cover development (cradle), raw material production and transport, manufacturing impacts, distribution, use, disposal/recycling (grave): cradle-to-grave.

LCD goes further than LCA, seeking a closed loop: designing a process from the outset. LCA and LCD should involve integrating with other producers, services, the supply chain, waste processing, etc. (Thiébat, 2019).

Gene modification, genetic engineering, nanotechnology and AI

Genetic engineering/gene modification (GM) is a contentious area of technology. GM opens up vast potential but must be tightly monitored and controlled. GM can speed the process of developing new crops, livestock, enzymes, energy supply, greenhouse gases control, etc. (Ferry and Gatehouse, 2009). GM and GMOs could play a vital role in Anthropocene EM; however, some environmentalists target GM as 'Frankenstein science' and probably a majority of citizens view it as a threat. Real risks have been flagged but benefits tend to be downplayed (Lin, 2013). GM also generates controversy over raw materials, intellectual property and transfer of the technology.

Nanotechnology generates fear and opposition because some think it might get out of control. Nanotechnology is already being used in things like water purification and pollution control (Pal, 2022; Shanker et al., 2022). There are similar worries over artificial intelligence (AI) that technology may get out of control. There might be some justification for fear because what impacts these technologies could have on humans, Earth and biota is largely unassessed (Katsikides, 2017; Lovelock, 2019; Katsikides, 2020; Lum, 2021).

Intermediate technology / appropriate technology

Intermediate/appropriate technology is that which is accessible, affordable, suitable, sustainable and does not bind users to a foreign or distant supplier of inputs and parts and which improves their lives and resilience. Its roots lie with P.D. Dunn, E.F. Schumacher and the Intermediate Technology Development Group.

Appropriate tends to be seen as meaning 'low tech' as opposed to 'high/advanced tech'. But the point is that technology suits culture, social, economic and environmental conditions and needs, uses accessible energy and materials and is controlled by people involved, whether low or high (Bihouix, 2020). Technology is linked with social and economic conditions;

Covid-19 has prompted changes in transport, shopping, telecom and building requirements that are being reflected in technology changes. IT technology enabled the adaptation to home working. Fashions change, a shift in the last decade to vegetarianism/veganism cut demand for meat and leather. There may be problems coordinating and managing scattered low tech developments and in getting economies of scale and obtaining high productivity and consistent quality of products.

SMART technology

Self-monitoring, analysis and reporting technology (SMART) uses data, machine learning and artificial intelligence to automate adaptive technology remotely – essentially the integration of data gathering, telecommunications and controls. It can conserve energy, improve traffic flows, make waste disposal more controllable, etc. The approach also has disadvantages: it could be vulnerable (e.g. to terrorism, malfunction or solar flares), might lead to loss of citizen privacy, etc. Applications so far have mainly been in urban settings (Clark II and Cooke, 2016; Kolhe et al., 2019; Ming Hu, 2020; Tomar and Kaur, 2020; Agarwal et al., 2022). Interlinking green and SMART to improve agriculture, assist resilience, cut carbon emissions and to help achieve SD are developing fields (Poonia et al., 2019; Lange and Santarius, 2020; Tomar and Kaur, 2020). SMART may mean more responsive and sensitive EM and assist SD and adaptation (Lange and Santarius, 2020).

Space/extraterrestrial technology

In the 1960s few guessed the tremendous value space technology would have for EM, communications, earth sciences, disaster management, etc. In 2023 many citizens still blindly condemn space technology, voicing the money would be 'better spent on social welfare'. The need to respond to extraterrestrial threats like asteroid strikes and the value of off-Earth studies for aiding monitoring and understanding terrestrial changes is great. Satellite remote sensing and mobile phones and GPS contribute hugely to human wellbeing. Knowledge about Mars, Venus, etc., helps in modelling Earth atmospheric/environmental change and improves awareness of risks. Space technology should allow the avoidance of many devastating asteroid, comet and meteoroid strikes in the near future.

The value of space technology will probably expand and in the very long term it will be a way for dispersal and survival of terrestrial life (Olla, 2009). However, careless technology has already polluted Earth orbits with debris threatening future launches and satellites. Already space research may be hindered by these debris, and, if orbits decay, things and people on the ground are threatened by debris strikes. EM may have to include near-Earth space clear-up soon.

Green attitudes, social media and other communications

EM has the challenge of promulgating rational, reliable environmental argument and action, and if need be encouraging caution when answers are not available.

Green attitudes

Popular views and environmental activism can assist EM or damage the environment, biota and humans. It is crucial that citizens value the environment and if need be stand up for it (Francis,

2015). While groups like Extinction Rebellion may be criticised, they do trigger a questioning of the status quo and may prompt thoughtful practical change. Attitudes are not uniform across society by age, class/affluence or spatially and EM must be aware of this (Tankha, 2017; Coskun, 2018; Aberšek and Flogie, 2022). The dangers for EM are to not act or to do so ineffectively or be accused of 'crying wolf' so future efforts will be viewed with suspicion.

Ethics, religious beliefs, politics and life experiences shape attitudes. Buddhist societies are commonly said to be more supportive of environmental care (Cooper and James, 2017) and some blame resources exploitation on Judeo-Christian attitudes that seem to place humans before nature. One problem is that environmental improvement may be slow and insidious so EM actions may not be linked to results as far as citizens and governments are concerned; the modern world has a short attention span: few governments plan five years ahead and many citizens lose interest after very short periods.

Attitudes change: Haberlein (2012) mentions US dams in the 1930s greeted as means to reduce flood damage and generate electricity, which they did; unfortunately, people saw the flood reduction and moved into at-risk areas, and flood damage increased. Attitudes are complex, difficult to predict and variable even within a group (Telešiene and Gross, 2017; Berry, 2018). When experts get things wrong, support for green issues may become lukewarm or there may be a backlash. An attitude commonly displayed is NIMBY ('not in my back yard'): the idea may be acceptable but someone else has to take the threat/side-effects.

Attacks on the oil industry miss that it has been an investor in future green technology, that it helped end whale slaughter (by substituting for whale products), was cleaner than the coal it replaced and supported real improvement of livelihoods and civilisation for nearly a century. EM issues are complex and seldom have neat 100% solutions. Haberlein (2012) proposed 'cognitive fix': e.g. give people flood risk maps and let them decide where to settle. Those dealing with EM tend not to have psychological skills, yet frequently need them. Interdisciplinary EM can be a shibboleth – in practice elusive. Many citizens favour 'wait and see' a few 'better safe than sorry' (the precautionary principle), but most are indifferent or unaware of challenges or opportunities.

Environmental psychology and environmental education Attitudes are influenced by education, parents, siblings, friends, media, religion, politics, etc. Behaviour, sense of identity and worldview are explored by environmental psychology (Steg et al., 2013; Scott et al., 2016; Lahiri, 2019; Krasney, 2020; UNESCO, 2020; *Journal of Environmental Education* (Taylor & Francis) – print ISSN: 0095-8964, online ISSN: 1940-1892). Until recently environmental education had limited support and often failed to focus on EM issues.

Communications and social media

Communication can be censored and manipulated. The world now has mobile phones even in remote areas and there are interesting new uses by farmers, banks, traders, NGOs, etc. Before about 150 years ago communication was by mail, newspapers and word of mouth, though the telegraph was spreading. Since the 1920s things accelerated and today often instant, affordable global communication is available to most people. Ideas now spread fast and attitudes change rapidly and may be affected by voices from outside a region or country. Where once social attitudes were stable they may now be changeable and confused. Social media can quickly mislead citizens and some are aware of this and willing to make use of it.

Science and authority was once held in respect and citizens responded to regulations and advice, but they are today likely to question and oppose almost any authority (Jewett, 2020). Ideas of justice, civilisation, progress and reasoned argument backed by free speech were common between 1900 and 1980, but today citizens and lobby groups commonly ignore, censor ('outing') or shout-down any views, evidence or arguments they dislike; sometimes the minority become a dominant voice. Democracies are less than efficient and may be slow to yield effective change. Authoritarian societies are controlled by the State and by official attitudes, but there may still be a diversity of media: radio, TV, magazines and so on. Authoritarian governments may find it easier to enforce EM if they choose to (Antonopoulos and Veglis, 2013; Craig, 2019).

There is currently popular fascination with conspiracy theories and this may make rational decision making more difficult and waste resources. Media and social media influence citizen attitudes, can destabilise governance and allow populist politicians or individuals to influence. This might hinder wise environmental policies, so media awareness is important for EM (Kahle and Gurel-Ata, 2014).

Social media The internet and social media keep people in touch and enable free exchange of information but can be misused and distorted by companies, states, hackers, etc. Internet users may not show responsibility and thus do damage but they can also aid research and monitoring (citizen science) (Narula et al., 2019). As Andersen (2014) pointed out, environmental issues are now in popular discourse; however, the focus tends to be more on problems than opportunities or solutions and opinion is often distorted and pious rather than useful (McFarland Taylor, 2019). Social media has proved valuable in disaster situations: *Twitter*® or *Facebook*®, etc., can spread wildfire, storm, tsunami or earthquake information and warnings rapidly, top-down (official to citizens) or bottom-up (citizen to citizen and to officials) (Cox et al., 2015; Comunello and Mulargia, 2018).

Marketing and advertising Marketing is activity aimed to sell a product, service or cause. Advertising is the promotion of a product, service or cause. Marketing and advertising (drawing on psychology and behavioural studies) seek to influence attitudes of consumers. At the core of problem consumerism and promotion of unnecessary and environmentally damaging technology adoption marketing and advertising are important challenges for EM.

Marketing has adopted some green outlook but much is not genuine (it is greenwashing) (Van Triip, 2013; Sheehan and Atkinson, 2015; Xinghua Li, 2016; Ottman, 2017; Coskun, 2018; Malyan and Duhan, 2019). Marketing and advertising create/respond to fashion and seek to shape it to their ends (McKenzie-Mohr et al., 2012; Kurisu, 2015; Farzana et al., 2018). Many organisations have media advisors or enforcers to ensure a viewpoint and avoid legal issues. EM must be media and social media wise and use advisors.

Fake news and misinformation False or misleading information circulates as fake news. A fun pastime for pranksters but with potential for damaging misuse. Part of gossip, fake news is as old as human culture and is used by politicians, religious bodies, the military, etc. Laced with propaganda it can effectively spread misinformation. Fake news manifested in the 2020s with conspiracy theorist opposition to Covid-19 controls and vaccinations, spreading views widely ('going viral') (Berman, 2020; Farmer, 2020). Fake/false news can be divided into misinformation (false but not necessarily spread to cause harm); mal-information (genuine information that is spread to cause harm); disinformation (false and spread with harmful intent).

Some now advocate 'prebunking' (spreading preparatory ideas and information) to prepare citizens for debate, rather than wait for false news that has to be 'debunked'. Fake news has become a force in politics and a dark shadow to responsible media. False news may mean that when a real situation appears citizens dismiss it. Fake news is not easy to control (Vaidhyanathan, 2021). Approaches like peer review can in theory check but are slow and imperfect and could also hold back new views and be dominated by conservative groups (peers, established but not necessarily objective). Separating true from false is difficult in many environmental debates, and democracy based on consensus is endangered by fake news and those who disseminate it are essentially unaccountable (Ireton and Posetti, 2018; Higdon, 2020; Zimdars and Mcleod, 2020; Anstead, 2021). EM must treat all information with caution and check sources for veracity. Statistics are often poorly conducted, presented and interpreted, yet even in academic works get past peer review.

Aggressive, dystopic climate lobbyists may claim tipping points will be passed within a few years based on limited proof or selected data. And climate change deniers equally misrepresent (Koonin, 2021). Influencers intend to persuade not inform: what is needed is objective, transparent, realistic information and decision making. Even well-meaning activism based on poor understanding or distorted views is damaging.

Even science suffers from fake news driven by pressure to publish, propaganda bodies or managerialism. Researchers are often encouraged to publish prematurely with weak, unoriginal and even contrived papers, rather than thoughtful, original study (driven by promotion and tenure pressures, governments keen to show active researchers, etc.). Selected key words help ensure supportive peers make checks. Often papers are the product of group authors, not one or a few original analytical thinkers, and there are states which assist publication using editing by hired academics. With publishers fearful of sales, some journals accept papers with cosmetic review or none at all – the problem is difficult to control. Science (and perhaps other disciplines) thus becomes debased and distorted. Also, new concepts may be resisted by established conservative researchers and managers slowing progress.

Fashion

Fashion is a manner of doing something, belief, behaviour, style, custom, trend, etc. It tends towards the short-lived, and is largely narcissistic and irrational. Consequently it is wasteful because items are deemed obsolete/outmoded/unfashionable before they become unserviceable and are discarded and new variants replace them. Some manufacturers design in obsolescence or update software to encourage or force regular sales. Fashion is largely about people's dreams and boredom and to be fashionable suggests consumers have awareness, status, taste and the ability to spend. Fashion has vast impact on people and the environment, much negative. Consumerism drives much of the carbon emissions and may be the root of an EM challenge. Consumer protection bodies can help change demands.

Fashion can be born of necessity or be prompted by commerce, attitudes (habits, vanity, etc.), and politics. Environmental concern has fashions: in the 1970s worries about population, pollution, technology and nuclear weapons; in the 1990s developing nations and SD; and today carbon emissions.

Fashion prompts good or bad consumption and is linked to consumerism. Fashion affects habits, goods, technology, transport, diet, clothing, music, etc. Fast fashion is a recent trend encouraging very rapid changes in demand, supplying affordable products as soon as a market is seen, especially clothing and accessories (handbags, etc.). A major branch of fashion is clothing which plays a significant role in many economies, provides employment but is very damaging to the environment (Siegle, 2011). Tourism is also influenced by fashion, and provides employment; some towns grew to service it, but it can damage the environment.

Green fashion (eco-fashion)

Environmental and social problems have prompted green fashion (Strähle, 2017). Producers and consumers like the idea of green fashion (there is a fashion for green fashion) – it promotes an attractive image, might reduce impacts but can be exaggerated or false (greenwashing) and is intended to sell more. Animal welfare activism has convinced or scared manufacturers and consumers to avoid fur and leather. The clothing (garments) industry is starting to use organic dyes, reduce use of bleach, avoid risky types of plastics, and perhaps cut water use. The drivers are genuine eco-consciousness, legislation and consumer pressure.

In 2023 the clothing industry was far from green in production or disposal: globally roughly 1.25 million tonnes of garments were discarded annually, some becoming second-hand clothing, some shredded for uses like house insulation, but the landfill amount was huge. Clothing manufacture and its supply chains also cause CO_2 emissions; water pollution with dyes, detergents, bleach, microfibres; plus pollution associated with transportation. Clothing manufacture has a poor health and safety record and competes with farming and city populations for water and natural products (Anguelov, 2016; Shabbir, 2019). Green (or eco-) fashion seeks to reduce negative environmental impacts (Muthu and Gardetti, 2016). An aspect of green fashion is sustainable fashion (Fletcher and Tham, 2015). While a shift from artificial fibres to flax, cotton or bamboo may look green to consumers it can mean competition with food crops for land and does little to reduce water demand and pollution during cultivation and manufacture.

A fashion absent in some nations, controlled in some, but rife in many is gambling. Huge sums are involved and the end product, unlike the recreational advertising image, may be citizen hardship and crime. Media and communication developments have offered new opportunities for the gambling industry.

Alternatives to consumerism and materialism have been proposed but require massive attitude changes or unpopular rules (Schlosberg and Craven, 2019). It would be good to encourage advertising that promotes green and more altruistic images, not status, material possessions and ill-advised consumption. However, draconian efforts to control alcohol in the USA and in Amazonia (by Ford) during the Prohibition Era failed.

Politics

Politics is tricky to pin down; it merges into ethics, political economy, political ecology, governance, policy studies, etc. One definition is: the process of making collective choices that express society's values (Conca, 2021), or co-operation of the concerned and control of rogues. Books about environmental politics tend to be conceptual, not really applicable (Dauvergne, 2012; Harris, 2014; Pattberg and Zelli, 2015; Dobson, 2016; Doyle et al., 2016; Stevenson, 2017; Carter, 2018; Kraft, 2018; Kütting and Herman, 2018). There are branches of politics: politics of growth, nationalistic politics, politics of trade, green/environmental politics, resources politics, global politics, etc. Whatever it focuses on, the core issue seems to be seeking good governance. Governance is the process of governing or overseeing the control and direction of something (such as a country, sector or an organisation). Governance ideally requires measures/processes that are transparent, accountability of decision makers, law/stability, communications, enforcement, co-ordination, capacity to respond to needs, etc. Whether it needs equitability, empowerment and democracy is arguable (Hansen-Magnusson and Vetterlein, 2020; Yifei Li and Shapiro, 2020).

The goal of EM is wise governance for the future, the Anthropocene (Koontz et al., 2010). Governance is often overshadowed by disagreements, so conflict avoidance and resolution and issues of justice are vital to EM and politics (Ostrom, 2015; Clarke and Peterson, 2016; Dodds et al., 2017). Politics and EM share many problems and demands and politics has had longer to develop coping strategies.

Most politicians have a short-term view (less than 5 years) and their own agenda, or do not have any. In an ideal world a policy (long term) is decided, ideally apolitically, and then programmes supporting and enforcing that policy are put in motion (medium term) and projects within programmes work at local and short-term scales (Vig and Kraft, 2021). Politics is shaped by politicians' and citizens' attitudes and these are as likely to be selfish as altruistic. Individuals, groups or countries have to be persuaded (or sometimes forced) to work co-operatively without exploiting others. Politics is unpredictable, and often ill-informed. Politics shapes the stage on which EM is played but for many involved in environmental science is something alien and to be avoided.

Progress with green politics since 1970 has mainly come from established politics becoming aware of environmental issues and then supporting/exploiting them. Governance by elected green parties is rare, greening is mainly by influencing others. A shift in politics can have massive impact (good or bad) on the environment, as demonstrated by a few 2020s populist leaderships. Since roughly 2018 populism has spread. Not new, it has recently manifested as a diverse political approach (right- or left-wing), often authoritarian. It strives to appeal to people who feel that their concerns are disregarded by established groups (including scientists). This backlash against experts and advice/regulation may prompt government to allow deforestation, global warming, ignore advice on Covid-19 precautions, etc., and voice or try to follow simplistic solutions to challenges (Huber, 2020). Some environmentalists might be called populist and seek fast changes (McCarthy, 2019). Populist action might sometimes provide one route to effective policies, which are otherwise discussed, delayed or shelved (Beeson, 2019). But it has led to damage in Amazonia and setbacks to C-emissions mitigation and biodiversity conservation. EM should seek understand populism and may occasionally need to work with it. However, there is risk in 'tyranny of the masses'. Expert knowledge, including that of environmental scientists, should be used wisely to inform decisions.

Things get more complicated when EM issues require co-operation between more than one political system. Regional unions – the EU, ASEAN, etc. – have helped, aiding trust, cross-border co-ordination, monitoring and development of regulations. The development of ocean resource exploitation technology and the loss of ice in the Arctic, Antarctic and Greenland look likely to lead to stresses on co-operation between political systems. Resource demands and the growth of the Indian and Chinese populations and economies may pose challenges.

Some resources have attracted the attention of politics; there are many publications on oil politics, water politics (especially the sharing and coordinated management of multistate rivers), the politics of pollution, ocean resources and climate change (Giddens, 2009; Vogler, 2016). Water is a common cause of disputes, controlling it can bring political power and prompt social organisation (Wittfogel, 1957; Conca, 2021).

Political ecology

Political ecology might help guide politics and governance towards a systems approach aiming at co-evolution of humans and nature (Clapp and Dauverge, 2011; Cadman et al., 2015; Fiorino, 2018; Robbins, 2020).

Green politics

Green political parties today seldom control a majority of votes in any country. But various other parties have embraced (or use) green (Wall, 2010; Newell, 2020). Most greens oppose nuclear weapons and nuclear generation, and some support the politics of survival. Degrowth, as opposed to economic growth/development, is professed by some; surprisingly rarely is there concern about population. Green politics is much influenced by NGOs like Greenpeace, Friends of the Earth, etc.

Politics for the Anthropocene

Politics for the Anthropocene will need to support Earth systems governance. The Earth is now 'post-nature'; humans will have to practice stewardship but few citizens have grasped this (Purdy, 2015; Biermann and Lövbrand, 2019; Dryzek and Pickering, 2019).

It is vital to improve human adaptability, improve resilience, reduce vulnerability and avoid major war (Beeson, 2019). Nationalism, capitalism and consumerism look unlikely to disappear, but they might feed into something more environmentally friendly (Arias-Maldonado and Trachtenberg, 2019; Fremaux, 2019). It is too soon to see what shape Anthropocene national and international politics will take. Perhaps something like the Swiss canton system: autonomous regions which have links to a trusted coordinating centre which they contribute to. New politico-economic challenges have appeared: debt-trapping, purchase of resources without claiming sovereignty, forms of neo-colonialism, populism, undeclared and deniable warfare, etc. There are hints that SMART approaches may be used to monitor citizen responses to governance.

Summary

- Too little attention has been given to improving adaptation. A Global Commission on Adaptation was launched in 2018.
- The Earth has started a transition, to the Anthropocene, a shift from humans using resources with little concern for consequences to a novel challenge: learning to co-evolve with Earth. If the transition cannot be made then civilisation and much of Earth's biota may face disaster.
- There have been great advances in technology and it might help with the transition.
- Attitudes determine whether technology will be wisely used, population managed, pollution controlled, whether catastrophic warfare will happen.
- EM must consider much in addition to environmental science: politics, education, media and social media will contribute to wise governance. EM must work with these and not be distracted by them.

Further reading

Books

Ashby, M.F. (2013) *Materials and the Environment: eco-informed material choice* 2nd edn. Elsevier (Butterworth-Heinemann), Waltham, MA, USA.

Beeson, M. (2019) *Environmental Populism: the politics of survival in the Anthropocene.* Springer (Palgrave Macmillan), Singapore, SG.

Carlson, R.H. (2010) *Biology Is Technology: the promise, peril, and new business of engineering life.* Harvard University Press, Cambridge, MA, USA.

Carlson, R.H. and Koonin, S.E. (2021) *Unsettled: what climate science tells us, what it doesn't, and why it matters.* BenBella Books, Dallas, TX, USA.

Ellis, E.C. (2018) *Anthropocene: a very short introduction.* Oxford University Press, Oxford, UK.

Haberlein, T.A. (2012) *Navigating Environmental Attitudes.* Oxford University Press, Oxford, UK.

Marinova, D., Annandale, D. and Phillimore, J. (Eds) (2006) *The International Handbook on Environmental Technology Management.* Edward Elgar, Cheltenham, UK.

UNDP (2020) *Human Development Report 2020: the next frontier – human development and the Anthropocene.* United Nations Development Program (UNDP), New York, NY, USA.

Journals

- *Environmental Politics* (Taylor & Francis). ISSN print: 0964-4016; ISSN online: 1743-8934. www.tandfonline.com/action/journalInformation?journalCode=fenp20 (accessed 17/11/22).
- *Global Environmental Politics* (MIT). ISSN print: 1526-3800; ISSN online: 1536-0091. https://direct.mit.edu/glep (accessed 17/11/22).
- *Journal of Political Ecology* (University of Arizona Libraries). ISSN online: 1073-0451. www.talktalk.co.uk/speedcheck/ (accessed 17/11/22).
- *Industrial Ecology* (Wiley). ISSN online: 1530-9290. https://onlinelibrary.wiley.com/journal/15309290 (accessed 01/11/22).
- *Anthropocene* (Elsevier). ISSN online: 2213-3054. www.sciencedirect.com/journal/anthropocene (accessed 17/11/22).
- *The Anthropocene Review* (Sage). ISSN print: 2053-0196; ISSN online: 2053-020X. https://journals.sagepub.com/home/anr (accessed 17/11/22).
- *Anthropocene Science* (Springer). ISSN online: 2731-3980. https://mail.google.com/mail/u/0/#inbox?compose=new (accessed 17/11/22).
- *Journal of Green Science and Technology* (American Scientific Publishers). ISSN print: 2164-7585; ISSN online: 2164-7607. www.aspbs.com/jgst/inst-auth_jgst.htm (accessed 18/11/22).
- *Green Technology, Resilience and Sustainability* (Springer). ISSN online: 2731-3425. www.springer.com/journal/44173 (accessed 18/11/22).
- *Green Energy and Environmental Technology* (IntechOpen). ISSN online: 2754-6314. www.intechopen.com/journals/7 (accessed 17/11/22).
- Green Energy & Environment (Elsevier). ISSN print: 2468-0257. www.sciencedirect.com/journal/green-energy-and-environment (accessed 17/11/22).
- *Biotechnology Journal* (Wiley). ISSN online: 1860-7314. https://onlinelibrary.wiley.com/journal/18607314 (accessed 18/11/22).
- *Journal of Biotechnology* (Elsevier). ISSN print: 0168-1656. www.sciencedirect.com/journal/journal-of-biotechnology (accessed 18/11/22).

www sources

- ❏ Top 10 green technology innovations: https://sustainabilitymag.com/top10/top-10-green-technology-innovations (accessed 16/11/22).
- ❏ Ten examples of green technology: https://tecamgroup.com/10-examples-of-green-technology/ (accessed 17/11/22).
- ❏ Anthropocene (Natural History Museum, London, UK): www.nhm.ac.uk/discover/what-is-the-anthropocene.htm (accessed 18/11/22).
- ❏ Anthropocene (Smithsonian): www.smithsonianmag.com/science-nature/what-is-the-anthropocene-and-are-we-in-it-164801414/ (accessed 17/11/22).
- ❏ Top green social media networks: https://thegreenmarketoracle.com/2011/03/05/top-green-social-networks/ (accessed 17/11/22).

Part V

The future

Chapter 15

The way ahead

Chapter overview

- Challenges and new opportunities
- Looking at the future: the Anthropocene
- EM going forwards
- Politics and ethics to support EM
- Closing note
- Summary
- Further reading
- www sources

Challenges and new opportunities

Environmental ethics, environmental law and EM have to evolve to serve diverse and changing socio-political systems (Ryder et al., 2021). There is still a need to better integrate physical and social sciences, and find effective problem-solving approaches and adaptable forms of environmental governance (Ehlers et al., 2006; Biermann, 2014; Dalbotten et al., 2014; Bonneuil and Fressoz, 2017; Kotzé, 2017; Schmitz, 2017; Arias-Maldonado and Trachtenberg, 2019; Clark and Szerszynski, 2020; Clement, 2021; Padilla, 2021).

The means and approaches to handle ongoing challenges are yet to be adequately identified (Galaz, 2014; Biello, 2016; Berners-Lee, 2021). Business management has future scenario prediction and assessment techniques (visioning) to aid in the identification of best strategies. EM is assisted by futures study, history, paleoecology and archaeology (Marshall and Connor, 2016; Centeno et al., 2023). The future may not unfold in ways like the past but past awareness does provide ideas, approaches, warnings, and prompt contingency planning.

In 2023 focus is on global warming, especially carbon emissions control, with some consideration of energy supply and pollution reduction. Other challenges get less attention and deserve far more (Spring, 2020; Allen, 2022). The need for resilience and adaptability are vital and need more attention. Urbanisation, vertical farming, high-tech agriculture, water recycling and non-polluting energy might free areas from human pressure that perhaps could be returned to nature. This technology does not cover cereal and potato staples, these remain field crops using 1st and 2nd Green Revolution approaches. As oil is phased out staples will need new approaches; perhaps chemical fertiliser can be substituted by green manures (like growing clover or alfalfa in rotation with cereal); farm mechanisation poses fuel challenges post-oil. If biochar proves as effective as hoped, agriculture could be more productive with less fertiliser demand and lower pollution, and might even be carbon neutral. Improved rain-fed agriculture, hydroponics and drip-irrigation should be supported, together with reduction of land degradation.

Biodiversity is being lost at an alarming rate; and its value in ensuring environmental stability and providing many crucial benefits for humans is largely undetermined. Even excluding moral obligations, it makes sense to conserve biodiversity to keep open future options (Wilson, 1992). It is clear that enforcement of conservation needs strengthening.

The value of biotechnology must be carefully weighed against risks but it should not be dismissed. There have already been cases of non-GM innovation which have had unwanted and unexpected economic and social impacts; e.g., the adoption of high fructose corn syrup by the food and drinks industries hit sugar producers badly between 1983 and 1984. Biotechnology might make it possible to produce things such as cocoa butter, or even alternatives to palm oil, which would cut pressure for land clearance and pollution but hit countries that rely on commodity exports. Assessment of potential and threats of innovations needs to be improved.

Looking at the future: the Anthropocene

The establishment of the Anthropocene may help strengthen and focus EM (Schwägerl, 2014). Much study so far is speculative, but serves to prompt thought about dealing with coming challenges and opportunities (Kolbert, 2006, 2014; Kress and Stine, 2017). Stern (2009) made an important point: danger lies in a poor *response* to a challenge, not just the challenge. Dystopian doom-assumptions might already have prompted ineffective and wasteful measures.

The problem with futures studies is for users to separate useful from unhelpful speculation. Sometimes futures assessments change what might otherwise have happened; *The Limits to Growth* debate prompted realisation that there were probably threats which may not have dawned otherwise for decades. A number of bodies do regular, even annual assessments which show trends and refresh baseline data. One of the resources which needs attention is food. It is generally agreed that, since 1990, agricultural production has slowed, while population growth is still running at over 70 million a year. The challenge is to feed growing numbers without degrading the environment or depending on petroleum inputs (Smil, 2001). Agricultural improvement to date may be hiding gradually accumulating damage and a threshold might be suddenly reached. Food is vulnerable, as in the past, to sudden disruption from events like volcanic eruption. Storage and responses need consideration – a global fall in temperatures of several degrees might be softened by use of lots of plastic growing-tunnels, something that can quite quickly be prepared.

Future predictions are biased towards pessimism (less often optimism), but even when unbiased, are fraught with uncertainties. It is unwise to assume the coming decades will be a dystopia and it is foolish to assume technology and governance will easily resolve problems. The wise approach is prudence, precaution, resilience and adaptability.

Dystopian future

It is argued the Anthropocene will be marked by Earth's 'sixth mass extinction' and already it is clear biodiversity loss is massive.

Going into survival mode is not wise. Aiming for humans just to survive will miss chances to advance civilisation (Lueddeke, 2019); there is a need to embrace change and seek to prosper, something dystopians miss. However, sudden nuclear war, catastrophic environmental change, uncontrollable migration or sudden natural disaster might necessitate survival strategies (McGuire, 2005; Dartnell, 2014). A risk is that a doom focus and loss of hope, or fatalism, or indifference, leads people to inaction or outright denial of the need for remedial actions. Also, a doom focus may trigger fatalistic inaction, quick-and-dirty technology, knee-jerk responses and risk taking, rather than careful policies. Consumerism is a problem and high ongoing human migration may hinder EM. It is unwise to allow marketing and consumerism to drive technology and cause environmental challenges, but this will be difficult to control. The default position is probably business-as-usual, i.e. carry on as normal with slow change and less than vigorous responses to problems already manifest.

Utopian future

To assume a dystopian future is undesirable, as is adoption of the opposite extreme. Techno-utopians (many citizens) feel science will find ways to overcome problems – they are over-optimistic. Adaptation will require human habits to change, especially consumer habits. The shift to the Anthropocene is the chance for humans to develop a new relationship with nature. There is now growing knowledge of how ecosystems and human socio-economic systems work so it should be possible to manage development better. Ecomodernist utopian extremists seek to take advantage of a warmer future climate to advance civilisation (ectopia). These hopes echo notions professed by Thomas Moore in *Utopia* (1516), by Francis Bacon in *New Atlantis* (1626) and by Tomasso Campanella in *The City of the Sun* (1623): civilisation, work, good governance and technology yield a better future life. Reality is more likely to be a struggle to achieve co-evolution and stable livelihoods (Thompson, 2005; Hamilton, 2017; Garforth, 2018; Harvey, 2019; Costanza et al., 2020; Thaler, 2022). Management of global and regional ecosystems and human systems will be a challenge and for decades population will be too high to sustain (overshoot).

During overshoot there will be many challenges, not simply global warming or global population. Some nations in 40 years will have large, young, restless populations; post-oil energy supply may be a struggle; and land degradation has been long neglected.

Realistic futures?

Humans have created AI and the Anthropocene might not see a human/nature co-evolution. It is possible human/AI/nature or AI/nature will appear. Utopia might be achieved but perhaps not by humans alone (Lovelock, 2019).

Humans must make the most of opportunities and better cope with challenges (Collier, 2010; Costanza and Kubiszewski, 2014; Denny, 2017; Rees, 2018; Riede and Sheets, 2020).

Overshoot is a period of living beyond the sustainable limits, but hopefully it will be possible to reduce demands and population in a non-draconian way to a level offering long-term security and prosperity. Deciding what target population each nation has will be fraught with difficulty. How long overshoot allows for problem solving is difficult to say but it will probably be less than a century (Sachs, 2015).

The UN 2030 SDGs and discussions by nations on the UN Framework for Climate Change led in 2015 to the Paris Agreement, where 196 governments decided to work to keep the world's average temperature from rising to 2°C above what it had been before the industrial revolution, and preferably to keep it below 1.5°C. In 2023 those aims seem unlikely and warming will perhaps exceed 1.5°C by 2050.

The planetary boundary model offers a tool providing targets for policy makers, giving a clear indication of the magnitude and direction of change. It also provides benchmarks and direction for science. This may help humans think more deeply and urgently about planetary limits and the critical actions to take. Critics argue boundaries are unpredictable and fail to consider population growth adequately, and should be treated with caution.

EM going forwards

After the late 1980s it became easier to exchange information and to co-operate on environmental care, but by 2023 relations between Russia, China and the West had deteriorated; there was spread of nuclear weapons to more countries and an expansion in overall number of warheads (Figure 15.1). The challenges of the Anthropocene are perhaps going to be aggravated by a decline in international relations, ongoing terrorism and migration.

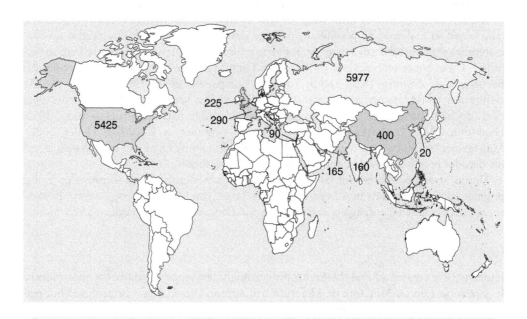

Figure 15.1 Nuclear weapons in 2023 (estimated). One nation (South Africa) renounced its weapons in 1989 and two more (Iran and Saudi Arabia) look set to become nuclear powers. World total in 2023 = roughly 12,755.

Various sources suggest the following nations are increasing and/or modernising/upgrading stockpiles and means of delivery: UK, Russia, China, India, Pakistan and North Korea. The only country to renounce weapons – South Africa – renounced about six atomic bombs in 1989. Sources vary a great deal – there may have been a global peak in 1986 (how many weapons are dismantled or stored is unknown) but yield of weapons has probably increased as more H-bombs appear and replace A-bombs. Some claim Israel has 75 to 400 and North Korea > 50. In the 1970s the UK may have had 520. Clearly figures are not reliable or precise. The totals in this figure do not distinguish the proportion of far more powerful H-bombs. The latter are possessed by the USA, Russia, the UK, France, China, India (?) and North Korea (?).

India and China are industrialising, undergoing economic growth and have expanding populations. By 2030 globally women will control more wealth, Asia and Africa will dominate consumption, and some nations will have aging and declining populations while others will be youthful and expanding. Western influence may well diminish.

Politics and ethics to support EM

More authoritarian politics can force through change, which in a democracy may be delayed or difficult to achieve due to selfish, irrational or indolent politicians and voters. Two nations which have made progress with EM are Cuba and Singapore, both quite authoritarian regimes. A relatively poor and authoritarian country, Dominica, has been enacting effective conservation and anti-land degradation measures since the 1990s. The People's Republic of China has implemented EM legislation and developed environmental policies. So, relatively non-liberal governments have taken proactive EM measures. Democrats would argue that people must have the right to make mistakes and learn; however, environmental problems affect more than one nation, mistakes can be difficult or impossible to resolve and there is little time for mistakes and 'learning'.

Anthropocene politics and ethics will not be easy (Purdy, 2015; Dryzek and Pickering, 2019). The current generation hold the environment in trust for future generations; in law and business, trust funds require a firm and impartial administrator to enforce things – international institutions do not yet provide that. Paying today to benefit future people is difficult to sell. Many environmentalists and green politicians adopt a decentralised, local focus, which could slow coping with environmental challenges. Also, many issues demand a large-scale approach to their solution; for example, investment in costly research, global problem avoidance, etc. Faced with challenges some nations may become ruthless and more nationalistic looking after their own interests, and ignoring any global obligations or costs.

One way of countering environmental inertia might be to adopt a World Charter on the Rights of Nature (efforts at the 1992 Earth Summit and those more recently have failed). Ecosystem services must be better incorporated into economics and law (Powe, 2007; Whitehead, 2014; Guttmann, 2018; Hiltner, 2020; Purseglove, 2020). To be anticipatory EM will need to work with political analysts as well as ecologists, specialists in ethics and law, etc.

EM operates as a multidisciplinary multi-layered process dealing with human–environment interaction (Hamilton et al., 2015; Hamilton, 2017; Thomas et al., 2020).

(a)

(b)

Figure 15.2 a) Singapore metro/underground railway. An efficient rapid transit system. b) View of Singapore metropolitan area, a city state with a dense population and a tropical environment, which has made impressive progress with urban EM. In a number of fields the city is among world leaders, notably in efforts to control car traffic and provide adequate popular public transport.   Source: Photos by author, 2004.

Closing note

There has been progress: in the 1950s few people were aware there was any environmental problem; in the 1970s many saw environmental care as a luxury and some suspected it might be green imperialism. In 1972 few or no nations had environmental ministries. By 1992 the UN Conference on Environment and Development attracted a huge attendance, and there were few delegates willing to state publicly that environmental care was not a vital component of development. By 1995 virtually every country had an environmental ministry or agency; most newspapers and television channels had environmental correspondents. Interest in SD was well enough established by 2002 for a major international conference to be held by the UN. By 2004 global warming was seen as a threat demanding international action, with some nations prepared to spend vast sums responding to it. By 2015 the UN was setting SDGs and popular environmental concern had grown.

EM has developed far beyond what might have been expected before 1970 when ecology was a virtually unknown word. If a similar degree of progress takes place in the coming 50 years then prospects might be reasonable. Recent populist politicians have shown how easy it is to set EM back, but citizen environmental awareness has hugely increased worldwide. Unthinking indiscriminate opposition to modern technology will not resolve environmental challenges, may exacerbate some and could cause widespread starvation. A 3rd phase – Doubly Green Revolution – is needed to refocus technology to cut environmental damage, not the rejection of technology. There must be more action to counter worsening soil degradation and environmental pollution. High petroleum input cereal farming with fertiliser pollution feeds the world today but has to be rethought. Water supplies and climate change will be serious challenges. Promising solutions are mainly those already developed that can be implemented fast and at affordable cost. Technology that needs 20 years to install or develop has to be assigned second priority. To back up wind and solar energy generation, small modular nuclear reactors may be the BPEO, even if unpopular. It is important to avoid policies that may be popular that divert efforts and funds from priority challenges, opportunities and effective solutions (politicians will not like this).

Sometimes privatisation has been an improvement, but it should be monitored and, if necessary, coordinated. So too should the process of globalisation. While pressing ahead with available technology there is also a need to ensure that governments and research bodies do not neglect 'blue-sky topics' in favour only of applied (too often meaning politically or commercially useful) studies: in the long term there will be new (unknown at present) challenges. There is a need to address infrequent but potentially catastrophic risks, which governments ignore or argue they cannot afford to waste funds upon. Funds for studies of atmospheric gas levels were hard to come by in the 1950s–1970s, yet without the studies and long-term monitoring ongoing, CO_2, O_3 and methane changes would be impossible to understand and extrapolate. Until recently comet or asteroid strikes were given little thought. In 1991, asteroid 1191B narrowly missed Earth (and it was big enough to have caused a global catastrophe). Then there was the collision of fragments of comet Shoemaker-Levy 9 with Jupiter (presenting Earth-size explosions on impact) and the discovery of the 2km-diameter asteroid 1997xf11, which it is hoped will miss Earth in 2028. Recent space missions indicate that technology can be easily developed to give fairly good protection against most incoming asteroids. For the first time in Earth's history a major threat can be realistically reduced. However, there are other worrying threats; the problem is that they are not currently perceived as important.

Managing the fate of the Earth prompts ethical issues: EM could be in the position of selecting one of various possible alternative futures, and in so doing prevent other possibilities. Humans will need to establish what ethical obligations they have to nature and see these 'rights' of nature are respected. It is important that EM applies the carpenter's rule: 'measure twice and cut once' (be very cautious). And more has to be done to embed the precautionary outlook in the world's administrations and politics.

Challenges lie ahead (not in priority order):

- large global population;
- eco challenges* (human): global warming, land degradation, pollution and biodiversity loss;
- eco challenges* (natural): volcanic, pandemic, solar flares, asteroids, and so on;
- nuclear war and/or chemical and biological weapons;
- totalitarianism/populism/nationalism – survival of the ruthless;
- corporate refusal to adhere to EM advice;
- economic collapse;
- breakdown of governance/social unrest;
- decline of respect for technology and learning;
- need to improve food supply, security and reserves;
- coordination and enforcement of EM at transnational scales – international bodies can really only promote and advise at present.

* Natural and human causes interact

There are numerous guides to future survival, useful as prompts for ideas (Torp and Andersen, 2020; Kolbert, 2021; Dixson-Declève et al., 2022). It is too soon to say whether human civilisation and survival prospects are good or not. Frank (2018) considered the Drake 'equation'. This thought model explores the chance of extraterrestrial life appearing in the universe. Recent finds of many exoplanets suggests life beyond Earth might be common. But Fermi posed a paradox some time ago: 'if life is common, how come none has made contact?' Frank mused that life appears but might rarely transition to stable co-evolution with a planet: collapse/failure to survive and maintain civilisation might be the rule. Hence, there are few or no advanced extraterrestrial civilisations to make contact with. Earth's history and archaeology already show that few civilisations have lasted more than a couple of hundred years, and now we have fearsome weapons. Planets also change naturally – an environment may not be stable and fixed: Mars is thought to have once been wet and had a denser atmosphere but is now a cold low pressure hyper-desert; Venus may have had oceans but it now has a surface hot enough to melt lead. Humans cannot expect the Earth to be the same forever; the future consists of moving targets which humans have to adapt to; resilience and adaptability are paramount.

Various future possibilities might be recognised:

i) *Pass Earth carrying capacity with too great a population* resulting in *sudden major collapse*. A bad scenario but perhaps slow eventual recovery.
ii) *Pass Earth carrying capacity (overshoot) and experience a gradual collapse*. Stress for a while and unpleasant but perhaps survival and eventually steady state. There must also be adequate response to natural and socio-economic threats/disasters.
iii) *Pass Earth carrying capacity resulting in catastrophic collapse*. No recovery. Disaster for humans and biota.

iv) *Major nuclear war – the exchange resulting in blast, radiation fall-out and widespread fire probably causing global nuclear winter.* Human and ecological disaster.

v) *Catastrophic natural challenge which humans fail to adequately mitigate or adapt to such as a large eruption or asteroid strike or an insidious but major shift in a critical global ecosystem.* Human and ecological disaster.

Organisms do not survive in the long run through strength or intelligence but resilience and adaptation (and perhaps luck). Humans can use intelligence to enhance resilience and adaptability *if they choose to*. At present the Earth's carrying capacity is uncertain but there is clearly overshoot. Even the best scenarios probably involve an unpleasant and uncertain journey.

Summary

- Some of the problems faced by EM are clear, but many will be unexpected. Watchfulness, monitoring and adaptive strategies are vital.
- Action on global warming focuses on carbon emissions control, largely neglects gases other than CO_2, may not be effective and fast enough, could prove economically damaging and diverts attention and resources from other challenges.
- The uncertainties in predicting what might happen places a premium on boosting resilience, cutting vulnerability and increasing adaptability.
- The goal is a transition to sustainable human–Earth co-evolution with minimum serious hardship en route.
- Overshoot of Earth's carrying capacity is happening now; the global population is above a sustainable level. A decline in birth rates in a coordinated non-draconian manner is needed.
- A global warming of 2.7°C by 2100 seems likely and would be most undesirable.
- A steady state co-evolution may be decades away. EM has to steer humans towards it, being proactive and precautionary.
- Nuclear war is a serious threat.

Further reading

Books

Brown, L.R. (2005) *Outgrowing the Earth: the food security challenge in an age of falling water tables and rising temperatures.* Routledge (Earthscan), London, UK.

Denny, M. (2017) *Making the Most of the Anthropocene: facing the future.* John Hopkins University Press, Baltimore, MD, USA.

Hamilton, C. (2017) *Defiant Earth: the fate of humans in the Anthropocene.* Polity Press, Cambridge, UK.

Harvey, M. (2019) *Utopia in the Anthropocene: a change plan for a sustainable and equitable world.* Routledge, Abingdon, UK.

Van Ginkel, H., Barrett, B., Court, J. and Velasquez, J. (Eds) (2005) *Human Development and the Environment: challenges for the United Nations in the new millennium.* United Nations University Press, Tokyo, JP.

Journals

- *Futures* (Elsevier). ISSN print: 0016-3287, ISSN online: 1873-6378. www.sciencedirect.com/journal/futures (accessed 02/01/23).
- *Future* (MDPI). ISSN online: 2813-2882. www.mdpi.com/journal/future (accessed 02/01/23).
- *International Journal of Futures Studies* (Tamkang University TW). ISSN print: 1027-6084. https://jfsdigital.org/ (accessed 03/01/23).
- *ENDLESS: International Journal of Future Studies* (Global Writing Academica Researching & Publishing). ISSN online: 2775-9180. https://endless-journal.com/index.php/endless (accessed 03/02/23).
- European Journal of Futures Research (Springer). ISSN online: 2195-2248. https://eujournalfuturesresearch.springeropen.com/ (accessed 02/01/23).
- *Foresight* (Emerald). ISSN online: 1463-6689. www.emerald.com/insight/publication/issn/1463-6689 (accessed 03/01/23).

www sources

- World Future Society. https://docs.google.com/document/u/0/ (accessed 03/01/23).
- Forum for the Future. www.forumforthefuture.org/ (accessed 02/01/23).
- The Futurist Society. www.gov.uk/renew-adult-passport (accessed 02/01/23).
- Future Earth. https://mail.google.com/mail/u/0/#inbox?compose=new (accessed 03/01/23).

Bibliography

Abate, R.S. and Warner, E.A.K. (Eds) (2013) *Climate Change and Indigenous Peoples: the search for legal remedies*. Edward Elgar, Cheltenham, UK.

Abbasi, S.A. and Abbasi, T. (2017) *Ozone Hole: past, present, future*. Springer, New York, NY, USA.

Abbati, M. (2019) *Communicating the Environment to Save the Planet: a journey into eco-communication*. Springer, Cham, CH.

Abdul, S. and Khan, R. (Eds) (2019) *Green Practices and Strategies in Supply Chain Management*. IntechOpen, London, UK.

Abdullah, A.M. (2007) *Introduction to Environmental Management System*. PENERBIT, Universiti Teknologi Malaysia, Skudai, MY.

Abdul-Matin, I. (2010) *Green Deen: what Islam teaches about protecting the Planet*. Berrett-Koehler Publisher, San Francisco, CA, USA.

Aberšek, B. and Flogie, A. (2022) *Human Awareness, Energy and Environmental Attitudes*. Springer, Cham, CH.

Abraham, M.A. (Ed) (2017) *Encyclopedia of Sustainable Technologies*. Elsevier, Amsterdam, NL.

Abram, S., Waltorp, K., Ortar, N. and Pink, S. (Eds) (2022) *Energy Futures: Anthropocene challenges, emerging technologies and everyday life*. Walter de Gruyter, Berlin, GR.

Abu Bakar, M.A.A., Amirul, S.M. and Aralas, S. (2016) Environmental management system and tools of Malaysian listed firms and its financial performance. *Journal of Global Business and Social Entrepreneurship* 3(5), 63–74.

Abunnasr, Y., Ryan, R.L. and Infield, E.M. (Eds) (2018) *Planning for Climate Change: a reader in green infrastructure and sustainable design for resilient cities*. Routledge, Abingdon, UK.

Acar, S. and Yeldan, A.E. (Eds) (2019) *Handbook of Green Economics*. Elsevier (Academic Press), London, UK.

Acevedo, M.F. (2012) *Data Analysis and Statistics for Geography, Environmental Science, and Engineering*. CRC Press, Boca Raton, FL, USA.

Acevedo, M.F. (2015) *Real-Time Environmental Monitoring: sensors and systems*. CRC Press, Boca Raton, FL, USA.

Achillas, C., Bochtis, D.D., Aidonis, D. and Folinas, D. (2019) *Green Supply Chain Management*. Routledge (Earthscan), Abingdon, UK.

Achour, N., Pantzartzis, E., Pascale, F. and Price, A.D.F. (2015) Integration of resilience and sustainability: from theory to application. *International Journal of Disaster Resilience in the Built Environment* 6(3), 347–362.

Bibliography

Ackerman, D. (2014) *The Human Age: the world shaped by us*. W.W. Norton, New York, NY, USA.

Adams, C.J. and Gruen, L. (Eds) (2022) *Ecofeminism: feminist intersections with other animals and the Earth* 2nd edn. Bloomsbury, London, UK.

Adams, W.M. (2003) *Future Nature: a vision of conservation* (revised edn). Earthscan, London, UK.

ADB. (2016) *River Basin Management Planning in Indonesia: policy and practice*. Asian Development Bank, Manila, PH.

Adger, W.N., Lorenzoni, I. and O'Brien, K.L. (Eds) (2009) *Adapting to Climate Change: thresholds, values, governance*. Cambridge University Press, Cambridge, UK.

Adger, W.N., Kelly, P.M. and Nguyen H.N. (2001) *Living with Environmental Change: social vulnerability, adaptation and resilience in Vietnam*. Routledge, London, UK.

Adler, F.R. and Tanner, C.J. (2013) *Urban Ecosystems: ecological principles for the built environment*. Cambridge University Press, Cambridge, UK.

Adoue, C. (2011) *Implementing Industrial Ecology: methodological tools and reflections for constructing a sustainable development*. CRC Press, Boca Raton, FL, USA.

Agarwal, P., Mittal, M., Ahmed, J. and Idrees, S.M. (Eds) (2022) *Smart Technologies for Energy and Environmental Sustainability*. Springer, Cham, CH.

Agrawal, S.B., Agrawal, M. and Anita Singh, A. (Eds) (2021) *Tropospheric Ozone: a hazard for vegetation and human health*. Cambridge Scholars Publishing, Newcastle-upon-Tyne, UK.

Agrawala, S. and Fankhauser, S. (Eds) (2008) *Economic Aspects of Adaptation to Climate Change: costs, benefits and policy instruments*. OECD, Paris, FR.

Agyeman, J. (2013) *Introducing Just Sustainabilities: policy, planning and practice*. Zed Books, London, UK.

Ahamad, A., Singh, P. and Tiwary, D. (Eds) (2022) *Plastic and Microplastic in the Environment: management and health risks*. Wiley-Blackwell, Oxford, UK.

Ahlhorn, F. (2017) *Integrated Coastal Zone Management: status, challenges and prospects*. Springer, Wiesbaden, GR.

Ahmad, N. and Jolly, S. (2019) *Climate Refugees in South Asia: protection under international legal standards and state practices in South Asia*. Springer, Singapore, SG.

Ahraman, C. and Sari, I.U. (Eds) (2017) *Intelligence Systems in Environmental Management: theory and applications*. Springer, Cham, CH.

Aitkenhead-Peterson, J. and Volder, J.A. (Eds) (2020) *Urban Ecosystem Ecology* (first published 2010). Wiley, New York, NY, USA.

Ajibade, I.J. and Siders, A.R. (Eds) (2021) *Global Views on Climate Relocation and Social Justice: navigating retreat*. Routledge, Abingdon, UK.

Ajith Sankar, R.N. (2015) *Environmental Management: vocabulary*. Oxford University Press India, Delhi, IN.

Akmam, W. (2017) *Arsenic Mitigation in Rural Bangladesh: a policy-mix for supplying safe water in badly affected areas of Meherpur District*. Springer, Tokyo, JP.

Alberti, M. (2015) Eco-evolutionary dynamics in an urbanizing planet. *Trends in Ecology & Evolution* 30(2), 114–126.

Albuquerque, U.P., Ramos, M.A., Ferreira, W.S. jnr. and Muniz de Medeiros, P. (2017) *Ethnobotany for Beginners* (original Portuguese edn 2005). Springer Nature, Cham, CH.

Alcamo, J. (Ed) (2009) *Environmental Futures: the practice of environmental scenario analysis*. Elsevier, Amsterdam, NL.

Alcantara, E. (Ed) (2013) *Remote Sensing: techniques, applications and technologies*. Nova Science Publishers, Hauppauge, NY, USA.

Aldrich, D.P. (2012) *Building Resilience: social capital in post-disaster recovery*. University of Chicago Press, Chicago, IL, USA.

Aldy, J.E. and Stavins, R.N. (Eds) (2010) *Post-Kyoto International Climate Policy: implementing architectures for agreement*. Cambridge University Press, Cambridge, UK.

Alex, G. (2020) *Environmental Impact Assessment (EIA) Simplified. @ GrG*. Blue Rose Publishers, Delhi, IN.

Alfredson, K. and Engeland, K. (Eds) (2020) *Hydrology and Water Resources Management in a Changing World*. IWA Publishing, London, UK.

Ali, M. (2013) *Sustainability Assessment: context of resource and environmental policy*. Elsevier, Amsterdam, NL.

Ali, S.S., Kaur, R. and Marmolejo Saucedo, J.A. (2019) *Best Practices in Green Supply Chain Management: a developing country perspective*. Emerald Publishing, Bingley, UK.

Al-Jayyousi, O.R. (2012) *Islam and Sustainable Development: new worldviews*. Ashgate Publishing (Gower), Farnham, UK.

Allan, C. and Stankey, G.H. (Eds) (2009) *Adaptive Environmental Management: a practitioner's guide*. Springer (with CSIRO), Collingwood, VI, AU.

Allen, J.P. (1991) *Biosphere 2: the human experiment*. Penguin, New York, NY, USA.

Allen, V.M. (2022) *Eight Billion Reasons Population Matters: the defining issue in the 21st century*. FriesenPress, Altona, Manitoba, CN.

Almusaed, A. (Ed) (2016) *Landscape Ecology: the influences of land use and anthropogenic impacts of landscape creation*. InTech, Rijeka, HR.

Alongi, D.M. (2018) *Blue Carbon: coastal sequestration for climate change mitigation*. Springer, Cham, CH.

Aloqaily, A. (2018) *Cross Country Pipeline Risk Assessments and Mitigation Strategies*. Elsevier, Amsterdam, NL.

Altieri, M.A. (2018) *Agroecology: the science of sustainable agriculture* 2nd edn. CRC Press, Boca Raton, FL, USA.

Alvarez, R.A. (2016) *Hurricane Mitigation for the Built Environment*. CRC Press, Boca Raton, FL, USA.

Alvarez, W. and Asaro, F. (1990) What caused the mass extinction? An extraterrestrial impact. *Scientific American* 263(4), 76–84.

Alves, F., Leal Filho, W. and Azeiteiro, U. (Eds) (2018) *Theory and Practice of Climate Adaptation*. Springer, Cham, CH.

Al-Zu'bi, M. and Radovic, V. (2019) *SDG 11 – Sustainable Cities and Communities: towards inclusive, safe, and resilient settlements*. Emerald Publishing, Bingley, UK.

Anbumozhi, V., Kimura, F. and Thangavelu, S.M. (Eds) (2020) *Supply Chain Resilience: reducing vulnerability to economic shocks, financial crises, and natural disasters*. Springer, Singapore, SG.

Andeobu, L., Wibowo, S. and Grandhi, S. (2021) A systematic review of e-waste generation and environmental management of Asia Pacific countries. *International Journal of Environmental Research & Public Health* 18(17), article No. 9051 (18 pp.).

Andersen, A.G. (2014) *Media, Environment and the Network Society*. Palgrave Macmillan, New York, NY, USA.

Andersen, S.O., Taddonio, K.N. and Sarma, K.M. (2007) *Technology Transfer for the Ozone Layer: lessons for climate change*. Routledge (Earthscan), London, UK.

Anderson, C.R., Bruil, J., Chappell, M.J., Kiss, C. and Pimbert, M.P. (2021) *Agroecology Now! Transformations towards more just and sustainable food systems*. Palgrave Macmillan (Springer), Cham, CH.

Bibliography

Anderson, D.A. (2019) *Environmental Economics and Natural Resource Management* 5th edn. Routledge, Abingdon, UK.

Anderson, F.R. (2011) *NEPA in the Courts: a legal analysis of the National Environmental Policy Act* (first published 1973). Routledge (Earthscan), London, UK.

Andresen, S., Boasson, E.L. and Hønneland, G. (Eds) (2012) *International Environmental Agreements: an introduction*. Routledge, Abingdon, UK.

Andrews, G.J., Crooks, V.A., Pearce, J.R. and Messina, J.P. (Eds) (2021) *Covid-19 and Similar Futures: pandemic geographies*. Springer, Cham, CH.

Angang Hu (2017) *China: Innovative Green Development* 2nd edn. Springer, Singapore, SG.

Angelsen, A., Brockhaus, M., Sunerlin, W.D. and Verchot, L.V. (Eds) (2012) *Analysing REDD+: challenges and choices*. Center for International Forestry Research, Bogor Barat, ID.

Angelsen, A., Martius, C., de Sy, V., Duchelle, A.E., Larson, A.M. and Pham, T.T. (Eds) (2018) *Transforming REDD+: lessons and new directions*. Center for International Forestry Research, Bogor Barat, ID.

Anguelov, N. (2016) *The Dirty Side of the Garment Industry: fast fashion and its negative impact*. CRC Press, Boca Raton, FL, USA.

Angus, I. and Butler, S. (2011) *Too Many People? Population, immigration, and the environmental crisis*. Haymarket Books, Chicago, IL, USA.

Anon (2012) *Ash Dieback Disease: Chalara fraxinea*. House of Commons Library, London, UK.

Anstead, N. (2021) *What Do We Know and What Should We Do About Fake News?* Sage, London, UK.

Anthopoulos, L. (Ed) (2019) *Smart City Emergence: cases from around the world*. Elsevier, Amsterdam, NL.

Anton, D.K. (Ed) (2021) *International Environmental Law* (2 vols). Edward Elgar, Cheltenham, UK.

Antonopoulos, N. and Veglis, A. (2013) *Green Websites: organizations portals newspapers magazines & TV*. Createspace, Scotts Valley, CA, USA.

Antunes, J. (Ed) (2021) *River Basin Management – sustainability issues and planning strategies*. IntechOpen, London, UK.

Apostolopoulou, E. (2022) *Cambridge Institute for Sustainability Leadership (CISL) – Policy Briefing: China's Belt and Road Initiative: sustainability in the New Silk Road*. Cambridge University, Cambridge, UK.

Apostolopoulou, E. and Cortes-Vazquez, J.A. (Eds) (2018) *The Right to Nature: social movements, environmental justice and neoliberal natures*. Routledge, London, UK.

ApSimon, H., Pearce, D. and Ozdemiroglu, E. (Eds) (2013) *Acid Rain in Europe: counting the cost* 2nd edn. Routledge (Earthscan), Abingdon, UK.

Aravind, J., Kamaraj, M., Devi, M.P. and Rajakumar, S. (Eds) (2021) *Strategies and Tools for Pollutant Mitigation: avenues to a cleaner environment*. Springer, Cham, CH.

Arce-Gomez, A., Donovan, J.D. and Bedggood, R.E. (2015) Social impact assessments: developing a consolidated conceptual framework. *Environmental Impact Assessment Review* 50(1), 85–94.

Arcidiacono, A. and Ronchi, S. (Eds) (2021) *Ecosystem Services and Green Infrastructure: perspectives from spatial planning in Italy*. Springer, Cham, CH.

Aresta, M. and Dibenedetto, A. (Eds) (2021) *The Carbon Dioxide Revolution: challenges and perspectives for a global society*. Springer Nature, Cham, CH.

Arias-Maldonado, M. and Trachtenberg, Z. (Eds) (2019) *Rethinking the Environment for the Anthropocene: political theory and socionatural relations in the new geological epoch.* Routledge, Abingdon, UK.

Armeni, C. (2016) Participation in environmental decision-making: reflecting on planning and community benefits for major wind farms. *Journal of Environmental Law* 28(3), 415–441.

Armstrong, A.K., Krasny, M.E. and Schuldt, J.P. (Eds) (2021) *Communicating Climatic Change: a guide for educators.* Cornell University Press, Ithaca, NY, USA.

Aronsson, T. and Löfgren, K.-G. (Eds) (2010) *Handbook of Environmental Accounting.* Edward Elgar, Cheltenham, UK.

Aronsson, T. and Löfgren, K.-G. (2012) *Handbook of Environmental Accounting* 2nd edn. Edward Elgar, Cheltenham, UK.

Aronsson, T., Johansson, P.-O. and Löfgren, K.-G. (1997) *Welfare Measurement, Sustainability and Green National Accounting: a growth theoretical approach.* Edward Elgar, London, UK.

Arora, S., Kumar, A., Ogita, S. and Yau, Y.Y. (Eds) (2022) *Innovations in Environmental Biotechnology.* Springer, Singapore, SG.

Arora-Jonsson, S. (2013) *Gender, Development and Environmental Governance: theorizing connections.* Routledge, New York, NY, USA.

Arts, B.J.M., van Bommel, S., Ros-Tonen, M. and Verschoor, G. (Eds) (2012) *Forest People Interfaces: understanding community forestry and biocultural diversity.* Wageningen Academic Publishers, Wageningen, NL.

Arts, B.J.M., Ingram, V. and Brockhaus, M. (Eds) (2020) *The Performance of REDD+: from global governance to local.* MDPI, Basel, CH.

Arya, A. (2017) *Biotechnology and Environmental Management.* Discovery Publishing House Pvt., New Delhi, IN.

Ascensão, F., Fahrig, L., Clevenger, A.P., Corlett, R.T., Jaeger, J.A.G., Laurance, W.F. and Pereira, H.M. (2018) Environmental challenges for the Belt and Road Initiative. *Nature Sustainability* 1(May), 206–209.

Ash, A.J., Stafford Smith, M., Parry, M., Waschka, M., Guitart, D., Palutikof, J.P. and Boulter, S.L. (Eds) (2013) *Climate Adaptation Futures.* Wiley-Blackwell, Chichester, UK.

Ashby, M.F. (2013) *Materials and the Environment: eco-informed material choice* 2nd edn. Elsevier (Butterworth-Heinemann), Waltham, MA, USA.

Ashley, J.M. (2016) *Food Security in the Developing World.* Elsevier (Academic Press), London, UK.

Ashley, P. and Boyd, W.E. (2006) *Quantitative and Qualitative Approaches to Research in Environmental Management.* Southern Cross University, Lismore, NSW, AU (Epublications@SCU).

Ashton Drew, C., Wiersma, Y.F. and Huettmann, F. (Eds) (2011) *Predictive Species and Habitat Modeling in Landscape Ecology: concepts and applications.* Springer, New York, NY, USA.

Asiri, N., Khan, T. and Kend, M. (2020) Environmental management accounting in the Middle East and North Africa region: significance of resource slack and coercive isomorphism. *Journal of Cleaner Production* 267, article No. 121870 (17 pp.).

Athithan, S. (2021) *Coastal Aquaculture and Mariculture.* CRC Press, Boca Raton, FL, USA.

Atkinson, A., Dávila, J.D. and Mattingley, M. (2020) *The Challenge of Environmental Management in Urban Areas.* Routledge (Taylor & Francis), Abingdon, UK.

Atkinson, G., Braathen, N.A., Groom, B. and Mourato, S. (2018) *Cost-Benefit Analysis and the Environment: further developments and policy use.* OECD Publishing, Paris, FR.

Atlas, R.M. and Maloy, S. (Eds) (2014) *One Health: people, animals, and the environment.* ASM Press, Washington, DC, USA.

Attenborough, D. (2020) *A Life on Our Planet: my witness statement and vision for the future.* Grand Central Publishing, New York, NY, USA.

Attfield, R. (Ed) (2016) *The Ethics of the Environment.* Routledge, Abingdon, UK.

Attfield, R. (2018) *Environmental Ethics: a very short introduction.* Oxford University Press, Oxford, UK.

Awange, J.L. (2017) *GNSS Environmental Sensing: revolutionizing environmental monitoring* 2nd edn. Springer, Cham, CH.

Awange, J.L. and Kyalo Kiema, J.B. (2013) *Environmental Geoinformatics: monitoring and management.* Springer-Verlag, Berlin, GR.

Ayyub, B.M. (Ed) (2018) *Climate-Resilient Infrastructure.* American Society of Civil Engineers, Reston, VA, USA.

Badenoch, N. (2002) *Transboundary Environmental Governance: principles and practice in mainland Southeast Asia* (Report). World Resources Institute, Washington, DC, USA.

Bahadur, A.V. and Tanner, T. (2022) *Resilience Reset: creating resilient cities in the global south.* Routledge, Abingdon, UK.

Bahro, R. (1982) *Socialism and Survival.* Heretic Books, London, UK.

Bahro, R. (1984) *From Red to Green: interviews with 'New Left Review'.* Verso, London, UK.

Bahro, R. (1986) *Building the Green Movement.* New Society Publishers, London, UK.

Bailey, R.G. (2009) *Ecosystem Geography: from ecoregions to sites* 2nd edn. Springer, Dordrecht, NL.

Bai Tian. (2019) *GIS Technology Applications in Environmental and Earth Sciences.* CRC Press, Boca Raton, FL, USA.

Bak, O. (Ed) (2021) *Sustainable and Green Supply Chains and Logistics: case study collection.* Kogan Page, London, UK.

Baker, R. (2020) *Environmental Management in the Tropics: an historical perspective* (first published 1992). CRC Press, Boca Raton, FL, USA.

Baldarelli, M.-G. and Del Baldo, M. (Eds) (2020) *A Journey in Social and Environmental Accounting, Accountability and Society.* Cambridge Scholars Publishing, Newcastle-upon-Tyne, UK.

Baldarelli, M.-G., Del Baldo, M. and Nesheva-Kiosseva, N. (Eds) (2017) *Environmental Accounting and Reporting: theory and practice.* Springer, Cham, CH.

Baldauf, C. (Ed) (2020) *Participatory Biodiversity Conservation: concepts, experiences, and perspectives.* Springer, Cham, CH.

Balducci, A., Chiffi, D. and Curci, F. (Eds) (2020) *Risk and Resilience: socio-spatial and environmental challenges.* Springer, Cham, CH.

Ballantyne, R. and Packer, J. (Eds) (2013) *International Handbook on Ecotourism.* Edward Elgar, Cheltenham, UK.

Bals, L. and Tate, W. (Eds) (2017) *Implementing Triple Bottom Line Sustainability into Global Supply Chains* 2nd edn. Routledge (Greenleaf Publishing), Abingdon, UK.

Bandarage, A. (2013) *Sustainability and Well-Being: the middle path to environment, society and the economy.* Palgrave Macmillan, Basingstoke, UK.

Bank, M.S. (Ed) (2022) *Microplastic in the Environment: pattern and process.* Springer, Cham, CH.

Bankoff, G. and Hihorst, D. (2022) *Why Vulnerability Still Matters: the politics of disaster risk creation*. Routledge (Taylor & Francis), Abingdon, UK.

Bankoff, G., Frerks, G. and Hilhorst, D. (Eds) (2004) *Mapping Vulnerability: disasters, development and people*. Routledge (Earthscan), London, UK.

Barbier, E.B. (2011) *Capitalizing on Nature: ecosystems as natural assets*. Cambridge University Press, Cambridge, UK.

Barbier, E.B. and Hanley, N. (2009) *Pricing Nature: cost-benefit analysis and environmental policy*. Edward Elgar, Cheltenham, UK.

Barbose, P. (Ed) (2020) *Urban Ecology: its nature and challenges*. CABI, Wallingford, UK.

Bardsley, D.K. and Hugo, G.J. (2010) Migration and climate change: examining thresholds of change to guide effective adaptation decision-making. *Population and Environment* 32(2–3), 238–262.

Barlow, M. and Clarke, T. (2003) *Blue Gold: the fight to stop the corporate theft of the world's water*. Routledge (Earthscan), London, UK.

Barnett, J. and Campbell, J. (2009) *Climate Change and Small Island States: power, knowledge and the South Pacific*. Routledge (Earthscan), London, UK.

Barnsley, M.J. (2007) *Environmental Modeling: a practical introduction*. CRC Press, Boca Raton, FL, USA.

Barrow, C.J. (2003) *Environmental Change and Human Development: controlling nature?* Arnold, London, UK.

Barrow, C.J. (2012) Socioeconomic adaptation to environmental change: towards sustainable development. In: J.A. Matthews (Ed), *The Sage Handbook of Environmental Change: Vol. 2* (pp. 426–446). Sage, London, UK.

Bartelmus, P. and Seifert, E.K. (Ed) (2019) *Green Accounting* (first published 2003). Routledge, Abingdon, UK.

Baskin, J. (2019) *Geoengineering, the Anthropocene and the End of Nature*. Springer (Palgrave Macmillan), Cham, CH.

Bassham, G. (2020) *Environmental Ethics: the central issues*. Hackett Publishing, Indianapolis, IN, USA.

Bastante-Ceca, M.J., Fuentes-Bargues, J.L., Hufnagel, L., Mihai, F.-C. and Latu, C. (Eds) (2020) *Sustainability Assessment at the 21st Century*. IntechOpen, London, UK.

Bates, A. (2010) *The Biochar Solution: carbon farming and climate change*. New Society Publishers, Gabriola Island, BC, CA.

Bates, A. and Draper, K. (2019) *Burn: using fire to cool the Earth*. Chelsea Green Publishing, White River Junction, VT, USA.

Batra, G., Uitto, J.I. and Feinstein, O.N. (2022) *Environmental Evaluation and Global Development Institutions: a case study of the Global Environmental Facility*. Routledge, Abingdon, UK.

Baughen, S. (2007) *International Trade and the Protection of the Environment*. Routledge-Cavendish, Abingdon, UK.

Baxter, B. (1999) *Ecologism: Introduction*. Georgetown University Press, Washington, DC, USA.

Bazeley, P. (2020) *Qualitative Data Analysis: practical strategies*. Sage, Los Angeles, CA, USA.

Beal, B. (2013) *Corporate Social Responsibility: definition, core issues, and recent developments*. Sage, Los Angeles, CA, USA.

Beall, J., Guha-Khasnobis, B. and Ravi Kanbur, R. (Eds) (2010) *Urbanization and Development*. OUP Oxford, Oxford, UK.

Beaumont, P. (1989) *Environmental Management and Development in Drylands*. Routledge, London, UK.

Bebbington, J., Gray, R. and Gray, S. (Eds) (2010) *Social and Environmental Accounting.* Sage, Thousand Oaks, CA, USA.

Bebbington, J., Larrinaga, C., O'Dwyer, B. and Thomson, I. (Eds) (2021) *Routledge Handbook of Environmental Accounting.* Routledge, Abingdon, UK.

Beck, P. (1986) *The International Politics of Antarctica.* St Martin's Press, New York, NY, USA.

Beeson, M. (2019) *Environmental Populism: the politics of survival in the Anthropocene.* Springer (Palgrave Macmillan), Singapore, SG.

Behera, B.K. and Prasad, R. (2020) *Environmental Technology and Sustainability: physical, chemical and biological technologies for a clean environmental management.* Elsevier, Amsterdam, NL.

Behnassi, M., Gupta, H., El Habi, M. and Ramachandran, G. (Eds) (2021) *Social-Ecological Systems (SES): from risks and insecurity to viability and resilience.* Springer, Cham, CH.

Behrens, K. (2010) Exploring African holism with respect to the environment. *Environmental Values* 19(4), 465–484.

Behrman, S. and Kent, A. (Eds) (2018) *Climate Refugees: beyond the legal impasse?* Routledge (Earthscan), Abingdon, UK.

Beier, G., Kiefer, J. and Knopf, J. (2020) Potentials of big data for corporate environmental management: a case study from the German automotive industry. *Journal of Industrial Ecology* 10, article No. 1111/jiec.13062 (14 pp.).

Belal, A., Cooper, S. and de Souza Freire, F. (Eds) (2017) *Advances in Environmental Accounting & Management: social and environmental accounting in Brazil.* Emerald Publishing, Bingley, UK.

Belcham, A.R. (2014) *Manual of Environmental Management.* Routledge, Abingdon, UK.

Bell, C.L., Hazel, P. and Slade, C. (1982) *Project Evaluation in Regional Perspective: a study of an irrigation project in northwestern Malaysia.* Johns Hopkins University Press, Baltimore, MD, USA.

Bell, C.L., Voorhees, J. and Woellner, R.A. (Eds) (2020) *International Environmental Risk Management: a systems approach.* CRC Press, Boca Raton, FL, USA.

Bell, M.M. (2012) *An Invitation to Environmental Sociology* 4th edn. Sage, Thousand Oaks, CA, USA.

Bell, S. (2018) *Urban Water Sustainability: constructing infrastructure for cities and nature.* Routledge, Abingdon, UK.

Bell, S. and Morse, S. (Eds) (2020) *Routledge Handbook of Sustainability Indicators.* Routledge, Abingdon, UK.

Bell, W. (2017) *Foundations of Futures Studies: Vol. 1: History, purposes, and knowledge.* Routledge, Abingdon, UK.

Belton, M.J.S., Morgan, T.H., Samarasinha, N.H. and Yeomans, D.K. (Eds) (2004) *Mitigation of Hazardous Comets and Asteroids.* Cambridge University Press, Cambridge, UK.

Belz, F.-M. and Peattie, K. (2014) *Sustainability Marketing: a global perspective* 2nd edn. Wiley, Chichester, UK.

Bendell, J. and Read, R. (Eds.) (2021) *Deep Adaptation.* Polity Press, Cambridge, UK.

Bendix, J., Beck, E., Bräuning, A., Makeschin, F., Mosandl, R., Scheu, S. and Wilcke, W. (Eds) (2013) *Ecosystem Services, Biodiversity and Environmental Change in a Tropical Mountain Ecosystem of South Ecuador.* Springer, Heidelberg, GR.

Benjaminsen, T.A. and Svarstad, H. (Eds) (2021) *Political Ecology: a critical engagement with global environmental issues.* Springer, Cham, CH.

Benkeblia, N. (Ed) (2019) *Agroecology, Ecosystems, and Sustainability.* CRC Press, Boca Raton, FL, USA.

Benna, U.G. and Garba, S.B. (Eds) (2016) *Population Growth and Rapid Urbanization in the Developing World*. IGI Global, Hershey, PA, USA.

Benner, S., Lax, G., Crutzen, P.J., Pöschl, U., Leieveld, J. and Brauch, H.G. (Eds) (2021) *Paul J. Crutzen and the Anthropocene: a new epoch in Earth's history*. Springer, Cham, CH.

Bennett, G. (1991) The history of the Dutch National Environmental Policy Plan. *Environment* 33(7), 6–9, 31–33.

Benson, M.H. and Craig, R.K. (2017) *The End of Sustainability: resilience and the future of environmental governance in the Anthropocene*. University Press of Kansas, Lawrence, KS, USA.

Bentlage, J. and Weiß, P. (2006) *Environmental Management Systems and Certification*. Baltic University Press, Uppsala, SE.

Beran, M.A. (2013) *Carbon Sequestration in the Biosphere: processes and prospects*. Springer, Berlin and Heidelberg, GR.

Berardi, P.C. and de Brito, R.P. (2015) Drivers of environmental management in the Brazilian context. *Brazilian Administration Revue* 12(1), 109–128.

Bercht, A.L. (2021) How qualitative approaches matter in climate and ocean change research: uncovering contradictions about climate concern. *Global Environmental Change* 70, 1–14.

Berg, G. (1969) *Circumpolar Problems: habitat, economy, and social relations in the Arctic: a symposium for anthropological research in the north, September, 1969*. Pergamon Press, Oxford, UK.

Bergin, T. (2018) *An Introduction to Data Analysis: quantitative, qualitative and mixed methods*. Sage, Los Angeles, CA, USA.

Bergmann, M., Gutow, L. and Klages, M. (Eds) (2015) *Marine Anthropogenic Litter*. Springer, Cham, CH.

Berkes, F. (2021) *Advanced Introduction to Community-Based Conservation*. Edward Elgar, Cheltenham, UK.

Berkes, F. and Bierkes, M.K. (2009) Ecological complexity, fuzzy logic, and holism in indigenous knowledge. *Futures* 41(1), 6–12.

Berkhard, B. and Maes, J. (Eds) (2017) *Mapping Ecosystem Services*. Pensoft Publishers, Sofia, BG.

Berman, J.M. (2020) *Anti-vaxxers: how to challenge a misinformed movement*. MIT Press, Cambridge, MT, USA.

Berners-Lee, M. (2010) *How Bad are Bananas? The carbon footprint of everything*. Profile Books Ltd., London, UK.

Berners-Lee, M. (2021) *There is no Planet B: a handbook for the make or break years*. Cambridge University Press, Cambridge, UK

Berrone, P. (2016) *Green Lies: how greenwashing can destroy a company (and how to go green without the wash)*. CreateSpace Independent Publishing Platform, Scotts Valley, CA, USA.

Berry, K.A. and Mollard, E. (Eds) (2010) *Social Participation in Water Governance and Management: critical and global perspectives*. Routledge, Abingdon, UK.

Berry, R.J. (2018) *Environmental Attitudes Through Time*. Cambridge University Press, Cambridge, UK.

Betts, A. (2021) *The Wealth of Refugees: how displaced people can build economies*. Oxford University Press, Oxford, UK.

Betzold, C. and Weiler, F. (2019) *Development Aid and Adaptation to Climate Change in Developing Countries*. Springer (Palgrave Macmillan), Cham, CH.

Beuron, T.A., da Rosa, L.R., Madruga, G., Garlet, V., Avila, L.V., Guarda, F.G.K., de Freitas Terra, C.C. and Balsan, L.A.G. (2020) Contributions of an environmental management system for sustainable development at a Brazilian university. *Environmental Quality Management* 29(4), 103–113.

Beven, K. (2018) *Environmental Modelling: an uncertain future?* CRC Press, Boca Raton, FL, USA.

Beyers, B. and Wackernagel, M. (2019) *Ecological Footprint: managing our biocapacity budget*. New Society Publishers, Gabriola Island, BC, CA.

Bharagava, R.N. (Ed) (2017) *Environmental Pollutants and their Bioremediation Approaches*. CRC Press, Boca Raton, FL, USA.

Bharagava, R.N. (Ed) (2020) *Emerging Eco-friendly Green Technologies for Wastewater Treatment*. Springer, Singapore, SG.

Bharati, L., Sharma, B.R. and Smatkhtin, V. (Eds) (2020) *The Ganges River Basin: status and challenges in water, environment and livelihoods* (first published 2016). Routledge, Abingdon, UK.

Bhatnagar, A.R. (2022) *Desertification and Land Degradation: concept to combating*. CRC Press, Boca Raton, FL, USA.

Bhatta, B. (2021) *Global Navigation Satellite Systems: new technologies and applications* 2nd edn. Routledge, Abingdon, UK.

Bhattacharyya, P., Pathak, H. and Pal, S. (2020) *Climate Smart Agriculture: concepts, challenges, and opportunities*. Springer, Singapore, SG.

Bhinekawati, A. (2017) *Corporate Social Responsibility and Sustainable Development: social capital and corporate development in developing countries*. Routledge (Gower Books), London, UK.

Bicknell, J., Dodman, D. and Satterthwaite, D. (Eds) (2009) *Adapting Cities to Climate Change: understanding and addressing the development challenges*. Routledge (Earthscan), London, UK.

Biehl, J. and Staudenmaier, P. (2001) *Ecofascism: lessons from the German experience*. AK Press, Chico, CA, USA.

Biello, D. (2016) *The Unnatural World: the race to remake civilization in Earth's newest age*. Scribner, New York, NY, USA.

Biermann, F. (2014) *Earth System Governance: world politics in the Anthropocene*. MIT Press, Cambridge, MA, USA.

Biermann, F. and Lövbrand, E. (Eds) (2019) *Anthropocene Encounters: new directions in green political thinking*. Cambridge University Press, Cambridge, UK.

Biermann, F., Pattberg, P. and Zelli, F. (Eds) (2010) *Global Climate Governance Beyond 2012: architecture, agency and adaptation*. Cambridge University Press, Cambridge, UK.

Bigger, P. and Webber, S. (2021) Green structural adjustment in the World Bank's resilient city. *Annals of the American Association of Geographers* 111(1), 35–51.

Biggs, R., Schlüter, M. and Schoon, M.L. (Eds) (2015) *Principles for Building Resilience: sustaining ecosystem services in social-ecological systems*. Cambridge University Press, Cambridge, UK.

Biggs, R., de Vos, A., Preiser, R., Clements, H., Maciejewski, K. and Schlüter, M. (Eds) (2022) *The Routledge Handbook of Research Methods for Socio-Ecological Systems*. Routledge, Abingdon, UK.

Bihouix, P. (2020) *The Age of Low Tech: towards a technologically sustainable civilization* (translated from French edn by Editions du Seuil, 2014). Bristol University Press, Bristol, UK.

Birch, E.L. and Wachter, S.M. (Eds) (2011) *Global Urbanization*. University of Pennsylvania Press, Philadelphia, PA, USA.

Birkes, F., Doubleday, N. and Armitage, D. (Eds) (2007) *Adaptive Co-Management: collaboration, learning, and multi-level governance*. UBC Press, Vancouver, BC, CA.

Birkhofer, K., Diehl, E., Andersson, J., Ekroos, J., Früh-Müller, A., Machnikowski, F., Mader, V.L., Nilsson, L., Sasaki, K., Rundlöf, M., Wolters V. and Smith, H.G. (2015) Ecosystem services: current challenges and opportunities for ecological research. *Frontiers in Ecology and Evolution* 2, article No. 87 (12 pp.).

Birkmann, J., Kienberger, S. and Alexander, D. (Eds) (2014) *Assessment of Vulnerability to Natural Hazards: a European perspective*. Elsevier, San Diego, CA, USA.

Birley, M.H. (2011) *Health Impact Assessment: principles and practice*. Routledge (Earthscan), London, UK.

Blaikie, P.M. (1985) *The Political Economy of Soil Erosion in Developing Countries*. Longman, Harlow, UK.

Blaikie, P.M. (1988) The explanation of land degradation in Nepal. In: J. Ives and D.C. Pitt (Eds), *Deforestation: social dynamics in watersheds and mountain ecosystems* (pp. 132–158). Routledge, London, UK.

Blaikie, P.M. and Unwin, T. (Eds) (1988) *Environmental Crisis in Developing Countries* (IBG Developing Areas Research Group, Monograph No. 5). Institute of British Geographers, London, UK.

Blair, G.S., Henrys, P., Leeson, A., Watkins, J., Eastoe, E., Jarvis, S. and Young, P.J. (2019) Data science of the natural environment: a research roadmap. *Frontiers in Environmental Science* 7(121), doi: 10.3389/fenvs.2019.00121 (11 pp.).

Blakeney, M. (2019) *Food Loss and Food Waste: causes and solutions*. Edward Elgar, Cheltenham, UK.

Blakley, J.A.E. and Franks, D. (Eds) (2021) *Handbook of Cumulative Impact Assessment*. Edward Elgar, Cheltenham, UK.

Blanco-Canqui, H. and Lal, R. (2010) *Principles of Soil Conservation and Management*. Springer, Dordrecht, NL.

Blench, R. (1998) *Biodiversity Conservation and its Opponents*. ODI Natural Resource Perspectives No. 32, 4 pp. Overseas Development Institute, London, UK.

Bles, M. and Sevenster, M. (2010) *Shadow Prices Handbook: valuation and weighting of emissions and environmental impacts*. CE Publications, Delft, NL.

Blewitt, J. and Tilbury, D. (2014) *Searching for Resilience in Sustainable Development: learning journeys in conservation*. Routledge (Earthscan), Abingdon, UK.

Bliss, L.C., Heal, O.W. and Moore, J.J. (Eds) (1981) *Tundra Ecosystems: a comparative analysis*. Cambridge University Press, Cambridge, UK.

Blokdyk, G. (2018) *Environmental Audit: the ultimate step-by-step guide*. CreateSpace Independent Publishing Platform, Scotts Valley, CA, USA.

Blokdyk, G. (2020a) *Risk Evaluation and Mitigation Strategies: a complete guide*. Emereo Publishing, Brisbane, QL, AU.

Blokdyk, G. (2020b) *Vulnerability Assessment a Complete Guide* 3rd edn. Emereo Publishing, Brisbane, QL, AU.

Boas, I. (2015) *Climate Migration and Security: securitisation as a strategy in climate change politics*. Routledge, New York, NY, USA.

Boasson, E.L., Andresen, S. and Hønneland, G. (2012) *International Environmental Agreements: an introduction*. Routledge, Abingdon, UK.

Bobek, V. (Ed) (2020) *Smart Urban Development*. IntechOpen, London, UK.

Bobrowsky, P.T. and Rickman, H. (Eds) (2007) *Comet/Asteroid Impacts and Human Society: an interdisciplinary approach*. Springer-Verlag, Berlin, GR.

Bodansky, D., Brunée, J. and Rajamani, L. (2014) *International Climate Change Law*. Oxford University Press, Oxford, UK.

Boddu, M., Gaayam, T. and Annamdas, V.G.M. (2011) *A review on inter basin transfer of water*. IPWE 2011, Proceedings of the 4th International Perspective on Water Resources & the Environment, January 4–6, 2011, National University of Singapore (NUS), Singapore, SG.

Bodiguel, L. and Cardwell, M. (Eds) (2010) *The Regulation of Genetically Modified Organisms: comparative approaches*. Oxford University Press, Oxford, UK.

Boer, B. (2015) *The Mekong: a socio-legal approach to river basin development*. Routledge (Earthscan), London, UK.

Boer, B., Hirsch, P., Johns, F., Saul, B. and Scurrah, N. (2015) *The Mekong: a socio-legal approach to riverbasin development*. Earthscan, Abingdon, UK.

Bogardi, J.J., Gupta, J., Wasantha Nandalal, K.D., Salamé, L., van Nooijen, R.R.P., Kumar, N., Tingsanchali, T., Bhaduri, A. and Kolechkina, A.G. (Eds) (2021) *Handbook of Water Resources Management: discourses, concepts and examples*. Springer Nature, Cham, CH.

Böhm, J., Palmié, M. and Gassmann, O. (2019) *Smart Cities: introducing digital innovation to cities*. Emerald Publishing, Bingley, UK.

Boillier, D. (2002) *Silent Theft: the plunder of our common wealth*. Routledge, London, UK.

Boiral, O., Heras-Saizarbitoria, I. and Brotherton, C. (2020) Improving environmental management through indigenous peoples' involvement. *Environmental Science & Policy* 103, 10–20.

Bojie Fu and Bruce Jones, K. (Eds) (2013) *Landscape Ecology for Sustainable Environment and Culture*. Springer, Basingstoke, UK.

Bolay, J.-C., Hostettler, S. and Hazboun, E. (Eds) (2014) *Technologies for Sustainable Development: a way to reduce poverty?* Springer, Cham, CH.

Bolgov, M., Gottschalk, L., Krasovskaia, I. and Moore, R.J. (Eds) (2002) *Hydrological Models for Environmental Management*. Kluwer, Dordrecht, NL.

Bollier, D. and Helfrich, S. (Eds) (2012) *The Wealth of the Commons: a world beyond market and state*. Levellers Press, Amhurst and Florence, MA, USA.

Bolognesi, T. (2018) *Modernization and Urban Water Governance: organizational change and sustainability in Europe*. Springer (Palgrave Macmillan), London, UK.

Bolwell, D. (2019) *Governing Technology in the Quest for Sustainability on Earth*. Routledge, Abingdon, UK.

Bonanno, G. and Orlando-Bonaca, M. (Eds) (2022) *Plastic Pollution and Marine Conservation: approaches to protect biodiversity and marine life*. Elsevier (Academic Press), London, UK.

Bonneuil, C. and Fressoz, J.-B. (2017) *The Shock of the Anthropocene: the Earth, history and us* (1st edn French released 2016). Verso, London, UK.

Bookchin, M. (1990) *Remaking Society*. Black Rose Books, Montreal, QC, CA.

Booker, C. (2010) *The Real Global Warming Disaster: is the obsession with 'climate change' turning out to be the most costly scientific blunder in history?* Continuum Publishing, London, UK.

Boon, P. and Raven, P. (Eds) (2012) *River Conservation and Management*. Wiley-Blackwell, Oxford, UK.

Boons, F. and Howard-Grenville, J.A. (Eds) (2009) *The Social Embeddedness of Industrial Ecology*. Edward Elgar, Cheltenham, UK.

Booth, C.A. and Charlesworth, S.M. (Eds) (2014) *Water Resources in the Built Environment: management issues and solutions*. Wiley-Blackwell, Oxford, UK.

Borgatti, S.P., Everett, M.G. and Johnson, J.C. (2013) *Analyzing Social Networks*. Sage, Thousand Oaks, CA, USA.

Bormann, B.T., Brookes, M.H., Ford, E.D., Kiesler, A.R., Oliver, C.D. and Weigand, J.F. (1993) *A Broad Strategic Framework for Sustainable-Ecosystem Management* (Eastside Forest Ecosystem Health Assessment). USDA Forest Service, Washington, DC, USA.

Börner, K. (2021) *Atlas of Forecasts: modeling and mapping desirable futures*. MIT Press, Cambridge, MA, USA.

Borowitz, M.J. (2017) *Open Space: the global effort for open access to environmental satellite data*. MIT Press, Cambridge, MA, USA.

Borsekova, K. and Nijkamp, P. (Eds) (2019) *Resilience and Urban Disasters: surviving cities*. Edward Elgar, Cheltenham, UK.

Boserüp, E. (1965) *The Conditions of Agricultural Growth: the economics of agrarian change under population pressure*. Allen and Unwin, London, UK.

Boström, M., Klintman, M. and Micheletti, M. (2011) *Eco-Standards, Product Labelling and Green Consumerism*. Palgrave Macmillan, London, UK.

Bothmer, V. and Daglis, I.A. (2007) *Space Weather: physics and effects*. Springer (Praxis), Berlin, GR.

Botterill, L.C. and Cockfield, G. (Eds) (2013) *Drought, Risk Management, and Policy: decision-making under uncertainty*. CRC Press, Boca Raton, FL, USA.

Bouchery, Y., Corbett, C.J., Fransoo, C.J. and Tan, T. (Eds) (2016) *Sustainable Supply Chains: a research-based textbook on operations and strategy*. Springer, Cham, CH.

Bouma, J. and van Beukering, P. (Eds) (2015) *Ecosystem Services: from concept to practice*. Cambridge University Press, Cambridge, UK.

Bourg, D. and Erkman, S. (Eds) (2017) *Perspectives on Industrial Ecology*. Routledge, Abingdon, UK.

Bouwer, L.M. and Aerts, J.C.J.H. (2006) Financing climate change adaptation. *Disasters* 30(1), 49–63.

Bowen, F. (2014) *After Greenwashing: symbolic corporate environmentalism and society*. Cambridge University Press, Cambridge, UK.

Bower, F.A. (2017) *Stratospheric Ozone and Man* (2 vols). Taylor & Francis, Abingdon, UK.

Bower, F.A. and Ward, R.B. (Eds) (2018) *Stratospheric Ozone and Man: Vol. 1*. CRC Press, Boca Raton, FL, USA.

Boyle, A.E. and Redgwell, C. (2021) *Birnie, Boyle, and Redgwell's International Law and the Environment* 4th edn. Oxford University Press, Oxford, UK.

Bragdon, J.H. (2021) *Economies that Mimic Life: from biomimicry to sustainable prosperity*. Routledge, Abingdon, UK.

Brake, M. (Ed) (2020) *Infrastructureatlas, daten und facten über öffentliche Räume und netze*. Drukhaus Kaufman, Lahr, GR (map drawn by Appenzeller, L., Hecher, S. and Sack, J.).

Brar, S.K., Hegde, K. and Pachapur, V. (Eds) (2019) *Tools, Techniques and Protocols for Monitoring Environmental Contaminants*. Elsevier, Amsterdam, NL.

Brasseur, G.P. (2020) *The Ozone Layer: from discovery to recovery*. University of Chicago Press, Chicago, IL, USA.

Brauch, H.G., Oswald Spring, U., Mesjasz, C., Grin, J., Dunay, P., Behera, N.C., Chourou, B., Kameri-Mbote, P. and Liotta, P.H. (Eds) (2008) *Globalization and Environmental Challenges: reconceptualizing security in the 21st century*. Springer-Verlag Berlin, GR.

Brauer, J. (2009) *War and Nature: the environmental consequences of war in a globalized world*. Rowman & Littlefield (AltaMira Press), Lanham, MD, USA.

Brears, R.C. (2020) *Nature-Based Solutions to 21st Century Challenges*. Routledge, Abingdon, UK.

Brears, R.C. (2021a) *Water Resources Management: innovative and green solutions*. Walter De Gruyter, Berlin, GR.

Brears, R.C. (2021b) *The Palgrave Handbook of Climate Resilient Societies*. Springer (Palgrave Macmillan), Cham, CH.

Brebbia, C.A. (Ed) (2011) *River Basin Management VI*. WIT Press, Southampton, UK.

Brebbia, C.A. and Bjornlund, H. (Eds) (2014) *Sustainable Irrigation and Drainage: Vol. 5: Management, technologies and policies*. WIT Press, Southampton, UK

Brennan, R. (Ed) (2016) *Environmental Risk Assessment and Management*. Syrawood Publishing House, New York, NY, USA.

Brereton, P. (2019) *Environmental Literacy and New Digital Audiences*. Routledge, Abingdon, UK.

Bressers, H., Bressers, N. and Larrue, C. (Eds) (2016) *Governance for Drought Resilience: land and water drought management in Europe*. Springer, Cham, CH.

Bréthaut, C. and Pflieger, G. (2020) *Governance of a Transboundary River: the Rhône*. Springer (Palgrave Macmillan), Cham, CH.

Breuste, J., Pauleit, S., Haase, D., Sauerwein, M. and Lay, M. (2021) *Urban Ecosystems: function, management and development*. Springer, Berlin, GR.

Bricker, D. and Ibbitson, J. (2019) *Empty Planet: the shock of global population decline*. Penguin (Random House), New York, NY, USA.

Bridge, G., Perreault, T. and McCarthy, J. (Eds) (2015) *The Routledge Handbook of Political Ecology*. Routledge, New York, NY, USA.

Brierley, G.J. and Fryirs, K.A. (Eds) (2008) *River Futures: an integrative scientific approach to river repair*. Island Press, Washington, DC, USA.

Briggs, S.L.K. (2017) *ISO 1400: 2015 Environmental Management Systems: a practical guide for SMEs*. American National Standards Institute (ANSI), New York, NY, USA.

Brightman, M. and Lewis, J. (Eds) (2017) *The Anthropology of Sustainability: beyond development and progress*. Palgrave, London, UK.

Bril, H., Kell, G. and Rasche, A. (Eds) (2020) *Sustainable Investing: a path to a new horizon*. Routledge, Abingdon, UK.

Brimblecombe, P., Hara, H., Houle, D. and Novak, M. (Eds) (2007) *Acid Rain – deposition to recovery*. Springer, Dordrecht, NL.

Brimicombe, A. (2020) *GIS, Environmental Modeling and Engineering* (first published 2010). CRC Press, Boca Raton, FL, USA.

British Ecological Society (2014) *A Guide to Data Management in Ecology and Evolution*. British Ecological Society, London, UK.

British Standards Institution (1994) *Environmental Management Systems BS7750*. British Standards Institution, Manchester, UK.

British Standards Institution (1996) *Environmental Management Systems: specifications with guidance for use*. British Standards Institution, Manchester, UK.

Broietti, C., Flach, L., Rover, S. and de Souza, J.A.S. (2018) Public expenditure and the environmental management of Brazilian municipalities: a panel data model. *International Journal of Sustainable Development & World Ecology* 25(7), 630–641.

Bromley, D.W. and Cernea, M.M. (1989) *The Management of Common Property Resources: some conceptual and operational fallacies* (World Bank Discussion Paper No. 57). World Bank, Washington, DC, USA.

Broniewicz, E. (Ed) (2011) *Environmental Management in Practice.* IntechOpen Publishing, London, UK.

Brooks, S. and Olive, A. (Eds) (2018) *Transboundary Environmental Governance Across the World's Longest Border.* Michigan State University Press, East Lansing, MI, USA.

Brouwer, F.M. and Ittersum, M.K. (Eds) (2010) *Environmental and Agricultural Modelling: integrated approaches for policy impact assessment.* Springer, Dordrecht, NL.

Brown, K. (2016) *Resilience, Development and Global Change.* Routledge, Abingdon, UK.

Bruckman, V.J., Varol, E.A., Uzun, B.B. and Lui, J. (Eds) (2016) *Biochar: a regional supply chain approach in view of climate change mitigation.* Cambridge University Press, Cambridge, UK.

Bruges, J. (2010) *The Biochar Debate: charcoal's potential to reverse climate change and build soil fertility.* Chelsea Green Publishing, River Junction, VT, USA.

Brukmeir, K. (2019) *Global Environmental Governance: social-ecological perspectives.* Springer (Palgrave Macmillan), Cham, CH.

Brusset, E. (2020) *Second Circle: the science and art of social impact assessment.* Urbane Publicaton, Abbey Park, UK.

Bryan, F.J. and Manly, J. (2008) *Statistics for Environmental Science and Management* 2nd edn. CRC Press, Boca Raton, FL, USA.

Bryant, C.R., Sarr, M.A. and Délusca, K. (Eds) (2016) *Agricultural Adaptation to Climate Change.* Springer, Cham, CH.

Bryant, E. (2014) *Tsunami: the underrated hazard* 3rd edn. Springer-Praxis, Cham, CH.

Bryant, R.L. (2015) *The International Handbook of Political Ecology.* Edward Elgar, Cheltenham, UK.

Buchholz, W. and Rübbelke, D. (2019) *Foundations of Environmental Economics.* Springer, Cham, CH.

Bucur, D. (Ed) (2016) *River Basin Management.* InTech, Rijeka, HR.

Buechler, S. and Hanson, A.-M.S. (2016) *A Political Ecology of Women, Water and Global Environmental Change.* Routledge, Abingdon, UK.

Bulkeley, H. (2013) *Cities and Climate Change.* Routledge, Abingdon, UK.

Bunker, B. (2018) *The Mythology of Global Warming: climate change fiction vs. scientific facts.* Moonshine Cove Publishing, Abbeville, SC, USA.

Burayidi, M.A., Allen, A., Twigg, J. and Wamsler, C. (Eds) (2020) *The Routledge Handbook of Urban Resilience.* Routledge, Abingdon, UK.

Burdge, R.J. (2015) *A Community Guide to Social Impact Assessment* 4th edn. Society and Natural Resources Press, Huntsville, TX, USA.

Burgos-Ayala, A., Jiménez-Aceituno, A., Torres-Torres, A.M., Rozas-Vásquez, D. and Lam D.P.M. (2020) Indigenous and local knowledge in environmental management for human-nature connectedness: a leverage points perspective. *Ecosystems and People* 16(1), 290–303.

Burnette, G.A. (2022) *Managing Environmental Data: principles, techniques, and best practices.* CRC Press, Boca Raton, FL, USA.

Burnett-Hall, R. and Jones, B. (Eds) (2012) *Burnett-Hall on Environmental Law.* Sweet & Maxwell, Mytholmroyd, UK.

Burns, T.A. Jnr. and Caniglia, B.S. (Eds) (2017) *Environmental Sociology: the ecology of late modernity* 2nd edn. Mercury Academic, Norman, OK, USA.

Burton, I., Kates, R. and White, G. (1977) *The Environment as Hazard.* Oxford University Press, Oxford, UK.

Bury, J. and Grunewald, W. (2015) *The GMO Revolution.* LannooCampus Publishers, Tielt, BE.

Bibliography

Buscher, B., Dressler, W. and Fletcher, R. (Eds) (2014) *Nature Inc.: environmental conservation in the neoliberal age (critical green engagements)*. University of Arizona Press, Tucson, AZ, USA.

Butler, D.R. (Ed) (2022) *The Anthropocene*. Routledge, Abingdon, UK.

Byron, F. (2019) *Environmental Remote Sensing*. Callisto Reference, Berlin, GR.

Cadman, T., Eastwood, L., Lopez-Casero Michaelis, F., Maraseni, T.N., Pittock, J. and Sarker, T. (2015) *The Political Economy of Sustainable Development: policy instruments and market mechanisms*. Edward Elgar, Cheltenham, UK.

Cahill, L.B. (2015) *Environmental Health and Safety Audits: a compendium of thoughts and trends*. Rowman & Littlefield, Washington, DC, USA.

Cahir, F., Clark, I.D. and Clarke, P. (2018) *Aboriginal Biocultural Knowledge in Southeastern Australia: perspectives of early colonists*. CSIRO Publishing, Clayton, VI, AU.

Caldwell, C.D. and Wang, S. (Eds) (2020) *Introduction to Agroecology*. Springer, Cham, CH.

Cameletti, M. and Finazzi, F. (Eds) (2018) *Quantitative Methods in Environmental and Climate Research*. Springer, Dordrecht, NL.

Camilleri, M.A. (2017) *Corporate Sustainability, Social Responsibility and Environmental Management: an introduction to theory and practice with case studies*. Springer International, Cham, CH.

Campbell, B. (2017) *Human Ecology: the story of our place in nature from prehistory to the present* 2nd edn. Routledge, Abingdon, UK.

Campbell, J.B., Wynne, R.H., Mesev, V., Treitz, P. and Lawrence, R. (2011) *Introduction to Remote Sensing* 5th edn. Guilford Press, New York, NY.

Campeol, G. (Ed) (2020) *Strategic Environmental Assessment and Urban Planning: methodological reflections and case studies*. Springer, Cham, CH.

Caniglia, B.S., Vallée, M. and Frank, B. (Eds) (2017) *Resilience, Environmental Justice and the City*. Routledge (Earthscan), Abingdon, UK.

Canter, L.W. (2015) *Cumulative Effects Assessment and Management: principles, processes and practices*. EIA Press, Washington, DC, USA.

Capri, E. and Alix, A. (Eds) (2018) *Sustainable Use of Chemicals in Agriculture*. Elsevier (Academic Press), Cambridge, MA, USA.

Caprotti, F. (2015) *Eco-Cities and the Transition to Low Carbon Economies*. Palgrave Macmillan, Basingstoke, UK.

Caputo, S. (2022) *Small Scale Soil-Less Urban Agriculture in Europe*. Springer, Cham, CH.

Cardoso, C.A.S. (2018) *Extractive Reserves in Brazilian Amazonia: local resource management and the global political economy* (first published 2002). Routledge, Abingdon, UK.

Carey, J. (2015) The 9 limits of our planet… and how we've raced past 4 of them. Ideas.TED.com 05/03/2015. https://ideas.ted.com/the-9-limits-of-our-planet-and-how-weve-raced-past-them/ (accessed 27/01/21).

Carilllo, F.J. and Koch, G. (Eds) (2021) *Knowledge for the Anthropocene: a multidisciplinary approach*. Edward Elgar, Cheltenham, UK.

Carlarne, C.P., Gray, K.R. and Tarasofsky, R. (Eds) (2016) *The Oxford Handbook of International Climate Change Law*. Oxford University Press, Oxford, UK.

Carley, M. and Christie, I. (1992) *Managing Sustainable Development*. Earthscan, London, UK.

Carlson, R.H. (2010) *Biology is Technology: the promise, peril, and new business of engineering life*. Harvard University Press, Cambridge, MA, USA.

Carolan, M. (2020) *Society and the Environment: pragmatic solutions to ecological issues* 3rd edn. Routledge, Abingdon, UK.

Carrol, A. (2019) *New Woman Ecologies: from arts and crafts to the Great War and beyond.* University of Virginia Press, Charlottesville, VA, USA.

Carroll, B., Fothergill, J., Murphy, J. and Turpin, T. (2019) *Environmental Impact Assessment Handbook: a practical guide for planners, developers and communities* 3rd edn. ICE Publishing, London, UK.

Carson, R. (1962) *Silent Spring.* Houghton Mifflin, Boston, MA, USA.

Carter, N. (2018) *The Politics of the Environment: ideas, activism, policy* 3rd edn. Cambridge University Press, Cambridge, UK.

Carter, N. and Mol, A.P.J. (Eds) (2007) *Environmental Governance in China.* Routledge, Abingdon, UK.

Cartwright, A., Parnell, S., Oelofse, G. and Ward, S. (Eds) (2012) *Climate Change at the City Scale: impacts, mitigation and adaptation in Cape Town.* Routledge (Earthscan), Abingdon, UK.

Cartwright, J. (1989) Conserving nature, decreasing debt. *Third World Quarterly* 11(2), 114–127.

Carvill, M., Butler, G. and Evans, G. (2021) *Sustainable Marketing: how to drive profits with purpose.* Bloomsbury Publishing, London, UK.

Cassin, J., Matthews, J.H. and Gunn, L. (Eds) (2022) *Nature-Based Solutions and Water Security: an action agenda for the 21st century.* Elsevier, Amsterdam, NL.

Cater, C.A., Garrod, B. and Low, T. (Eds) (2015) *The Encyclopedia of Sustainable Tourism.* CABI, Wallingford, UK.

Cathelat, B. (2019) *Smart Cities: shaping the society of 2030.* UNESCO, Paris, FR.

Cato, M.S. (2013) *The Bioregional Economy: land, liberty and the pursuit of happiness.* Routledge, London, UK.

Cavalier, D., Hoffman, C. and Cooper, C. (2020) *The Field Guide to Citizen Science: how you can contribute to scientific research and make a difference.* Timber Press (Workman Publishing Co.), New York, NY, USA.

Cays, J. (2020) *An Environmental Life Cycle Approach to Design: LCA for designers and the design market.* Springer, Cham, CH.

Centeno, P., Callaghan, P. and Patterson, T. (2023) *How Worlds Collapse: what history, systems and complexity teach us about our modern world and fragile future.* Routledge, Abingdon, UK.

Cerqueti, R., Cinelli, M. and Minervini, L.F. (2021) Municipal waste management: A complex network approach with an application to Italy. *Waste Management* 126, 597–607.

Chabbi, A. (2020) *Grassland Management for Sustainable Agroecosystems.* MDPI Ag, Basle, CH.

Chadwick, M.A. and Francis, R.A. (2013) *Urban Ecosystems: understanding the human environment.* Routledge (Earthscan), London, UK.

Chakraborty, C. (Ed) (2021) *Green Technological Innovation for Sustainable Smart Societies: post pandemic era.* Springer, Cham, CH.

Chambers, R.H. (1994a) The origins and practice of participatory rural appraisal. *World Development* 22(7), 953–969.

Chambers, R.H. (1994b) Participatory rural appraisal (PRA): analysis of experience. *World Development* 22(9), 1253–1268.

Chambers, R.H. (2013) *Rural Development: putting the last first* (first published 1983). Routledge, Abingdon, UK.

Chambers, W.B. (2008) *Interlinkages and the Effectiveness of Multilateral Environmental Agreements.* UNU Press, Tokyo, JP.

Chand, G. and Kumar, S. (Eds) (2016) *Crop Diseases and Their Management: integrated approaches*. CRC Press (Apple Academic Press), Boca Raton, FL, USA.

Chandler, D. and Coaffee, J. (Eds) (2017) *The Routledge Handbook of International Resilience*. Routledge, London, UK.

Chandler, D., Grove, K. and Wakefield, S. (Eds) (2020) *Resilience in the Anthropocene: governance and politics at the end of the world*. Routledge, Abingdon, UK.

Chandler, R. (1988) *Understanding the New Age*. Ward, London, UK.

Chandra, R. (Ed) (2016) *Environmental Waste Management*. CRC Press, Boca Raton, FL, USA.

Chandrappa, R. and Das, D.B. (2012) *Solid Waste Management: principles and practice*. Springer-Verlag, Berlin, GR.

Chapin, F.S., Matson, P.A. and Vitousek, P.M. (2011) The ecosystem concept. In: F.S. Chapin., P.A. Matson and Vitousek, P.M. (Eds), *Principles of Terrestrial Ecosystem Ecology* 2nd edn (pp. 3–32). Springer, New York, NY, USA.

Chapman, M.D. (1989) The political ecology of fisheries depletion in Amazonia. *Environmental Conservation* 16(4), 331–337.

Chaube, H. (2018) *Plant Disease Management: principles and practices* (first published 1991). CRC Press, Boca Raton, FL, USA.

Chaurasia, A., Hawksworth, D.L. and Pessoa de Miranda, M. (Eds) (2020) *GMOs: implications for biodiversity conservation and ecological processes*. Springer, Cham, CH.

Chemhuru, M. (Ed) (2019) *African Environmental Ethics: a critical reader*. Springer Nature, Cham, CH.

Chen, J.P., Wang, L.K., Wang, M.-H.S., Hung, Y.T. and Shammas, N.K. (Eds) (2017) *Remediation of Heavy Metals in the Environment*. CRC Press, Boca Raton, FL, USA.

Cheshmehzangi, A. and Dawodu, A. (2019) *Sustainable Urban Development in the Age of Climate Change: people – the cure or curse*. Springer (Palgrave Macmillan), Singapore, SG.

Cheshmehzangi, A., Dawodu, A. and Sharifi, A. (2022) *Sustainable Urbanism in China*. Routledge, New York, NY, USA.

Chicharo, L., Müller, F. and Fohrer, N. (Eds) (2015) *Ecosystem Services and River Basin Ecohydrology*. Springer, Dordrecht, NL.

Child, B. (2019) *Sustainable Governance of Wildlife and Community-Based Natural Resource Management: from economic principles to practical governance*. Routledge (Earthscan), Abingdon, UK.

Chitty, G. (Ed) (2017) *Heritage, Conservation and Communities: engagement, participation and capacity building*. Routledge, Abingdon, UK.

Chiu, E. (2022) Environmental implications of the Belt and Road Initiative: geopolitics and climate change. *The Yale Review of International Studies*. http://yris.yira.org/comments/5879 (accessed 18/01/23).

Chmielowski, W.Z. (2016) *Fuzzy Control in Environmental Engineering*. Springer, Cham, CH.

Cho, C.H., Giordano-Spring, S., Belal, A., Maurice, J. and Cooper, S. (Eds) (2018) *Sustainability Accounting: education, regulation, reporting and stakeholders*. Emerald Publishing, Bingley, UK.

Chompunth, C. (2013) Public participation in environmental management in constitutional and legal frameworks. *American Journal of Applied Sciences* 10(1), 73–80.

Chowdhary, P., Kumar, V., Kumar, S. and Hare, V. (Eds) (2023) *Environmental Management Technologies: challenges and opportunities*. CRC Press, Boca Raton, FL, USA.

Christie, E. (2008) *Finding Solutions for Environmental Conflicts: power and negotiation.* Edward Elgar, Cheltenham, UK.

Christoff, P. and Eckersley, R. (2013) *Globalization and the Environment.* Rowman & Littlefield, Lanham, MD, USA.

Chua, L. and Fair, H. (2019) Anthropocene. In: F. Stein (Ed), *Cambridge Encyclopedia of Anthropology* (pp. 1–19). Cambridge University Press, Cambridge, UK.

Church, J.A., Woodworth, P.L., Aarup, T. and Stanley Wilson, W. (Eds) (2010) *Understanding Sea-Level Rise and Variability.* Wiley-Blackwell, Oxford, UK.

Chuvieco, E. (2020) *Fundamentals of Satellite Remote Sensing: an environmental approach* 3rd edn. CRC Press, Boca Raton, FL, USA.

CIEL (2022) *Sowing a Plastic Planet: how microplastics in agrochemicals are affecting our soils, our food, and our future.* Center for International Environmental Law, Washington, DC, USA.

Ciracy-Wantrup, S.V. (1952) *Resource Conservation: economics and policies.* University of California Press, Berkeley, CA, USA.

Cirino, E. (2021) *Thicker than Water: the quest for solutions to the plastic crisis.* Island Press, Washington, DC, USA.

Clapp, J. and Dauvergne, P. (2011) *Paths to a Green World: the political economy of the global environment* 2nd edn. MIT Press, Cambridge, MA, USA.

Clark, C.K.L., Pengzhi, L. and Hong, X. (2021) *Water Environment Modeling.* CRC Press, Boca Raton, FL, USA.

Clark, C.W. (2006) *The Worldwide Crisis in Fisheries: economic models and human behaviour.* Cambridge University Press, Cambridge, UK.

Clark, N. and Szerszynski, B. (2020) *Planetary Social Thought: the Anthropocene challenge to the social sciences.* Polity Press, Cambridge, UK.

Clark II, W.W. and Cooke, G. (2016) *Smart Green Cities: toward a carbon neutral world.* Routledge (Gower), Abingdon, UK.

Clarke, T. and Peterson, T.R. (2016) *Environmental Conflict Management.* Sage, Thousand Oaks, CA. USA.

Clayton, S. and Myers, G. (2015) *Conservation Psychology: understanding and promoting human care for nature* 2nd edn. Wiley-Blackwell, Oxford, UK.

Clement, S. (2021) *Governing the Anthropocene: novel ecosystems, transformation and environmental policy. Springer* (Palgrave Macmillan), Cham, CH.

Clements, F.E. (1916) *Plant Succession: an analysis of the development of vegetation* (Publication No. 242). Carnegie Institute, Washington, DC, USA.

Clift, B. (2021) *Comparative Political Economy: states, markets and global capitalism* 2nd edn. Palgrave Macmillan (Bloomsbury), London, UK.

Clift, R. and Druckman, A. (Eds) (2016) *Taking Stock of Industrial Ecology.* Springer, Cham, CH.

Closmann, C.E. (Ed) (2009) *War and the Environment: military destruction in the modern age.* Texas A&M University Press, College Station, TX, USA.

Clynes, T. (2015) *The Boy who Played with Fusion: extreme science, extreme parenting, and how to make a star.* Houghton Mifflin Harcourt, Boston, MA, USA.

Coaffee, J.C. (2019) *Futureproof: how to build resilience in an uncertain world.* Yale University Press, New Haven, CT, USA.

Coaffee, J.C. and Lee, P. (2016) *Urban Resilience: planning for risk, crisis and uncertainty.* Bloomsbury Publishing, London, UK.

Coenen, F., Huitema, D. and O'Toole, L.J. jnr. (Eds) (2012) *Participation and the Quality of Environmental Decision Making*. Springer, Berlin, GR.

Coghlan, A. (2019) *An Introduction to Sustainable Tourism*. Goodfellow Publishers, Oxford, UK.

Cohen, B.J. (2022) *Rethinking International Political Economy*. Edward Elgar, Cheltenham, UK.

Cohen, E., McKay, A. and Wolfe, P. (2013) *Sustainability Reporting for SMEs: competitive advantage through transparency*. Routledge, London, UK.

Cohen, F., Hepburn, C.J. and Teytelboym, A. (2019) Is natural capital really substitutable? *Annual Review of Environment and Resources* 44, 425–448.

Cohen, M.J. (Ed) (2021) *Advances in Food Security and Sustainability*. Elsevier (Academic Press), Cambridge, MA, USA.

Cohen, S. and Dong, G. (2021) *The Sustainable City*. Columbia University Press, New York, NY, USA.

Cohen, Y.H. and Reich, Y. (2016) *Biomimetic Design Method for Innovation and Sustainability*. Springer, Cham, CH.

Coll, M. and Wajnberg, E. (Eds) (2017) *Environmental Pest Management: challenges for agronomists*. Wiley, Oxford, UK.

Collier, P. (2010) *The Plundered Planet: why we must – and how we can – manage nature for global prosperity*. Oxford University Press, Oxford, UK.

Collings, D.A. (2014) *Stolen Future, Broken Promise: the human significance of climate change*. Open Humanities Press, Ann Arbor Press, MI, USA.

Collins, A. and Flynn, A. (2015) *The Ecological Footprint: new developments in policy and practice*. Edward Elgar, Cheltenham, UK.

Collins, P., Menone, M.L., Metcalfe, C.D. and Tundisi, J.G. (Eds) (2020) *The Paraná River Basin: managing water resources to sustain ecosystem services*. Routledge, Abingdon, UK.

Commins, S.K., Lofchie, M.F. and Payne, R. (Eds) (1986) Africa's *Agrarian Crisis: the roots of famine*. Lynne Rienner, Boulder, CO, USA.

Comunello, F. and Mulargia, S. (2018) *Social Media in Earthquake-Related Communication: shake networks*. Emerald Publishing, Bingley, UK.

Conca, K. (2021) *Advanced Introduction to Water Politics*. Edward Elgar, Cheltenham, UK.

Conde, E. and Sánchez, S.I. (Eds) (2016) *Global Challenges in the Arctic Region: sovereignty, environment and geopolitical balance*. Routledge, Abingdon, UK.

Conroy, C. and Litvinoff, M. (1988) *The Greening of Aid: sustainable livelihoods in practice*. Earthscan, London, UK.

Conroy, M.J. and Perterson, J.T. (2013) *Decision Making in Natural Resource Management: a structured, adaptive approach*. Wiley-Blackwell, Oxford, UK.

Constantini, E.A.C. (Ed) (2017) *Manual of Methods for Soil and Land Evaluation* (first published 2009). CRC Press, Boca Raton, FL, USA.

Constanza, R., Graumlich, L.J. and Steffen, W. (Eds) (2007) *Sustainability and Collapse? An integrated history and future of people on Earth*. MIT Press, Cambridge, MA, USA.

Conway, G.R. and Barbier, E.B. (1990) *After the Green Revolution: sustainable agriculture for development*. Earthscan, London, UK.

Cook, B.I. (2019) *Drought: an interdisciplinary perspective*. Columbia University Press, New York, NY, USA.

Cook, J.A., Cylke, O., Larson, D.F., Nash, J.D. and Steadman-Edwards, P. (Eds) (2010) *Vulnerable Places, Vulnerable People: trade liberalization, rural poverty and the environment*. Edward Elgar/WWF/World Bank, Cheltenham, UK.

Cooper, C. (2016) *Citizen Science: how ordinary people are changing the face of discovery.* Overlook Press, New York, NY, USA.

Cooper, D.E. and James, S.P. (2017) *Buddhism, Virtue and Environment* 2nd edn. Routledge, Abingdon, UK.

Cooper, J. and Sheets, P. (Eds) (2012) *Surviving Sudden Environmental Change: answers from archaeology.* University Press of Colorado, Boulder, CO, USA.

Copeland, B.R. (2014) *Recent Developments in Trade and the Environment.* Edward Elgar, Cheltenham, UK.

Copenhagen Institute for Futures Studies (2020) *Visions of a Greener World: solutions for tackling climate change.* CIFS, Copenhagen, DK.

Copithorne, M.D. (1991) *Circumpolar Environmental Management and Regulation: from coexistence to cooperation – international law and organizations in the post-Cold War era.* Martinus Nijhoff, Dordrecht, NL.

Cory, J., Bradshaw, A. and Ehrlich, P.R. (2015) *Killing the Koala and Poisoning the Prairie: Australia, America and the environment.* University of Chicago Press, Chicago, IL, USA.

Coskun, A. (2018) *Understanding Green Attitudes: driving green consumerism through strategic sustainability marketing.* IGI Global, Hershey, PA, USA.

Cossu, R. and Stegmann, R. (2018) *Solid Waste Landfilling: concepts, processes, technology.* Elsevier, Amsterdam, NL.

Costanza, R. (2020) Valuing natural capital and ecosystem services toward the goals of efficiency, fairness, and sustainability. *Ecosystem Services* 43, 2212–2216.

Costanza, R. and Kubiszewski, I. (Eds) (2014) *Creating a Sustainable and Desirable Future: insights from 45 global thought leaders.* World Scientific Publishing, Singapore, SG.

Costanza, R., de Groot, R., Braat, L., Kubiszewski, I., Fioramonti, L., Sutton, P., Farber, S. and Grasso, M. (2017) Twenty years of ecosystem services: how far have we come and how far do we still need to go? *Ecosystem Services*, 28: 1–16.

Costanza, R., Erikson, J.D., Farley, J. and Kubiszewski, I. (Eds) (2020) *Sustainable Wellbeing Futures: a research and action agenda for ecological economics.* Edward Elgar, Cheltenham, UK.

Coulson, R.N. and Tchakerian, M.D. (2010) *Basic Landscape Ecology.* KEL Partners, College Station, TX, USA.

Cox, R., Phaedra, C. and Pezzullo, C. (2015) *Environmental Communication and the Public Sphere* 4th edn. Sage, Thousand Oaks, CA, USA.

Cracknell, A.P. and Varotsos, C. (2012) *Remote Sensing and Atmospheric Ozone: human activities versus natural variability.* Springer, Cham, CH.

Craig, G. (2019) *Media, Sustainability and Everyday Life.* Springer (Palgrave Macmillan), London, UK.

Craik, N., Jefferies, C.S.G., Seck, S.L. and Stephens, T. (Eds) (2018) *Global Environmental Change and Innovation in International Law.* Cambridge University Press, Cambridge, UK.

Cramer, A. and Karabell, Z. (2010) *Sustainable Excellence: the future of business in a fast-changing world.* Macmillan (Rodale), New York, NY, USA.

Crampton, P., MacKay, D.J.C., Ockenfels, A. and Stoft, S. (Eds) (2017) *Global Carbon Pricing: the path to climate cooperation.* MIT Press, Cambridge, MS, USA.

Cranganu, C. (2021) *Climate Change, Torn Between Myth and Fact.* Cambridge Scholars Publishing, Newcastle-upon-Tyne, UK.

Crawford, R. (2011) *Life Cycle Assessment in the Built Environment.* Spon Press, Abingdon, UK and New York, NY, USA.

Criollo, R., Malheiros, T. and Alfaro, J.F. (2018) Municipal environmental management indicators: a bottom-up approach applied to the Colombian context. *Social Indicators Research* 141, 1037–1054.

Crist, E. (2019) *Abundant Earth: toward an ecological civilization*. University of Chicago Press, Chicago, IL, USA.

Croner Publications (1997) *Croner's Environmental Policy and Procedures*. Croner Publications, Kingston upon Thames, UK (also available as CD-ROM).

Cross, K., Tondera, K., Rizzo, A., Andrews, L, Pucher, B., Istenič, D., Karres, N. and McDonald, R. (Eds) (2021) *Nature Based Solutions for Wastewater Treatment*. IWA Publications, London, UK.

Crowther, D. and Seifi, S. (2021) *The Palgrave Handbook of Corporate Social Responsibility* 2nd edn. Springer (Palgrave Macmillan), London, UK.

Cubbage, F.W. (2022) *Natural Resource Leadership and Management: a practical guide for professionals*. Routledge, Abingdon, UK.

Cugurullo, F. (2021) *Frankenstein Urbanism: eco, smart and autonomous cities, artificial intelligence and the end of the city*. Routledge, Abingdon, UK.

Curran, M.A. (Ed) (2012) *Life Cycle Assessment Handbook: a guide for environmentally sustainable products*. Wiley, Hoboken, NJ, USA.

Curran, M.A. (Ed) (2017) *Goal and Scope Definition in Life Cycle Assessment*. Springer, Dordrecht, NL.

Cutaia, F. (2016) *Strategic Environmental Assessment: integrating landscape and urban planning*. Springer, Cham, CH.

Czúcz, B., Molnár, Z.S., Horváth, F. and Botta-Dukát, Z. (2008) The natural capital index of Hungary. *Acta Botanica Hungarica* 50(suppl.), 161–177.

Dabashi, H. (Ed) (2011) *The Green Movement in Iran*. Routledge, Abingdon, UK.

da Cal Seixas, S.R. and de Moraes Hoefel, J.L. (2021) *Environmental Sustainability: sustainable development goals and human rights*. CRC Press, Boca Raton, FL, USA.

da Costa de Sousa, C., Lins, R.R., de Lima Albuquerque, J., da Silva Correia Neto, J., Silva, I.M.M., de Souza, E.R., Bignetti, V.M.B. and da Nóbrega Marinho, G.G. (2021) Green cities: an analysis of the environmental management strategy in the City of Paragominas – Pará – Brazil. *International Journal of Development Research* 11, article No. 21865 (5 pp.).

Dada, A., Stanoevska, K. and Gómez, J.M. (Eds) (2013) *Organizations' Environmental Performance Indicators: measuring, monitoring, and management*. Springer-Verlag, Berlin, GR.

Dahiya, A. (Ed) (2015) *Bioenergy: biomass to biofuels*. Elsevier (Academic Press), London, UK.

Dahlman, C.J. (2012) *The World Under Pressure: how China and India are Influencing the global economy and environment*. Stanford University Press, Stanford, CA, USA.

Dalal-Clayton, B. and Sadler, B. (2005) *Strategic Environmental Assessment: a sourcebook and reference guide to international experience*. Routledge (Earthscan), London, UK.

Dalbotten, D., Roehrig, G. and Hamilton, P. (Eds) (2014) *Future Earth: advancing civic understanding of the Anthropocene* (AGU). Wiley, Hoboken, NJ, USA.

Dale, L. (2022) *Climate Change Adaptation: an Earth Institute sustainability primer*. Columbia University Press, New York, NY, USA.

Dalezios, N.R. (Ed) (2017) *Environmental Hazards: methodologies for risk assessment and management*. IWA Publishing, London, UK.

Dameris, M. and Fabian, P. (2014) *Ozone in the Atmosphere: basic principles, natural and human impacts*. Springer, Berlin, GR.

D'Amico, A.R., Figueira, J.E.C., Cândido, J.F. Jnr. and Drumond, M.A. (2020) Environmental diagnoses and effective planning of Protected Areas in Brazil: is there any connection? *PLOS One* 15(12), e0242687.

Daniell, K.A., Squires, V.R. and Milner, H.M. (Eds) (2014) *River Basin Management in the Twenty-First Century: understanding people and place*. CRC Press, Boca Raton, FL, USA.

Darby, S. and Sear, D. (Eds) (2008) *River Restoration: managing the uncertainty in restoring physical habitat*. Wiley, Chichester, UK.

Darling, F.F. and Dasmann, R.F. (1969) The ecosystem view of human society. *Impact of Science on Society* 19(2), 109–121.

Darnhofer, I., Gibbon, D. and Dedieu, B. (Eds) (2012) *Farming Systems Research into the 21st Century: the new dynamic*. Springer, Cham, CH.

Dartnell, L. (2014) *The Knowledge: how to rebuild our world from scratch*. Penguin Press, New York, NY, USA.

Darwin, C. (1859) *The Origin of Species by Means of Natural Selection: or the preservation of favoured races in the struggle of life*. John Murray, London, UK.

Das, J. (2019) *Reporting Climate Change in the Global North and South: journalism in Australia and Bangladesh*. Routledge, Abingdon, UK.

da Silva, G.C.S. and de Medeiros, D.D. (2004) Environmental management in Brazilian companies. *Management of Environmental Quality* 15(4), 380–388.

Datta, R. and Meena, R.S. (Eds) (2021) *Soil Carbon Stabilization to Mitigate Climate Change*. Springer, Singapore, SG.

Dauvergne, P.P. (Ed) (2012) *Handbook of Global Environmental Politics* 2nd edn. Edward Elgar, Cheltenham, UK.

Dauvergne, P.P. and Lister, J. (2011) *Timber*. Polity Press, Cambridge, UK.

David, V. (2017) *Data Treatment in Environmental Sciences: multivaried approach*. Elsevier (ISTE Press), London, UK.

Davidson, N. (2019) The Anthropocene Epoch: have we entered a new phase of planetary history? *The Guardian* (UK) 30/5/2019. www.theguardian.com/environment/2019/may/30/anthropocene-epoch-have-we-entered-a-new-phase-of-planetary-history (accessed 05/01/21).

Davies, T. and Rosser, N. (Eds) (2022) *Landslide Hazards, Risks, and Disasters* 2nd edn. Elsevier, Amsterdam, NL.

Davis, D.K. (2007) *Resurrecting the Granary of Rome: environmental history and French colonial expansion in North Africa*. Ohio University Press, Athens, OH, USA.

Day, K.A. (2005) *China's Environment and the Challenge of Sustainable Development*. Routledge, Abingdon, UK.

D'Ayala, P., Hein, P. and Beller, W. (1990) *Sustainable Development and Environmental Management of Small Islands* (MAB Series No. 5). Parthenan Publishing Group (for UNESCO), Paris, FR.

Dayton, M., Reynolds, E., Bates, M.E., Morgan, H., Clarke, S.S. and Linkov, I. (2018) Resilience and sustainability: similarities and differences in environmental management applications. *Science of the Total Environment* 613–614, 1275–1283.

de Andrade, J.G.P. (2011) Interbasin water transfers: the Brazilian experience and international case comparisons. *Water Resources Management* 25(8), 1915–1934.

Debnath, S. (2019) *Environmental Accounting, Sustainability and Accountability*. Sage (India), New Delhi, IN.

De Boef, W.S., Subedi, A., Peroni, M., Thijssen, M. and O'Keeffe, E. (Eds) (2013) *Community Biodiversity Management: promoting resilience and the conservation of plant genetic resources*. Routledge (Earthscan), Abingdon, UK.

DeCarlo, J. (2011) *Fair Trade and How It Works*. The Rosen Publishing Group, New York, NY, USA.

Defu He and Yongming Luo (Eds) (2021) *Microplastics in Terrestrial Environments: emerging contaminants and major challenges*. Springer, Cham, CH.

DeHaas, H., Castles, S. and Miller, M.J. (2020) *The Age of Migration: international population movements in the modern world* 6th edn. Springer (Red Globe Press), London, UK.

De Haas, H. (2023) *How Migration Really Works: a practical guide to the most divisive issue in politics*. Random House Penguin (Viking Books), New York, NY, USA.

Dehm, J. (2021) *Reconsidering REDD+: authority, power and law in the green economy*. Cambridge University Press, Cambridge, UK.

de Jong, W. and van Ommen, J.R. (Eds) (2015) *Biomass as a Sustainable Energy Source for the Future: fundamentals of conversion processes*. Wiley, Hoboken, NJ, USA.

Delang, C.O. and Yuan, Z. (2015) *China's Grain for Green Program: a review of the largest ecological restoration and rural development program in the world*. Springer, Cham, CH.

Delang, C.O. (2016) *China's Water Pollution Problems*. Routledge, Abingdon, UK.

De la Rosa, J.M. (Ed) (2020) *Biochar as Soil Amendment: impact on soil properties and sustainable resource management*. MDPI, Basel, CH.

De Lima, M. (Ed) (2016) *Geographic Information Systems*. ML Books International, New Delhi, IN.

Delingpole, J. (2012) *Watermelons: how environmentalists are killing the planet, destroying the economy and stealing your children's future*. Biteback Publishing, London, UK.

DellaSala, D.A. and Goldstein, M.I. (Eds) (2018) *Encyclopedia of the Anthropocene* (5 vols). Elsevier, Oxford, UK.

De Lucia, V. (2020) *The 'Ecosystem Approach' in International Environmental Law: genealogy and biopolitics*. Routledge, Abingdon, UK.

Demirbas, A. (2010) *Methane Gas Hydrates*. Springer, London, UK.

De Moor, T. (2015) *The Dilemma of the Commoners: understanding the use of common-pool resources in long-term perspective*. Cambridge University Press, New York, NY, USA.

Denny, M. (2017) *Making the Most of the Anthropocene: facing the future*. John Hopkins University Press, Baltimore, MD, USA.

Dentch, M.P. (2016) *The ISO 14001:2015 Implementation Handbook: using the process approach to build an environmental management system*. ASQ Quality Press, Milwaukee, WI, USA.

De Nys, E., Engle, N. and Rocha Magalhães, A. (Eds) (2016) *Drought in Brazil: proactive management and policy*. CRC Press, Boca Raton, FL, USA.

de Ohveira, J.A.P. (2017) Networks for environmental management involving public and quasi-public organisations for market development towards sustainability in Rio De Janeiro, Brazil. In: I. Demirag (Ed), *Corporate Social Responsibility, Accountability and Governance: global perspectives* (pp. 237–243; first published 2005). Routledge, London, UK.

Depoe, S.P., Delicath, J.W. and Elsenbeer, M.-F.A. (Eds) (2011) *Communication and Public Participation in Environmental Decision Making*. SUNY Press, New York, NY, USA.

DePriest, D. (2013) *A GPS User Manual: working with Garmin receivers*. AuthorHouse, Bloomington, IN, USA.

Desai, A. and Mital, A. (2020) *Sustainable Product Design and Development*. CRC Press, Boca Raton, FL, USA.

Desai, B.H. (2014) *International Environmental Governance – towards UNEPO*. Brill Nijhoff, Leiden, NL.

Despommier, D. (2010) *The Vertical Farm: feeding the world in the 21st century.* Macmillan (St Martin's Press), New York, NY, USA.

Dessler, A. and Parson, E.A. (2016) *The Science and Politics of Global Climate Change: a guide to the debate* (first published 2010). Cambridge University Press, Cambridge, UK.

Detraz, N. (2017) *Gender and the Environment – gender and global politics.* Polity Press, Cambridge, UK.

Deutz, P., Lyons, D.I. and Bi, J. (Eds) (2015) *International Perspectives on Industrial Ecology.* Edward Elgar, Cheltenham, UK.

Devall, B. (Ed) (2021) *Living Deep Ecology: a bioregional journey.* Rowman & Littlefield Publishers, Lanham, MD, USA.

De Vente, J., Reed, M.S., Stringer, L.C., Valente, S. and Newig, J. (2016) How does the context and design of participatory decision making processes affect their outcomes? Evidence from sustainable land management in global drylands. *Ecology and Society* 21(2), article No. 24 (18 pp.).

Devere, H., Kelli, T.M. and Synnott, J.P. (Eds) (2017) *Peacebuilding and the Rights of Indigenous Peoples: experiences and strategies for the 21st century.* Springer, Cham, CH.

Devi, G.T.V. (2019) *Understanding Human Ecology: knowledge, ethics and politics.* Routledge, Abingdon, UK.

Devlin, J.F. (Ed) (2019) *Social Movements Contesting Natural Resource Development.* Routledge (Earthscan), Abingdon, UK.

de Vries, J., Schusters, M., Procee, P. and Mengers, H. (2001) *Environmental Management of Small and Medium Sized Cities in Latin America and the Caribbean.* Institute for Housing and Urban Development Studies (IHS), Washington, DC, USA.

de Zeeuw, H., and Drechsel, P. (Eds) (2015) *Cities and Agriculture: developing resilient urban food systems.* Routledge (Earthscan), Abingdon, UK.

Dhyani, S., Gupta, A.K. and Karki, M. (Eds) (2020) *Nature-Based Solutions for Resilient Ecosystems and Societies.* Springer, Singapore, SG.

Diamond, J.M. (2005) *Collapse: how societies choose to fail or succeed.* Penguin, London, UK.

Diamond, J.M. (2012) *The World Until Yesterday: what can we learn from traditional societies?* Penguin (Allen Lane), London, UK.

Diamond, J.M. (2020) *Upheaval: how nations cope with crisis and change.* Penguin (Allen Lane), London, UK.

Diamond, J.M. (2021) *The Last Tree on Easter Island.* Penguin, London, UK.

Dickensen, J.L. and Bonney, R. (Eds) (2012) *Citizen Science: public participation in environmental research.* Cornell University Press (Comstock), Ithaca, NY, USA.

Diehl, J.A. and Kauer, H. (Eds) (2021) *New Forms of Urban Agriculture: an urban ecology perspective.* Springer, Singapore, SG.

Diesendorf, M. and Hamilton, C. (Eds) (2020) *Human Ecology, Human Economy* 2nd edn. Routledge, Abingdon, UK.

Dinar, A. and Tsur, Y. (Eds) (2017) *Management of Transboundary Water Resources under Scarcity: a multidisciplinary approach.* World Scientific, Singapore, SG.

Di Nardo, A., Boccelli, D.L., Herrera, M., Creaco, E., Cominola, A., Sitzenfrei, R. and Taormina, R. (Eds) (2021) *Smart Urban Water Networks: solutions, trends and challenges.* MDPI, Basel, CH.

Dinesh Kumar, M., Ratna, R.V. and James, A.J. (Eds) (2019) *From Catchment Management to Managing River Basins: science, technology choices, institutions and policy.* Elsevier, Amsterdam, NL.

Divya, M.P. (2008) *Social Forestry and Agroforestry.* Vikram Jain Books, New Delhi, IN.

Dixon, J.A., James, D.E. and Sherman, P.B. (Eds) (1989) *The Economics of Dryland Management*. Earthscan, London, UK.

Dixson-Declève, S., Gaffney, O., Ghosh, J., Randers, J., Rockström, J. and Stocknes, P.E. (2022) *Earth for All: a survival guide for humanity* (a report of the Club of Rome). New Society Publishers, London, UK.

Di Zhou (2020a) China's environmental vertical management reform: an effective and sustainable way forward or trouble in itself? *Laws* 9(4), 1–27.

Dobson, A. (2016) *Environmental Politics: a very short introduction*. Oxford University Press, Oxford, UK.

Do Carmo, J.S.A. (Ed) (2021) *River Basin Management: sustainability issues and planning strategies*. IntechOpen, London, UK.

Do Carmo Azevedo, G.M., Fialho, A. and Eugénio, T. (Eds) (2022) *Modern Regulations and Practices for Social and Environmental Accounting*. IGI Global Publishers, Hershey, PA. USA.

Dodds, F. and Bartram, J. (Eds) (2016) *The Water, Food, Energy and Climate Nexus: challenges and an agenda for action*. Routledge (Earthscan), Abingdon, UK.

Dodds, K., Hemmings, A.D. and Roberts, P. (Eds) (2017) *Handbook on the Politics of Antarctica*. Edward Elgar, Cheltenham, UK.

Doherty, B. (2016) *Ideas and Actions in the Green Movement* (first published 2002). Taylor & Francis, Abingdon, UK.

Döös, B.R. (1997) Can large-scale environmental migrations be predicted? *Global Environmental Change* 7(1), 41–61.

Dormann, C. (2020) *Environmental Data Analysis: an introduction with examples in R*. Springer, Cham, CH.

Doron, A. and Jeffrey, R. (2018) *Waste of a Nation: garbage and growth in India*. Harvard University Press, Cambridge, MA, USA.

Douglas, I., Anderson, P.M.L., Goode, D., Houck, M.C., Maddox, D., Nagendra, H. and Tan, P. Y. (Eds) (2021) *The Routledge Handbook of Urban Ecology* 2nd edn. Routledge, Abingdon, UK.

Douglas, I., Goode, D.D., Houck, M.C. and Wang, R. (Eds) (2011) *The Routledge Handbook of Urban Ecology*. Routledge, Abingdon, UK.

Douglas, I. and James, P. (2015) *Urban Ecology: an introduction*. Routledge, Abingdon, UK.

Doumelzel, V. (2023) *The Seaweed Revolution: how seaweed has shaped our past and can save our future* (trans. from French). Hero Press, London, UK.

Dove, M.R. and Carpenter, C. (Eds) (2007) *Environmental Anthropology: a historical reader*. Wiley, Chichester, UK.

Doxiadis, C. (1977) *Ecology and Ekistics*. Elek Books, London, UK.

Doyle, T., McEachern, D. and MacGregor, S. (2016) *Environment and Politics*. Routledge, Abingdon, UK.

Drakakis-Smith, D.W. (2000) *Third World Cities*. Routledge, London, UK.

Drechsler, M. (2020) *Ecological-Economic Modelling for Biodiversity Conservation*. Cambridge University Press, Cambridge, UK.

Drengson, A. and Devall, B. (Eds) (2016) *Ecology of Wisdom: writings by Arne Naess*. Penguin Classics, London, UK.

Dryzek, J.S. and Pickering, J. (2019) *The Politics of the Anthropocene*. Oxford University Press, Oxford, UK.

Dukes, E.F. and Hirsch, S. (2014) *Mountaintop Mining in Appalachia: understanding stakeholders and change in environmental conflict*. Ohio University Press, Athens, OH, USA.

Dupont, H. (Ed) (2012) *Environmental Management: systems, sustainability, and current issues*. Nova Science Publishers, New York, NY, USA.

Dusik, J. and Sadler, B. (Eds) (2016) *European and International Experiences of Strategic Environmental Assessment: recent progress and future prospects*. Routledge (Earthscan), Abingdon, UK.

Dyball, R. and Newell, B.A.T. (2015) *Understanding Human Ecology: a systems approach to sustainability*. Routledge (Earthscan), Abingdon, UK.

Earle, A., Öjendal, J., Swain, A., Jägerskog, A. and Cascão, A.E. (2015) *Transboundary Water Management and the Climate Change Debate*. Routledge (Earthscan), Abingdon, UK.

Earnest, P. (Ed) (2015) *The Global Positioning System*. Nova Science Publishers Inc., Hauppauge, NY, USA.

Eastwood, L.E. (2019) *Negotiating the Environment: civil society, globalisation and the UN*. Routledge, Abingdon, UK.

Ebadi, A.G., Toughani, M., Najafi, A. and Babaee, M. (2020) A brief overview on current environmental issues in Iran. *Central Asian Journal of Environmental Science and Technology Innovation* 1(1), 1–11.

Eccleston, C.H. (2011) *Environmental Impact Assessment: a guide to best professional practices*. CRC Press, Boca Raton, FL, USA.

Eccleston, C.H. (2014) *The EIS Book: managing and preparing environmental impact statements*. CRC Press, Boca Raton, FL, USA.

Eccleston, C.H. (Ed) (2020) *NEPA and Environmental Planning: tools, techniques, and approaches for practitioners* (first published 2008). CRC Press, Boca Raton, FL, USA.

Eckersley, R. (1988) The road to ecotopia? Socialism versus environmentalism. *The Ecologist* 18(4–5), 142–148.

Edden, R.J., Hatfield, J.L., Yadav, S.S., Lotze-Campen, H. and Hall, A.E. (Eds) (2011) *Crop Adaptation to Climate Change*. Wiley-Blackwell, Chichester, UK.

Edelman, D.J., Schuster, M. and Said, J. (2017) Environmental management in Latin America, 1970–2017. *Current Urban Studies* 5(3), 305–331.

Edgell, D.L. Snr. (2020) *Managing Sustainable Tourism: a legacy for the future*. Routledge, Abingdon, UK.

Edington, J.M. (2017) *Indigenous Environmental Knowledge: reappraisal*. Springer International, Cham, CH.

Edwards, C.A., Aracon, N.Q. and Sherman, R.L. (Eds) (2021) *Vermiculture Technology: earthworms, organic wastes, and environmental management*. CRC Press, Boca Raton, FL, USA.

Ehlers, E., Kraft, T. and Moss, C. (Eds) (2006) *Earth System Science in the Anthropocene: emerging issues and problems*. Springer, Berlin, GR.

Ehrlich, P.R. (1970) *The Population Bomb*. New Ballentine Books, New York, NY, USA.

Ehrlich, P.R., Ehrlich, A.H. and Holdren, J.P. (1970) *Ecoscience: population, resources, environment*. Freeman, San Francisco, CA, USA.

Ehrlich, S.D. (2018) *The Politics of Fair Trade: moving beyond free trade and protection*. Oxford University Press, New York, NY, USA.

Eiglad, E. (Ed) (2015) *Social Ecology and Social Change*. AK Press, Chico, CA, USA.

Einarsson, Á. and Óladóttir, Á.D. (2020) *Fisheries and Aquaculture: the food security of the future*. Elsevier (Academic Press), London, UK.

Eisma-Osorio, R.-L., Kirk, E.A. and Steinberg Albin, J. (Eds) (2020) *The Impact of Environmental Law: stories of the world we want*. Edward Elgar, Cheltenham, UK.

Ekins, P. and Andersen, M.S. (Eds) (2009) *Carbon-Energy Taxation: lessons from Europe*. Oxford University Press, Oxford, UK.

Bibliography

Ekins, P. and Speck, S. (Eds) (2011) *Environmental Tax Reform (ETR): a policy for green growth*. Oxford University Press, Oxford, UK.

Ekundayo, E. (Ed) (2011) *Environmental Monitoring*. IntechOpen, London, UK.

Elachi, C. and van Zyl, J. (2021) *Introduction to the Physics and Techniques of Remote Sensing* 3rd edn. Wiley, Hoboken, NJ, USA.

El-Din, H. and Saleh, M. (Eds) (2019) *Municipal Solid Waste Management*. IntechOpen, London, UK.

El-Gendy, N.S. (Ed) (2021) *Sustainable Solutions for Environmental Pollution: Vol. 1: Waste management and value-added products*. Wiley (Scrivener), Hoboken, NY, USA.

Elias Bibri, S.E. (2018) *Smart Sustainable Cities of the Future: the untapped potential of big data and context-aware computing for advancing sustainability*. Springer, Cham, CH.

Elkington, J. and Knight, P. (1991) *The Green Business Guide*. Victor Gollancz, London, UK.

Elling, B. (2010) *Rationality and the Environment: decision-making in environmental politics and assessment*. Routledge (Earthscan), Abingdon, UK.

Ellis, E.C. (2018) *Anthropocene: a very short introduction*. Oxford University Press, Oxford, UK.

Elmqvist, T.M., Fragkias, J., Goodness, B., Guneralp, P.J., Marcotullio, R.I., McDonald, S., Parnell, M., Schewenius, M., Sendstad, K.C. and Wilkinson, S.C. (Eds) (2013) *Urbanization, Biodiversity and Ecosystem Services: challenges and opportunities: a global assessment*. Springer, Dordrecht, NL.

Eloi, L. (2020) *The New Environmental Economics: sustainability and justice*. Polity Press, Cambridge, UK.

Elton, C.S. (1958) *The Ecology of Invasions by Animals and Plants*. Chapman and Hall, London, UK.

Emanuel, K.A. (2007) *What We Know About Climate Change*. MIT Press, Cambridge, MA, USA.

Emery, W. and Camps, A. (2017) *Introduction to Satellite Remote Sensing: atmosphere, ocean, land and cryosphere applications*. Elsevier, Amsterdam, NL.

Emetere, M.E. (2019) *Environmental Modeling Using Satellite Imaging and Dataset Re-processing*. Springer, Cham, CH.

Emetere, M.E. (2020) *Numerical Methods in Environmental Data Analysis*. Elsevier, Amsterdam, NL.

Emetere, M.E. and Akinlabi, E.T. (2020) *Introduction to Environmental Data Analysis and Modeling*. Springer, Cham, CH.

Emmanuel, E.O. (2014) *Environmental Management Systems Guide*. CreateSpace Independent Publishing Platform, Scotts Valley, CA, USA.

Emmett, S. and Sood, V. (2010) *Green Supply Chains: an action manifesto*. Wiley, Oxford, UK.

Emmott, S. (2013) *Ten Billion: facing our future*. Penguin Canada, Toronto, ON, CA.

Endres, A. and Radke, V. (2018) *Economics for Environmental Studies: a strategic guide to micro- and macroeconomics*. Springer, Berlin, GR.

Engel, M., Pilarczyk, J., May, S.M., Brill, D. and Garrett, E. (Eds) (2020) *Geological Records of Tsunamis and Other Extreme Waves*. Elsevier, Amsterdam, NL.

Englander, J. (2021) *Moving to Higher Ground: rising sea level and the path forward*. The Science Bookshelf, New, NY, USA.

Epler Wood, M. (2017) *Sustainable Tourism on a Finite Planet: environmental, business and policy solutions*. Routledge (Earthscan), Abingdon, UK.

Epstein, M.J. and Buhovac, A.R. (2014) *Making Sustainability Work: best practices in managing and measuring corporate social, environmental and economic impact* 2nd edn. Routledge, Abingdon, UK.

Erdkamp, P., Manning, J.G. and Verboven, C. (Eds) (2021) *Climate Change and Ancient Societies in Europe and the Near East: diversity in collapse and resilience.* Springer (Palgrave Macmillan), Cham, CH.

Esakki, T. (Ed) (2017) *Green Marketing and Environmental Responsibility in Modern Corporations.* IGI Global, Hershey, PA, USA.

Eslamian, S. (Ed) (2016) *Urban Water Reuse Handbook.* CRC Press, Boca Raton, FL, USA.

Eslamian, S. and Eslamian, F.A. (Eds) (2017) *Handbook of Drought and Water Scarcity: environmental impacts and analysis of drought and water scarcity.* CRC Press, Boca Raton, FL, USA.

Espinosa, A. and Walker, J. (2006) Environmental management revisited: lessons from a cybernetic intervention in Colombia. *Cybernetics and Systems* 37(1), 75–92.

Etzion, D. and Aragon-Correa, J.A. (2016) Big data, management, and sustainability. *Organization & Environment* 29(2), 147–155.

Evanoff, R. (2014) *Bioregionalism and Global Ethics: a transactional approach to achieving ecological sustainability, social justice, and human well-being.* Routledge, Abingdon, UK.

Evans, J.P. (2012) *Environmental Governance.* Routledge, Abingdon, UK.

Evans, N. (2015) *Strategic Management for Tourism, Hospitality and Events* 2nd edn. Routledge, Abingdon, UK.

Everard, C. (2019) *Desert Locust Plagues: controlling the ancient scourge.* Bloomsbury Academic, London, UK.

Everard, M. (2017) *Ecosystem Services: key issue.* Routledge (Earthscan), Abingdon, UK.

Ewert, A.W., Baker, D.C. and Bissix, G.C. (2004) *Integrated Resource and Environmental Management: the human dimension.* CABI Publishing, Wallingford, UK.

Extinction Rebellion (2019) *This is Not a Drill: an Extinction Rebellion handbook.* Penguin, London, UK.

Ezban, M. (2020) *Aquaculture Landscapes: fish farms and the public realm.* Routledge, Abingdon, UK.

Fabian, P. and Demeris, M. (2014) *Ozone in the Atmosphere: basic principles, natural and human impacts.* Springer, Berlin, GR.

Fadly, D. (2020) Greening industry in Vietnam: environmental management standards and resource efficiency in SMEs. *Sustainability* 12(18), article No. 7455 (27 pp.).

Fagan, B.M. (2000) *The Little Ice Age: how climate made history 1300–1850.* Basic Books, New York, NY, USA.

Fagan, B.M. (2008) *The Great Warming: climate change and the rise and fall of civilisations.* Bloomsbury, London, UK.

Fagan, B.M. (2013) *The Attacking Ocean: the past, present, and future of rising sea levels.* Bloomsbury, London, UK.

Fagan, B.M. (2019) *Floods, Famines and Emperors: El Niño and the fate of civilizations.* Basic Books, New York, NY, USA.

Fagan, B.M. and Durrani, N. (2021) *Climate Chaos: lessons on survival from our ancestors.* Hachette, London, UK.

Fahimnia, B., Bell, M., Hensher, D.A. and Sarkis, J. (Eds) (2015) *Green Logistics and Transportation: a sustainable supply chain perspective.* Springer, Cham, CH.

Fairhead, J. and Leach, M. (1996) *Misreading the African Landscape: society and ecology in a forest–savanna mosaic.* Cambridge University Press, Cambridge, UK.

Falatoonitoosi, E., Hasan, M., Leman, Z. and Masoumik, S.M. (2016) *Green Supply Chain Management: product life cycle approach.* Reference, London, UK.

Falck-Zepeda, J., Gruère, G.P. and Sithole-Niang, I. (Eds) (2013) *Genetically Modified Crops in Africa: economic and policy lessons from countries south of the Sahara*. IFPRI, Washington, DC, USA.

Falconer, R.A. (2020) *Water Quality Modelling* (first published 1992). Routledge, Abingdon, UK.

Falk, D.A., McKenzie, D. and Miller, C. (Eds) (2011) *The Landscape Ecology of Fire*. Springer, Dordrecht, NL.

Fanfani, D. and Ruiz, A.M. (Eds) (2020a) *Bioregional Planning and Design: Vol. 1: Perspectives on a transitional century*. Springer, Cham, CH.

Fanfani, D. and Ruiz, A.M. (Eds) (2020b) *Bioregional Planning and Design: Vol. 2: Issues and practices for a bioregional regeneration*. Springer, Cham, CH.

FAO (2016) *RIMA II – Resilience Index Measurement and Analysis: analysing resilience for better targeting and action*. FAO, Rome, IT.

FAO (2017) *The Charcoal Transition: greening the charcoal value chain to mitigate climate change and improve local livelihoods*. FAO, Rome, IT.

FAO (2022) *Managing Risksto Build Climate-Smart and Resilient Agrifood Value Chains: the role of climate services*. FAO, Rome, IT.

FAO and ILO (2009) *The Livelihood Assessment Tool-kit: analysing and responding to the impact of disasters on the livelihoods of people*. FAO and ILO, Geneva, CH.

FAO, IFAD, UNICEF, WFP and WHO (2021) *The State of Food Security and Nutrition in the World 2021: transforming food systems for food security, improved nutrition and affordable healthy diets for all*. FAO, Rome, IT.

Fares, A. (Ed) (2021) *Climate Change and Extreme Events*. Elsevier, Amsterdam, NL.

Farid, M., Keen, M., Papaioannou, M., Parry, I., Pattillo, C.A. and Ter-Martirosyan, A. (2016) *After Paris: fiscal, macroeconomic, and financial implications of climate change*. International Monetary Fund, Washington, DC, USA.

Farmer, A. (1997) *Managing Environmental Pollution*. Routledge, London, UK.

Farmer, A. (2016) *Handbook of Environmental Protection and Enforcement: principles and practice*. Routledge (Earthscan), Abingdon, UK.

Farmer, A.M. (Ed) (2012) *Manual of European Environmental Policy*. Routledge, London, UK.

Farmer, L.S.J. (2020) *Fake News in Context*. Routledge, Abingdon, UK.

Farmer, T. and Barnes, J. (2018) Environment and society in the Middle East and North Africa: introduction. *International Journal of Middle East Studies* 50(3), 375–382.

Farvar, M.T. and Milton, J.P. (Eds) (1972) *The Careless Technology: ecology and international development*. Doubleday (The Natural History Press), New York, NY, USA.

Farzana, Q., Ramayah, T. and Jihad, M. (Eds) (2018) *Driving Green Consumerism through Strategic Sustainability Marketing*. IGI Global, Hershey, PA, USA.

Fatemi, M., Rezaei-Moghaddam, K., Wackernagel, M. and Shennan, C. (2018) Sustainability of environmental management in Iran: an ecological footprint analysis. *Iran Agricultural Research* 37(2), 53–68.

Fath, B.D. and Jørgensen, S.E. (2011) *Fundamentals of Ecological Modelling: applications in environmental management and research* 4th edn. Elsevier, Amsterdam, NL.

Fath, B.D. and Jørgensen, S.E. (Eds) (2021) *Managing Biological and Ecological Systems: environmental management handbook: Vol. 2*, 2nd edn. CRC Press, Boca Raton, FL, USA.

Faulseit, R.K. (Ed) (2016) *Beyond Collapse: archaeological perspectives on resilience, revitalization, and transformation in complex societies*. Southern Illinois University, Carbondale, IL, USA.

Faure, M. (Ed) (2018) *Elgar Encyclopedia of Environmental Law: Vol. 6*. Edward Elgar, Cheltenham, UK.

Faure, M., de Smedt, P. and Stas, A. (Eds) (2015) *Environmental Enforcement Networks: concepts, implementation and effectiveness*. Edward Elgar, Cheltenham, UK.

Fawaz, U. and Long, D.G. (2014) *Microwave Radar and Radiometric Remote Sensing*. University of Michigan Press, Ann Arbor, MI, USA.

Federal Ministry for Economic Cooperation and Development (BZM) (2017) *The Vulnerability Sourcebook: concept and guidelines for standardised vulnerability assessments*. BZM, Bonn, GR.

Federici, S. (2019) *Re-Enchanting the World: feminism and the politics of the commons*. PM Press, Oakland, CA, USA.

Feeney, D., Berkes, F., McCay, B.J. and Acherson, J.M. (1990) The tragedy of the commons – 22 years later. *Human Ecology* 18(1), 1–19.

Fekete, A. and Fiedrich, F. (Eds) (2018) *Urban Disaster Resilience and Security: addressing risks in societies*. Springer, Cham, CH.

Fennell, D.A. (2020) *Ecotourism* 5th edn. Routledge, Abingdon, UK.

Ferguson, R.J. (2021) *Greening China's New Silk Roads: the sustainable governance of Belt and Road*. Edward Elgar, Cheltenham, UK.

Fernandez, L. and Carson, R.T. (Eds) (2003) *Both Sides of the Border: transboundary environmental management Iisues facing Mexico and the United States*. Kluwer Academic, New York, NY, USA.

Fernández, P.V. (2017) Municipal governance, environmental management and disaster risk reduction in Chile. *Bulletin of Latin American Research* 36(4), 440–458.

Fernández, R.A., Zubelzu, S. and Martínez, M. (Eds) (2017) *Carbon Footprint and the Industrial Life Cycle: from urban planning to recycling*. Springer, Cham, CH.

Ferranti, P., Berry, E.M. and Anderson, J.R. (Eds) (2018) *Encyclopedia of Food Security and Sustainability* (3 vols). Elsevier, Amsterdam, NL.

Ferrari, E. and Rae, A. (2019) *GIS for Planning and the Built Environment: an introduction to spatial analysis – planning, environment, cities*. Bloomsbury Publishing, London, UK.

Ferreira, C.S.S., Kalantari, Z., Hartmann, T. and Pereira, P. (Eds) (2022) *Nature-Based Solutions for Flood Mitigation: environmental and socio-economic aspects*. Springer, Cham, CH.

Ferrier, R.C. and Jenkins, A. (Eds) (2010) *Handbook of Catchment Management*. Blackwell, Oxford, UK.

Ferrier, R.C. and Jenkins, A. (Eds) (2021) *Handbook of Catchment Management* 2nd edn. Wiley, Hoboken, NJ, USA.

Ferry, N. and Gatehouse, A.M.R. (Eds) (2009) *Environmental Impact of Genetically Modified Crops*. CABI, Wallingford, UK.

Fiadjoe, A. (2004) *Alternative Dispute Resolution: a developing world perspective*. Routledge-Cavendish, Abingdon, UK.

Fiddian-Qasmiyeh, E., Loescher, G., Long, K. and Sigona, N. (Eds) (2016) *The Oxford Handbook of Refugee and Forced Migration Studies*. Oxford University Press, Oxford, UK.

Field, C.B., Barros, V.R., Dokken, D.J., Mach, K.J., Mastrandrea, M.D., Bilir, T.E., Chatterjee, M., Ebi, K.L., Estrada, Y.O., Genova, R.C., Girma, B., Kissel, E.S., Levy, A.N., MacCracken, S., Mastrandrea, P.R. and White, L.L. (Eds.) (2014) *Climate Change 2014: impacts, adaptation, and vulnerability. Part A: global and sectoral aspects*. Contribution of Working Group II to the Fifth Assessment Report of the Intergovernmental Panel on Climate Change Cambridge University Press, Cambridge, UK.

Fields, B. and Renne, J.L. (2021) *Adaptation Urbanism and Resilient Communities: transforming streets to address climate change.* Routledge, Abingdon, UK.

Figueres, C. and Rivett-Carnac, T. (2020) *The Future We Choose: surviving the climate crisis.* Manilla Press, London, UK.

Fiksel, J. (2009) *Design for Environment: a guide to sustainable product development* 2nd edn. McGraw-Hill, New York, NY, USA.

Fiksel, J. (Ed) (2022) *Resilient by Design: creating businesses that adapt and flourish in a changing world* 2nd edn. Springer, Cham, CH.

Finamore, B. (2018) *Will China Save the Planet?* Polity Press, Cambridge, UK.

Finkbeiner, M. (Ed) (2011) *Towards Life Cycle Sustainability Management.* Springer, Dordrecht, NL.

Finsinger, J. and Marx, J.F. (1996) Eco-Management and Audit Scheme (EMAS): opportunities and risks for insurance companies. *International Journal of Environment and Pollution* 6(4–6), 491–499.

Fiorino, D.J. (2018) *A Good Life on a Finite Earth: the political economy of green growth.* Oxford University Press, New York, NY, USA.

Fischer, M. (Ed) (2021) *Environmental Management: ecosystems, competitiveness and waste management.* Nova Science Publishers, Hauppauge, NY, USA.

Fischer, T.B. (2007) *Theory & Practice of Strategic Environmental Assessment: towards a more systematic approach.* Routledge (Earthscan), Abingdon, UK.

Fischer, T.B. and González, A. (Eds) (2021) *Handbook on Strategic Environmental Assessment.* Edward Elgar, Cheltenham, UK.

Fischer-Kowalski, M., Krausmann, F., Haberl, H. and Winiwarter, V. (Eds) (2016) *Social Ecology: society-nature relations across time and space.* Springer, Cham, CH.

Fiscus, D.A. and Fath, B.D. (2018) *Foundations for Sustainability: a coherent framework of life–environment relations.* Elsevier (Academic Press), London, UK.

Fish, R.D. (2011) Environmental decision making and an ecosystems approach: Some challenges from the perspective of social science. *Progress in Physical Geography* 35(5), 671–680.

Fisher, E. (2017) *Environmental Law: a very short introduction.* Oxford University Press, Oxford, UK.

Fisher, E., Lange, B. and Scotford, E. (2013) *Environmental Law: text, cases & materials.* Oxford University Press, Oxford, UK.

Fitzgerald, J. (2020) *Greenovation: urban leadership on climate change.* Oxford University Press, Oxford, UK.

Flannery, T. (2007) *The Weather Makers: how man is changing the climate and what it means for life on Earth* (first published 2005). Grove/Atlantic, New York, NY, USA.

Fletcher, K. and Tham, M. (Eds) (2015) *Routledge Handbook of Sustainability and Fashion.* Routledge (Earthscan), Abingdon, UK.

Fluss, H. and Frim, L. (2022) *Prometheus and Gaia: technology, ecology and anti-humanism.* Anthem Press, London, UK.

Foley, G. (1986) *Charcoal Making in Developing Countries.* Routledge (Earthscan), London, UK.

Foltz, R.C., Baharuddin, A. (Haji) and Denny, F. (Eds) (2003) *Islam and Ecology: a bestowed trust.* Harvard University Press, Cambridge, MA, USA.

Fonseca, A. (Ed) (2022) *Handbook of Environmental Impact Assessment.* Edward Elgar, Cheltenham, UK.

Ford, A. (2009) *Modeling the Environment* 2nd edn. Island Press, Washington, DC, USA.

Ford, J.D. and Barrang-Ford, L. (Eds) (2011) *Climate Change Adaptation in Developed Nations: from theory to practice*. Springer, Dordrecht, NL.

Forman, R.T.T. (2014) *Urban Ecology: science of cities*. Cambridge University Press, Cambridge, UK.

Fort, H. (2020) *Ecological Modelling and Ecophysics: agricultural and environmental applications*. IOP Publishing, Bristol, UK.

Fowler, A. and Malunga, C. (Eds) (2010) *NGO Management: the Earthscan companion*. Routledge (Earthscan), Abingdon, UK.

Fowler, C. (2016) *Seeds on Ice: Svalbard and the Global Seed Vault*. Easton Studio Press, Norwalk, CT, USA.

Fox, M. (1989) A call for a spiritual renaissance. *Green Letter* 5(1), 4, 16–17.

Frache, F., Grigore, G., Stancu, A. and McQueen, D. (2020) *Values and UK Corporate Responsibility: CSR and sustainable development*. Palgrave Macmillan, London, UK.

Franchetti, M.J. and Apul, D. (2013) *Carbon Footprint Analysis: concepts, methods, implementation, and case studies*. CRC Press, Boca Raton, FL, USA.

Francis, Pope. (2015) *Encyclical letter: Laudato si' of the Holy Father Francis* ('on the care of the common home' by Pope Francis) 1st edn. Vatican City Publication, VC.

Francis, R.A., Millington, J.D.A. and Chadwick, M.A. (Eds) (2016) *Urban Landscape Ecology: science, policy and practice*. Routledge (Earthscan), London, UK.

Frank, A. (2018) *Light of the Stars: the fate of the Earth*. W.N. Norton, New York, NY, USA.

Frankel, D., Lawrence, S. and Webb, J. (Eds) (2013) *Archaeology in Environment and Technology: intersections and transformations*. Routledge, Abingdon, UK.

Frankopan, P. (2023) *The Earth Transformed: an untold history*. Bloomsbury, London, UK.

Franks, B., Hanscomb, S. and Johnston, S.F. (2018) *Environmental Ethics and Behavioural Change*. Routledge, Abingdon, UK.

Frederic, A. and Humphreys, P. (Eds) (2008) *Encyclopedia of Decision Making and Decision Support Technologies*. IGI Global (Information Science Reference), Hershey, PA, USA.

Freedman, E., Hiles, S.S. and Sachsman, D.B. (Eds) (2022) *Communicating Endangered Species: extinction, news and public policy*. Routledge, Abingdon, UK.

Freedman, E. and Neuzil, M. (Eds) (2017) *Biodiversity, Conservation and Environmental Management in the Great Lakes Basin*. Routledge (Earthscan), Abingdon, UK.

Freestone, D. (Ed) (2018) *Sustainable Development and International Environmental Law*. Edward Elgar, Cheltenham, UK.

Freire, P. (1970) *Pedagogy of the Oppressed*. Continuum International Publishing, New York, NY, USA.

Fremaux, A. (2019) *After the Anthropocene: green republicanism in a post-capitalist world*. Springer (Palgrave Macmillan), Cham, CH.

Freudenberger, K.S. (2011) *Rapid Rural Appraisal and Participatory Rural Appraisal: a manual for CRS field workers and partners*. Catholic Relief Services, Baltimore, MD, USA.

Fridell, G., Gross, Z. and McHugh, S. (2021) *The Fair Trade Handbook: building a better world, together*. Fernwood Publishing, Winnipeg, MB, CA.

Friis, R.H. (2019) *Essentials of Environmental Health* 3rd edn. Jones & Bartlett, Burlington, MA, USA.

Fris, C. and Nielsen, J.Ø. (Eds) (2019) *Telecoupling: exploring land-use change in a globalised world*. Springer (Palgrave Macmillan), Cham, CH.

Fuchs, S. and Thaler, T. (Eds) (2018) *Vulnerability and Resilience to Natural Hazards*. Cambridge University Press, Cambridge, UK.

Fuhrer, J. and Gregory, P.J. (Eds) (2014) *Climate Change Impact and Adaptation in Agricultural Systems*. CBI, Wallingford, UK.

Bibliography

Funk, C.C. (2021) *Drought, Flood, Fire: how climate change contributes to catastrophes.* Cambridge University Press, Cambridge, UK.

Funk, C.C. and Shukla, S. (2020) *Drought Early Warning and Forecasting: theory and practice.* Elsevier, Amsterdam, NL.

Funtowicz, S.O. and Ravetz, J.R. (1992) The good, the true and the post-modern. *Futures* 24(11), 963–976.

Furness, R.W. and Rainbow, P.S. (Eds) (2018) *Heavy Metals in the Marine Environment.* CRC Press, Boca Raton, FL, USA.

Fusheng Li, Awaya, Y., Kageyama, K. and Yongfen Wei (2022) *River Basin Environment: evaluation, management and conservation.* Springer, Singapore, SG.

Füssel, H.-M. (2005) *Vulnerability in Climate Change Research: a comprehensive conceptual framework.* University of California International and Area Studies Bresia Paper No 6. http://repositories.cdlib.org/ucias/breslauer/6 (accessed 23/05/23).

Gade, A.M. (2018) *Islam and the Environment.* Oneworld Publications, London, UK.

Gade, A.M. (2019) *Muslim Environmentalisms: religious and social foundations.* Columbia University Press, New York, NY, USA.

Galanakis, C.M. (Ed) (2021a) *Food Security and Nutrition.* Elsevier (Academic Press), London, UK.

Galanakis, C.M. (Ed) (2021b) *Food Losses, Sustainable Postharvest and Food Technologies.* Elsevier (Academic Press), London, UK.

Galaz, V. (Ed) (2014) *Global Environmental Governance, Technology and Politics: the Anthropocene gap.* Edward Elgar, Cheltenham, UK.

Galen, D., Newman, G.D. and Qiao, Z. (Eds) (2022) *Landscape Architecture for Sea Level Rise: innovative global solutions.* Routledge, New York, NY, USA.

Gallagher, K.P. (Ed) (2008) *Handbook on Trade and the Environment.* Edward Elgar, Cheltenham, UK.

Gallagher, N., Myers, L. and Chouinard, Y. (Eds) (2016) *Tools for Grassroots Activists: best practices for success in the environmental movement.* Patagonia Books (Independent Publishers Group), Chicago, IL, USA.

Ganpat, W. and Isaac, W.-A. (Eds) (2017) *Environmental Sustainability and Climate Change Adaptation Strategies.* IGI, Hershey, PA, USA.

Garbolino, E. and Voiron-Canicio, C. (Eds) (2020) *Ecosystem and Territorial Resilience: a geoprospective approach.* Elsevier, Amsterdam, NL.

Garcia, E.J. and Vale, B. (2017) *Unravelling Sustainability and Resilience in the Built Environment.* Routledge, Abingdon, UK.

Garcia, J.M. (Ed) (2014) *Climate & Environmental Protection: international funding.* Nova Science Publishing, Hauppauge, NY, USA.

Gardiner, D.K. (2018) *Environmental Pollution in China: what everyone needs to know.* Oxford University Press, New York, NY, USA.

Gardner, S.M., Ramsden, S.J. and Hails, R.S. (Eds) (2019) *Agricultural Resilience: perspectives from ecology and economics.* Cambridge University Press, Cambridge, UK.

Garforth, L. (2018) *Green Utopias: environmental hope before and after nature.* Polity Press, Cambridge, UK.

Garrick, D.E., Anderson, G.R.M. and Pittock, J. (Eds) (2014) *Federal Rivers: managing water in multi-layered political systems.* Edward Elgar, Cheltenham, UK.

Garvey, J. (2008) *The Ethics of Climate Change: right and wrong in a warming world.* Continuum International Publishing, London, UK.

Gasparatos, A. and Willis, K.J. (Eds) (2015) *Biodiversity in the Green Economy.* Routledge, Abingdon, UK.

Gates, B. (2021) *How to Avoid a Climate Disaster: the solutions we have and the break-throughs we need*. Penguin Random House (Alfred A. Knopf), New York, NY, USA.

Gates, B. (2022) *How to Prevent the Next Pandemic*. Penguin Random House (Alfred A. Knopf), New York, NY, USA.

Gatt, C. (2018) *An Ethnography of Global Environmentalism: becoming friends of the Earth*. Routledge, New York, NY, USA.

Geall, S. (Ed) (2013) *China and the Environment: the green revolution*. Zed Books, London, UK.

Geary, W.L., Bode, M., Doherty, T.S., Fulton, E.A., Nimmo, D.G., Tulloch, A.I.T., Tulloch, V.J.D. and Ritchie, E.G. (2020) A guide to ecosystem models and their environmental applications. *Nature Ecology & Evolution* 4, 1459–1471.

Geertz, C. (1971) *Agricultural Involution: the process of ecological change in Indonesia*. University of California Press, Berkeley, CA, USA.

GEF (2021) *Reflecting on 30 Years of the GEF*. Global Environmental Facility, Washington, DC, USA.

Gemenne, F. and McLeman, R.A. (Eds) (2018) *Routledge Handbook of Environmental Displacement and Migration*. Routledge, Abingdon, UK.

Gemenne, F., Zickgraf, C. and Ionesco, D. (Eds) (2016) *The State of Environmental Migration 2016: a review of 2015*. University of Liege Press, Liege, FR.

Gemitzi, A., Koutsias, N. and Lakshmi, V. (2020) *Advanced Environmental Monitoring with Remote Sensing Time Series Data and R*. CRC Press, Boca Raton, FL, USA.

Genletti, D., Cortinovis, C., Zardo, L. and Esmail, A. (2019) *Planning for Ecosystem Services in Cities*. Springer Open, Dordrecht, NL.

George, G. and Schillebeeckx, S.J.D. (Eds) (2018) *Managing Natural Resources: organizational strategy, behaviour and dynamics*. Edward Elgar, Cheltenham, UK.

Gerber, P.J., Steinfeld, H., Henderson, B., Mottet, A., Opio, C., Dijkman, J., Falcucci, A. and Tempio, G. (2013) *Tackling Climate Change Through Livestock: a global assessment of emissions and mitigation opportunities*. FAO, Rome, IT.

Gerged, A.M., Cowton, C.J. and Beddewela, E.S. (2017) Towards sustainable development in the Arab Middle East and North Africa region: a longitudinal analysis of environmental disclosure in corporate annual reports. *Business Strategy & Environment* 27(4), 572–587.

Gerraester, T. and Hester, T. (Eds) (2018) *Climate Engineering and the Law: regulation and liability for solar radiation management and carbon dioxide removal*. Cambridge University Press, Cambridge, UK.

Gerrard, M.B. and Wannier, G.E. (Eds) (2013) *Threatened Island Nations: legal implications of rising seas and a changing climate*. Cambridge University Press, Cambridge, UK.

Gerten, D. and Bergmann, S. (Eds) (2011) *Religion in Environmental and Climate Change: suffering, values, lifestyles*. Bloomsbury Publishing, London, UK.

Ghassemi, F. and White, I. (2012) *Inter-Basin Water Transfer: case studies from Australia, United States, Canada, China and India* (first published 2007). Cambridge University Press, Cambridge, UK.

Gheorghe, A.V. (Ed) (2005) *Integrated Risk and Vulnerability Management Assisted by Decision Support: relevance and impact on governance*. Springer, Dordrecht, NL.

Giampietro, M., Aspinall, R.J., Ramos-Martin, J. and Bukkens, S.G.F. (Eds) (2014) *Resource Accounting for Sustainability Assessment: the nexus between energy, food, water and land use*. Routledge, Abingdon, UK.

Gibson, P. and Power, C.H. (2013) *Introductory Remote Sensing: principles and concepts*. Routledge, Abingdon, UK.

Gibson, R.B. (Ed) (2016) *Sustainability Assessment: applications and opportunities.* Routledge (Earthscan), Abingdon, UK.

Giddens, A. (2009) *The Politics of Climate Change.* Polity Press, London, UK.

Gidley, J.M. (2017) *The Future: a very short introduction.* Oxford University Press, Oxford, UK.

Gilespie, A. (2011) *Conservation, Biodiversity and International Law.* Edward Elgar, Cheltenham, UK.

Gilio-Whitaker, D. (2020) *As Long as Grass Grows: the indigenous fight for environmental justice, from colonization to Standing Rock.* Beacon Press, Boston, MA, USA.

Gille, Z. and Lepawsky, J. (Eds) (2022) *The Routledge Handbook of Waste Studies.* Abingdon, UK.

Ginty, A. (2021) *Climate Change Solutions and Environmental Migration: the injustice of maladaptation and the gendered 'silent offset' economy.* Routledge, Abingdon, UK.

Giorgi, E., Cattaneo, T., Herrera, A.M.F. and del Tarango, V. S.A. (Eds) (2022) *Design for Vulnerable Communities.* Springer, Cham, CH.

Giri, C.P. (Ed) (2012) *Remote Sensing of Land Use and Land Cover: principles and applications.* Routledge, London, UK.

Giridhar, M.V.S.S. (Ed) (2019) *Sustainable Water and Environmental Management.* BS Publications, Hyderabad, IN.

Gitzen, R.A., Millspaugh, J.J., Cooper, A.B. and Licht, D.S. (Eds) (2014) *Design and Analysis of Long-Term Ecological Monitoring Studies.* Cambridge University Press, Cambridge, UK.

Giupponi, C. and Jakeman, A.J. (Eds) (2006) *Sustainable Management of Water Resources: an integrated approach.* Edward Elgar, Cheltenham, UK.

Glaeser, B. (Ed) (2013) *Learning From China? Development and environment in third world countries.* Routledge, Abingdon, UK.

Glaser, M., Ratter, B.M.W., Krause, G. and Welp, M. (Eds) (2012) *Human-Nature Interactions in the Anthropocene: potentials of social-ecological systems analysis.* Routledge, New York, NY, USA.

Glaßmeier, K.H., Soffel, H. and Negendank, J. (Eds) (2009) *Geomagnetic Field Variations.* Springer-Verlag, Berlin, GR.

Glasson, J. and Therivel, R. (2019) *Introduction to Environmental Impact Assessment* 5th edn. Routledge, Abingdon, UK.

Gliessman, S.R. (2014) *Agroecology: the ecology of sustainable food systems* 3rd edn. CRC Press, Boca Raton, FL, USA.

Gliessman, S.R. (Ed) (2019) *Agroecosystem Sustainability: developing practical* strategies. CRC Press, Boca Raton, FL, USA.

Glotfelty, C. and Quesnel, E. (Eds) (2014) *The Biosphere and the Bioregion: essential writings of Peter Berg.* Routledge, London, UK.

Godfrey, L. (Ed) (2021) *Waste Management Practices in Developing Countries.* MDPI, Basle, CH.

Goeden, G.B. (1979) Biogeographic theory as a management tool. *Environmental Conservation* 6(1), 27–32.

Goel, M., Satyanarayana, T., Sudhakar, M. and Agrawal, D.P. (Eds) (2021) *Climate Change and Green Chemistry of CO_2 Sequestration.* Springer, Singapore, SG.

Goh, E.A., Zailani, S. and Abd Wahid, N. (2006) A study on the impact of environmental management system (EMS) certification towards firms' performance in Malaysia. *Management of Environmental Quality: An International Journal* 17(1), 73–93.

Gokten, P.O. and Gokten, S. (2018) *Sustainability Assessment and Reporting*. IntechOpen, London, UK.

Göktepea, O., Altmb, E. and Kasımoğlucc, M. (2014) A strategic environmental management model: Salt Lake case. *Procedia – Social and Behavioral Sciences* 150, 310–319.

Golding, E.L. (2017) *A History of Technology and Environment: from stone tools to ecological crisis*. Routledge, Abingdon, UK.

Goldsmith, E., Allan, R., Allaby, M., Davol, J. and Lawrence, S. (1972) *Blueprint for Survival*. Penguin, Harmondsworth, UK.

Gonen, R. (2021) *The Waste-Free World: how the circular economy will take less, make more and save the Planet*. Penguin (Random House), New York, NY, USA.

Gonenc, I.E., Wolfin, J.P. and Russo, R.C. (Eds) (2015) *Sustainable Watershed Management*. CRC Press, Boca Raton, FL, USA.

Goodenough, A. and Hart, A. (2017) *Applied Ecology: monitoring, managing, and conserving*. Oxford University Press, Oxford, UK.

Goodman, A. and Mesa, N. (2022) *Collaborating for Climate Resilience*. Routledge, Abingdon, UK.

Goonetilleke, A., Tan Yigitcanlar, Ayoko, G.A. and Egodawatta, P. (2014) *Sustainable Urban Water Environment: climate, pollution and adaptation*. Edward Elgar, Cheltenham, UK.

Gore, A. (2006) *An Inconvenient Truth: the planetary emergency of global warming and what we can do about it*. Rodale, New York, NY, USA.

Gore, A. (2013) *The Future: six drivers of global change*. Random House, New York, NY, USA.

Gorman, M. (1979) *Island Ecology*. Chapman and Hall, London, UK.

Gottlieb, R.S. (Ed) (2010) *Religion and the Environment*. Routledge, Abingdon, UK.

Gould, K.A. and Lewis, T.L. (2009) *Twenty Lessons in Environmental Sociology* 3rd edn. Oxford University Press, New York, NY, USA.

Gould, S.J. (1984) Toward the vindication of punctuational change, in W.A. Berggren and J.A. Van Couvering (Eds) *Catastrophes and Earth History: the new uniformitarianism*. Princeton University Press, Princeton, NJ, USA.

Graaf-van Dinther, R. (Ed) (2021) *Climate Resilient Urban Areas: governance, design and development in coastal delta cities*. Springer (Palgrave Macmillan), Cham, CH.

Grant, J. (2020) *Greener Marketing*. Wiley, Chichester, UK.

Grant, W.E. and Swannak, T.M. (2008) *Ecological Modeling: a common sense approach to theory and practice*. Blackwell, Malden, MA, USA.

Grasso, M. and Giugni, M. (Eds) (2022) *The Routledge Handbook of Environmental Movements*. Routledge, Abingdon, UK.

Graves, P.E. (2007) *Environmental Economics: a critique of benefit-cost analysis*. Rowman & Littlefield, Lanham, MA, USA.

Gray, G.A. and Gray, W.G. (2017) *Introduction to Environmental Modelling*. Cambridge University Press, Cambridge, UK.

Gray, R.H., Adams, C.A. and Owen, D. (2014) *Accountability, Social Responsibility and Sustainability: accounting for society and the environment*. Pearson, London, UK.

Gray, R.H., Bebbington, J. and Walters, D. (2002) *Accounting for the Environment*. M. Wiener Publishing, Princeton, NJ, USA.

Gray, S., Paolisso, M. and Jordan, R. (Eds) (2017) *Environmental Modeling with Stakeholders: theory, methods, and applications*. Springer, Dordrecht, NL.

Gray, W. (1993) *Coral Reefs and Islands: the natural history of a threatened paradise*. David and Charles, Newton Abbot, UK.

Grbich, C. (2012) *Qualitative Data Analysis: an introduction* 2nd edn. Sage, Los Angeles, CA, USA.

Greenberg, M.E. (2012) *The Environmental Impact Statement After Two Generations: managing environmental power.* Routledge, Abingdon, UK.

Greenberg, M.R. (2017) *Explaining Risk Analysis: protecting health and the environment.* Routledge (Earthscan), Abingdon, UK.

Greenspan, I., Handy, F. and Katz-Gerro, T. (2012) Environmental philanthropy: is it similar to other types of environmental behavior? *Organization & Environment* 25(2), 111–130.

Gregory, R., Failing, L., Harstone, M., Long, G., McDaniels, T. and Ohlson, D. (2012) *Structured Decision Making: a practical guide to environmental management choices.* Wiley-Blackwell, Oxford, UK.

Grillo, O. and Venora, G. (Eds) (2011) *Biodiversity Loss in a Changing Planet.* BoD – Books on Demand, Norderstedt, GR.

Grim, J.A. (2001) *Indigenous Traditions and Ecology – religions of the world and ecology.* Harvard University Press, Cambridge, MA, USA.

Grim, J.A. and Tucker, M.E. (2014) *Ecology and Religion.* Island Press, Washington, DC, USA.

Gross, M. and Heinrichs, H. (Eds) (2010) *Environmental Sociology: European perspectives and interdisciplinary challenges.* Springer, Dordrecht, NL.

Grove, K. (2018) *Resilience.* Routledge, Abingdon, UK.

Grunwald, A. (2019) *Technology Assessment in Practice and Theory.* Routledge, Abingdon, UK.

Grunewald, K. and Bastian, O. (Eds) (2012) *Ecosystem Services – concept, methods and case studies.* Springer-Verlag, Berlin, GR.

Grunewald, K. and Bastian, O. (Eds) (2015) *Ecosystem Services: concept, methods and case studies.* Springer-Verlag, Berlin, GR.

Gude, V.G., Venkataramana, G. and Kandiah, R. (Eds) (2020) *Sustainable Water: resources, management and challenges.* Nova Science Publishers, Hauppauge, NY, USA.

Guia-Pedrosa, P.V. (2016) *Environmental Management in the Philippines.* ADB-PPD Center, EMB-DENR, PCE-CSSD, Manila, PH.

Guizhen He, Yonglong Lu, Mol, A.P.J. and Beckers, T. (2012) Changes and challenges: China's environmental management in transition. *Environmental Development* 3, 25–38.

Gunderson, L.H., Allen, C.R. and Holling, C.S. (Eds) (2010) *Foundations of Ecological Resilience.* Island Press, Washington, DC, USA.

Guntenspergen, G.R. (Ed) (2014) *Application of Threshold Concepts in Natural Resource Decision Making.* Springer, New York, NY, USA.

Guntenspergen, G.R., Breuste, J. and Niemelä, J. (Eds) (2011) *Urban Ecology: patterns, processes, and applications.* Oxford University Press, Oxford, UK.

Gupta, A., Farjad, B., Wang, G., Hyung Eum and Dubé, M. (2021) *Integrated Environmental Modelling Framework for Cumulative Effects Assessment.* University of Calgary Press, Edmonton, AB, CA.

Gurjar, B.R., Molina, L.T. and Ojha, C.S.P. (Eds) (2010) *Air Pollution: health and environmental impacts.* CRC Press, Boca Raton, FL, USA.

Guttmann, R. (2018) *Eco-Capitalism: carbon money, climate finance, and sustainable development.* Springer (Palgrave Macmillan), Cham, CH.

Haab, T. and Whitehead, J.C. (Eds) (2014) *Environmental and Natural Resource Economics: an encyclopedia.* Greenwood, Santa Barbara, CA, USA.

Habbash, M. (2016) Corporate governance and corporate social responsibility disclosure: evidence from Saudi Arabia. *Social Responsibility Journal* 12(4): 740–754.

Haberlein, T.A. (2012) *Navigating Environmental Attitudes*. Oxford University Press, Oxford, UK.

Haddaway, N.R., Kohl, C., da Silva, N.R., Schiemann, J., Spök, A., Stewart, R., Sweet, R.B. and Wilhelm, R. (2017) A framework for stakeholder engagement during systematic reviews and maps in environmental management. *Environmental Evidence* 6(11), 1–14.

Hagan, S. (2015) *Ecological Urbanism: the nature of the city*. Routledge, London, UK.

Haigh, J.D. and Cargill, P. (2015) *The Sun's Influence on Climate*. Princeton University Press, Princeton, NJ, USA.

Haigh, M., Krecek, J., Kubin, E. and Höfer, T. (Eds) (2012) *Management of Mountain Watersheds*. Springer, Dordrecht, NL.

Hák, T., Moldan, B. and Dahl, A.L. (Eds) (2007) *Sustainability Indicators: a scientific assessment*. Island Press, Washington, DC, USA.

Hale, J.D., Grayson, N., Sadler, J.P., Hunt, D., Bouch, C., Sadler, J., Rogers, C.D.F. and Locret-Collet, M. (2018) *The Little Book of Ecosystem Services in the City*. Birmingham University, Birmingham, UK.

Halegoua, G.R. (2020) *Smart Cities*. MIT Press, Cambridge, MA, USA.

Halford, N.G. (2012) *Genetically Modified Crops*. World Scientific, Singapore, SG.

Hall, M. (2013) *The Severn Tsunami: the story of Britain's greatest natural disaster*. History Press, Cheltenham, UK.

Halldorsson, G., Sigurdsson, B.D. and Leena, F. (2015) *Soil Carbon Sequestration – for climate, food security and ecosystem services*. NORDEN, Copenhagen, DK.

Halvorsen, K.E., Schelly, C., Handler, R.M., Pischke, E.C. and Knowlton, J.L. (Eds) (2019) *A Research Agenda for Environmental Management*. Edward Elgar, Cheltenham, UK.

Hamilton, C. (2017) *Defiant Earth: the fate of humans in the Anthropocene*. Polity Press, Cambridge, UK.

Hamilton, C., Gemenne, F. and Bonneuil, C. (Eds) (2015) *The Anthropocene and the Global Environmental Crisis: rethinking modernity in a new Epoch*. Routledge (Earthscan), Abingdon, UK.

Hamin Infield, E.M., Abunnasr, Y. and Ryan, R.L. (2019) *Planning for Climate Change: a reader in green infrastructure and sustainable design for resilient cities*. Routledge, New York, NY, USA.

Hampton, P.S. (2015) *Workers and Trade Unions for Climate Solidarity: tackling climate change in a neoliberal world*. Routledge, Abingdon, UK.

Han, M. and Nguyen, D.C. (2018) *Hydrological Design of Multipurpose Micro-Catchment Rainwater Management*. IWA Publishing, London, UK.

Hanasz, P. (2018) *Transboundary Water Governance and International Actors in South Asia: the Ganges-Brahmaputra-Meghna basin*. Routledge (Earthscan), Abingdon, UK.

Hancock, G. (1989) *Lords of Poverty: the power, prestige, and corruption of the international aid business*. Grove Press, New York, NY, USA.

Hancock, R. (2018) *Islamic Environmentalism: activism in the United States and Great Britain*. Routledge, Abingdon, UK.

Hanley, N. and Barbier, E.B. (2009) *Pricing Nature: cost–benefit analysis and environmental policy*. Edward Elgar, Cheltenham, UK.

Hanna, K.S. (Ed) (2005) *Environmental Impact Assessment: practice and participation*. Oxford University Press Canada, North York, ON, CA.

Hanna, K.S. (Ed) (2022) *Routledge Handbook of Environmental Impact Assessment*. Routledge, Abingdon, UK.

Hanna, K.S. and Bullock, R.C.L. (2012) *Community Forestry: local values, conflict and forest governance*. Cambridge University Press, Cambridge, UK.

Hansen-Magnusson, H., and Vetterlein, A. (Eds) (2020) *The Rise of Responsibility in World Politics*. Cambridge University Press, Cambridge, UK.

Harari, Y.N. (2017) *Homo Deus: a history of tomorrow*. Harper-Collins, New York, NY, USA.

Harcourt, W. and Nelson, I.L. (Eds) (2015) *Practising Feminist Political Ecologies: moving beyond the 'green economy'*. Zed Books, London, UK.

Hardin, G. (1968) The tragedy of the commons. *Science* 162(3859), 1243–1248.

Hardin, G. (1974a) *The Ethics of a Lifeboat*. American Association for the Advancement of Science, Washington, DC, USA.

Hardin, G. (1974b) Lifeboat ethics: the case against helping the poor. *Psychology Today* 8(1), 38–43, 123–126.

Hardman, M. and Larkham, P.J. (2014) *Informal Urban Agriculture: the secret lives of guerrilla gardeners*. Springer, Cham, CH.

Haring, B. (2020) *Why Biodiversity Loss Is Not a Disaster*. Leiden University Press, Leiden, NL.

Harper, B. (2020) *Agroecosystems: an ecological perspective*. Callisto Reference, Forest Hills, NY, USA.

Harper, K. (2021) *Plagues Upon the Earth: disease and the course of human history*. Princeton University Press, Princeton, NJ, USA.

Harremoës, P., Gee, D., MacGarvin, M., Stirling, A., Keys, J., Wynne, B. and Guedes Vaz, S. (Eds) (2002) *The Precautionary Principle in the 20th Century: late lessons from early warnings*. Routledge (Earthscan), London, UK.

Harris, C.M. and Meadows, J. (1992) Environmental management in Antarctica: instruments and institutions. *Marine Pollution Bulletin* 25(9–12), 239–249.

Harris, F. (Ed) (2012a) *Global Environmental Issues* 2nd edn. Wiley-Blackwell, Oxford, UK.

Harris, J.M. and Roach, B. (2013) *Environmental and Natural Resource Economics: a contemporary approach* 3rd edn. M.E. Sharpe Inc., Armonk, NY, USA.

Harris, P.G. (2012b) *Environmental Policy and Sustainable Development in China*. Policy Press, Bristol, UK.

Harris, P.G. (Ed) (2014) *Routledge Handbook of Global Environmental Politics*. Routledge, Abingdon, UK.

Harris, U.S. (2019) *Participatory Media in Environmental Communication: engaging communities in the periphery*. Routledge, Abingdon, UK.

Harrison, R.M. and Hester, R.E. (Eds) (2014) *Geoengineering of the Climate System*. The Royal Society of Chemistry, Cambridge, UK.

Hart, B., Byron, N., Pollino, C.A., Stewardson, M. and Bond, N. (Eds) (2020) *Murray-Darling Basin, Australia: its future management*. Elsevier, Amsterdam, NL.

Harvey, M. (2019) *Utopia in the Anthropocene: a change plan for a sustainable and equitable world*. Routledge, Abingdon, UK.

Hasan, F. (2018) *Participatory Networks and the Environment: the BGreen Project in the US and Bangladesh*. Routledge (Taylor & Francis), Abingdon, UK.

Hashmi, M.Z. (Ed.) (2020) *Antibiotics and Antimicrobial Resistance Genes: environmental occurrence and treatment technologies*. Springer. Cham, CH.

Hassanien, A.E., Darwish, A., Gyampoh, B., Abdel-Monaim, A.T. and Anter, A.M. (Eds) (2021) *The Global Environmental Effects During and Beyond COVID-19: intelligent computing solutions*. Springer, Cham, CH.

Hauschild, M.Z. and Huijbregts, M.A.J. (Eds) (2015) *Life Cycle Impact Assessment.* Springer, Dordrecht, NL.

Hauschild, M.Z., Rosenbaum, R.K. and Olsen, S.I. (Eds) (2018) *Life Cycle Assessment: theory and practice.* Springer, Cham, CH.

Hausknost, D. and Hammond, M. (2020) Beyond the environmental state? The political prospects of a sustainability transformation. *Environmental Politics* 29(1), 1–16.

Hawken, P. (Ed) (2017) *Drawdown: the most comprehensive plan ever proposed to reverse global warming.* Penguin, New York, NY, USA.

Hawken, P. (2021) *Regeneration: ending the climate crisis in one generation.* Penguin, New York, NY, USA.

Hawksworth, D.L. (2009) *Methods and Practice in Biodiversity Conservation.* Springer, Dordrecht, NL.

Hawley, J. (Ed) (2015) *Why Women Will Save the Planet.* Zed Books, London, UK.

Hawley, J. (Ed) (2018) *Why Women Will Save the Planet* 2nd edn. Routledge (Earthscan), London, UK.

Head, L., Saltzman, K., Setten, G. and Stenseke, M. (Eds) (2020) *Nature, Temporality and Environmental Management: Scandinavian and Australian perspectives on peoples and landscapes.* Routledge, Abingdon, UK.

Healy, R.G., VanNijnatten, D.L. and López-Vallejo, M. (2020) Environmental management approaches and capacities. In: R.G. Healy., D.L. VanNijnatten and M. López-Vallejo (Eds), *Environmental Policy in North America: approaches, capacity, and the management of transboundary issues* (pp. 13–46) University of Toronto Press, Toronto, ON, CA.

Heberlein, T.A. (2012) *Navigating Environmental Attitudes.* Oxford University Press, New York, NY, USA.

Hecht, J.E. (2005) *National Environmental Accounting: bridging the gap between ecology and economy.* Routledge, Abingdon, UK.

Hecker, S., Bonn, A., Haklay, M., Bowser, A., Makuch, Z., Vogel, J. and Bonn, A. (Eds) (2018) *Citizen Science: innovation in open science, society and policy.* UCL Press, London, UK.

Hedemann-Robinson, M. (2019) *Enforcement of International Environmental Law: challenges and responses at the international level.* Routledge, Abingdon, UK.

Heidkamp, C.P. and Morrissey, J. (Eds) (2019) *Towards Coastal Resilience and Sustainability.* Routledge, Abingdon, UK.

Heimann, T. (2020) *Culture, Space and Climate Change: vulnerability and resilience in European coastal areas* 2nd edn (trans. from German). Routledge, Abingdon, UK.

Heinberg, R. (2010) *Peak Everything: waking up to the century of decline in Earth's resources.* Clairview Books, West Hoathly, UK.

Helke, P. (Ed) (2019) *Critical Terms in Futures Studies.* Springer, Cham, CH.

Helm, D. (2015) *Natural Capital: valuing the planet.* Yale University Press, New Haven, CT, USA.

Helming, K., Perez-Soba, M. and Tabbush, P. (Eds) (2008) *Sustainability Impact Assessment of Land Use Changes.* Springer-Verlag Berlin, GR.

Henderson, H. (1994) Paths to sustainable development: the role of social indicators. *Futures* 26(2), 125–137.

Henry, C. and Tubiana, L. (2018) *Earth at Risk: natural capital and the quest for sustainability.* Columbia University Press, New York, NY, USA.

Heras-Saizarbitoria, I. (Ed) (2018) *ISO 9001, ISO 14001, and New Management Standards.* Springer International, Cham, CH.

Herzog, H.J. (2018) *Carbon Capture*. MIT Press, Cambridge, MA, USA.

Hester, R.E. and Harrison, R.M. (Eds) (2009) *Electronic Waste Management: design, analysis and application*. Royal Society of Chemistry, Cambridge, UK.

Hester, R.E. and Harrison, R.M. (Eds) (2010a) *Ecosystem Services*. Royal Society of Chemistry Publishing, Cambridge, UK.

Hester, R.E. and Harrison, R.M. (Eds) (2010b) *Carbon Capture: sequestration and storage*. Royal Society of Chemistry Publishing, Cambridge, UK.

Heusemann, M. and Heusemann, J. (2011) *Techno-Fix: why technology won't save us or the environment*. New Society Publishers, Gabriola Island, BC, CA.

Hewitt, R.J. and Macleod, C.J.A. (2017) What do users really need? Participatory development of decision support tools for environmental management based on outcomes. *Environments* 4(4), article No. 88 (14 pp.).

Hewitt, R.J., Hernandez Jimenez, V., Moratalla, A., Martín, B., Bermejo, L. and Encinas, M. (Eds) (2017) *Participatory Modelling for Resilient Futures: action for managing our environment from the bottom-up*. Elsevier, Amsterdam, NL.

Hey, E. (2018) *Advanced Introduction to International Environmental Law*. Edward Elgar, Cheltenham, UK.

Heynen, N., McCarthy, J., Proudham, S. and Robbins, P. (Eds) (2008) *Neoliberal Environments: false promises and unnatural consequences*. Routledge, London, UK.

Heywood, I., Cornelius, S. and Carver, S. (2011) *An Introduction to Geographical Information Systems* 4th edn. Pearson (Prentice-Hall), Harlow, UK.

Hicks, R.L., Parks, B.C., Timmons Roberts, J. and Tierney, M.J. (2008) *Greening Aid? Understanding the environmental impact of development assistance*. Oxford University Press, Oxford, UK.

Higdon, N. (2020) *The Anatomy of Fake News: a critical news literacy education*. University of California Press, Oakland, CA, USA.

Higgins, M. (Ed) (2019) *Environmental Quality, Monitoring and Management*. Callisto Reference, Forest Hills, NY, USA.

Higgs, A.J. (1981) Island biogeographic theory and nature reserve design. *Journal of Biogeography* 8(2), 117–124.

Hill, A.C. and Martinez-Diaz, L. (2020) *Building a Resilient Tomorrow: how to prepare for the coming climate disruption*. Oxford University Press, Oxford, UK.

Hill, J. (2020) *Environmental, Social, and Governance (ESG) Investing: a balanced analysis of the theory and practice of a sustainable portfolio*. Elsevier (Academic Press), London, UK.

Hillary, R. (Ed) (2017) *Small and Medium-Sized Enterprises and the Environment: business imperatives* 2nd edn. Routledge, London, UK.

Hilson, G.M. (Ed) (2005) *The Socio-Economic Impacts of Artisanal and Small-Scale Mining in Developing Countries*. A.A. Balkema, Abingdon, UK.

Hiltner, K. (2020) *Writing a New Environmental Era: moving forward to nature*. Routledge, Abingdon, UK.

Hilty, J.A., Lidicker, W.Z. Jnr. and Merenlender, A.M. (Eds) (2006) *Corridor Ecology: the science and practice of linking landscapes for biodiversity conservation*. Island Press, Washington, DC, USA.

Hines, A. and Bishop, A. (2015) *Thinking About the Future: guidelines for strategic foresight* 2nd edn. Hinesight, Houston, TX, USA.

Hipel, K.W., Fang, L., Cullmann, J. and Bristow, M. (Eds) (2015) *Conflict Resolution in Water Resources and Environmental Management*. Springer, Cham, CH.

Hitchens, D.M.W.M., Clausen, J. and Fichter, K. (Eds) (2012) *International Environmental Management Benchmarks: best practice experiences from America, Japan, and Europe* 2nd edn. Springer, Berlin, GR.

Ho, S. (Ed) (2021) *A River Flows Through It: a comparative study of transboundary water disputes and cooperation in Asia.* Routledge, Abingdon, UK.

Hoalst-Pullen, N. and Patterson, M.W. (Eds) (2010) *Geospatial Technologies in Environmental Management.* Springer, Dordrecht, NL.

Hoang, T.C., Black, M.C., Knuteson, S.L. and Roberts, A.P. (2019) Environmental pollution, management, and sustainable development: strategies for Vietnam and other developing countries. *Environmental Management* 63, 433–436.

Hobohm, C. (Ed) (2021) *Perspectives for Biodiversity and Ecosystems.* Springer, Cham, CH.

Hobson, C., Bacon, P. and Cameron, R. (Eds) (2014) *Human Security and Natural Disasters.* Routledge (Earthscan), Abingdon, UK.

Hodge, I. (1995) *Environmental Economics: individual incentives and public choices.* Macmillan, Basingstoke, UK.

Hoffman, A.J. and Devereaux Jennings, P. (2018) *Re-engaging with Sustainability in the Anthropocene Era.* Cambridge University Press, Cambridge, UK.

Hofmann-Wellenhof, B., Lichtenegger, H. and Collins, J. (2010) *GPS: theory and practice.* Springer-Verlag, New York, NY, USA.

Holdgate, M.W. (1990) Antarctica: ice under pressure. *Environment* 32(6), 4–9, 30–33.

Holdridge, L.R. (1967) *Life Zone Ecology* 2nd edn. Tropical Science Center, San Jose, CR.

Holdridge, L.R. (1971) *Forest Environments in Tropical Life Zones: a pilot study.* Pergamon Press, Oxford, UK.

Hollender, J. and Breen, B. (2010) *The Responsibility Revolution: how the next generation of businesses will win.* Wiley (Jossey-Bass), San Francisco, CA, USA.

Holling, C.S. (1978) *Adaptive Environmental Assessment and Management* 2nd edn. Wiley, Chichester, UK.

Holling, C.S. (1986) Adaptive environmental management. *Environment: Science and Policy for Sustainable Development* 28(9), 39.

Hollo, E.J. (2017) *Water Resource Management and the Law.* Edward Elgar, Cheltenham, UK.

Holmberg, J. and Robèrt, K.H. (2000) Backcasting from non-overlapping sustainability principles: a framework for strategic planning. *International Journal of Sustainable Development and World Ecology* 74, 291–308.

Holzbecher, E. (2012) *Environmental Modeling in China: using MATLAB.* Springer-Verlag, Berlin, GR.

Honey, M. (2008) *Ecotourism and Sustainable Development: who owns paradise?* 2nd edn. Island Press, Washington, DC, USA.

Honey, M. and Frenkiel, K. (2021) *Overtourism: lessons for a better future.* Island Press, Washington, DC, USA.

Hong, S.-K., Wu, J., Kim, J.-E. and Nakagoshi, N. (Eds) (2010) *Landscape Ecology in Asian Cultures.* Springer, Tokyo, JP.

Hopkins, R. (2008) *The Transition Handbook: from oil dependency to local resilience.* Transition Books (Green Books), Dagenham, UK.

Hopkins, R. (2013) *The Power of Just Doing Stuff: how local action can change the world.* Transition Books (Green Books), Dagenham, UK.

Hopkins, R. (2019) *From What Is to What If: unleashing the power of imagination to create the future we want.* Chelsea Green Publishing, White River Junction, VT, USA.

Horn, E. and Bergthaller, H. (2020) *The Anthropocene: key issues for the humanities*. Routledge (Earthscan), Abingdon, UK.

Horne, A.C., Webb, A., Stewardson, M.J., Richter, B. and Acreman, M. (Eds) (2017) *Water for the Environment: from policy and science to implementation and management*. Elsevier (Academic Press), London, UK.

Horner, C.C. (2007) *The Politically Incorrect Guide to Global Warming and Environmentalism*. Regnery Publishing, Washington, DC, USA.

Horning, N.R. (2018) *The Politics of Deforestation in Africa: Madagascar, Tanzania, and Uganda*. Springer (Palgrave Macmillan), Cham, CH.

Hossein, A.M. (2011) *Practices of Irrigation & On-Farm Water Management: Vol. 2*. Springer, New York, NY, USA.

Hossein, A.M. (Ed) (2013) *Irrigation Management, Technologies & Environmental Impact*. Nova Science Publishers, Hauppauge, NY, USA.

Houbing Song, Srinivasan, R., Sookoor, T. and Jeschke, S. (2017) (Eds) *Smart Cities: foundations, principles, and applications*. Wiley, Hoboken, NJ, USA.

Hourdequin, M. (2015) *Environmental Ethics: from theory to practice*. Bloomsbury Academic, London, UK.

Howarth, C. (2019) *Resilience to Climate Change: communication, collaboration and co-production*. Springer (Palgrave Macmillan), Cham, CH.

Howe, C.W. (2011) *Interbasin Transfers of Water: economic issues and impacts* (first published 1971). Earthscan, New York, NY, USA.

Ho-Won, Jeong (2005) *Globalization and the Physical Environment*. Chelsea House Publishers, New York, NY, USA.

Hsu, S.-L. (2017) *The Case for a Carbon Tax: getting past our hang-ups to effective climate policy*. Island Press, Washington, DC, USA.

Huan Quingzhi (Ed) (2010) *Eco-Socialism as Politics: rebuilding the basis of our modern civilisation*. Springer, Dordrecht, NL.

Huapu Lu (2020) *Eco-Cities and Green Transport*. Elsevier, Amsterdam, NL.

Huber, R.A. (2020) The role of populist attitudes in explaining climate change skepticism and support for environmental protection. *Environmental Politics* 29(6), 373–386.

Hudson, I., Hudson, M. and Fridell, M. (2013) *Fair Trade, Sustainability and Social Change*. Palgrave Macmillan, New York, NY, USA.

Hufnagel, L. (Ed) (2018) *Ecosystem Services and Global Ecology*. IntechOpen, London, UK.

Huggins, A. (2017) *Multilateral Environmental Agreements and Compliance: the benefits of administrative procedures*. Routledge, Abingdon, UK.

Hughes, A.C. (2019) Understanding and minimizing environmental impacts of the Belt and Road Initiative. *Conservation Biology* 33(4), 883–894.

Hugo, M.L. (2018) *Environmental Management: an ecological guide to sustainable living in Southern Africa*. Ecopla, Pretoria, ZA.

Cao, Hui, Amiraslani, F., Jian Liu and Na Zhou (2015) Identification of dust storm source areas in West Asia using multiple environmental datasets. *Science of the Total Environment* 1(502), 224–235.

Hull, V. and Liu, J. (2018) Telecoupling: a new frontier for global sustainability. *Ecology and Society* 23(4), 41–49.

Hulme, M. (2009) *Why We Disagree About Climate Change: understanding controversy, inaction and opportunity*. Cambridge University Press, Cambridge, UK.

Hülsberg, W. (1988) *The German Greens: a social and political profile* (English trans. by G. Fagan). Verso, London, UK.

Hundloe, T. (2021) *Environmental Impact Assessment: incorporating sustainability principles*. Springer (Palgrave Macmillan), Cham, CH.

Hunt, D. and Johnson, C. (1995) *Environmental Management Systems: principles and practice*. McGraw-Hill, London, UK.

Ha, H. and Rose, J. (2017) Public participation and environmental governance in Singapore. *International Journal of Environment, Workplace and Employment* 4(3), 186–204.

Hurst, C.J. (Ed) (2018) *The Connections Between Ecology and Infectious Disease*. Springer International, Cham, CH.

Hussain, C.H. (Ed) (2021) *Environmental Management of Waste Electrical and Electronic Equipment*. Elsevier, Amsterdam, NL.

Hutham, F.C., Hussein, K.M., Tariq, T. and Yousif, A. (2021) Does environmental management accounting matter in promoting sustainable development? A study in Iraq. *Journal of Accounting Science* 5(2), 114–126.

Hutter, G., Neubert, M. and Ortlepp, R. (Eds) (2021) *Building Resilience to Natural Hazards in the Context of Climate Change: knowledge, integration, implementation and learning*. Springer, Wiesbaden, GR.

Huu Hao Ngo and Tian Cheng Zhang (2016) *Green Technologies for Sustainable Water Management*. American Society of Civil Engineers, Reston, VA, USA.

Hyndman, R.J. and Athanasopoulos, G. (2021) *Forecasting: principles and practice* 3rd edn. OTexts, Melbourne, VI, AU.

Iannone, A.P. (2017) *Practical Environmental Ethics*. Routledge, Abingdon, UK.

Iannuzzi, A. (2017) *Greener Products: the making and marketing of sustainable brands* 2nd edn. CRC Press, Boca Raton, FL, USA.

Idllalène, S. (2021) *Rediscovery and Revival in Islamic Environmental Law: back to the future of nature's trust*. Cambridge University Press, Cambridge, UK.

IGI Global (2016) *Natural Resources Management: concepts, methodologies, tools, and applications* (3 vols). Information Resources Management Association USA, Hershey, PA, USA.

Iglesias, A., Assimacopoulos, D. and Van Lanen, H.A.J. (Eds) (2019) *Drought: science and policy*. Wiley-Blackwell, Hoboken, NJ, USA.

Imeson, A. (2012) *Desertification, Land Degradation and Sustainability*. Wiley-Blackwell, Oxford, UK.

Independent Commission on International Development Issues (1980) *North–South: a programme for survival*. Pan, London, UK.

Ingegnoli, V., Lombardo, F. and La Torre, G. (Eds) (2022) *Environmental Alteration Leads to Human Disease: a planetary health approach*. Springer, New York, NY, USA.

Inkinen, T., Yigitcanlar, T. and Wilson, M. (Eds) (2021) *Smart Cities and Innovative Urban Technologies*. Routledge, Abingdon, UK.

Inoguchi, T., Newman, E. and Paoletto, G. (Eds) (2005) *Cities and the Environment: new approaches for eco-societies*. United Nations University Press, Tokyo, JP.

International Chamber of Commerce (1989) *Environmental Auditing* (ICC Publication No. 468). International Chamber of Commerce, Paris, FR.

Inyang, H.I. and Daniels, J.L. (Eds) (2009) *Environmental Monitoring*. EOLSS Publishers, Oxford, UK.

Ioannau, C. (2012) *SWOT Analysis: an easy to understand guide* Kindle edn. Amazon, USA.

Iqbal, A., Afroze, S. and Shipin, O. (2011) *Cumulative Effect Assessment*. Lambert Academic Publishing, Warsaw, PL.

Iqbal, M. (Ed) (2005) *Islamic Perspectives on Sustainable Development.* Palgrave Macmillan, London, UK.

Ireton, C. and Posetti, J. (2018) *Journalism, Fake News & Disinformation: handbook for journalism education and training.* UNESCO, Paris, FR.

Irfan, M. (2015) *Consumer Awareness About Green Marketing & Effect on Buying Behaviour.* Lambert Academic Publishing. Chisinau, MD, USA.

Irwin, A. (1995) *Citizen Science: a study of people, expertise and sustainable development.* Routledge, London, UK.

Isaacs, N. (2020) *Every Woman's Guide to Saving the Planet.* ABC Books, Springfield, MO, USA.

Islam, K.M.B. and Nomani, Z.M. (Eds) (2022) *Environment Impact Assessment: precept & practice.* CRC Press, Boca Raton, FL, USA.

Islam, M. (2013) *Development, Power, and the Environment: neoliberal paradox in the age of vulnerability.* Taylor & Francis, Abingdon, UK.

Islam, M., Yamaguchi, R., Sugiawan, Y. and Managi, S. (2019) Valuing natural capital and ecosystem services: a literature review. *Sustainability Science* 14, 159–174.

Islam, T. and Ryan, J. (2016) *Hazard Mitigation in Emergency Management.* Elsevier, Amsterdam, NL.

Itard, L., van Bueren, E.M., Visscher, H. and van Bohemen, H. (Eds) (2011) *Sustainable Urban Environments: an ecosystem approach.* Springer, Dordrecht, NL.

IUCN, UNEP and WWF (1980) *World Conservation Strategy: living resource conservation for sustainable development.* International Union for Conservation of Nature and Natural Resources, Gland, CH.

Jabbour, C.J.C. (2015) Environmental training and environmental management maturity of Brazilian companies with ISO14001: empirical evidence. *Journal of Cleaner Production* 96, 331–338.

Jabbour, C.J.C., da Silva, E.M., Paivac, E.L. and Santos, F.C.A. (2012) Environmental management in Brazil: is it a completely competitive priority? *Journal of Cleaner Production* 21, 11–22.

Jabbour, C.J.C., Santos, F.C.A. and Nagano, M.S. (2010) Contributions of HRM throughout the stages of environmental management: methodological triangulation applied to companies in Brazil. *The International Journal of Human Resource Management* 21(7), 1049–1089.

Jabbour, C.J.C., Santos, F.C.A., Fonseca, S.A. and Nagano, M.S. (2013) Green teams: understanding their roles in the environmental management of companies located in Brazil. *Journal of Cleaner Production* 46, 58–66.

Jackson, M.C. (2003) *Systems Thinking: creative holism for managers.* Wiley, Chichester, UK.

Jacobs, S., Dendoncker, J. and Keune, H. (Eds) (2013) *Ecosystem Services: global issues, local practices.* Elsevier, Dordrecht, NL.

Jäger, J. and Afifi, T. (Eds) (2010) *Environment, Forced Migration and Social Vulnerability.* Springer-Verlag, Berlin, GR.

Jager, N.W., Newig, J., Challies, E. and Kochskämper, E. (2020) Pathways to implementation: evidence on how participation in environmental governance impacts on environmental outcomes. *Journal of Public Administration Research and Theory* 30(3), 383–399.

Jain, H.K. (2010) *The Green Revolution: history, impacts and future.* Stadium Press, Houston, TX, USA.

Jain, R.K., Zengdi, C.C. and Domen, J.K. (Eds) (2016) *Environmental Impact of Mining and Mineral Processing: management, monitoring and auditing strategies.* Elsevier (BH), Oxford, UK.

Jakeman, A.J., Voinov, A.A., Rizzoli, A.E. and Chen, S.H. (Eds) (2008) *Environmental Modelling, Software and Decision Support: state of the art and new perspective.* Elsevier, Amsterdam, NL.

James, S.B. (2016) *Zen Buddhism and Environmental Ethics* 2nd edn. Routledge, Abingdon, UK.

Janssen, R. (2013) *Multiobjective Decision Support for Environmental Management* 3rd edn. Springer, Dordrecht, NL.

Jarratt-Snider, K. and Nielsen, M.O. (Eds) (2020) *Indigenous Environmental Justice – Indigenous Justice.* University of Arizona Press, Tucson, AZ, USA.

Jatoi, W.N., Mubeen, M., Ahmad, A., Cheema, M.A., Lin, Zhaohui and Hashmi, M.Z. (Eds) (2022) *Building Climate Resilience in Agriculture: theory, practice and future perspective.* Springer, Cham, CH.

Jayamani, C.V. (2012) *Environmental Management: from ancient to modern times.* New Century Publications, New Delhi, IN.

Jenkins, W., Tucker, M.E. and Grim, J. (Eds) (2016) *Routledge Handbook of Religion and Ecology.* Routledge (Earthscan), Abingdon, UK.

Jensen, D., Lierre, K. and Wilbert, M. (2021) *Bright Green Lies: how the environmental movement lost its way and what we can do about it.* Monkfish Book Publishing Company, Rhinebeck, NY, USA.

Jensen, M.E. and Bourgeron, P.S. (Eds) (2021) *A Guidebook for Integrated Ecological Assessments.* Springer-Verlag, New York, NY, USA.

Jerolleman, A. and Kiefer, J.J. (Eds) (2013) *Natural Hazard Mitigation.* CRC Press, Boca Raton, FL, USA.

Jewett, A. (2020) *Science under Fire: challenges to scientific authority in modern America.* Harvard University Press, Cambridge, MA, USA.

Jha, M.K. (Ed) (2010) *Natural and Anthropogenic Disasters: vulnerability, preparedness and mitigation.* Springer and Capitol Publishing (New Delhi, IN), Dordrecht, NL.

Jiahua Pan (2016) *China's Environmental Governing and Ecological Civilization* 2nd edn. Springer-Verlag, Heidelberg, GR.

Jianming Yang (2017a) *Environmental Management in Mega Construction Projects.* Springer, Singapore, SG.

Jiaping Wu, Junyu He and Christakos, G. (2022) *Quantitative Analysis and Modeling of Earth and Environmental Data: space-time and spacetime data considerations.* Elsevier, Amsterdam, NL.

Jia Shaofeng and Lie Xie (2017) *China's International Transboundary Rivers: politics, security and diplomacy of shared water resources.* Routledge (Earthscan), Abingdon, UK.

Jia Wang, S. and Liu, J. (2010) *China's Environment.* China International Press, Beijing, PRC.

Jia Wang, S. and Moriarty, P. (2018) *Big Data for Urban Sustainability: a human-centered perspective.* Springer International, Cham, CH.

Jingling Liu, Lulu Zhang and Zhijie Liu (2017a) *Environmental Pollution Control.* Walter de Gruyter, Berlin, GR.

Jing Wu and I.-Shin Chang (2020) *Environmental Management in China: policies and institutions.* Springer (with Chemical Industry Press), Singapore, SG.

Jingyuan Li and Tongjin Yang (Eds) (2016) *China's Eco-City Construction.* Springer, Berlin, GR and Social Sciences Academic Press (China), Beijing, PRC.

Jingzheng Ren (Ed) (2021a) *Multi-Criteria Decision Analysis for Risk Assessment and Management*. Springer, Cham, CH.

Jingzheng Ren (Ed) (2021b) *Methods in Sustainability Science: assessment, prioritization, improvement, design and optimization*. Elsevier, Amsterdam, NL.

Jingzheng Ren (Ed) (2022) *Advances of Footprint Family for Sustainable Energy and Industrial Systems*. Springer, Cham, CH.

Jingzhen Ren and Toniolo, S. (Eds) (2020) *Life Cycle Sustainability Assessment for Decision-Making: methodologies and case studies*. Elsevier, Amsterdam, NL.

Jodoin, S. (2017) *Forest Preservation in a Changing Climate: REDD+ and indigenous and community rights in Indonesia and Tanzania*. Cambridge University Press, Cambridge, UK.

Johansen, B.E. (2015) *Eco-Hustle! Global Warming, Greenwashing, and Sustainability*. Praeger, Santa Barbara, CA, USA.

Johnson, K.N., Swanson, F.J., Herring, M. and Greene, S. (Eds) (1999) *Bioregional Assessments: science at the crossroads of management and planning*. Island Press, Washington, DC, USA.

Johnson, S. (Ed) (2020) *Monitoring Environmental Contaminants*. Elsevier, Amsterdam, NL.

Jolliet, O., Saade-Sbeih, M., Shaked, S., Jolliet, A. and Crettaz, P. (2016) *Environmental Life Cycle Assessment*. CRC Press, Boca Raton, FL, USA.

Jones, H.G. and Vaughan, R.A. (2010) *Remote Sensing of Vegetation: principles, techniques, and applications*. Oxford University Press, New York, NY, USA.

Jones, K.E., Venter, O., Fuller, R.A., Allen, J.R., Maxwell, S.L., Negret, P.J. and Watson, J.E. (2018) One-third of global protected areas is under intense human pressure. *Science* 360, 788–791.

Jones, L. (2019) Resilience isn't the same for all: comparing subjective and objective approaches to resilience measurement. *Wiley Interdisciplinary Reviews: Climate Change* 10(1), 1–19.

Jones, R. (2016) *Violent Borders: refugees and the right to move*. Verso Books, London, UK.

Jørgensen, S.E. (2009) *Ecological Modelling: an introduction*. WIT Press, Southampton, UK.

Jørgensen, S.E. (2011a) *Fundamentals of Ecological Modelling* 4th edn. Elsevier, Amsterdam, NL.

Jørgensen, S.E. (2011b) *Handbook of Ecological Models Used in Ecosystem and Environmental Management*. CRC Press, Boca Raton, FL, USA.

Jørgensen, S.E. (2016a) *Introduction to Systems Ecology*. CRC Press, Boca Raton, FL, USA.

Jørgensen, S.E. (Ed) (2016b) *Ecological Model Types*. Elsevier, Amsterdam, NL.

Jørgensen, S.E. (2017) *Handbook of Environmental and Ecological Modeling*. CRC Press, Boca Raton, FL, USA.

Jørgensen, S.E. (Ed) (2019a) *Handbook of Ecological Models used in Ecosystem and Environmental Management* (first published 2011). CRC Press, Boca Raton, FL, USA.

Jørgensen, S.E. (Ed) (2019b) *A Systems Approach to the Environmental Analysis of Pollution Minimization*. CRC Press, Boca Raton, FL, USA.

Jørgensen, S.E. and Fath, B.D. (Eds) (2020) *Managing Water Resources and Hydrological Systems*. CRC Press, Boca Raton, FL, USA.

Jørgensen, S.E., Marques, J.C. and Nielsen, S.N. (2021) *Integrated Environmental Management: a transdisciplinary approach*. CRC Press, Boca Raton, FL, USA.

Joseph, A. (2011) *Tsunamis: detection, monitoring, and early-warning technologies*. Elsevier (AP), Amsterdam, NL.

Joseph, B., Lata Madhavi, A., Hemalatha, S. and Laxminarayana, P. (2018) *Practical Manual on Principles and Practices of Social Forestry*. BS Publications, Hyderabad, IN.

Joshi, A.B. (Ed) (2008) *Eco-labels: concerns and experiences.* ICFAI University Press, Hyderabad, IN.

Jungkunst, H.F., Goepel, J., Horvath, T., Ott, S., and Brown, M. (2021) New uses for old tools: reviving Holdridge Life Zones in soil carbon persistence research. *Journal of Plant Nutrition and Soil Science* 184(1), 5–11.

Jurin, R.R., Roush, D. and Danter, J. (2000) *Environmental Communication: skills and principles for natural resource managers, scientists, and engineers* 2nd edn. Springer, Dordrecht, NL.

Kääpä, P. (2018) *Environmental Management of the Media: policy, industry, practice.* Routledge, Abingdon, UK.

Kabisch, N., Korn, H., Stadler, J. and Bonn, A. (Eds) (2017) *Nature-Based Solutions to Climate Change Adaptation in Urban Areas: linkages between science, policy and practice.* Springer, Cham, CH.

Kaczynski, T.J. (2020) *Anti-Tech Revolution: Why and how* 2nd edn. Fitch & Madison Publishers, Scottsdale, AZ, USA.

Kah, S. and Akeneroye, T. (2020) Evaluation of social impact measurement tools and techniques: a systematic review of the literature. *Social Enterprise Journal* 16(4), 381–402.

Kahle, L.R. and Gurel-Ata, E. (2014) *Communicating Sustainability for the Green Economy.* Routledge, New York, NY, USA.

Kahn, M.E. (2021) *Adapting to Climate Change: markets and the management of an uncertain future.* Yale University Press, New Haven, CT, USA.

Kahraman, C. and Sari, İ.U. (Eds) (2017) *Intelligence Systems in Environmental Management: theory and applications.* Springer, Cham, CH.

Kaiser, W. and Meyer, J.-H. (Eds) (2016) *International Organizations and Environmental Protection: conservation and globalization in the twentieth century.* Berghahn Books, New York, NY, USA.

Kallis, G. and Bliss, S. (2019) Post-environmentalism: origins and evolution of a strange idea. *Journal of Political Ecology* 26(1), 466–485.

Kamei, T. (2012) Recent research on thorium molten-salt reactors from a sustainability viewpoint. *Sustainability* 4(10), 2399–2418.

Kandpal, P.C. (2019) *Environmental Governance in India: issues and challenges.* Sage India, New Delhi, IN.

Kanga, S., Singh, S.K., Meraj, G. and Majid, F. (Eds) (2022) *Geospatial Modeling for Environmental Management: case studies from South Asia.* CRC Press, Boca Raton, FL, USA.

Kanwat, M. and Kumar, P.S. (2011) *Participatory Rural Appraisal: tools and techniques for need assessment.* Agrotech Publishing Academy, Udaipur, IN.

Kaplan, D.M. (Ed) (2017) *Philosophy, Technology, and the Environment.* MIT Press, Cambridge, MA, USA.

Kapoor, R.T. and Shah, M.P. (Eds) (2022) *BioChar: applications for bioremediation of contaminated systems.* Walter de Gruyter, Berlin, GR.

Kapoor, R.T., Treichel, H. and Shah, M.P. (Eds) (2021) *Biochar and its Application in Bioremediation.* Springer, Singapore, SG.

Kapsar, K.E., Hovis, C.L., da Silva, R.F.B., Buchholtz, E.K., Carlson, A.R., Dou, Y., Du, Y., Furumo, P.R., Li, Y., Torres, A., Yang, D., Wan, H.Y., Zaehringer, J.G. and Liu, J. (2019) Telecoupling reassessed: the first five years. *Sustainability* 11(4), 1–13.

Kapustka, L.A. and Landis, W. (Eds) (2010) *Environmental Risk Assessment and Management from a Landscape Perspective.* Wiley, Hoboken, NJ, USA.

Bibliography

Karampelas, P. and Bourlai, T. (Eds) (2018) *Surveillance in Action: technologies for civilian, military and cyber surveillance*. Springer, Cham, CH.

Kareiva, P., Talis, H., Ricketts, T.H., Daley, G.C. and Poalsky, S. (Eds) (2011) *Natural Capital: theory and practice of mapping ecosystem services*. Oxford University Press, Oxford, UK.

Karpf, A. (2021) *How Women Can Save the Planet*. Hurst & Co., London, UK.

Kasperson, R.E. and Berberian, M. (Eds) (2011) *Integrating Science and Policy: vulnerability and resilience in global environmental change*. Routledge (Earthscan), London, UK.

Kaswamila, A. (Ed) (2016) *Land Degradation and Desertification: a global crisis*. InTech Open, Rijeka, HR.

Katsikides, S.A. (Ed) (2017) *An Exploration of Technology and its Social Impact*. Cambridge Scholars, Newcastle-upon-Tyne, UK.

Katsikides, S.A. (2020) *The Societal Impact of Technology* (first published 1998). Routledge, Abingdon, UK.

Katsoni, V. and Segarra-Oña, M. (Eds) (2019) *Smart Tourism as a Driver for Culture and Sustainability*. Springer, Cham, CH.

Katz, M. and Thornton, D. (2019) *Environmental Management Tools on the Internet: accessing the world of environmental information*. CRC Press, Boca Raton, FL, USA.

Kaushik, A., Kaushik, C.P. and Attri, S.D. (Eds) (2021) *Climate Resilience and Environmental Sustainability Approaches: global lessons and local challenges*. Springer, Singapore, SG.

Kayalica, M.O., Hakan, H. and Çağatay, S. (Eds) (2017) *Economics of International Environmental Agreements: a critical approach*. Routledge, Abingdon, UK.

Kaza, S. (2019) *Green Buddhism: practice and compassionate action in times*. Shambhala, Boulder, CO, USA.

Keen, P.L. and Monforts, M.H.M.M. (Eds) (2012) *Antimicrobial Resistance in the Environment*. Wiley-Blackwell, Hoboken, NJ, USA.

Keeso, A. (2014) *Big Data and Environmental Sustainability: a conversation starter*. Smith School Working Paper Series No. 14-04. Smith School of Environment and Enterprise, Oxford University, Oxford, UK.

Keil, R., Bell, D.V.J., Penz, P. and Fawcett, L. (Eds) (1998) *Political Ecology: global and local*. Routledge, London, UK.

Keith, D. (2013) *A Case for Climate Engineering*. MIT Press, Cambridge, MA, USA.

Keller, D.R. (2010) *Environmental Ethics: the big questions*. Wiley-Blackwell, Oxford, UK.

Kelley, B.R.P., Erickson, A.L., Mease, L.A., Battista, W., Kittinger, J.N. and Fujita, R. (2015) Embracing thresholds for better environmental management. *Philosophical Transactions of the Royal Society of London B Biological Sciences* 370(1659), article No. 20130267 (10 pp.).

Kelson, M., Kimball, E. and Leonard, A. (2016) *What is Green Spirituality?* GreenSpirit Press, Barnsley, UK.

Kemm, J. (2013) *Health Impact Assessment: past achievement, current understanding, and future progress*. Oxford University Press, Oxford, UK.

Kenis, A. and Mathijs, E. (2014) (De)politicising the local: the case of the Transition Towns Movement in Flanders (Belgium). *Journal of Rural Studies*, 34, 172–183.

Kennedy, J. (2023) *Pathogenesis: how germs made history*. Torva Penguin (Transworld), London, UK.

Kennet, M., Seeberg, A. and Berg, T.H. (Eds) (2012a) *Green Economics Methodology: an introduction*. Green Economics Institute, Los Angeles, CA, USA.

Kennet, M., Courea, E., Black, K., Bouquet, A. and Pipinyte, L. (Eds) (2012b) *Handbook of Green Economics: a practitioner's guide*. Green Economics Institute, Los Angeles, CA, USA.

Kennet, M., Bukaveckaitė, K. and Caporale Madi, M.A. (Eds) (2015) *The Greening of Global Banking and Finance*. The Green Economics Institute, Los Angeles, CA, USA.

Kerr, M. (2022) *Wilder; how rewilding is transforming conservation and changing the world*. Bloomsbury Publishing (Sigma), London, UK.

Kershaw, K.A. (1973) *Quantitative and Dynamic Plant Ecology* 2nd edn. Edward Arnold, London, UK.

Kershaw, T. (2017) *Climate Change Resilience in Urban Environments*. IOP Publishing, Bristol, UK.

Keskitalo, E.C.H. (2008) *Climate Change and Globalization in the Arctic: an integrated approach to vulnerability assessment*. Earthscan, London, UK.

Khalid, F. (2019) *Signs on the Earth: Islam, modernity and the climate crisis*. Kube Publishing, Markfield, UK.

Khalil, M.A.K. (Ed) (2000) *Atmospheric Methane: its role in the global environment*. Springer-Verlag, Berlin, GR.

Khan, M., Hussain, M. and Ajmal, M.M. (Eds) (2017) *Green Supply Chain Management for Sustainable Business Practice*. IGI Global, Hershey, PA, USA.

Khan, S.A.R. (Ed) (2019) *Global Perspectives on Green Business Administration and Sustainable Supply Chain Management*. IGI Global, Hershey, PA, USA.

Khan, S.I. and Adams III, T.E. (Eds) (2019) *Indus River Basin: water security and sustainability*. Elsevier, Dordrecht, NL.

Khatun, K., Imbach, P. and Zamora, J. (2013) An assessment of climate change impacts on the tropical forests of Central America using the Holdridge life zone (HLZ) land classification system. *iForest – Biogeosciences and Forestry* 6(4), 183–189.

Khdair, A. and Abu-Rumman, G. (2020) Sustainable environmental management and valorization options for olive mill byproducts in the Middle East and North Africa (MENA) region. *Processes* 8(6), 671.

Khorram, S., Nelson, S.A.C., Koch, F.H. and van der Wiele, C.F. (2012) *Remote Sensing*. Springer, New York, NY, USA.

Khurshid, M., Al-Aali, A., Ali Soliman, A. and Mohamad, A.S. (2014) Developing an Islamic corporate social responsibility model (ICSR). *Competitiveness Review* 24(4), 258–274.

Kidd, S., Plater, A. and Frid, C. (Eds) (2011) *The Ecosystem Approach to Marine Planning and Management*. Routledge, London, UK.

Kidokoro, T., Okata, J., Matsumura, S. and Shima, N. (Eds) (2008) *Vulnerable Cities: realities, innovations and strategies*. Springer, Tokyo, JP.

Kilby, P. (2019) *The Green Revolution: narratives of politics, technology and gender*. Routledge, Abingdon, UK.

Kinchy, A. (2012) *Seeds, Science, and Struggle: the global politics of transgenic crops*. MIT Press, Cambridge, MA, USA.

Kindon, S., Pain, R. and Kesby, M. (Eds) (2007) *Participatory Action Research Approaches and Methods: connecting people, participation and place*. Routledge, Abingdon, UK.

King, B. (2017) *States of Disease: political environments and human health*. University of California Press, Oakland, CA, USA.

King-Tak Ip (Ed) (2009) *Environmental Ethics: intercultural perspectives*. Brill (Editions Rodopi), Amsterdam, NL.

Kingston, S., Heyvaert, V. and Čavoški, A. (2017) *European Environmental Law*. Cambridge University Press, Cambridge, UK.

Kitchin, R. (2014) *The Data Revolution: big data, open data, data infrastructures and their consequences*. Sage, Los Angeles, CA, USA.

Kittikhoun, A. and Schmeier, S. (Eds) (2021) *River Basin Organizations in Water Diplomacy*. Routledge, Abingdon, UK.

Klein, D., Carazo, M.P., Doelle, M., Bulmer, J. and Higham, A. (Eds) (2017) *The Paris Agreement on Climate Change: analysis and commentary*. Oxford University Press, Oxford, UK.

Klijn, F., De Waal, R.W. and Oude Voshaar, J.H. (1995) Ecoregions and ecodistricts: ecological regionalizations for The Netherlands' environmental policy. *Environmental Management* 19(6), 797–813.

Klöpffer, W. and Grahl, B. (2014) *Life Cycle Assessment (LCA): a guide to best practice*. Wiley-VCH, Weinheim, GR.

Knight, D.W. and Shamseldin, A.Y. (2006) *River Basin Modelling for Flood Risk Mitigation*. Taylor & Francis, London, UK.

Kochskämper, E., Challies, E., Jager, N.W. and Newig, J. (Eds) (2019) *Participation for Effective Environmental Governance: evidence from the European Water Framework Directive Implementation*. Routledge, Abingdon, UK.

Koh Kheng-Lian (Ed) (2010) *Crucial Issues in Climate Change and the Kyoto Protocol: Asia and the world*. World Scientific Publishing, Singapore, SG.

Koh Kheng-Lian (2012) Transboundary and global environmental issues: the role of ASEAN. *Transnational Environmental Law* 1(1), 67–82.

Koivurova, T. (2013) *Introduction to International Environmental Law*. Routledge, Abingdon, UK.

Kokenes, S. (2008) *Debt-for-Nature Swaps – conservancy or cemetery? – NGOs reduce foreign debt/save environments*. VDM Verlag Dr Müller, Riga, LV.

Kokotsis, E. and Kirton, J. (2015) *The Global Governance of Climate Change: G7, G20, and UN leadership*. Ashgate, Farnham, UK.

Kolbert, E. (2006) *Field Notes from a Catastrophe: man, nature, and climate change*. Bloomsbury, New York, NY, USA.

Kolbert, E. (2014) *The Sixth Extinction: an unnatural history*. Henry Holt & Co., New York, NY, USA.

Kolbert, E. (2021) *Under a White Sky: the nature of the future*. Random House, New York, NY, USA.

Kolhe, M.L., Labhasetwar, P.K. and Suryawanshi, H.M. (Eds) (2019) *Smart Technologies for Energy, Environment and Sustainable Development*. Springer, Singapore, SG.

Komarov, B. (1981) *The Destruction of Nature in the Soviet Union*. Pluto Press, London, UK.

Komatina, D. (Ed) (2018) *Achievements and Challenges of Integrated River Basin Management*. IntechOpen, London, UK.

Kondrup, C., Mercogliano, P., Bosello, F., Mysiak, J., Scoccimarro, E., Rizzo, A., Ebrey, R., de Ruiter, M., Jeuken, A. and Watkiss, P. (Eds) (2022) *Climate Adaptation Modelling*. Springer, Cham, CH.

Koninklijk Instituut voor de Tropen (1990) *Environmental Management in the Tropics: an annotated bibliography 1985–89*. Royal Tropical Institute, Amsterdam, NL.

Konrad, J. and Shroder, T. (2011) *Fire on the Horizon: the untold story of the Gulf Oil disaster*. Harper-Collins, New York, NY, USA.

Koohi, E., Shobeiri, S.M., Koohi, E. and Meiboudi, H. (2014) Women's participation in environmental management and development promotion culture. *International Journal of Resistive Economics* 2(2), 49–61.

Koonin, S.E. (2021) *Unsettled: what climate science tells us, what it doesn't, and why it matters*. BenBella Books, Dallas, TX, USA.

Koontz, T.M., Steelman, T.A., Carmin, J.-A., Korfmacher, K.S., Moseley, C. and Thomas, C.W. (2010) *Collaborative Environmental Management: what roles for government* (first published 2004). Resources for the Future, Washington, DC, USA.

Kopnina, H. and Shoreman-Ouimet, E. (2015a) *Sustainability: key issues*. Routledge (Earthscan), Abingdon, UK.

Kopnina, H. and Shoreman-Ouimet, E. (Eds) (2015b) *Environmental Anthropology: future directions*. Routledge, Abingdon, UK.

Kopnina, H. and Shoreman-Ouimet, E. (Eds) (2017) *Routledge Handbook of Environmental Anthropology*. Routledge, Abingdon, UK.

Köster, S., Reese, M. and Zuo, J. (Eds) (2019) *Urban Water Management for Future Cities: technical and institutional aspects from Chinese and German perspective*. Springer, Cham, CH.

Kotlyakov, V.M. (1991) The Aral Sea basin: a critical environmental zone. *Environment* 33(1), 4–9; 36–38.

Kotzé, L. (Ed) (2017) *Environmental Law and Governance for the Anthropocene*. Bloomsbury Publishing (Hart), London, UK.

Kraft, M. (2018) *Environmental Policy and Politics* 7th edn. Routledge, Abingdon, UK.

Krämer, L. (2012) *Principles of Environmental Law*. Edward Elgar, Cheltenham, UK.

Krämer, L. (2016a) *EU Environmental Law* 8th edn. Sweet & Maxwell, Mytholmroyd, UK.

Krämer, L. (Ed) (2016b) *Enforcement of Environmental Law*. Edward Elgar, Cheltenham, UK.

Krasney, M.E. (2020) *Advancing Environmental Education Practice*. Cornell University Press, Ithaca, NY, USA.

Kreiser, L.A., Yábar Sterling, A., Herrera, P., Milne, J.E. and Ashiabor, H. (Eds) (2012) *Green Taxation and Environmental Sustainability*. Edward Elgar, Cheltenham, UK.

Kress, W.J. and Stine, J.K. (Eds) (2017) *Living in the Anthropocene: Earth in the age of humans*. Smithsonian Books, Washington, DC, USA.

Krieger, T., Panke, D. and Pregernig, M. (Eds) (2020a) *Environmental Conflicts, Migration and Governance*. Bristol University Press, Bristol, UK.

Krieger, T., Panke, D. and Pregernig, M. (Eds) (2020b) *Climate Change, Vulnerability and Migration*. Bristol University Press, Bristol, UK.

Krishna, K.R. (Ed) (2013) *Agroecosystems: soils, climate, crops, nutrient dynamics and productivity*. CRC Press, Boca Raton, FL, USA.

Krishna, K.R. (2021) *Agroecosystems: soils, climate, crops, nutrient dynamics and productivity*. Academic Press (Elsevier), Cambridge, MA, USA.

Krishnamoorthy, B. (2017) *Environmental Management: text and case studies* 3rd edn. Prentice-Hall India, Delhi, IN.

Krishnamurthy, R.R., Jonathan, M.P., Srinivasalu, S. and Glaeser, B. (Eds) (2019) *Coastal Management: global challenges and innovations*. Elsevier (Academic Press), London, UK.

Kristen, A. (2005) *China's Environment and the Challenge of Sustainable Development*. Taylor & Francis, Abingdon, UK.

Kroll, G.M. and Robbins, R.H. (Eds) (2009) *World in Motion: the globalization and the environment reader*. Rowman & Littlefield (Alta Mira Press), Lanham, MD, USA.

Kuhad, R.C. and Singh, A. (Eds) (2013) *Biotechnology for Environmental Management and Resource Recovery*. Springer, New Delhi, IN.

Kuhn, G.L. and Emery, J.R. (2013) *Watersheds: processes, management, and impact*. Nova Science Publishers, Hauppauge, NY, USA.

Kulenbekov, Z.E. and Asanov, B.D. (Eds) (2021) *Water Resource Management in Central Asia and Afghanistan: current and future environmental and water issues.* Springer, Cham, CH.

Kulkami, V. and Ramachandra, T.V. (2006) *Environmental Management.* The Energy and Resources Institute (TERI), New Delhi, IN.

Kull, C.A., Arnauld de Sartre, X. and Castro-Larrañaga, M. (2015) The political ecology of ecosystem services. *Geoforum* 61, 122–134.

Kumar, A. (2009) *Integrated Environmental Management.* Daya Publishing House, New Delhi, IN.

Kumar, A. and Droby, S. (Eds) (2021) *Food Security and Plant Disease Management.* Elsevier (Woodhead Publishing), Duxford, UK.

Kumar, B.M. and Ramachandran Nair, P.K. (Eds) (2011) *Carbon Sequestration Potential of Agroforestry Systems: opportunities and challenges.* Springer, Dordrecht, NL.

Kumar, K.C., Kumar, A. and Singh, A.K. (Eds) (2019a) *Climate Change and Agricultural Ecosystems: current challenges and adaptation.* Elsevier, Amsterdam, NL.

Kumar, M.D. (2010) *Managing Water in River Basins: hydrology, economics and institutions.* Oxford University Press, Oxford, UK.

Kumar, M.D., Reddy, R. and James, A.J. (Eds) (2019b) *From Catchment Management to Managing River Basins: science, technology choices, institutions and policy.* Elsevier, Amsterdam, NL.

Kumar, P. (Ed) (2019) *Mainstreaming Natural Capital and Ecosystem Services into Development Policy.* Routledge (Earthscan), Abingdon, UK.

Kumar, P. and Thiaw, I. (Eds) (2013) *Values, Payments and Institutions for Ecosystem Management: a developing country perspective.* Edward Elgar, Cheltenham, UK.

Kumar, S. (2016) *Municipal Solid Waste Management in Developing Countries.* CRC Press, Boca Raton, FL, USA.

Kumari, K. (Ed) (2022) *Persistent Organic Pollutants: gaps in management and associated challenges.* CRC Press, Boca Raton, FL, USA.

Kummer, S., Wakolbinger, T., Novoszel, L. and Geske, A.M. (Eds) (2022) *Supply Chain Resilience: insights from theory and practice.* Springer, Cham, CH.

Kunstadter, P., Bird, E.C.F. and Sabhasri, S. (Eds) (1989) *Man in the Mangroves: the socio-economic situation of human settlements in mangrove forests.* United Nations University Press, Tokyo, JP.

Kuo, L., Yeh, C-C and Yu, H-C. (2011) Disclosure of corporate social responsibility and environmental management: evidence from China. *Corporate Social Responsibility and Environmental Management* 19(5), 273–287.

Kurisu, K. (2015) *Pro-Environmental Behaviors.* Springer, Tokyo, JP.

Kurra, S. (2021) *Environmental Noise and Management: overview from past to present.* Wiley, Hoboken, NY, USA.

Kütting, G. and Herman, K. (Eds) (2018) *Global Environmental Politics: concepts, theories and case studies.* Routledge, Abingdon, UK.

Kutz, M. (2018) *Handbook of Environmental Engineering.* Wiley, Hoboken, NJ, USA.

Kuzilwa, J.A., Fold, N., Henningsen, A. and Larsen, M.N. (Eds) (2017) *Contract Farming and the Development of Smallholder Agricultural Businesses: improving markets and value chains in Tanzania.* Routledge (Earthscan), Abingdon, UK.

Kvam, R. (2018) *Social Impact Assessment: integrating social issues in development projects.* Inter-American Development Bank, New York, NY, USA.

Kwashirai, V.C. (2013) Environmental change, control and management in Africa. *Global Environment* 12, 166–196.

Lacuna-Richman, C. (2011) *Growing from Seed: an introduction to social forestry*. Springer, Dordrecht, NL.

Ladygina, N. and Rineau, F. (Eds) (2013) *Biochar and Soil Biota*. CRC Press, Boca Raton, FL, USA.

Lahiri, S. (Ed) (2019) *Environmental Education*. Studera Press, Delhi, IN.

Lahiri-Dutt, K. (Ed) (2018) *Between the Plough and the Pick: informal, artisanal and small-scale mining in the contemporary world*. Australian National University Press, Acton, ACT, AU.

Lal, R. and Stewart, B.A. (Eds) (2013) *Principles of Sustainable Soil Management in Agroecosystems*. CRC Press, Boca Raton, FL, USA.

Lal, R. and Stewart, B.A. (Eds) (2019) *Soil and Climate*. CRC Press, Boca Raton, FL, USA.

Lameed, G.A. (Ed) (2012) *Biodiversity Conservation and Utilization in a Diverse World*. IntechOpen, Rijeka, HR.

Landsberger, S. (2019) *Beijing Garbage: a city besieged by waste*. Amsterdam University Press, Amsterdam, NL.

Lang, T. and Yu, D. (1992) Free trade versus the environment: a debate. *Our Planet (UNEP)* 4(2), 12–13.

Lange, S. and Santarius, T. (2020) *Smart Green World? Making digitalization work for sustainability*. Routledge, Abingdon, UK.

Lannon, E.C. (Ed) (2013) *Drainage Basins and Catchment Management: classification, modelling and environmental assessment*. Nova Science Publishers, Hauppauge, NY, USA.

Larkin, A. (2013) *Environmental Debt: the hidden costs of a changing global economy*. Springer (Palgrave Macmillan), Cham, CH.

Larramendy, M.L. and Solonesky, S. (Eds) (2021) *Agroecosystems: very complex environmental systems*. IntechOpen, London, UK.

Latawiec, A. and Agol, D. (2015) *Sustainability Indicators in Practice*. De Gruyter Open, Warsaw, PL.

Latin, H.A. (2012) *Climate Change Policy Failures: why conventional mitigation approaches cannot succeed*. World Scientific Publishing, Singapore, SG.

Lautze, J. (2020) *The Zambezi River Basin: water and sustainable development*. Routledge (Earthscan), Abingdon, UK.

Lavender, S. and Lavender, A. (2015) *Practical Handbook of Remote Sensing*: a beginner's guide to the world of satellite data. CRC Press, Boca Raton, FL, USA.

Lawn, P. (2006) *Sustainable Development Indicators in Ecological Economics*. Edward Elgar, Cheltenham, UK.

Lawson, N. (2008) *An Appeal to Reason: a cool look at global warming*. Duckworth, London, UK.

Leach, G. and Mearns, R. (2009) *Beyond the Woodfuel Crisis: people, land and trees in Africa* 2nd edn. Routledge (Earthscan), London, UK.

Leach, M. (Ed) (2016) *Gender Equality and Sustainable Development*. Routledge, Abingdon, UK.

Leach, M. and Mearns, R. (Eds) (1996) *The Lie of the Land: challenging received wisdom on the African environment*. James Currey, London, UK.

Leach, M., Joekes, S. and Green, C. (1995) Editorial: gender relations and environmental change. *IDS Bulletin* 26(1), 1–8.

Leach, M., Scoons, I. and Stirling, A. (2010) *Dynamic Sustainabilities: technology, environment, social justice*. Routledge (Earthscan), Abingdon, UK.

Leal Filho, W. (Ed) (2016) *Innovation in Climate Change Adaptation*. Springer, Cham, CH.

Leal Filho, W. (Ed) (2019) *Handbook of Climate Change Resilience*. Springer, Cham, CH.

Leal Filho, W., Azeiteiro, U.M. and Alves, F. (Eds) (2016) *Climate Change and Health: improving resilience and reducing risks*. Springer, Cham, CH.

Leal Filho, W., Djekic, I., Smetana, S. and Kovaleva, M. (Eds) (2022) *Handbook of Climate Change Across the Food Supply Chain*. Springer, CH.

Leary, N., Conde, C., Kulkarni, J., Nyong, A., Adejuwon, J., Barros, V., Burton, I., Lasco, R. and Pulhin, J. (Eds) (2008a) *Climate Change and Vulnerability and Adaptation* (2 vols). Routledge (Earthscan), London, UK.

Leary, N., Conde, C., Kulkami, J., Nyong, A. and Pulhin, J. (Eds) (2008b) *Climate Change and Vulnerability*. Routledge (Earthscan), London, UK.

Lechner, A.M., Tan, C.M., Tritto, A., Horstmann, A., Teo, H.C., Owen, J.R., and Campos-Arceiz, A. (2019) *The Belt and Road Initiative: environmental impacts in Southeast Asia*. ISEAS, Yusof Ishak Institute, Singapore, SG.

Ledgerwood, G., Street, E. and Therivel, R. (1992) *Environmental Audit and Business Strategy*. Pitman, London, UK.

Lee, N. and George, C. (Eds) (2000) *Environmental Assessment in Developing and Transitional Countries: principles, methods, and practice*. Wiley, Chichester, UK.

Lee, N. and Kotler, P. (2005) *Corporate Social Responsibility: doing the most good for your company and your cause*. Wiley, Hoboken, NJ, USA.

Legrand, W., Sloan, P. and Chen, J.S. (2017) *Sustainability in the Hospitality Industry: principles of sustainable operations* 3rd edn. Routledge, Abingdon, UK.

Legun, K., Keller, J., Carolan, J.M. and Bell, M.M. (Eds) (2020) *The Cambridge Handbook of Environmental Sociology: Vol. 1*. Cambridge University Press, Cambridge, UK.

Lehmann, J. and Joseph, S. (Eds) (2015) *Biochar for Environmental Management: science, technology and implementation* 2nd edn. Routledge, Abingdon, UK.

Lehner, O.M. (Ed) (2017) *Routledge Handbook of Social and Sustainable Finance*. Routledge, Abingdon, UK.

Lelo, F., Ayieko, J.O. and Njeri Muhia, R. (2021) *Participatory Rural Appraisal Approaches: a resource for trainers and practitioners*. Moran Publishers, Nairobi, KE.

Lemaire, G., Kronberg, S., César De Faccio, P. and Recous, S. (Eds) (2019) *Agroecosystem Diversity: reconciling contemporary agriculture and environmental quality*. Elsevier (Academic Press), London, UK.

Le Masson, V. and Buckingham-Hatfield, S. (Eds) (2017) *Understanding Climate Change through Gender Relations*. Routledge, Abingdon, UK.

Lennox, C. and Short, D. (Eds) (2018) *Handbook of Indigenous Peoples' Rights*. Routledge, Abingdon, UK.

Lenton, R. and Muller, M. (Eds) (2012) *Integrated Water Resources Management in Practice: better water management for development*. Earthscan, London, UK.

Lenton, T. and Vaughn, N. (Eds.) (2013) *Geoengineering Responses to Climate Change: selected entries from The Encyclopedia of Sustainability Science and Technology*. Springer, New York, NY, USA.

Leonelli, G.C. (2021) *Transnational Narratives and Regulation of GMO Risks*. Bloomsbury Publishing, London, UK.

Leopold, A. (1949) *A Sand County Almanac*. Oxford University Press, London, UK.

Lepczyk, C.A., Boyle, O.D. and Vargo, T.L.V. (Eds) (2020) *Handbook of Citizen Science in Ecology and Conservation*. University of California Press, Oakland, CA, USA.

Leroux, M. (2005) *Global Warming – Myth or reality the erring ways of climatology.* Springer-Verlag (with Praxis), Berlin and Heidelberg, GR.

Letcher, T.M. (Ed) (2019) *Managing Global Warming: an interface of technology and human issues.* Elsevier (Academic Press), London, UK.

Letcher, T.M. (Ed) (2020) *Plastic Waste and Recycling: environmental impact, societal issues, prevention, and solutions.* Elsevier (Academic Press), London, UK.

Leung Ping Sun and Setboonsarng, S. (Eds) (2014) *Making Globalization Work Better for the Poor through Contract Farming.* Asian Development Bank, Manila, PH.

Lewis, B. (2013) *Small Dams: planning, construction and maintenance.* CRC Press, Boca Raton, FL, USA.

Lewis, D. (2014) *Non-Governmental Organizations, Management and Development* 3rd edn. Routledge, Abingdon, UK.

Lewis, D., Kanji, N. and Themudo, N.S. (2021) *Non-Governmental Organizations and Development* 2nd edn. Routledge, Abingdon, UK.

Lewis, H. and Demmers, M. (2013) Life cycle assessment and environmental management. *Australian Journal of Environmental Management* 3(2), 110–123.

Liang Zhang, Sisi Li, Loáiciga, H.A., Zhuang, Y. , and Du, Y. (2015) Opportunities and challenges of interbasin water transfers: a literature review with bibliometric analysis. *Scientometrics* 105(1), 279–294.

Liebig, M., Franzluebbers, A.J. and Follett, R.F. (Eds) (2012) *Managing Agricultural Greenhouse Gases: coordinated agricultural research through GRACEnet to address our changing climate.* Elsevier (Academic Press), London, UK.

Lin, A.C. (2013) *Prometheus Reimagined: technology, environment, and law in the twenty-first century.* University of Michigan Press, Ann Arbor, MI, USA.

Lindenmeyer, D.B. and Likens, G.E. (2018) *Effective Ecological Monitoring* 2nd edn. CSIRO, Clayton South, VI, AU.

Lindenmeyer, D.B., Burns, E., Thurgate, N.Y. and Lowe, A. (Eds) (2014) *Biodiversity and Environmental Change: monitoring, challenges and direction.* CSIRO, Clayton South, VI, AU.

Linebaugh, P. (2014) *Stop, Thief! The commons, enclosures, and resistance.* PM Press, Oakland, CA, USA.

Lin-Heng Lye (2007) Land use planning, environmental management, and the garden city as an urban development approach in Singapore. In: N.J. Chalifour, P. Kameri-Mbote, Lin-Heng Lye and J.R. Nolan (Eds), *Land Use Law for Sustainable Development* (pp. 374–396). Cambridge University Press, Cambridge, UK.

Link, S.J. (2020) *Forging Global Fordism: Nazi Germany, Soviet Russia, and the contest over the industrial order.* Princeton University Press, Princeton, NJ, USA.

Linkov, I. and Palma-Oliveira, J.M. (Eds) (2016) *Resilience and Risk: methods and application in environment, cyber and social domains.* Springer, Dordrecht, NL.

Linnenluecke, M.K. and Griffiths, A. (2015) *The Climate Resilient Organization: adaptation and resilience to climate change and weather extremes.* Edward Elgar, Cheltenham, UK.

Lipper, L. (2011) *Climate Change Mitigation Finance for Smallholder Agriculture: a guide book to harvesting soil carbon sequestration benefits.* Rome, IT.

Lipper, L., McCarthy, N., Zilberman, D., Asfaw, S. and Branca, G. (Eds) (2018) *Climate Smart Agriculture: building resilience to climate change.* Springer (FAO), Cham, CH.

Lippert, I. (2014) Latour's Gaia – not down to Earth? Social studies of environmental management for grounded understandings of the politics of human-nature relationships. In: A. Bamme, G. Gertzinger and Berger, T. (Eds), *Yearbook 2012 of the Institute for Advanced*

Studies on Science, Technology and Society (pp. 91–111). Munchen Wien, Profil Verlag Gmbh, GR.

Liu, J., Sun, W., and Hu, W. (Eds) (2017b) *The Development of Eco Cities in China.* Springer, Singapore, SG.

Liverman, D.M. and Vilas, S. (2006) Neoliberalism and the environment in Latin America. *Annual Revue of Environment and Resources* 31, 2.1–2.7.

Livermore, M.A. and Revesz, R.L. (Eds) (2013) *The Globalization of Cost-Benefit Analysis in Environmental Policy.* Oxford University Press, New York, NY, USA.

Llan, A.J.L., Brenner, S. and Al Hmaidi, M. (2013) Peace and pollution: an examination of Palestinian–Israeli trans-boundary hazardous waste management 20 years after the Oslo Peace Accords, *Journal of Peacebuilding & Development*, 8(1), 15–29.

Lobell, D.B. and Burke, M. (Eds) (2010) *Climate Change and Food Security: adapting agriculture to a warmer world.* Springer, Dordrecht, NL.

Lockyer, J. and Veteto, J.R. (Eds) (2013) *Environmental Anthropology Engaging Ecotopia: bioregionalism, permaculture, and ecovillages.* Berghahn Books, New York, NY, USA.

Lofrano, G. (Ed) (2012) *Green Technologies for Wastewater Treatment: energy recovery and emerging compounds recovery.* Springer, Dordrecht, NL.

Logo, D. (2010) *Environmental Compliance and Auditing.* Lambert Academic Publishing, Saarbrücken, GR.

Lohrberg, F., Lička, L., Scazzosi, L. and Timpe, A. (2016) *Urban Agriculture Europe.* Jovis, Berlin, GR.

Loh, Chin-Ling. (2021) *Sustainability Management of Public Listed Companies in Malaysia: a professional intelligence about key performance area of environmental economic and social aspects* Kindle edn. Amazon, Seattle, WS, USA.

Loh, Z. and Chia, A. (2010) Sustainability reporting in Singapore: environmental management in the Anthropocene. In: Lin-Heng Lye, G. Ofori, Lai Choo Malone-Lee, V.R. Savage and Yen-Peng Lee (Eds), *Sustainability Matters: environmental management in Asia* (pp. 161–188). World Scientific Publishing, Singapore, SG.

Lombardo, J.S. and Buckridge, D.L. (Eds) (2007) *Disease Surveillance: a public health informatics approach.* Wiley, Chichester, UK.

Lomborg, B. (2001) *The Skeptical Environmentalist: measuring the real state of the World* (English trans). Cambridge University Press, Cambridge, UK.

Lomborg, B. (Ed) (2004) *Global Crisis: global solutions.* Cambridge University Press, Cambridge, UK.

Lomborg, B. (2009) *Cool it: the sceptical environmentalist's guide to global warming.* Alfred Knopf, New York, NY, USA.

Lomborg, B. (Ed) (2010) *Smart Solutions to Climate Change: comparing costs and benefits.* Cambridge University Press, Cambridge, UK.

Lomborg, B. (2020) *False Alarm: how climate change panic costs us trillions, hurts the poor, and fails to fix the planet.* Basic Books, New York, NY, USA.

Longley, P.A., Goodchild, M.F., Maguire, D.J. and Rhind, D.W. (2015) *Geographic Information Science & Systems* 4th edn. Wiley, Chichester, UK.

Lopez, M.I. and Suryomenggolo, J. (Eds) (2018) *Environmental Resources Use and Challenges in Contemporary Southeast Asia: tropical ecosystems in transition.* Springer, Singapore, SG.

Lopez, R.D. and Frohn, R.C. (2017) *Remote Sensing for Landscape Ecology: monitoring, modelling and assessment of ecosystems* 2nd edn. CRC Press, Boca Raton, FL, USA.

Lorenz, C. and Lal, R. (2018) *Carbon Sequestration in Agricultural Ecosystems.* Springer, Cham, CH.

Lostarnau, C., Oyarzún, J., Maturana, H., Soto, G., Señoret, M., Soto, M., Rötting, T.S., Amezaga, J.M. and Oyarzún, R. (2011) Stakeholder participation within the public environmental system in Chile: major gaps between theory and practice. *Journal of Environmental Management* 92, 2470–2478.

Lotus, H. (2016) *Ecohydrology and Environmental Watershed Management.* Syrawood Publishing House, New York, NY, USA.

Loughlin, S.C., Sparks, R.S.J., Sparks, S., Brown, S.K., Jenkins, S.F. and Vye-Brown, C. (Eds) (2015) *Global Volcanic Hazards and Risk.* Cambridge University Press, Cambridge, UK.

Lovejoy, T. and Hannah, L.J. (Eds) (2019) *Biodiversity and Climate Change: transforming the biosphere.* Yale University Press, New Haven, CT, USA.

Lovelock, J.E. (1979) *Gaia: a new look at life on Earth.* Oxford University Press, Oxford, UK.

Lovelock, J.E. (1988) *The Ages of Gaia: a biography of our living Earth.* Oxford University Press, Oxford, UK.

Lovelock, J.E. (1992) *Gaia: the practical science of planetary medicine.* Gaia Books, London, UK.

Lovelock J.E. (2006) *The Revenge of Gaia: why the Earth is fighting back – and how we can still save humanity.* Allen Lane, London, UK.

Lovelock, J.E. (2009) *The Vanishing Face of Gaia: a final warning.* Penguin, London, UK.

Lovelock, J.E. (2019) *Novacene: the coming age of hyperintelligence.* MIT Press, Cambridge, MA, USA.

Lovelock, J.E. and Margulis, L. (1973) Atmospheric homoeostasis by and for the biosphere: the Gaia hypothesis. *Tellus* 26(1), 2.

Lovett, J.C. and Ockwell, D.G. (Eds) (2010) *A Handbook of Environmental Management.* Edward Elgar, Cheltenham, UK.

Loy, D., Dorje, G. and Stanley, J. (Eds) (2009) *A Buddhist Response to the Climate Emergency.* Wisdom Publications, Somerville, MA, USA.

Lucivero, F. (2020) Big data, big waste? A reflection on the environmental sustainability of big data initiatives. *Science and Engineering Ethics* 26, 1009–1030.

Ludwig, F., Kabat, P., von Schaik, H. and van der Valk, M. (Eds) (2009) *Climate Change Adaptation in the Water Sector.* Routledge (Earthscan), London, UK.

Lueddeke, G.R. (2019) *Survival: One Health, One Planet, One Future.* Routledge, Abingdon, UK.

Lum, H. (Ed) (2021) *Human Factors Issues and the Impact of Technology on Society.* IGI Global, Hershey, PA, USA.

Lunan, D. (2014) *Incoming Asteroid! What could we do about it?* Springer, New York, NY, USA.

Lusted, M.A. (Ed) (2018) *Extreme Weather Events.* Greenhaven Publishing, New York, NY, USA.

Lutz, W., Butz, W.P. and Samir, K.C. (Eds) (2014) *World Population and Human Capital in the Twenty-First Century: an overview.* Oxford University Press, Oxford, UK.

Lynas, M. (2007) *Six Degrees: our future on a hotter planet.* Harper-Collins (Fourth Estate), London, UK.

Lynas, M. (2018) *Seeds of Science: why we got it so wrong on GMOs.* Bloomsbury, London, UK.

Lynch, A.H. and Veland, S. (2018) *Urgency in the Anthropocene.* MIT Press, Cambridge, MA, USA.

Lynch, T., Glotfelty, C. and Armbruter, K. (2012) *The Bioregional Imagination: literature, ecology, and place.* University of Georgia Press, Athens, GA, USA.

Lyon, T. (Ed) (2010) *Good Cop/Bad Cop: environmental NGOs and their strategies toward business.* Routledge, Abingdon, UK.

Lyster, R., MacKenzie, C. and McDermott, C. (Eds) (2013) *Law, Tropical Forests and Carbon: the case of REDD+*. Cambridge University Press, Cambridge, UK.

Lyubchich, V., Gel, Y., Kilbourne, H.K., Miller, T.J., Newlands, N.K. and Smith, A.B. (Eds) (2021) *Evaluating Climate Change Impacts*. CRC Press, Boca Raton, FL, USA.

MacArthur, R.H. and Wilson, E.O. (1967) *The Theory of Island Biogeography*. Princeton University Press, Princeton, NJ, USA.

MacCracken, M.C. (2012) *Beyond Mitigation: potential options for counter-balancing the climatic and environmental consequences of the rising concentrations of greenhouse gases.* World Bank, Washington, DC, USA.

MacCracken, M.C., Moore, F. and Topping, J.C. Jnr. (Eds) (2008) *Sudden and Disruptive Climate Change: exploring the real risks and how we can avoid them*. Routledge (Earthscan), London, UK.

MacDonald, C. (2008) *Green Inc. An environmental insider reveals how a good cause has gone bad*. The Lyons Press (Pequot Press), Guilford, CT, USA.

MacGregor, S. (Ed) (2017) *Routledge Handbook of Gender and Environment*. Routledge, Abingdon, UK.

MacKenzie, D. (1998) Waste not. *New Scientist* 159(2149), 26–30.

MacKinnon, J.B. (2021) *The Day the World Stops Shopping: how ending consumerism gives us a better life and a greener world*. Penguin (Random House), New York, NY, USA.

McAdam, J. (Ed) (2010) *Climate Change and Displacement: multidisciplinary perspectives*. Hart Publishing, Oxford, UK.

McCarthy, J. (2019) Authoritarianism, populism, and the environment: comparative experiences, insights, and perspectives, *Annals of the American Association of Geographers*, 109(2), 301–313.

McCarthy, J.J., Canziani O.F., Leary N.A., Dokken D.J. and White K.S. (Eds) (2001) *Climate Change 2001: impacts adaptation, and vulnerability contribution of Working Group II to the Third Assessment Report of the IPCC*. Cambridge University Press, Cambridge, UK.

McCool, S.F. and Bosak, K. (Eds) (2016) *Reframing Sustainable Tourism*. Springer, Dordrecht, NL.

McCormick, J.F. (1992) *The Global Environmental Movement*. Belhaven, London, UK.

McCracken, J., Pretty, J. and Conway, G.G. (1988) *An Introduction to Rapid Rural Appraisal for Agricultural Development*. IIED, London, UK.

McDonnell, M.J., Hahs, A.K. and Breuste, J. (Eds) (2009) *Ecology of Cities and Towns: a comparative approach*. Cambridge University Press, Cambridge, UK.

McFarland Taylor, S. (2019) *Ecopiety: green media and the dilemma of environmental virtue*. New York University Press, New York, NY, USA.

McGuire, B. (2005) *Surviving Armageddon: solutions for a threatened planet*. Oxford University Press, Oxford, UK.

McHale, R.K. (Ed) (2012) *Landsat & Its Valuable Role in Satellite Imagery of Earth*. Nova Science Publishers, Hauppauge, NY, USA.

McHarg, I. (1969) *Design with Nature*. Doubleday (Natural History Press), Garden City, NY, USA.

McIntosh, B.S., Ascough II, J.C., Twery, M., Chew, J., Elmahdi, A., Haase, D., Harou, J.J., Hepting, D., Cuddy, S., Jakeman, A.J., Chen, S., Kassahun, A., Lautenbach, S., Matthews, K., Merrittj, W., Quinn, N.T.W., Rodriguez-Roda, I., Sieber, S., Stavenga, M., Sulis, A., Ticehurst, J., Volk, M., Wrobel, M., van Delden, H., El-Sawah, S., Rizzoli, A. and Voinov, A. (2011) Environmental decision support systems (EDSS) development: challenges and best practices. *Environmental Modelling & Software* 26, 1389–1402.

McIntyre, J.R., Ivanaj, V. and Ivanaj, S. (Eds) (2018) *CSR and Climate Change Implications for Multinational Enterprises*. Edward Elgar, Cheltenham, UK.

McKenzie, C. (2020) *GATT and Global Order in the Postwar Era*. Cambridge University Press, Cambridge, UK.

McKenzie, D. (2011) *Fostering Sustainable Behavior: an introduction to community-based social marketing* 3rd edn. New Society Publishers, Gabriola Island, BC, CA.

McKenzie-Mohr, D., Lee, N.R., Wesley Schultz, P. and Kotler, P. (2012) *Social Marketing to Protect the Environment: what works*. Sage, Thousand Oaks, CA, USA.

McKinnon, A.C., Browne, M., Piecyck, M. and Whiteing, A.E. (Eds) (2010) *Green Logistics: improving the environmental sustainability of logistics*. Kogan Page, London, UK.

McKinnon, A.C., Browne, M., Piecyk, H. and Whiteing, A. (Eds) (2015) *Green Logistics: improving the environmental sustainability of logistics* 2nd edn. Kogan Page, London, UK.

McLaren, A.E. (2013) *Preservation and Cultural Heritage in China*. Common Ground Publishing, Champaign, IL, USA.

McLeman, R.A. (2013) *Climate and Human Migration: past experiences, future challenges*. Cambridge University Press, New York, NY.

McLeman, R.A. and Gemenne, F. (Eds) (2018) *Routledge Handbook of Environmental Displacement and Migration*. Routledge (Earthscan), Abingdon, UK.

McLeman, R.A., Schade, J. and Faist, T. (Eds) (2016) *Environmental Migration and Social Inequality*. Springer, Cham, CH.

McMahon, E.T. (2010) *Conservation Communities: creating value with nature, open space and agriculture*. Urban Land Institute, London, UK.

McNulty, S. (Ed) (2023) *Future Forests: adaptation to climate change*. Elsevier, Philadelphia, PA, USA.

McPhaden, M.J., Santoso, A. and Cai, J. (Eds) (2002) *El Niño Southern Oscillation in a Changing Climate*. Wiley (AGU), Hoboken, NY, USA.

Ma, Y. (2017) Vertical environmental management: a panacea to the environmental enforcement gap in China?, *Chinese Journal of Environmental Law*, 1(1), 37–68.

Macaes, B. (2020) *Belt and Road: a Chinese world order*. C. Hurst & Co., London, UK.

Mace, G.M., Schreckenberg, K. and Poudyal, M. (Eds) (2018) *Ecosystem Services and Poverty Alleviation: trade-offs and governance*. Routledge (Earthscan), Abingdon, UK.

Macey, J. (2013) *Spiritual Ecology: the cry of the Earth*. The Golden Sufi Center, Point Reyes, CA, USA.

Macleod, C.J.A., Blackstock, K., Brown, K., Eastwood, A., Gimona, A., Prager, K. and Irvine, R.J. (2016) *Adaptive Management: an overview of the concept and its practical application in the Scottish context*. The James Hutton Institute, Edinburgh, UK.

Macphee, R.D.E. (2018) *End of the Megafauna: the fate of the world's hugest, fiercest, and strongest animals*. W.W. Norton, New York, NY, USA.

Macpherson, E.J. and O'Donnell, E. (2015) Desafíos para la gestión ambiental en Chile: una perspectiva Australiana [Challenges for environmental water management in Chile: an Australian perspective]. *Revista de Derecho Administrativo Economico* 21, 171–203.

Macrory, R. (2014) *Regulation, Enforcement and Governance in Environmental Law*. Hart Publishing, Oxford, UK.

Macura, B., Suškevičs, M., Garside, R., Hannes, K., Rees, R. and Rodela, R. (2019) Systematic reviews of qualitative evidence for environmental policy and management: an overview of different methodological options. *Environmental Evidence* 8(24), article No. 24 (11 pp.).

Madeley, J. (2002) *Food for All: the need for a new agriculture*. Zed Books, London, UK.

Maes, J. and Burkhard, B. (Eds) (2017) *Mapping Ecosystem Services*. Pensoft Publishers, Sofia, BG.

Magalhães Pires, J.C. (Ed) (2019) *Carbon Capture and Storage*. MDPI, Basel, CH.

Magalhães Pires, J.C. and da Cunha Goncalves, A.L. (Eds) (2019) *Bioenergy with Carbon Capture and Storage: using natural resources for sustainable development*. Elsevier (Academic Press), London, UK.

Magdalena, Z. (Ed) (2021) *Adapting and Mitigating Environmental, Social, and Governance Risk in Business*. IGI Global, Hershey, PA, USA.

Magee, T. (2013) *A Field Guide to Community Based Adaptation*. Routledge (Earthscan), Abingdon, UK.

Magnan, A.K., Schipper, E.L.F. and Duvat, K.E. (2020) Frontiers in climate change adaptation science: advancing guidelines to design adaptation pathways. *Current Climate Change Reports* 6, 166–177.

Magnussen, A. (2020) *Practical Vulnerability Management: a strategic approach to managing cyber risk*. William Pollock, San Francisco, CA, USA.

Maha, B. (Ed) (2017) *Examining the Role of Environmental Change on Emerging Infectious Diseases and Pandemics*. IGI Global, Hershey, PA, USA.

Mahaffey, J. (2014) *Atomic Accidents: a history of nuclear meltdowns and disasters: from the Ozark Mountains to Fukushima*. Pegasus Books, New York, NY, USA.

Maizatun, M. (2011) *Environmental Law in Malaysia* 4th edn. Kluwer Law, Dordrecht, NL.

Major, D.C. and Juhola, S. (2021) *Climate Change Adaptation in Coastal Cities: a guidebook for citizens, public officials and planners*. Helsinki University Press, Helsinki, FI.

Malik, B. (2018) *Measuring Environmental Attitude and Pro-Environmental Behavior*. Lambert Academic Publishing, Sunnyvale, CA, USA.

Maliwal, G.L. (2006) *Handbook of Environmental Management*. Agrotech Publishing Academy, Udaipur, IN.

Maliwal, P.L. (2020) *Text Book of Rainfed Agriculture and Watershed Management*. Scientific Publishers, Jodhpur, IN.

Maljean-Dubois, S. (Ed) (2017) *The Effectiveness of Environmental Law*. Intersentia, Cambridge, UK.

Maltby, E. (1986) *Waterlogged Wealth: why waste the world's wet places?* Earthscan, London, UK.

Malthus, T.R. (1798) *An Essay on the Principle of Population as it Affects the Future Improvement of Mankind*. J. Johnson, London, UK.

Malyan, R.S. and Duhan, P. (Eds) (2019) *Green Consumerism: perspectives, sustainability, and behavior*. CRC Press, Boca Raton, FL, USA.

Mambretti, S. and Brebbia, C.A. (Eds) (2012) *Urban Water*. WIT Press, Southampton, UK.

Management, Information Resources Association(Ed) (2018) *Hydrology and Water Resource Management: breakthroughs in research and practice*. IGI Global, Hershey, PA, USA.

Managi, S. (Ed) (2013) *The Economics of Biodiversity and Ecosystem Services*. Routledge, Abingdon, UK.

Managi, S. and Kuriyama, K. (2017) *Environmental Economics*. Routledge, London, UK.

Manaia, C.M., Donner, E., Vaz-Moreira, I. and Hong, P. (Eds) (2020) *Antibiotic Resistance in the Environment: a worldwide overview*. Springer, Cham, CH.

Mann, M.E. (2021) *The New Climate War: the fight to take back our planet*. Hachette Book Group, New York, NY, USA.

Manou, D., Thorp, T.M., Baldwin, A., Mihr, A. and Cubie, D. (Eds) (2017) *Climate Change, Migration and Human Rights: law and policy perspectives*. Routledge (Earthscan), Abingdon, UK.

Mansvelt, J. (Ed) (2011) *Green Consumerism: an A-to-Z guide*. Sage Publications, Thousand Oaks, CA, USA.

Manya, J.J. and Gasco, G. (Eds) (2021) *Biochar as a Renewable-Based Material: with applications in agriculture, the environment and energy*. World Scientific Publishing Europe, London, UK.

Manyuchi, M.M., Mbohwa, C., Muzenda, E. and Sukdeo, N. (2021) *Environmental Impact Assessments and Mitigation*. CRC Press, Boca Raton, FL, USA.

Manzello, S.L. (Ed) (2020) *Encyclopedia of Wildfires and Wildland-Urban Interface (WUI) Fires*. Springer, Cham, CH.

Mapedza, E., Tsegai, D., Brüntrup, M. and Mcleman, R. (2019) *Drought Challenges: policy options for developing countries*. Elsevier, Amsterdam, NL.

Marboe, I. (Ed) (2021) *Legal Aspects of Planetary Defence*. Brill/Nijhoff, Leiden, NL.

Marciano, C., Romano, G. and Fiorelli, M.S. (Eds) (2021) *Best Practices in Urban Solid Waste Management: ownership, governance, and drivers of performance in a zero waste framework*. Emerald Group Publishing, Bingley, UK.

Marcus, A.A. (2015) *The Future of Technology Management and the Business Environment: lessons on innovation, disruption, and strategy execution*. Pearson Educational, London, UK.

Marfe, G. and Di Stefano, C. (Eds) (2020) *Hazardous Waste Management and Health Risks*. Bentham Science Publishers, Singapore, SG.

Margaris, N.S. (1987) Desertification in the Aegean Islands. *Ekistics* 54(323/324), 132–137.

Margerum, R.D. (2008) A typology of collaboration efforts in environmental management. *Environmental Management* 41(4), 487–500.

Mariciano, C., Romano, G. and Fiorelli, M.S. (2021) *Best Practices in Urban Solid Waste Management: ownership, governance, and drivers of performance in a zero waste framework*. Emerald Group Publishing, Bingley, UK.

Markandya, A., Galarraga, I. and Sainz de Murieta, E. (Eds) (2016) *Routledge Handbook of the Economics of Climate Change Adaptation*. Routledge, Abingdon, UK.

Marques, J.C., Jørgensen, S.E. and Nielsen, S.N. (2015) *Integrated Environmental Management: a transdisciplinary approach*. CRC Press, Boca Raton, FL, USA.

Marsden, S. (2008) *Strategic Environmental Assessment in International and European Law: a practitioner's guide*. Routledge (Earthscan), Abingdon, UK.

Marsden, S. (2017) *Environmental Regimes in Asian Subregions: China and the Third Pole*. Edward Elgar, Cheltenham, UK.

Marsden, S. and Brandon, E. (2015) *Transboundary Environmental Governance in Asia: practice and prospects with the UNECE agreements*. Edward Elgar, Cheltenham, UK.

Marsh, G.P. (1864) *Man and Nature: or physical geography as modified by human action*. Charles Scribner, New York, NY, USA.

Marshall, J.P. and Connor, L.H. (Eds) (2016) *Environmental Change and the World's Futures: ecologies, ontologies and mythologies*. Routledge (Earthscan), Abingdon, UK.

Masterson, J.H., Peacock, W.G., Van Zandt, S.S., Grover, H., Schwarz, L.F. and Cooper, J.T. (Eds) (2014) *Planning for Community Resilience: a handbook for reducing vulnerability*. Island Press, Washington, DC, USA.

Martens, P. and Chiung Ting, Chang (Eds) (2010) *The Social and Behavioural Aspects of Climate Change: linking vulnerability, adaptation and mitigation*. Routledge (Greenleaf Publishing), Abingdon, UK.

Martin, A. (2017) *Just Conservation: biodiversity, wellbeing and sustainability*. Routledge (Earthscan), Abingdon, UK.

Martin, D.L. and Schouten, J. (2011) *Sustainable Marketing*. Pearson, London, UK.

Martin, P. and Kennedy, A. (Eds) (2015) *Implementing Environmental Law*. Edward Elgar, Cheltenham, UK.

Martin, P.S. and Klein, R.G. (Eds) (1984) *Quaternary Extinctions*. University of Arizona Press, Tucson, AZ, USA.

Martin, P.S., Du Plessis, A., Lin Zhiping Li, Qin Tianbao, and Le Bouthillier, Y. (Eds) (2012) *Environmental Governance and Sustainability*. Edward Elgar, Cheltenham, UK.

Martin, R. (2012) *SuperFuel: thorium, the green energy source for the future*. St Martin's Press, New York, NY, USA.

Martinez, J.L., Munoz-Acevedo, A. and Rai, M. (Eds) (2019) *Ethnobotany: local knowledge and traditions*. CRC Press, Boca Raton, FL, USA.

Martinez-Alier, J. and Muradian, R. (Eds) (2015) *Handbook of Ecological Economics*. Edward Elgar, Cheltenham, UK.

Martinez Romera, B. and Broberg, M. (Eds) (2021) *The Third Pillar of International Climate Change Policy: on 'loss and damage' after the Paris Agreement*. Routledge, Abingdon, UK.

Martin-Guay, M.-O., Paquette, A., Dupras, J. and Rivest, D. (2018) The New Green Revolution: sustainable intensification of agriculture by intercropping. *Science of the Total Environment* 615, 767–772.

Martin-Ortega, J., Gordon, I.J., Ferrier, R.C. and Khan, S. (Eds) (2015) *Water Ecosystem Services: a global perspective*. Cambridge University Press, Cambridge, UK.

Maser, C. (2015) *Interactions of Land, Ocean and Humans: a global perspective*. CRC Press, Boca Raton, FL, USA.

Maser, C. and Pollio, C.A. (2012) *Resolving Environmental Conflict* 2nd edn. CRC Press, Boca Raton, FL, USA.

Mason, L.R. (2019) *People and Climate Change: vulnerability, adaptation, and social justice*. Oxford University Press, Oxford, UK.

Massei, G., Rocchi, L., Paolotti, L., Greco, S. and Boggia, A. (2014) Decision support systems for environmental management: a case study on wastewater from agriculture. *Journal of Environmental Management* 146, 491–504.

Matejicek, L. (Ed) (2010) *Environmental Modelling with GPS*. Nova Science Publishers Inc., Hauppauge, New York, NY, USA.

Mathur, H.M. (Ed) (2018) *Assessing the Social Impact of Development Projects: experience in India and other Asian countries*. Springer, Cham, CH.

Matlock, M.D. and Morgan, R.A. (2010) *Ecological Engineering Design: restoring and conserving ecosystem services*. Wiley, Hoboken, NJ, USA.

Matsushita, M. and Schoenbaum, T.J. (Eds) (2016) *Emerging Issues in Sustainable Development: international trade law and policy relating to natural resources, energy, and the environment*. Springer (Japan), Tokyo, JP.

Mattson, I., Hofmann, M.W., Mieth, D. and Arneth, M. (2011) *Islam, Christianity & the Environment*. The Royal Aal Al-Bayt Institute for Islamic Thought (RABIIT), 20 Sa'ed Bino Road, Dabuq, PO Box 950361, Amman, JO.

Maurya, S.P., Yadav, A.K. and Singh, R. (2022) *Modeling and Simulation of Environmental Systems: a computation approach*. CRC Press, Boca Raton, FL, USA.

Maware, M. (2013) *Environment and Natural Resource Conservation and Management in Mozambique*. Langaa RPCID, Mankon, CM.

Maware, M. (2014) *Environmental Conservation through Ubuntu and Other Emerging Perspectives*. Langaa Research and Publishing CIG, Mankon, CM.

Max-Neef, M.A. (1992a) *From the Outside Looking In: experiences in 'barefoot economics'* (first published in 1982 by Dag Hammarskjold Foundation). Zed Press, London, UK.

Max-Neef, M.A. (Ed) (1992b) *Real-Life Economics: understanding wealth creation.* Routledge, London, UK.

Maxwell, D. (2015) *Valuing Natural Capital: future proofing business and finance.* Routledge, Abingdon, UK.

Mayer, B. (2016) *The Concept of Climate Migration: advocacy and its prospects.* Edward Elgar, Cheltenham, UK.

Mayer, B. (2018) *The International Law on Climate Change.* Cambridge University Press, Cambridge, UK.

Mayer, B. and Crépeau, F. (Eds) (2017) *Research Handbook on Climate Change, Migration and the Law.* Edward Elgar, Cheltenham, UK.

Mazo, A.F. and López, M.A.M. (Eds) (2021) *Food Security Issues and Challenges.* Nova Science Publishers, Hauppauge, NY, USA.

Meadows, D.H., Meadows, D.L., Randers, J. and Behrens, W.W. III (1972) *The Limits to Growth (a report to the Club of Rome's project on the predicament of mankind).* Universal Books, New York, NY, USA.

Meadows, D.H., Meadows, D.L. and Randers, J. (1992) *Beyond the Limits: global collapse or sustainable future?* Earthscan, London, UK.

Meadows, D.H., Randers, J. and Meadows, D.L. (2004) *Limits to Growth: the 30-year update.* Routledge (Earthscan), London, UK.

Measham, T. and Lockie, S. (Eds) (2012) *Risk and Social Theory in Environmental Management.* CSIRO, Collingwood, VI, AU.

Medeiros, E. (Ed) (2020) *Territorial Impact Assessment.* Springer, Cham, CH.

Megdal, S.B., Varady, R.G. and Eden, S. (Eds) (2013) *Shared Borders, Shared Waters: Israeli-Palestinian and Colorado River basin water challenges.* CRC Press, Boca Raton, FL, USA.

Megdal, S.B., Varady, R.G., Gude, V.G., Venkataramana, G. and Kandiah, R. (Eds) (2020) *Sustainable Water: resources, management and challenges.* Nova Science Publishers, Hauppauge, NY, USA.

Meisel, J.H. and Puttaswamaiah, K. (Eds) (2017) *Cost-Benefit Analysis: with reference to environment and ecology.* Routledge, London, UK.

Meissener, M. and Lindner, C. (Eds) (2016) *Global Garbage: urban imaginaries of waste, excess, and abandonment.* Routledge, Abingdon, UK.

Melack, J.M., Sadro, S., Sickman, J.O. and Dozier, J. (2021) *Lakes and Watersheds in the Sierra Nevada of California: responses to environmental change.* University of California Press, Oakland, CA, USA.

Melianda, E. (2009) *Vulnerability and Development of a Tsunami-Affected Coast: a case study of Banda Aceh, Indonesia.* Lambert Academic Publishing, Sunnyvale, CA, USA.

Mendez, V.E. (2013) *Agroecology: as a transdisciplinary, participatory and action-oriented approach.* CRC Press, Boca Raton, FL, USA.

Menon, S. and Pillai, P.A. (2012) *Watershed Management: concepts & experiences.* ICFAI Books, Hyderabad, IN.

Mensah, I. (2019) *Environmental Management Concepts and Practices for the Hospitality Industry.* Cambridge Scholars Publishing, Newcastle-upon-Tyne, UK.

Merchant, C. (2020) *The Anthropocene & the Humanities: from climate change to a new age of sustainability.* Yale University Press, New Haven, CT, USA.

Merriam, S.B. and Tisdell, E.J. (2015) *Qualitative Research: a guide to design and implementation* 4th edn. Jossey-Bass, San Francisco, CA, USA.

Merrill, R.T. (2010) *Our Magnetic Earth: the science of geomagnetism.* The University of Chicago Press, Chicago, IL, USA.

Metcalfe, C.D., Collins, P., Menone, M.L. and Tundisi, J.G. (Eds) (2020) *The Paraná River Basin: managing water resources to sustain ecosystem services.* Routledge (Earthscan), Abingdon, UK.

Metcalf, G.E. (2019) *Paying for Pollution: why a carbon tax is good for America.* Oxford University Press, New York, NY, USA.

Metzger, J. and Lindblad, J. (2021) *Dilemmas of Sustainable Urban Development: a view from practice.* Routledge, New York, NY, USA.

Meyer, K. and Newman, P. (2022) *Planetary Accounting: quantifying how to live within planetary limits at different scales of human activity.* Springer, Singapore, SG.

Michaelides, E.E. (2018) *Energy, the Environment, and Sustainability.* CRC Press, Boca Raton, FL, USA.

Mieila, M. (2013) *Sustainable Development Indicators: a review of paradigms.* IGI Global, Hershey, PA, USA.

Mihailovic, D.T. and Lalic, B. (Eds) (2010) *Advances in Environmental Modeling and Measurements.* Nova Science Publishers, Hauppauge, NY, USA.

Mikhail, A. (2012) *Water on Sand: environmental histories of the Middle East and North Africa.* Oxford University Press, New York, NY, USA.

Miles, M.B., Huberman, A.M. and Saldana, J. (2019) *Qualitative Data Analysis: a methods sourcebook* 4th edn. Sage, Los Angeles, CA, USA.

Milford, J.B., Ramaswami, A. and Small, M.J. (2005) *Integrated Environmental Modeling: pollutant transport, fate, and risk in the environment.* Wiley, Hoboken, NJ, USA.

Miller, M., Barden, P. and Candlin, L. (2005) *Environmental Data Management.* IEMA, London, UK.

Miller, T. (2017a) *Greenwashing Culture.* Routledge, Abingdon, UK.

Miller, T. (2017b) *Storming the Wall: climate change, migration, and homeland security.* City Lights Books, San Francisco, CA, USA.

Minaei, N. (Ed) (2022) *Smart Cities: critical debates on big data, urban development and social environmental sustainability.* CRC Press, Boca Raton, FL, USA.

Minaverry, C.M. and Caceres, V.L. (2016) Contributions for discussion: environmental management instruments in Buenos Aires Province, Argentina: an interdisciplinary approach. *Estudios Socio-Jurídicos* 18(1), 57–78.

Mines, R.O. (Ed) (2014) *Environmental Engineering: principles and practice.* Wiley, Chichester, UK.

Ming Hu (2020) *Smart Technologies and Design for Healthy Built Environments.* Springer, Cham, CH.

Ministerie VROM (1989) *National Environmental Policy Plan: to choose or lose?* SDU, The Hague, NL.

Mingxin Guo, Zhongqi He and Uchimiya, S.M. (Eds) (2016) *Agricultural and Environmental Applications of Biochar: advances and barriers.* SSSA, Madison, WI, USA.

Miralles-Wilhelm, F. (2021) *Nature-based Solutions in Agriculture: sustainable management and conservation of land, water and biodiversity.* FAO, VA, USA.

Mirumachi, N. (2017) *Transboundary Water Politics in the Developing World.* Earthscan, Abingdon, UK.

Mishra, V.N., Rai, P.K. and Singh, P. (Eds) (2022) *Geospatial Technology for Landscape and Environmental Management: sustainable assessment and planning.* Springer, Singapore, SG.

Mitchell, A. (2020) *ESRI Guide to GIS Analysis: Vol. 1.* ESRI Press, Redlands, CA, USA.

Mitchell, B. (1997) *Resource and Environmental Management.* Addison Wesley Longman, Harlow, UK.

Miyake, F., Usoskin, I.G. and Poluianov, S. (2019) *Extreme Solar Particle Storms: the hostile Sun*. Institute of Physics Publishing, Bristol, UK.

Mmbali, O. (2013) *Empowering the Community for Effective Environment Management: the case of the Kakamega Forest, Kenya*. GRIN Verlag, Munich, GR.

Mohan, D.M. and Saikia, M.D. (2012) *Watershed Management*. PHI Learning, New Delhi, IN.

Mohanraj, R., Kumaraswamy, K. and Ramkumar, M. (Eds) (2015) *Environmental Management of River Basin Ecosystems*. Springer, Cham, CH.

Mokhnacheva, D., Ionesco, D. and Gemenne, F. (2016) *The Atlas of Environmental Migration*. Routledge (Earthscan), Abingdon, UK.

Moksness, E., Dahl, E. and Støttrup, J. (2009) *Integrated Coastal Zone Management*. Wiley-Blackwell, Oxford, UK.

Mokhtsim, N. and Salleh, K.O. (2016) An appraisal of environmental management strategies in Malaysia: towards achieving sustainable development goals. *International Journal of Applied Environmental Sciences* 11(6), 1569–1579.

Molden, D. (Ed) (2007) *Water for Food, Water for Life: a comprehensive account of water management in agriculture*. Routledge (Earthscan), Abingdon, UK.

Mölders, N. (2012) *Land-Use and Land-Cover Changes: impact on climate and air quality*. Springer, Dordrecht, NL.

Momtaz, S. and Kabir, S.M.Z. (2018) *Evaluating Environmental and Social Impact Assessment in Developing Countries* 2nd edn. Elsevier, Amsterdam, NL.

Monbiot, G. (2009) *Heat: how to stop the planet from burning*. Penguin, London, UK.

Monbiot, G. (2014) *Feral: rewilding the land, the sea, and human life*. University of Chicago Press, Chicago, IL, USA.

Monbiot, G. (2022) *Regenesis: feeding the world without devouring the planet*. Penguin (Allan Lane), London, UK.

Monirul Alam, G.M., Erdiaw-Kwasie, M.O., Nagy, G.J. and Filho, W.L. (Eds) (2021) *Climate Vulnerability and Resilience in the Global South: human adaptations for sustainable futures*. Springer, Cham, CH.

Montgomery, D.R. (2007) *Dirt: the erosion of civilizations*. University of California Press, Berkeley, CA, USA.

Moon, J. (2014) *Corporate Social Responsibility: a very short introduction*. Oxford University Press, Oxford, UK.

Moore, J.W. (Ed) (2016) *Anthropocene or Capitalocene? Nature, history, and the crisis of capitalism*. PM Press, Oakland, CA, USA.

Moosa, I.A. and Ramiah, V. (2014) *The Costs and Benefits of Environmental Regulation*. Edward Elgar, Cheltenham, UK.

Morain, S.A. and Budge, A.M. (Eds) (2012) *Environmental Tracking for Public Health Surveillance*. CRC Press, Boca Raton, FL, USA.

Moran, A. (Ed) (2015) *Climate Change the Facts*. Stockade Books, Woodsville, NH, USA.

Moran, E.F. (2010) *Environmental Social Science: human – environment interactions and sustainability*. Wiley-Blackwell, Oxford, UK.

Morgan, C.L. (2011) *Vulnerability Assessments: a review of approaches*. IUCN, Gland, CH.

Morgan, R.P.C. and Nearing, M.A. (Eds) (2011) *Handbook of Erosion Modelling*. Wiley-Blackwell, Oxford, UK.

Morre, S.M. (2018) *Subnational Hydropolitics: conflict, cooperation, and institution-building in shared river basins*. Oxford University Press, New York, NY, USA.

Morris, B. (2012) *Pioneers of Ecological Humanism*. Book Guild Ltd., Kibworth, UK.

Bibliography

Morris, W. (1891) *News from Nowhere*. Reeves and Turner, London, UK (more recent edn published by Routledge and Kegan Paul, London, UK).

Morrison-Saunders, A. (2018) *Advanced Introduction to Environmental Impact Assessment*. Edward Elgar, Cheltenham, UK.

Morrison-Saunders, A., Bond, A. and Howitt, R. (Eds) (2013) *Sustainability Assessment: pluralism, practice and progress*. Routledge, Abingdon, UK.

Morrison-Saunders, A., Bond, A. and Pope, J. (Eds) (2015) *Handbook of Sustainability Assessment*. Edward Elgar, Cheltenham, UK.

Morton, O. (2015) *The Planet Remade: how geoengineering could change the world*. Granta Books, London, UK.

Moser, S.C. and Boykoff, M.T. (Eds) (2013) *Successful Adaptation to Climate Change: linking science and policy in a rapidly changing world*. Routledge, Abingdon, UK.

Mostafavi, M. and Doherty, G. (Eds) (2010) *Ecological Urbanism*. Lars Müller Publishers, Zurich, CH.

Moustafaev, J. (2015) *Project Scope Management: a practical guide to requirements for engineering, product, construction, it and enterprise projects*. CRC Press (Auerback), Boca Raton, FL, USA.

Muddiman, S. (2019) *Ecosystem Services: economics and policy*. Springer (Palgrave Macmillan), Cham, CH.

Mueller-Dombois, D. (1975) Some aspects of island ecosystem analysis. In: F.B. Golley and E. Medina (Eds), *Tropical Ecological Systems: trends in terrestrial and aquatic research* (pp. 353–366). Springer-Verlag, Berlin, GR.

Mueller-Dombois, D., Bridges, K.W. and Carson, H.L. (Eds) (1981) *Island Ecosystems: Ecological organization in selected Hawaiian communities*. Hutchinson Ross, Stroudsburg, PA, USA.

Mukh, A. (2010) *Frontiers in PRA and PLA: PRA & PLA in applied research*. Academic Foundation, New Delhi, IN.

Mukherjee, M. and Shaw, R. (Eds) (2021) *Ecosystem-Based Disaster and Climate Resilience: integration of blue-green infrastucture in sustainable development*. Springer, Singapore, SG.

Mukherjee, N., Hugé, J., William J., Sutherland, W.J., McNeill, J., Van Opstal, M., Dahdouh-Guebas, F. and Koedam, N. (2015) The Delphi technique in ecology and biological conservation: applications and guidelines. *Methods in Ecology and Evolution* 6(9), 1097–1109.

Mukherjee, R. (2002) *Environmental Management and Awareness Issues*. Sterling Publishers, New Delhi, IN.

Mukonza, C., Hinson, R.E., Adeola, O., Adisa, I., Mogaji, E. and Kirgiz, A.C. (Eds) (2021) *Green Marketing in Emerging Markets: strategic and operational perspectives*. Elsevier (Palgrave Macmillan), London, UK.

Mullen, J. (Ed) (2018) *Rural Poverty, Empowerment and Sustainable Livelihoods* 2nd edn. Routledge, Abingdon, UK.

Muller, A. (Ed) (2016) *Environmental Pollution and Management*. Syrawood Publishing House, New York, NY, USA.

Müller, R. (Ed) (2012) *Stratospheric Ozone Depletion and Climate Change*. Royal Society of Chemistry Publishing, Cambridge, UK.

Mulvihill, P. and Harris Ali, S. (2016) *Environmental Management: critical thinking and emerging practices*. Routledge, Abingdon, UK.

Mupambwa, H.A., Nciizah, A.D., Nyambo, P., Muchara, B. and Gabriel, N.N. (Eds) (2022) *Food Security for African Smallholder Farmers*. Springer, Singapore, SG.

Muralikrishna, I.V. and Manickam, V. (2017) *Environmental Management: science and engineering for industry*. Elsevier (Butterworth-Heinemann), Oxford, UK.

Murray, M. (2016) *Participatory Rural Planning: exploring evidence from Ireland* 2nd edn. Routledge (Ashgate), Abingdon, UK.

Musavengane, R., Tantoh, H.B. and Simatele, D. (2019) A comparative analysis of collaborative environmental management of natural resources in sub-Saharan Africa: a study of Cameroon and South Africa. *Journal of Asian and African Studies* 54(4), 512–532.

Muthu, S.S. (Ed) (2014) *Social Life Cycle Assessment: an insight*. Springer, Singapore, SG.

Muthu, S.S. (Ed) (2016) *The Carbon Footprint Handbook*. CRC Press, Boca Raton, FL, USA.

Muthu, S.S. (Ed) (2019) *Development and Quantification of Sustainability Indicators*. Springer-Verlag, Singapore, SG.

Muthu, S.S. (Ed) (2020) *Carbon Footprints: case studies from the building, household, and agricultural sectors*. Springer, Singapore SG.

Muthu, S.S. (Ed) (2021a) *Life Cycle Sustainability Assessment (LCSA)*. Springer, Singapore, SG.

Muthu, S.S. (Ed) (2021b) *LCA Based Carbon Footprint Assessment*. Springer, Singapore, SG.

Muthu, S.S. (Ed) (2021c) *Microplastic Pollution*. Springer, Singapore, SG.

Muthu, S.S. and Gardetti, M.A. (Eds) (2016) *Green Fashion: Vol. 1*. Springer, Singapore SG.

Muthuraman, L. and Ramaswamy, S. (2018) *Solid Waste Management*. MJP Publishers, New Delhi, IN.

Mysiak, J., Henrikson, H.J., Sullivan, J.C., Bromley, J. and Pahl-Wostl, C. (Eds) (2010) *The Adaptive Water Resource Management Handbook*. Earthscan, London, UK.

Naess, A. (1988) Deep ecology and ultimate premises. *The Ecologist* 18(4–5), 128–132.

Naiman, R.J. (1992) *Watershed Management: balancing sustainability and environmental change*. Springer-Verlag, London, UK.

Najam, A. and Meléndez-Ortiz, R. (Eds) (2007) *Envisioning a Sustainable Development Agenda for Trade and Environment*. Palgrave Macmillan, New York, NY, USA.

Nakashima, D., Krupnik, I. and Rubis, J. (2018) *Indigenous Knowledge for Climate Change Assessment and Adaptation*. Cambridge University Press, Cambridge, UK.

Nanwal, R.K. (2019) *Rainfed Agriculture and Watershed Management*. New India Publishing Agency, New Delhi, IN.

Nanwal, R.K. and Rajanna, G.A. (2017) *Rainfed Agriculture*. New India Publishing Agency, New Delhi, IN.

Narayanasamy, N. (2009) *Participatory Rural Appraisal: principles, methods and application*. Sage Publications India, New Delhi, IN.

Narula, S.N., Rai, S. and Sharma, A. (Eds) (2019) *Environmental Awareness and the Role of Social Media*. IGI Global, Hershey, PA, USA.

Nash, C.E. (2011) *The History of Aquaculture*. Wiley-Blackwell, Ames, IO, USA.

Nash, S. (2019) *Negotiating Migration in the Context of Climate Change: international policy and discourse*. Bristol University Press, Bristol, UK.

Nasreen, A., Naz, A. and Fatima, Z. (2019) Women participation in environmental management in the urban and rural areas of District Lahore. *Pakistan Social Sciences Review* 3(1), 429–442.

Nath, A.J., Sileshi, G.W. and Das, A.K. (2020) *Bamboo: climate change adaptation and mitigation*. CRC Press, Boca Raton, FL, USA.

National Academies (2012) *Disaster Resilience: a national imperative*. Policy and Global Affairs, Committee on Science, Engineering, and Public Policy, Committee on Increasing National Resilience to Hazards and Disasters, National Academies Press, Washington, DC, USA.

National Academy of Sciences (2009) *Science and Decisions: advancing risk assessment*. The National Academy Press, Washington, DC, USA.

National Research Council of the National Academies (2015) *Climate Intervention: reflecting sunlight to cool Earth*. The National Academies Press, Washington, DC, USA.

Navarra, A. and Tubiana, L. (Eds) (2013) *Regional Assessment of Climate Change in the Mediterranean: Vol. 2: agriculture, forests and ecosystem services and people*. Springer, Dordrecht, NL.

Nazaroff, W.W. and Alvarez-Cohen, L. (2021) *Environmental Engineering Science*. Wiley, New York, NY, USA.

Nazrul Islam, M. and Jørgensen, S.E. (Eds) (2020) *Environmental Management of Marine Ecosystems*. Taylor & Francis, Abingdon, UK.

Ndongo, S. (2014) *The Fair Trade Scandal: marketing poverty to benefit the rich*. Ohio University Press, Athens, OH, USA.

Nelson, M.K. and Shilling, D. (Eds) (2018) *Traditional Ecological Knowledge: learning from indigenous practices for environmental sustainability*. Cambridge University Press, Cambridge, UK.

Nelson, V.C. (2011) *Introduction to Renewable Energy*. CRC Press, Boca Raton, FL, USA.

Neteler, M. and Mitasova, H. (Eds) (2008) *Open Source GIS: a GRASS GIS approach*. Springer, Boston, MA, USA.

Newall, P. and Timmons Roberts, J. (Eds) (2017) *The Globalization and Environment Reader*. Wiley-Blackwell, Oxford, UK.

Newby, H. (1990) *Environmental change and the social sciences*. Paper to 1990 Annual Meeting of the British Association for the Advancement of Science, 22 pp.

Newell, P. (2012) *Globalization and the Environment: capitalism, ecology and power*. Polity Press, Cambridge, UK.

Newell, P. (2020) *Global Green Politics*. Cambridge University Press, Cambridge, UK.

Newell, P. and Timmons Roberts, J. (Eds) (2016) *The Globalization and Environment Reader*. Wiley, Oxford, UK.

Newitz, A. (2013) *Scatter, Adapt and Remember: how humans will survive a mass extinction*. DoubleDay, New York, NY, USA.

Newman, P. and Jennings, I. (2008) *Cities as Sustainable Ecosystems: principles and practices*. Island Press, Washington, DC, USA.

Newton, D.E. (2014) *GMO Food: a reference handbook*. ABC-CLIO, Santa Barbara, CA, USA.

Newton, D.E. (2021) *GMO Food: a reference handbook* 2nd edn. ABC-CLIO, Santa Barbara, CA, USA.

Ngee-Choon Chia, and Sock-Yong Phang (2006) Motor vehicle taxes as an environmental management instrument: the case of Singapore. In A. Muller and T. Sterner (Eds), *Environmental Taxation in Practice* (Chapter 22). Routledge, Abingdon, UK.

Nguyen, A.T. (2010) *GIS-Based Sustainable Landscape Planning and Management: theory, case studies, methodology and pilot project*. VDM Verlag, Saarbrücken, GR.

Nguyen Hoang Tien, Phan Phung Phu, Nguyen Tien Phuc, Le Doa Minh Duc and Tran Duy Thuc (2020) Sustainable development and environmental management in Vietnam. *International Journal of Research in Finance and Management* 3(1), 72–79.

Nhamo, G. and Inyang, E. (2011) *Framework and Tools for Environmental Management in Africa*. African Books Collective, Oxford, UK and CODESRIA, Dakar, SN.

Niaounakis, M. (2017) *Management of Marine Plastic Debris*. Elsevier, Oxford, UK.

Nicastro, R. and Carillo, P. (2021) Food loss and waste prevention strategies from farm to fork. *Sustainability* 13, article No. 5443 (23 pp.).

Nielsen, S.N. (2020) *Sustainable Development Indicators: an exergy-based approach*. CRC Press, Boca Raton, FL, USA.

Nijkamp, P., Poof, J. and Sahin, M. (2012) *Migration Impact Assessment: new horizons*. Edward Elgar, Cheltenham, UK.

Nikolakis, W. and Innes, J.L. (Eds) (2020) *The Wicked Problem of Forest Policy: a multidisciplinary approach to sustainability in forest landscapes*. Cambridge University Press, Cambridge, UK.

Nikolaou, I.E. and Evangelinos, K.I. (2010) A SWOT analysis of environmental management practices in the Greek mining and mineral industry. *Resources Policy* 35, 226–234.

Ninan, K.N. (Ed) (2014) *Valuing Ecosystem Services: methodological issues and case studies*. Edward Elgar, Cheltenham, UK.

Ninan, K.N. (Ed) (2020) *Environmental Assessments Scenarios, Modelling and Policy*. Edward Elgar, Cheltenham, UK.

Ninan, K.N. and Inoue, M. (Eds) (2017) *Building a Climate Resilient Economy and Society: challenges and opportunities*. Edward Elgar, Cheltenham, UK.

Nissen, B. (Ed) (2002) *Unions in a Globalized Environment: changing borders, organizational boundaries and social roles*. M.E. Sharp Inc., Armonk, NY, USA.

Nitivattananon, V. and Noonin, C. (2008) Thailand urban environmental management: case of environmental infrastructure and housing provision in the Bangkok Metropolitan Region. In: T. Kidokoro, J. Okata, S. Matsumura and N. Shima (Eds), *Vulnerable Cities: realities, innovations and strategies* (pp. 187–208; cSUR-UT Series: Library for Sustainable Urban Regeneration). Springer, Tokyo, JP.

Njoku, E.G. (Ed) (2014) *Encyclopedia of Remote Sensing*. Springer, New York, NY, USA.

Noble, B.F. (2020) *An Introduction to Environmental Assessment*. Oxford University Press, Oxford, UK.

Nobukazu Nakagoshi and Sun-Kee Hong (Eds) (2018) *Landscape Ecology for Sustainable Society*. Springer, Cham, CH.

Nolan, D. (Ed) (2020) *Environmental and Resource Management Law* 7th edn. LexisNexis, New York, NY, USA.

Nolt, J. (2015) *Environmental Ethics for the Long Term: an introduction*. Routledge, Abingdon, UK.

Nomani, M.Z.M. (2015) Environmental security and sustainability in West Asia. In A. Rahman (Ed), *West Asia In Transition: issues, perspectives & global concerns* (pp. 234–252). Academic Excellence, Delhi, IN.

NRC (2010) *Defending Planet Earth: near-Earth-object surveys and hazard mitigation strategies*. National Academy of Sciences, Washington, DC, USA.

Nugent, C. (2017) *Asteroid Hunters*. Simon & Schuster, New York, NY, USA.

Nuguti, E. (2015) *Project Monitoring and Evaluation: introduction and the logical framework approach*. Ekon Publishers, Doetinchem, NL.

Nunan, F. (Ed) (2019) *Governing Renewable Natural Resources: theories and frameworks*. Routledge (Earthscan), Abingdon, UK.

Núñez, J.E. (2017) *Sovereignty Conflicts and International Law and Politics: a distributive justice issue*. Routledge, Abingdon, UK.

Nüsser, M. (Ed) (2014) *Large Dams in Asia: contested environments between technological hydroscapes and social resistance*. Springer, Dordrecht, NL.

Nyborg, K. (2012) *The Ethics and Politics of Environmental Cost-Benefit Analysis*. Routledge, Abingdon, UK.

O'Brian, R. and Williams, M. (2020) *Global Political Economy: evolution and dynamics* 6th edn. Bloomsbury, London, UK.

O'Bryan, K. (2018) *Indigenous Rights and Water Resource Management: not just another stakeholder*. Routledge, Abingdon, UK.

O'Donnel, E. (2019) *Legal Rights for Rivers: competition, collaboration and water governance*. Routledge, Abingdon, UK.

O'Donnell, A. and Wodon, Q. (Eds) (2015) *Climate Change Adaptation and Social Resilience in the Sundarbans*. Routledge (Earthscan), Abingdon, UK.

O'Faircheallaigh, C. and Ali, S. (Eds) (2008) *Earth Matters: indigenous peoples, the extractive industries and corporate social responsibility*. Taylor & Francis (Greenleaf), London, UK.

O'Higgins, T. and Al-Kalbani, M.S. (2015) *A Systems Approach to Environmental Management: it's not easy being green*. Dunedin Academic Press, Dunedin, NZ.

O'Sills, E., Atmadja, S.S., de Sassi, C., Duchelle, A.E., Kweka, D.L., Resosudarmo, A.A.P. and Sunderlin W.D. (Eds) (2014) *REDD+ on the Ground: a case book of subnational initiatives across the globe*. Center for International Forestry Research, Bogor Barat, ID.

Oba, G. (2021) *African Environmental Crisis: a history of science for development*. Routledge, Abingdon, UK.

Obermann, M. (Ed) (2014) *Sustainable Agroecosystems in Climate Change Mitigation*. Wageningen Academic Publishers, Wageningen, NL.

Oberthur, S. and Ott, H.E. (1999) *The Kyoto Protocol: international climate policy for the 21st century* (English edn). Springer-Verlag, Berlin, GR.

Odenwald, S. (2015) *Solar Storms: 2000 years of human calamity*. CreateSpace Independent Publishing Platform, Scotts Valley, CA, USA.

Odum, E.P. (1975) *Ecology: the link between the natural and the social sciences* 2nd edn. Holt, Rinehart and Winston, London, UK.

Odum, E.P. (1983) *Systems Ecology: an introduction*. Wiley, New York, NY, USA.

OECD (1991) *Environmental Indicators*. Organization for Economic Cooperation and Development, Paris, FR.

OECD (2012) *Strategic Environmental Assessment in Development Practice: a review of recent experience*. OECD, Paris, FR.

OECD (2018) *Cost-Benefit Analysis and the Environment: further developments and policy use*. OECD, Paris, FR.

Oelbermann, M. (Ed) (2014) *Sustainable Agroecosystems in Climate Change Mitigation*. Wageningen Academic Publishers, Wageningen, NL.

Ogaji, J. (2019) *Alternative Dispute Resolution in the Niger Delta*. Lambert Academic Publishers, Saarbrücken, GR.

Ogden, L., Heynen, N., Oslender, U., West, P. Kassam, K.-A. and Robbins, P. (Eds) (2013) Global assemblages, resilience, and Earth stewardship in the Anthropocene. *Frontiers in Ecology and the Environment* 11(7), 341–347.

Oguge, N., Ayal, D., Adeleke, L. and da Silva, I. (Eds) (2021) *African Handbook of Climate Change Adaptation*. Springer, Cham, CH.

Ojha, C.S.P. (Ed) (2017) *Sustainable Water Resources Management*. American Society of Civil Engineers, Reston VA, USA.

Okereke, C. (2010) *Global Justice and Neoliberal Environmental Governance: ethics, sustainable development and international co-operation (environmental politics)*. Taylor & Francis (Routledge), London, UK.

Oliver-Smith, A. and Shen Xiaomeng (2009) *Linking Environmental Change, Migration & Social Vulnerability*. United Nations University, ENH, Tokyo, JP.

Olla, P. (Ed) (2009) *Space Technologies for the Benefit of Human Society and Earth*. Springer, Cham, CH.

Olson, S.S. (2019) *International Environmental Standards Handbook*. CRC Press, Boca Raton, FL, USA.

Olsson, E.G.A. and Gooch, P. (Eds) (2019) *Natural Resource Conflicts and Sustainable Development*. Routledge (Earthscan), Abingdon, UK.

Ondrasek, G. (Ed) (2020) *Drought: detection and solutions*. IntechOpen, London, UK.

Orr, S.K. (2013) *Environmental Policymaking and Stakeholder Collaboration: theory and practice*. Routledge, Abingdon, UK.

Orsini, A. and Morin, J.-F. (Eds) (2021) *Essential Concepts of Global Environmental Governance* 2nd edn. Routledge, Abingdon, UK.

Osborn, F. (1948) *Our Plundered Planet*. Little, Brown and Co., Boston, MA, USA.

Osborn, F. (1953) *Limits of the Earth*. Little, Brown and Co., Boston, MA, USA.

Osório, K.F. (2020) *Financial Environmental Management: the secret to business growth*. Alcance, Mid Valley, CA, USA.

Ossola, A., Schifman, L., Herrmann, D.L., Garmestani, A.S., Schwarz, K. and Hopton, M.E. (2018) The provision of urban ecosystem services throughout the private-social-public domain: a conceptual framework. *Cities Environment* 11(1), 1–15.

Ostrom, E. (2015) *Governing the Commons: the evolution of institutions for collective action*. Cambridge University Press, Cambridge, UK.

Ottman, J. (2017) *The New Rules of Green Marketing: strategies, tools, and inspiration for sustainable branding* (first published 2011). Routledge (Greenleaf), Abingdon, UK.

Oulton, W. (Ed) (2009) *Investment Opportunities for a Low-Carbon World*. Indiana University Press, Bloomington, IN, USA.

OXFAM International (2007) *Adapting to Climate Change: what's needed in poor countries and who should pay*. Oxfam Briefing Paper No. 104. OXFAM International Secretariat, Oxford, UK.

Ozge, E. and Yalciner, O. (Eds) (2012) *Green and Ecological Technologies for Urban Planning: creating smart cities*. IGI Global, Hershey, PA, USA.

Pacey, A. and Bray, F. (2021) *Technology in World Civilization: a thousand-year history* (revised edn). MIT Press, Cambridge MA, USA.

Paddock, L.C., Markell, D.L., Goldstein, S.M. and Bryner, N.S. (Eds) (2017) *Compliance and Enforcement of Environmental Law*. Edward Elgar, Cheltenham, UK.

Padilla, L.-A. (2021) *Sustainable Development in the Anthropocene: towards a new holistic and cosmopolitan paradigm*. Springer, Cham, CH.

Paegelow, M. and Olmedo, M.T.C. (Eds) (2008) *Modelling Environmental Dynamics: advances in geomatic solutions*. Springer-Verlag, Berlin, GR.

Pain, S.W. (2018) *Safety, Health and Environmental Auditing: a practical guide* 2nd edn. CRC Press, Boca Raton, FL, USA.

Paksoy, T., Huber, S. and Huber, S. (Eds) (2018) *Lean and Green Supply Chain Management: optimization models and algorithms*. Springer, Cham, CH.

Pal, K. (Ed) (2022) *Green Nanomaterials: sustainable technologies and applications*. CRC Press (Apple), Boca Raton, FL, USA.

Palinkas, L.A. (2020) *Global Climate Change, Population Displacement, and Public Health: the next wave of migration*. Springer, Cham, CH.

Paloviita, A. and Järvelä, M. (Eds) (2017) *Climate Change Adaptation and Food Supply Chain Management*. Routledge, Abingdon, UK.

Pandey, A., Larroche, C., Gnansounou, E., Khanal, S.K., Dussap, C.-G. and Ricke, S.C. (Eds) (2019) *Biomass, Biofuels, Biochemicals: alternative feedstocks and conversion processes for the production of liquid and gaseous biofuels* 2nd edn. Elsevier (Academic Press), London, UK.

Pandy, A., Tyagi, R.D. and Varjani, S. (Eds) (2021) *Biomass, Biofuels, Biochemicals: circular bioeconomy-current developments and future outlook.* Elsevier, Amsterdam, NL.

Pant, D., Nadda, A.K., Pant, K.K. and Agarwal, A.K. (Eds) (2021) *Advances in Carbon Capture and Utilization.* Springer Nature, Cham, CH.

Panya, N., Poboon, C., Phoochinda, W. and Teungfun, R. (2018) The performance of the environmental management of local governments in Thailand. *Kasetsart Journal of Social Sciences* 39(1), 33–41.

Papale, P. (Ed) (2015) *Volcanic Hazards, Risks and Disasters.* Elsevier, Amsterdam, NL.

Papantoniou, A. and Fitzmaurice, M. (Eds) (2017) *Multilateral Environmental Treaties.* Edward Elgar, Cheltenham, UK.

Pariatamby, A., Hamid, F.S. and Bhatti, M.S. (Eds) (2019) *Sustainable Waste Management Challenges in Developing Countries.* IGI Global, Hershey, PA, USA.

Park, P. (2013) *International Law for Energy and the Environment* 2nd edn. CRC Press, Boca Raton, FL, USA.

Park, S. and Kramarz, T. (Eds) (2019) *Global Environmental Governance and the Accountability Trap.* MIT Press, Cambridge, MA, USA.

Parkinson, C.L. (2010) *Coming Climate Crisis? Consider the past, beware the big fix.* Rowman & Littlefield, Lanham, MA, USA.

Parris, K.M. (2016) *Ecology of Urban Environments.* Wiley-Blackwell, Oxford, UK.

Parry, I.W.H., Keen, M. and de Mooij, R.A. (2012) *Fiscal Policy to Mitigate Climate Change: a guide for policymakers.* International Monetary Fund, Washington, DC, USA.

Parsons, M., Fisher, K. and Crease, R.P. (2021) *Decolonising Blue Spaces in the Anthropocene: freshwater management in Aotearoa, New Zealand.* Springer Nature (Palgrave Macmillan), Cham, CH.

Passarini, F. and Ciacci, L. (Eds) (2021) *Life Cycle Assessment (LCA) of Environmental and Energy Systems.* MDPI, Basel, CH.

Pasteur, K. (2011) *From Vulnerability to Resilience: a framework for analysis and action to build community resilience.* Practical Action Publishing, Rugby, UK.

Pathak, H. (2015) *A Handbook of Environmental Audit.* CreateSpace Independent Publishing Platform, Scotts Valley, CA, USA.

Paton, D. (Ed) (2015) *Wildfire Hazards, Risks, and Disasters.* Elsevier, Amsterdam, NL.

Pattberg, P.H. and Zelli, F. (Eds) (2015) *Encyclopedia of Global Environmental Governance and Politics.* Edward Elgar, Cheltenham, UK.

Pattberg, P.H. and Zelli, F. (Eds) (2016) *Environmental Politics and Governance in the Anthropocene: institutions and legitimacy in a complex world.* Routledge, Abingdon, UK.

Patton, M.Q. (2019) *Blue Marble Evaluation: premises and principles.* Guilford Press, New York, NY, USA.

Pawłowski, L., Litwińczuk, Z. and Guomo Zhou (Eds) (2020) *The Role of Agriculture in Climate Change Mitigation.* CRC Press, Boca Raton, FL, USA.

Pearce, D.W., Atkinson, G. and Mourato, S. (2006) *Cost-Benefit Analysis and the Environment: recent developments.* OECD, Paris, FR.

Pearce, F. (2015) *The New Wild: why invasive species will be nature's salvation.* Beacon Press, Boston, MA, USA.

Pearce, F. (2019) *When the Rivers Run Dry: the global water crisis and how to solve it.* Granta Books, London, UK.

Pearce, F. (2021) *A Trillion Trees: how we can reforest our world.* Granta Books, London, UK.

Pearse, G. (2012) *Greenwash: big brands and carbon scams.* Schwartz Media Pty., Collingwood, VI, AU.

Peattie, K. (1995) *Environmental Marketing Management: meeting the green challenge.* Pitman, London, UK.

Peden, D., Awulachew, S.B., Molden, D. and Smahktin, V. (Eds) (2012) *The Nile River Basin: water, agriculture, governance and livelihoods.* Routledge, London, UK.

Pedersen, J.S. and Wilkinson, A. (Eds) (2019) *Big Data: promise, application and pitfalls.* Edward Elgar, Cheltenham, UK.

Peet, R., Robbins, P. and Watts, M. (Eds) (2011) *Global Political Ecology.* Routledge, New York, NY, USA.

Pekmezovic, A., Walker, G. and Walker, J. (Eds) (2019) *Sustainable Development Goals: harnessing business to achieve the SDGs through finance, technology and law reform.* Wiley, Chichester, UK.

Pelling, M. (2011) *Adaptation to Climate Change: from resilience to transformation.* Routledge, Abingdon, UK.

Pender, G. and Faulkner, H. (Eds) (2011) *Flood Risk Science and Management.* Wiley-Blackwell, Chichester, UK.

Pentreath, R.J. (2000) Strategic environmental management: time for a new approach. *Science of the Total Environment* 249(1–3): 3–11.

Pepper, D. (1996) *Modern Environmentalism: an introduction.* Routledge, London, UK.

Perdew, L. (2018) *The California Drought.* ABDO Publishing Company, North Mankato, MN, USA.

Perdicoúlis, A., Durning, B. and Palframan, L. (Eds) (2012) *Furthering Environmental Impact Assessment: towards a seamless connection between EIA and EMS.* Edward Elgar, Cheltenham, UK.

Pereira, H.M. and Navarro, L.M. (Eds) (2015) *Rewilding European Landscapes.* Springer, Cham, CH.

Pereira, P. (Ed) (2019) *Soil Degradation, Restoration and Management in a Global Change Context.* Elsevier (Academic Press), Cambridge, MA, USA.

Perkins, P.E. (1995) An overview of international institutional mechanisms for environmental management with reference to Arctic pollution. *Science of the Total Environment* 161(9), 849–857.

Perman, R., Ma, Yue and McGilvray, J. (1996) *Natural Resource and Environmental Economics.* Longman, Harlow, UK.

Pernetta, J. (1993) *Monitoring Coral Reefs for Global Change.* IUCN, Cambridge, UK.

Perreault, T., Bridge, G. and McCarthy, J. (Eds) (2015) *The Routledge Handbook of Political Ecology.* Routledge, London, UK.

Perrings, C. (2014) *Our Uncommon Heritage: biodiversity change, ecosystem services, and human wellbeing.* Cambridge University Press, Cambridge, UK.

Perz, S.G. (Ed) (2019) *Collaboration Across Boundaries for Social-Ecological Systems Science: experiences around the world.* Springer (Palgrave Macmillan), Cham, CH.

Peters, A., Koechlin, L., Förster, T. and Zinkernagel, G.F. (Eds) (2009) *Non-State Actors as Standard Setters.* Cambridge University Press, Cambridge, UK.

Peters, R. (Ed) (2016) *Natural Hazard Preparedness and Mitigation.* Syrawood Publishing House, New York, NY, USA.

Peterson, T.R. and Clarke, T. (2016) *Environmental Conflict Management.* Sage, Los Angeles, CA, USA.

Petorelli, N. (2019) *Satellite Remote Sensing and the Management of Natural Resources.* Oxford University Press, Oxford, UK.

Petropoulos, G. and Srivastava, P.K. (Eds) (2021) *GPS and GNSS Technology in Geosciences.* Elsevier, Amsterdam, NL.

Pettorelli, N., Durant, S.M. and du Toit, J.T. (Eds) (2019) *Rewilding.* Cambridge University Press, Cambridge, UK.

Pham, P.H. and Doane, D.L. (2021) *Climate Change, Gender Roles and Hierarchies: socio-economic transformation in an ethnic minority community in Viet Nam.* Routledge, Abingdon, UK.

Philander, S.G. (Ed) (2008) *Encyclopedia of Global Warming and Climate Change.* Sage, Los Angeles, CA, USA.

Philips, A. (2013) *Designing Urban Agriculture: a complete guide to the planning, design, construction, maintenance and management of edible landscapes.* Wiley, Hoboken, NJ, USA.

Phillips, B.D., Thomas, D.S.K., Fothergill, A. and Blinn-Pike, L. (Eds) (2010) *Social Vulnerability to Disasters.* CRC Press, Boca Raton, FL, USA.

Phillips, M. and Barinaga, E. (2021) *Climate Adaptation: accounts of resilience, self-sufficiency and systems change.* Arkbound, Bristol, UK.

Phillips, M. and Rumens, N. (Eds) (2015) *Contemporary Perspectives on Ecofeminism.* Routledge (Earthscan), London, UK.

Philpott, C. (2011) *Green Spirituality: one answer to global environmental problems and world poverty.* AuthorHouse, Milton Keynes, UK.

Piccolo, A. (Ed) (2012) *Carbon Sequestration in Agricultural Soils: a multidisciplinary approach to innovative methods.* Springer, Heidelberg, GR.

Pichon, L. (1993) *Environmental Management for Hotels: the industry guide to best practice.* Butterworth-Heinemann, Oxford, UK.

Pickard, D. (2022) *Urban Agriculture for Improving the Quality of Life: examples from Bulgaria.* Springer, Cham, CH.

Pielke, R.A. (Ed) (2013) *Climate Vulnerability: understanding and addressing threats to essential resources* (4 vols). Elsevier (Academic Press), London, UK.

Piera, J. and Ceccaroni, L. (Eds) (2016) *Analyzing the Role of Citizen Science in Modern Research.* IGI Global, Hershey, PA, USA.

Pierre-Louis, K. (2012) *Green Washed: why we can't buy our way to a green planet.* Ig Pub, New York, NY, USA.

Pigou, A.C. (1920) *The Economics of Welfare* 1st edn. Macmillan, London, UK.

Piguet, E., Pécoud, A. and de Guchteneire, P. (Eds) (2011) *Migration and Climate Change.* Cambridge University Press/UNESCO Publishing, New York, NY, USA.

Piguet, E. and Laczko, F. (Eds) (2013) *People on the Move in a Changing Climate: the regional impact of environmental change on migration.* Springer, Dordrecht, NL.

Pilkey, K.C. and Pilkey, O.H. (2019) *Sea Level Rise: a slow tsunami on America's shores.* Duke University Press, Durham, NC, USA.

Pilkey, O.H., Pilkey-Jarvis, L. and Pilkey, K.C. (2016) *Retreat from a Rising Sea: hard choices in an age of climate change.* Columbia University Press, New York, NY, USA.

Pindyck, R.S. (2022) *Climate Future: averting and adapting to climate change.* Oxford University Press, Oxford, UK.

Pinto, G.M.C., Pedroso, B., Moraes, J., Pilatti, L.A. and Picinin, C.T. (2018) Environmental management practices in industries of Brazil, Russia, India, China and South Africa (BRICS) from 2011 to 2015. *Journal of Cleaner Production,* 198, 1251–1261.

Pirlot, A. (2017) *Environmental Border Tax Adjustments and International Trade Law: fostering environmental protection.* Edward Elgar, Cheltenham, UK.

Pittock, J., Meng, J., Geiger, M. and Chapagain, A.K. (Eds) (2009) *Interbasin Water Transfers and Water Scarcity in a Changing World – a solution or pipe dream?* WWF Germany,

Frankfurt, GR. 61 pp. https://assets.panda.org/downloads/pipedreams18082009.pdf (accessed 12/02/22).

Pogue, D. (2021) *How to Prepare for Climate Change: a practical guide to surviving the chaos*. Simon & Schuster, New York, NY, USA.

Pollak, H.N. (2010) *A World Without Ice*. Avery Hill, London, UK.

Poonia, R.C., Gao, Xiao-Zhi, Linesh, R., Sugam, S. and Sonali, V. (Eds) (2019) *Smart Farming Technologies for Sustainable Agricultural Development*. IGI Global, Hershey, PA, USA.

Popp, J.S., Jahn, M.M., Matlock, M.D. and Kemoer, N.P. (Eds) (2012) *The Role of Biotechnology in a Sustainable Food Supply*. Cambridge University Press, Cambridge, UK.

Porter, A.L., Cunningham, S.W., Banks, J., Roper, A.T., Mason, T.W. and Rossini, F.A. (2011) *Forecasting and Management of Technology* 2nd edn. Wiley, Hoboken, NJ, USA.

Porter-Bolland, L., Ruiz-Mallén, I., Camacho-Benavides, C. and McCandless, S.R. (Eds) (2013) *Community Action for Conservation: Mexican experiences*. Springer, New York, NY, USA.

Pörtner, H.-O., Roberts, D.C., Tignor, M., Poloczanska, E.S., Mintenbeck, K., Alegría, A., Craig, M., Langsdorf, S., Löschke, S., Möller, V., Okem, A. and Rama, B. (Eds.) (2022) *Climate Change 2022: impacts, adaptation, and vulnerability* (Contribution of Working Group II to the Sixth Assessment Report of the Intergovernmental Panel on Climate Change). Cambridge University Press, Cambridge, UK.

Postel, S. (1999) *Pillar of Sand: can the irrigation miracle last?* W.W. Norton, New York, NY, USA.

Potschin, M., Haines-Young, R.H., Fish, R. and Turner, K. (Eds) (2016) *Routledge Handbook of Ecosystem Services*. Routledge (Earthscan), London, UK.

Pourmokhtari, N. (2021) *Iran's Green Movement: everyday resistance, political contestation and social mobilization*. Taylor & Francis, Abingdon, UK.

Powe, N.A. (2007) *Redesigning Environmental Valuation: mixing methods within stated preference techniques*. Edward Elgar, Cheltenham, UK.

Prasad, A. (2018) *Environmental Performance Auditing in the Public Sector: enabling sustainable development*. Routledge, Abingdon, UK.

Prasad, M.N.V. and Vithanage, M. (Eds) (2019) *Electronic Waste Management and Treatment Technology*. Elsevier (Butterworth-Heinemann), Oxford, UK.

Preeg, E.H. (2008) *India and China: an advanced technology race and how the United States should respond*. Manufacturers Alliance/MAPI, Arlington, VA, USA.

Prell, C., Hubacek, K. and Reed, M. (2009) Stakeholder analysis and social network analysis in natural resource management. *Society & Natural Resources* 22(6), 501–518.

Pretty, J. and Bharucha, Z.P. (2018) *Sustainable Intensification of Agriculture: greening the world's food economy*. Routledge, London, UK.

Price, T.J. (2014) *Environmental Management Systems: how to boost your organization's environmental performance*. CreateSpace Independent Publishing Platform, Scotts Valley, CA, USA.

Price, T.J. (2016) *Environmental Management Systems: an easy to use guide to boosting your organization's environmental performance* 2nd edn. CreateSpace Independent Publishing Platform, Scotts Valley, CA, USA.

Primrose, S.B. (2020) *Biomimetics: nature-inspired design and innovation*. Wiley, Oxford, UK.

Pritchard, S.B. and Zimring, C.A. (2020) *Technology and the Environment in History*. Johns Hopkins University Press, Baltimore, MD, USA.

Pritwani, K. (2016) *Sustainability of Business in the Context of Environmental Management.* TERI Press, New Delhi, IN.

Proux, A. (2022) *Fen, Bog and Swamp: a short history of peatland destruction and its role in the climate crisis.* Scribner, New York, NY, USA.

Prud' Homme, A. (2011) *The Ripple Effect: the fate of fresh water in the twenty-first century.* Scribner, New York, NY, USA.

Pryde, R.R. (1991) *Environmental Management in the Soviet Union.* Cambridge University Press, Cambridge, UK.

Publications Office of the European Union (2020) *Nature-Based Solutions for Climate Mitigation: analysis of EU-funded projects.* Publications Office of the European Union, Brussels, BE.

Purdy, J. (2015) *After Nature: a politics for the Anthropocene.* Harvard University Press, Cambridge, MA, USA.

Purseglove, J. (2020) *Working With Nature: saving and using the world's wild places.* Profile Books, London, UK.

Pyne, S.J. (2021) *The Pyrocene: how we created an age of fire, and what happens next.* University of California Press, Oakland, CA, USA.

Quante, M. and Colijn, F. (Eds) (2016) *North Sea Region Climate Change Assessment.* Springer, Cham, CH.

Quoquab, F., Thurasamy, R. and Mohammad, J. (Eds) (2017) *Driving Green Consumerism through Strategic Sustainability Marketing.* IGl Global, Hershey, PA, USA.

Rabe, B.G. (2018) *Can We Price Carbon?* MIT Press, Cambridge, MS, USA.

Rackley, S.A. (2017) *Carbon Capture and Storage* 2nd edn. Elsevier (Butterworth-Heinemann), Oxford, UK.

Rackley, S.A., Sewel, A., Clery, D., Dowson, G., Styring, P., Andrews, G., McCord, S., Knops, P., de Richter, R., Ming, Tingzhen, Li, Wei and Tyka, M. (2023) *Negative Emissions Technologies for Climate Change Mitigation.* Elsevier, Amsterdam, NL.

Radkau, J. (2012) *Wood: a history* (first published 2007). Polity Press, Cambridge, UK.

Raga, J. (2017) *Environmental Management for Hotels: a comprehensive guide for sustainable operation.* Society Publishing, Burlington, ON, Canada.

Rahimpour, M.R., Farsi, M. and Makarem, M.A. (Eds) (2020) *Advances in Carbon Capture: methods, technologies and applications.* Elsevier (Woodhead Publishing), Oxford, UK.

Rai, P.K., Mishra, V.N. and Singh, P. (Eds) (2022) *Geospatial Technology for Landscape and Environmental Management: sustainable assessment and planning.* Springer, Singapore, SG.

Rai, P.K., Singh, P. and Mishra, N. (Eds) (2021) *Recent Technologies for Disaster Management and Risk Reduction: sustainable community resilience & responses.* Springer, Cham, CH.

Rai, S.C. (2012) *Ecotourism and Biodiversity Conservation.* Nova Science Publishers, Hauppauge, NY, USA.

Rajaram, V., Siddiqui, F.Z., Agrawal, S. and Khan, M.E. (2016) *Solid and Liquid Waste Management: waste to wealth.* PHI Learning, Delhi, IN.

Ralebitso-Senior, T.K. and Orr, C.H. (Eds) (2016) *Biochar Application: essential soil microbial ecology.* Elsevier, Amsterdam, NL.

Ramachandran, B., Justice, C.O. and Abrams, M.J. (Eds) (2011) *Land Remote Sensing and Global Environmental Change: NASA's Earth observing system and the science of ASTER and MODIS.* Springer, New York, NY, USA.

Raman, N.S., Gajbhiye, A.R. and Khandeshwar, S.R. (2014) *Environmental Impact Assessment.* I K International Publishing House, New Delhi, IN.

Ramanathan, A.L., Sabarathinam, C., Arriola, F., Prasanna, M.V., Kumar, P. and Jonathan, M.P. (Eds) (2021) *Environmental Resilience and Transformation in Times of Covid-19: climate change effects on environmental functionality*. Elsevier, Amsterdam, NL.

Ramiah, V. and Gregoriou, G.N. (Eds) (2015) *Handbook of Environmental and Sustainable Finance*. Academic Press, Cambridge, MA, USA.

Ramkumar, M., Kumaraswamy, K. and Mohanraj, R. (Eds) (2015) *Environmental Management of River Basin Ecosystems*. Springer, Cham, CH.

Ramkumar, M., Kumaraswamy, K. and Mohanraj, R. (Eds) (2016) *Environmental Management of River Basin Ecosystems*. Springer, Cham, CH.

Ramkumar, M., James, A., Menier, D. and Kumaraswamy K. (Eds) (2018) *Coastal Zone Management: global perspectives, regional processes, local issues*. Elsevier, Amsterdam, NL.

Ramlogan, R. (1996) Environmental refugees: a review. *Environmental Conservation* 23(1), 81–88.

Rampino, M.R. (2017) *Cataclysms: a new geology for the twenty-first century*. Columbia University Press, New York, NY, USA.

Ramutsindela, M., Spierenburg, M. and Wels, H. (2017) *Sponsoring Nature: environmental philanthropy for conservation* 2nd edn. Routledge (Earthscan), Abingdon, UK.

Rankin, A. (2020) *Jainism and Environmental Politics*. Routledge, Abingdon, UK.

Rapiah, M., Zuriana, C. and Jamil, M. (2020) The influence of environmental management accounting practices on environmental performance in small-medium manufacturing in Malaysia. *International Journal of Environment and Sustainable Development* 19(4), 378–392.

Rathi, A.K.A. (2021) *Handbook of Environmental Impact Assessment: concepts and practice*. Cambridge Scholars Publishing, Newcastle-upon-Tyne, UK.

Rathore, H.S. and Nollet, L.M.L. (Eds) (2012) *Pesticides: evaluation of environmental pollution*. CRC Press, Boca Raton, FL, USA.

Rathoure, A.K. (Ed) (2020) *Zero Waste: management practices for environmental sustainability*. CRC Press, Boca Raton, FL, USA.

Räthzel, N. and Uzzel, D. (Eds) (2013) *Trade Unions in the Green Economy: working for the environment*. Routledge (Earthscan), London, UK.

Rawhouser, H., Cummings, M. and Newbert, S.L. (2019) Social impact measurement: current approaches and future directions for social entrepreneurship research. *Entrepreneurship Theory and Practice* 43(1), 82–115.

Reason, P. and Bradbury, H. (Eds) (2008) *The Sage Handbook of Action Research: participative inquiry and practice*. Sage, London, UK.

Reay, D., Smith, P. and van Amstel, A. (Eds) (2010) *Methane and Climate Change*. Routledge (Earthscan), Abingdon, UK.

Redclift, M.E. (1987) *Sustainable Development: exploring the contradictions*. Methuen, London, UK.

Reddy, M.A. (2016) *Geoinformatics for Environmental Management*. BS Publications, Petersfield, UK.

Reddy, M.V. and Wilkes, K. (Eds) (2015) *Tourism in the Green Economy*. Routledge (Earthscan), Abingdon, UK.

Reddy, P.P. (2015) *Climate Resilient Agriculture for Ensuring Food Security*. Springer New Delhi, IN.

Reddy, R., Kurian, M. and Ardakanian, R. (2015) *Life-Cycle Cost Approach for Management of Environmental Resources: a primer*. Springer, Cham, CH.

Reddy, V.R., Syme, G. and Tallapragada, C. (2019) *Integrated Approaches to Sustainable Watershed Management in Xeric Environments: a training manual.* Elsevier, New York, NY, USA.

Rees, G. (2008) *The Remote Sensing Data Book.* Cambridge University Press, Cambridge, UK.

Rees, M. (2018) *On the Future: prospects for humanity.* Princeton University Press, Princeton, NJ, USA.

Regis, E. (2019) *Golden Rice: the imperilled birth of a GMO superfood.* Johns Hopkins University Press, Baltimore, MD, USA.

Rego, F.C., Bunting, S.C., Strand, E.K. and Godinho-Ferreira, P. (2019) *Applied Landscape Ecology.* Wiley, Oxford, UK.

Reichental, J. (2020) *Smart Cities for Dummies.* Wiley, Hoboken, NJ, USA.

Ren, J. and Toniolo, S. (Eds) (2019) *Life Cycle Sustainability Assessment for Decision-Making: methodologies and case studies.* Elsevier, Amsterdam, NL.

Renn, O. (2010) *Risk Governance: coping with uncertainty in a complex world.* Routledge (Earthscan), Abingdon, UK.

Reynolds, K.M., Hessburg, P.F. and Bourgeron, P.S. (Eds) (2014a). *Decision Support for Environmental Management: applications of the ecosystem management decision support system.* Springer, Cham, CH.

Reynolds, K.M., Hessburg, P.F. and Bourgeron, P.S. (Eds) (2014b) *Making Transparent Environmental Management Decisions: applications of the ecosystem management decision support system.* Springer, Berlin, GR.

Rich, N. (2019) *Losing Earth: the decade we could have stopped climate change.* Picador, London, UK.

Richards, J.-A. and Bradshaw, S. (2017) *Uprooted by Climate Change: responding to the growing risk of displacement.* Oxfam, Oxford, UK.

Richter, M. and Weiland, U. (Eds) (2012) *Applied Urban Ecology: a global framework.* Wiley-Blackwell, Oxford, UK.

Riede, F. and Sheets, P. (Eds) (2020) *Going Forward by Looking Back: archaeological perspectives on socio-ecological crisis, response and collapse.* Berghahn Books, New York, NY, USA.

Rifkin, J. (2011) *The Third Industrial Revolution: how lateral power is transforming energy, the economy and the world.* St Martin's Press, New York, NY, USA.

Rip, A. (2015) Technology assessment. In: J.D. Wright (Ed.), *International Encyclopedia of the Social & Behavioral Sciences* 2nd edn (pp 125–128). Elsevier, Amsterdam, NL.

Rist, L., Felton, A., Samuelsson, L., Sandström, C. and Rosvall, O. (2013) A new paradigm for adaptive management. *Ecology and Society* 18(4), article No. 63 (24 pp.).

Rivera, F.I. and Kapucu, N. (2015) *Disaster Vulnerability, Hazards and Resilience: perspectives from Florida.* Springer, Heidelberg, GR.

Rivera, J.E., Chang Hoon Oh, Oetzal, J. and Clement, V. (2022) *Business Adaptation to Climate Change.* Cambridge University Press, Cambridge, UK.

Roaf, S., Crichton, D. and Nicol, F. (2009) *Adapting Buildings and Cities for Climate Change: a 21st century survival guide* 2nd edn. Elsevier, Oxford, UK.

Robbins, D. and Wennerstein, J.R. (2017) *Rising Tides: climate refugees in the twenty-first century.* Indiana University Press, Bloomington, IN, USA.

Robbins, P. (2020) *Political Ecology: a critical introduction* 3rd edn. Wiley, Hoboken, NJ, USA.

Roberts, J. (2020) Political ecology. In: F. Stein, S. Lazar, M. Candea, H. Diemberger, J. Robbins, A. Sanchez and R. Stasch (Eds), *The Cambridge Encyclopedia of Anthropology* (pp. 1–17). Cambridge University Press, Cambridge UK.

Robinson, E. (2022) *An Integrated Approach to Environmental Management.* States Academic Press, New York, NY, USA.

Robinson, N.A., Wang Xi, Harmon, L. and Wegmueller, S. (Eds) (2013) *Dictionary of Environmental and Climate Change Law.* Edward Elgar, Cheltenham, UK.

Rocha-Santos, T., Costa, M.F. and Mouneyrac, C. (Eds) (2022) *Handbook of Microplastics in the Environment.* Springer, Cham, CH.

Rockström, J. and Gaffney, O. (2021) *Breaking Boundaries: the science of our planet.* Dorling Kindersley, London, UK.

Rockström, J., Steffen, W.L., Noone, K., Persson, Å. and Chapin III, F.S. (2009a) Planetary boundaries: exploring the safe operating space for humanity. *Ecology & Society* 14(2), 32.

Rockström, J., Wani, S.P. and Oweis, T.Y. (Eds) (2009b) *Rainfed Agriculture: unlocking the potential.* CABI, Wallingford, UK.

Rodenbiker, J. (2021) Adapting participatory research methods for reflexive environmental management. *Qualitative Research* 22(4), 559–577.

Rodríguez, L. and Sanchez, T. (2011) *Designing and Building Mini and Micro Hydropower Schemes: a practical guide.* Practical Action Publications, Rugby, UK.

Rodriguez-Bachiller, A. and Glasson, J. (2004) *Expert Systems and Geographic Information Systems for Impact Assessment.* CRC Press, Boca Raton, FL, USA.

Rodgers, P., Straughton, E., Winchester, L. and Pieraccini, M. (Eds) (2011) *Contested Common Land: environmental governance past and present.* Routledge (Earthscan), London, UK.

Rogers, D.T., Bregman, J.I. and Edell, R.D. (2016) *Environmental Compliance Handbook* 2nd edn. CRC Press, Boca Raton, FL, USA.

Rogers, H. (2010) *Green Gone Wrong: how our economy is undermining the environmental revolution.* Scribner, New York, NY, USA.

Roggema, R. (Ed) (2016) *Sustainable Urban Agriculture and Food Planning.* Routledge (Earthscan), Abingdon, UK.

Rolston III, H. (2012) *A New Environmental Ethics: the next millennium for life on Earth.* Routledge, New York, NY, USA.

Roothaan, A. (2019) *Indigenous, Modern and Postcolonial Relations to Nature: negotiating the environment.* Routledge, Abingdon, UK.

Rosales, J. (2020) *Environmental Impact Assessment.* Arcler Education Inc., Burlington, ON, CA.

Rose, A. (2018) *Defining and Measuring Economic Resilience from a Societal, Environmental and Security Perspective.* Springer, Cham, CH.

Rosenberg, M. (2015) *Strategy and Sustainability: a hardnosed and clear-eyed approach to environmental sustainability for business.* Palgrave Macmillan, London, UK.

Rosenzweig, C., Solecki, W.D., Romero-Lankao, P., Mehrotra, S., Dhakal, S. and Ibrahim, S.A. (Eds) (2018) *Climate Change and Cities: second assessment report of the Urban Climate Change Research Network.* Cambridge University Press, Cambridge, UK.

Ross, A., Sherman, K.P., Delcore, H.D., Snodgrass, J.G. and Sherman, D. (2010) *Indigenous Peoples and the Collaborative Stewardship of Nature: knowledge binds and institutional conflicts.* Left Coast Press, Walnut Creek, CA, USA.

Rossini, F. (2020) *Integrated Impact Assessment* (first published 1983). Routledge, Abingdon, UK.

Røste, O.B. (2021) *Norway's Sovereign Wealth Fund: sustainable investment of natural resource revenues.* Springer, Cham, CH.

Rotherham, I.D. and Lambert, R.A. (Eds) (2017) *Invasive and Introduced Plants and Animals: human perceptions, attitudes and approaches to management.* Routledge (Earthscan), London, UK.

Rothschild, R.E. (2019) *Poisonous Skies: acid rain and the globalization of pollution.* University of Chicago Press, Chicago, IL, USA.

Rougier, J., Sparks, S. and Hill, L. (Eds) (2018) *Risk and Uncertainty Assessment for Natural Hazards.* Cambridge University Press, Cambridge, UK.

Rowe, P.G. and Limin Hee (2019) *A City in Blue and Green: the Singapore story.* Springer Open, Singapore, SG.

Rowell, A. and van Zeben, J. (2020) *A Guide to EU Environmental Law.* University of California Press, Berkeley, CA, USA.

Roy, H.E., Pocock, M.J.O., Preston, C.D., Roy, D.B., Savage, J., Tweddle, J.C. and Robinson, L.D. (2012) *Understanding Citizen Science & Environmental Monitoring.* Final Report on behalf of UK-EOF. NERC Centre for Ecology & Hydrology and Natural History Museum, London, UK.

Roy, W.R. (2021) *Radioactive Waste Management in the 21st Century.* World Scientific Publishing, Singapore, SG.

Rudderian, W.F. and Kutzback, J.E. (1991) Plateau uplift and climatic change. *Scientific American* 264(3), 42–50.

Rukhsana, H., Alam, A. and Satpati, L. (Eds) (2021) *Habitat, Ecology and Ekistics: case studies of human-environment interactions in India.* Springer, Basingstoke, UK.

Runyan, C. and D'Odorico, P. (2016) *Global Deforestation.* Cambridge University Press, New York, NY, USA.

Ruokonen, E. and Temmes, A. (2019) The approaches of strategic environmental management used by mining companies in Finland. *Journal of Cleaner Production* 210, 466–476.

Rustamov, R.B. and Samadova, N.E. (2017) *Global Management & Geo-Spatial Information System Applications.* Nova Science Publishers Inc., Hauppauge, NY, USA.

Ruth, M. (2008) *Smart Growth and Climate Change.* Routledge (Earthscan), London, UK.

Ruthenberg, I.-M. (2001) *A Decade of Environmental Management in Chile.* World Bank Environment Department, Washington, DC, USA.

Ryan, J.M. (2020) *Covid-19: social consequences and cultural adaptations.* Routledge, Abingdon, UK.

Ryan, M. (2016) *Human Value, Environmental Ethics and Sustainability: the precautionary ecosystem health principle.* Rowman & Littlefield, London, UK.

Ryder, S., Powlen, K., Laituri, M., Malin, S.A., Sbicca, J. and Stevis, D. (Eds) (2021) *Environmental Justice in the Anthropocene: from (un)just presents to just futures.* Routledge, Abingdon, UK.

Rydin, Y. (2012) *Governing for Sustainable Urban Development.* Routledge (Earthscan), London, UK.

Sabatier, P., Vedlitz, A., Lubell, M., Trachtenberg, Z.M., Focht, W. and Matlock, M.D. (Eds) (2005) *Swimming Upstream: collaborative approaches to watershed management.* MIT Press, Cambridge, MA, USA.

Sachs, J. (2015) *The Age of Sustainable Development.* Columbia University Press, New York, NY, USA.

Sachsman, D.B. and Valenti, J.M. (Eds) (2020) *Routledge Handbook of Environmental Journalism* 1st edn. Routledge, London, UK.

Sadiku, M.N.O. (2022) *Emerging Green Technologies* (first published 2020). CRC Press, Boca Raton, FL, USA.

Sadler, B., Aschemann, R., Dusik, J., Fischer, T., do Rosário Partidário, M. and Verheem, R. (Eds) (2011) *Handbook of Strategic Environmental Assessment*. Routledge (Earthscan), Abingdon, UK.

Sadler, B., Dusik, J., Fischer, T., do Rosário Partidario, M., Verheem, R. and Aschemann, R. (Eds) (2015) *Handbook of Strategic Environmental Assessment*. Routledge, Abingdon, UK.

Sahni, P. (2008) *Environmental Ethics in Buddhism: a virtues approach*. Routledge, London, UK.

Sala, O.E., Meyerson, L.A. and Parmesan, C. (Eds) (2009) *Biodiversity Change and Human Health: from ecosystem services to spread of diseases*. Island Press, Washington, DC, USA.

Sale, K. (2000) *Dwellers in the Land: the bioregional vision*. The University of Georgia Press, Athens, GA, USA.

Salih, M.A. (Ed) (2009) *Climate Change and Sustainable Development: new challenges for poverty reduction*. Edward Elgar, Cheltenham, UK.

Salinger, J., Sivakumar, M.V.K. and Motha, R.P. (Eds) (2005) *Increasing Climate Variability and Change: reducing the vulnerability of agriculture and forestry*. Springer, Dordrecht, NL.

Saljnikov, E., Mueller, L., Lavrishchev, A. and Eulenstein, F. (Eds) (2022) *Advances in Understanding Soil Degradation*. Springer, Cham, CH.

Salomone, R., Clasadonte, M.T., Proto, M. and Raggi, A. (Eds) (2013) *Product-Oriented Environmental Management Systems (POEMS): improving sustainability and competitiveness in the agri-food chain with innovative environmental management tools*. Springer, Dordrecht, NL.

Salpeteur, M., Calvet-Mir, L., Diaz-Revirego, I. and Reyes-Garcia, V. (2017) Networking the environment: social network analysis in environmental management and local ecological knowledge studies. *Ecology and Society* 22(1), article No. 41 (22 pp.).

Samah, M.A.A. and Kamarudin, M.K.A. (Eds) (2022) *Environmental Management and Sustainable Development: case studies and solutions from Malaysia*. Springer, Cham, CH.

Samimia, A.J., Ahmadpour, M. and Ghaderic, S. (2012) Governance and environmental degradation in MENA Region. *Procedia – Social and Behavioral Sciences* 62, 503–507.

Samson, L. (2018) *Epitaph for the Ash: in search of recovery and renewal*. Harper-Collins (4th Estate), London, UK.

Sànchez-Marrè, M., Gibert, K., Sojda, R.S., Steyer, J.P., Struss, P., Rodríguez-Roda, I., Comas, J., Brilhante, V. and Roehl, E.A. (2008) Intelligent environmental decision support systems. In: A.J. Jakeman, A.A. Voinov, A.E. Rizzoli and S.H. Chen (Eds), *Environmental Modelling, Software and Decision Support: state of the art and new perspectives* (pp. 119–144). Elsevier, Amsterdam, NL.

Sánchez-Triana, E. and Ahmed, K. (Eds) (2008) *Strategic Environmental Assessment for Policies: an instrument for good governance*. The World Bank, Washington, DC, USA.

Sand, P.H. (Ed) (2015) *The History and Origin of International Environmental Law*. Edward Elgar, Cheltenham, UK.

Sand, P.H. (Ed) (2019) *International Environmental Agreements*. Edward Elgar, Cheltenham, UK.

Sandbrook, C., Roe, D. and Walpole, M. (Eds) (2013) *Biodiversity Conservation and Poverty Alleviation: exploring the evidence for a link*. Wiley-Blackwell, Chichester, UK.

Sanders, M.H. and Sanders, C.E. (2020) *Nuclear Waste Management Strategies: an international perspective.* Elsevier (Academic Press), London, UK.

Sanderson, H., Hilden, M., Russel, D., Pehna-Lopes, G. and Capriolo, A. (Eds) (2021) *Adapting to Climate Change in Europe: exploring sustainable pathways – from local measures to wider policies.* Elsevier, Amsterdam, NL.

Sandham, L.A., Chabalala, J.J. and Spaling, H.H. (2019) Participatory rural appraisal approaches for public participation in EIA: lessons from South Africa. *Land* 8(10), 1–16.

Sandler, R.L. (2017) *Environmental Ethics: theory in practice.* Oxford University Press, Oxford, UK.

Sands, P.H. and Peel, J. (2012) *Principles of International Environmental Law* 4th edn. Cambridge University Press, Cambridge, UK.

Sang, N. (Ed) (2020) *Modelling Nature-Based Solutions: integrating computational and participatory scenario modelling for environmental management and planning.* Cambridge University Press, Cambridge, UK.

Sanjay, R.K.J., Gnanaseelan, C., Mujumdar, M., Kulkarni, A. and Chakraborty, S. (Eds) (2020) *Assessment of Climate Change Over the Indian Region: a report of the Ministry of Earth Sciences (MoES), Government of India.* Springer Nature, Singapore, SG.

Santos, R. (Ed) (2019) *Geoengineering: counteracting climate change.* Greenhaven Publishing, New York, NY, USA.

Sapountzaki, S. (Ed) (2022) Risk mitigation, vulnerability management, and resilience under disasters. *Sustainability* 14(6), special issue (ISSN 2071-1050).

Saravanan, K. and Sakthinathan, G. (Eds) (2022) *Handbook of Green Engineering Technologies for Sustainable Smart Cities.* CRC Press, Boca Raton, FL, USA.

Sarkar, A., Sensarma, S.R. and vanLoon, G.W. (Eds) (2019) *Sustainable Solutions for Food Security: combating climate change by adaptation.* Cham, CH.

Sarkar, D., Datta, R., Mukherjee, A. and Hannigan, R. (Eds) (2016) *An Integrated Approach to Environmental Management.* Wiley, Hoboken, NJ, USA.

Sarkis, J. (Ed) (2019) *Handbook on the Sustainable Supply Chain.* Edward Elgar, Cheltenham, UK.

Sarkis, J. and Dou, Y. (2013) *Green Supply Chain Management: a concise introduction.* Routledge, Abingdon, UK.

Sarmah, A.K. and Barceló, D. (2021) *Biochar: fundamentals and applications in environmental science and protection.* Elsevier (Academic Press), Cambridge, MA, USA.

Sassa, K., Fukuoka, H., Wang, F. and Gonghui Wang (Eds) (2007) *Progress in Landslide Science.* Springer, Berlin and Heidelberg, GR.

Sassa, K., Rouhban, B., Briceño, S., McSaveney, M. and Bin He (Eds) (2013) *Landslides: global risk preparedness.* Springer-Verlag, Berlin and Heidelberg, GR.

Sato, T., Helgeson, J. and Chabay, I. (Eds) (2018) *Transformations of Social-Ecological Systems: studies in co-creating integrated knowledge toward sustainable futures.* Springer, Singapore, SG.

Sattler, C., Schröter, B., Meyer, A., Giersch, G., Meyer, C. and Matzdorf, B. (2016) Multilevel governance in community-based environmental management: a case study comparison from Latin America. *Ecology and Society* 21(4), article No. 24 (14 pp.).

Savory, A. and Butterfield, J. (2016) *Holistic Management: a commonsense revolution to restore our environment* 3rd edn. Island Press, Washington, DC, USA.

Savarimuthu, X., Usha Rao, S.J. and Reynolds, M.F. (Eds) (2022) *Go Green for Environmental Sustainability: an interdisciplinary exploration of theory and applications.* CRC Press, Boca Raton, FL, USA.

Sayles, J.S., Garcia, M.M., Hamilton, M., Alexander, S.M., Baggio, J.A., Fischer, A.P., Ingold, K., Meredith, G.R. and Pittman, J. (2019) Social-ecological network analysis for sustainability sciences: a systematic review and innovative research agenda for the future. *Environmental Research Letters* 14(9), article No. 093003 (18 pp.).

Scales, I.R. (Ed) (2014) *Conservation and Environmental Management in Madagascar.* Routledge (Earthscan), Abingdon, UK.

Scarce, R. (2016) *Eco-Warriors: understanding the radical environmental movement* (first published 1990). Routledge, Abingdon, UK.

Scavone, G.M. (2005) Environmental management accounting: current practice and future trends in Argentina. In: P.M. Rikhardsson, M. Bennett, J.J. Bouma and S. Schaltegger (Eds), *Implementing Environmental Management Accounting: status and challenge* (Eco-Efficiency in Industry and Science). Springer, Dordrecht, NL.

Scavone, G.M. (2006) Challenges in internal environmental management reporting in Argentina. *Journal of Cleaner Production* 14(14), 1276–1285.

Schäli, J. (2022) *The Mitigation of Marine Plastic Pollution in International Law: facts, policy and legal implications.* Brill Nijhoff, Leiden, NL.

Schaltegger, S., Burritt, R. and Bennett, M. (Eds) (2006) *Sustainability Accounting and Reporting.* Springer, Dordrecht, NL.

Schaltegger, S., Bennett, M., Burritt, R.L. and Jasch, C. (Eds) (2008) *Environmental Management Accounting for Cleaner Production.* Springer, Dordrecht, NL.

Schellnhuber, H.J. (2009) *Climate Change as a Security Risk.* Routledge (Earthscan), London, UK.

Schenck, R. and White, P. (Eds) (2014) *Environmental Life Cycle Assessment: measuring the environmental performance of products.* American Center for Life Cycle Assessment, Washington DC, USA.

Scheub, U., Haiko Pieplow, Schmidt, H.-P. and Draper, K. (2016) *Terra Preta: how the world's most fertile soil can help reverse climate change and reduce world hunger.* Greystone Books, Vancouver, BC, CA.

Schipper, E.L.F. (2007) *Climate Change Adaptation and Development: exploring the linkages.* Tyndall Centre Working Paper No. 107. University of East Anglia, Norwich, UK.

Schipper, E.L.F., Schipper, L. and Burton, I. (2009) *The Earthscan Reader on Adaptation to Climate Change.* Routledge (Earthscan), Abingdon, UK.

Schipper, L. (2015) *A Comparative Overview of Resilience Measurement Frameworks Analyzing Indicators and Approaches.* Overseas Development Institute, London, UK.

Schlesinger, M.E., Kheshgi, H.S., Smith, J., de la Chesnaye, F.C., Reilly, J.M., Wilson, T. and Kolstad, C. (Eds) (2007) *Human-Induced Climate Change: an interdisciplinary assessment.* Cambridge University Press, Cambridge, UK.

Schlosberg, D. and Craven, L. (2019) *Sustainable Materialism: environmental movements and the politics of everyday life.* Oxford University Press, Oxford, UK.

Schmeier, S. (2013) *Governing International Watercourses: river basin organizations and the sustainable governance of internationally shared rivers and lakes.* Routledge (Earthscan), Abingdon, UK.

Schmelev, S. (Ed) (2016) *The Green Economics Reader: lectures in ecological economics and sustainability.* Springer, Dordrecht, NL.

Schmidt, B., Gemeinholzer, B. and Treloar, A. (2016) Open data in global environmental research: the Belmont Forum's open data survey. *PLOS One* 11(1), article No. e0146695 (18 pp.).

Schmidt, M., João, E. and Albrecht, E. (Eds) (2005) *Implementing Strategic Environmental Assessment.* Springer, Berlin, GR.

Schmidt, N. (Ed) (2019) *Planetary Defense: global collaboration for defending Earth from asteroids*. Springer, Cham, CH.

Schmitz, O.J. (2017) *The New Ecology: rethinking a science for the Anthropocene*. Princeton University Press, Princeton, NJ, USA.

Schmutz, S. and Sendzimir, J. (Eds) (2018) *Riverine Ecosystem Management: science for governing towards a sustainable future*. Springer, Cham, CH.

Schneider-Myerson, M. (2018) *The Influence of Climate Fiction: an empirical survey of readers*. Dover University Press, Durham, NC, USA.

Schnoor, J.L. and McAvoy, D.C. (2019) *Environmental Modeling: fate and transport of pollutants in water, air, and soil* 2nd edn. Wiley-Blackwell, Oxford, UK.

Schoch, R.M. (2021) *Forgotten Civilization: new discoveries on the solar-induced dark age* revised edn. Inner Traditions, Rochester, VT, USA.

Schoenmaker, D. and Schramade, W. (2018) *Principles of Sustainable Finance*. Oxford University Press, Oxford, UK.

Scholz, S.B., Sembres, T., Roberts, K., Whitman, T., Wilson, K. and Lehmann, J. (2014) *Biochar Systems for Smallholders in Developing Countries: leveraging current knowledge and exploring future potential for climate-smart agriculture*. World Bank, Washington, DC, USA.

Schrâter, D., Marc, J., Metzger, M.J., Cramer, W. and Leemans, R. (2004) Vulnerability assessment and analysing the human-environment system in the face of global environmental change. *The ESS Bulletin* 2(2), 11–17.

Schreckenberg, K., Mace, G. and Poudyal, M. (Eds) (2018) *Ecosystem Services and Poverty Alleviation: trade-offs and governance*. Routledge (Earthscan), Abingdon, UK.

Schreurs, M. and Papadakis, E. (2007) *The A to Z of the Green Movement*. Rowman & Littlefield (Scarecrow Press), Plymouth, UK.

Schröter, M., Bonn, A., Klotz, S., Seppelt, R. and Baessler, C. (Eds) (2019) *Atlas of Ecosystem Services: drivers, risks, and societal responses*. Springer International, Cham, CH.

Schultes, R.E. and von Reis, S. (Eds) (1995) *Ethnobotany: evolution of a discipline*. Chapman & Hall, London, UK.

Schultz, J. (2005) *The Ecozones of the World: the ecological divisions of the geosphere* 2nd edn. Springer, Berlin, GR.

Schulze, S. (2015) *China Environmental Law - sourcebook 2016* Chinese/English bilingual edn. CreateSpace Independent Publishing Platform, Scotts Valley, CA, USA.

Schumacher, E.F. (1973) *Small Is Beautiful: a study of economics as if people mattered*. Harper and Row, New York, NY, USA.

Schumann, G.L. and D'Arcy, C.J. (2012) *Hungry Planet: stories of plant disease*. American Phytopathological Society, Saint Paul, MN, USA.

Schwab, J. (Ed) (2009) *Planning the Urban Forest: ecology, economy, and community development*. Routledge, New York, NY, USA.

Schwägerl, C. (2014) *The Anthropocene: the human era and how it shapes our planet*. Synergetic Press, Santa Fe, NM, USA.

Schwartz, G. and Cohen, T.D. (2021) *Bright Green Future: how everyday heroes are re-imagining the way we feed, power, and build our world*. Harmon Street Press, Sarasota, FL, USA.

Schweinsberg, S. and Wearing, S. (2019) *Ecotourism: transitioning to the 22nd century* 3rd edn. Routledge, Abingdon, UK.

Sciaccaluga, G. (2020) *International Law and the Protection of 'Climate Refugees'*. Springer International (Palgrave), Cham, CH.

Scoones, I. (2015) *Sustainable Livelihoods and Rural Development.* Fernwood Publishing, Black Point, NS, CA.

Scotford, E. (2017) *The Strategic Environmental Assessment Directive: a plan for success?* Bloomsbury, London, UK.

Scott, A.C. (2018a) *Burning Planet: the story of fire through time.* Oxford University Press, Oxford, UK.

Scott, B.A., Amel, E.L., Koger, S.M. and Manning, C.M. (2016) *Psychology for Sustainability* 4th edn. Routledge, New York, NY, USA.

Scott, J. (2017) *Social Network Analysis* 4th edn. Sage, Thousand Oaks, CA, USA.

Scott, M. (2020) *Climate Change, Disasters, and the Refugee Convention.* Cambridge University Press, Cambridge, UK.

Scott, N.D. (2018b) *Food, Genetic Engineering and Philosophy of Technology: magic bullets, technological fixes and responsibility to the future.* Springer, Cham, CH.

Scott Cato, M. (2012) *The Bioregional Economy: land, liberty and the pursuit of happiness.* Taylor & Francis, London, UK.

Scow, K.M., Fogg, G.E., Hinton, D.E. and Johnson, M.L. (Eds) (2018) *Integrated Assessment of Ecosystem Health.* CRC Press, Boca Raton, FL, USA.

Scruton, R. (2012) *Green Philosophy: how to think seriously about the planet.* Atlantic Books, London, UK.

Scudder, T. (2012) *The Future of Large Dams: dealing with social, environmental, institutional and political costs* 2nd edn. Earthscan, London, UK.

Scudder, T. (2019) *Large Dams: long term impacts on riverine communities and free flowing rivers.* Springer, Singapore, SG.

Seenipandi, K., Rani, M., Sajjad, H., Kumar, P. and Rehman, S. (Eds) (2020) *Remote Sensing of Ocean and Coastal Environments.* Elsevier, Amsterdam, NL.

Seidl, A. (2011) *Finding Higher Ground: adaptation in the age of warming.* Beacon Press, Boston, MA, USA.

Seiegel, F.R. (2016) *Mitigation of Dangers from Natural and Anthropogenic Hazards: prediction, prevention and preparedness.* Springer, Cham, CH.

Sejian, V., Gaughan, J., Baumgard, L. and Prasad, C. (Eds) (2015) *Climate Change Impact on Livestock: adaptation and mitigation.* Springer, New Delhi, IN.

Sellers, K. (2015) *Product Stewardship: life cycle analysis and the environment.* CRC Press, Boca Raton, FL, USA.

Semple, E.C. (1911) *Influences of Geographic Environment.* Henry Holt, New York, NY, USA.

Sen, Z. (2018) *Flood Modeling, Prediction and Mitigation.* Springer, Cham, CH.

Sene, K. (2013) *Flash Floods: forecasting and warning.* Springer, Dordrecht, NL.

Senthil Kumar, A.V. and Kalpana, M. (Eds) (2019) *Fuzzy Expert Systems and Applications in Agricultural Diagnosis.* IGI Global, Hershey, PA, USA.

Sepkoski, D. (2020) *Catastrophic Thinking: extinction and the value of diversity from Darwin to the Anthropocene.* University of Chicago Press, Chicago, IL, USA.

Seraphin, H. and Nolan, E. (Eds) (2019) *Green Events and Green Tourism: an international guide to good practice.* Routledge, Abingdon, UK.

Serfilippi, E. and Giovannucci, D. (Eds) (2017) *Simpler resilience measurement: tools to diagnose and improve how households fare in difficult circumstances from conflict to climate change.* The Committee on Sustainability Assessment (COSA), Philadelphia, PA, USA.

Seto, K.C., Solecki, W.D. and Griffith, C.A. (Eds) (2016) *The Routledge Handbook of Urbanization and Global Environmental Change.* Routledge, Abingdon, UK.

Setthasakko, W. (2010) Barriers to the development of environmental management accounting: an exploratory study of pulp and paper companies in Thailand. *EuroMed Journal of Business*, 5(3), 315–331.

Seymour, N. (2018) *Bad Environmentalism: irony and irreverence in the ecological age.* University of Minnesota Press, Minneapolis, MN, USA.

Shabbir, M. (Ed) (2019) *Textiles and Clothing: environmental concerns and solutions.* Hoboken, NJ, USA.

Shah, E., Boelens, R. and Bruins, B. (2019a) *Contested Knowledges: water conflicts on large dams and mega-hydraulic development.* MDPI, Basel, CH.

Shah, S., Venkatramanan, V. and Prasad, R. (Eds) (2019b) *Sustainable Green Technologies for Environmental Management.* Springer, Singapore, SG.

Shakley, S., Ruysschaert, G., Zwart, K. and Glaser, B. (Eds) (2016) *Biochar in European Soils and Agriculture: science and practice.* Routledge, Abingdon, UK.

Shaltegger, S., Burritt, R. and Petersen-Dyggve, H.N. (2017) *An Introduction to Corporate Environmental Management: striving for sustainability.* Taylor & Francis (Routledge), Abingdon, UK.

Shamdasani, P.N. and Stewart, D.W. (2014) *Focus Groups: theory and practice* 3rd edn. Sage Publications, Thousand Oaks, CA, USA.

Shaner, W.W., Philipp, P.F. and Schmehl, W.R. (2021) *Farming Systems Research and Development: guidelines for developing countries.* Routledge, Abingdon, UK.

Shanker, U., Hussain, C.M. and Rani, M. (Eds) (2022) *Green Functionalized Nanomaterials for Environmental Applications.* Elsevier, Amsterdam, NL.

Shapiro, J. (2015) *China's Environmental Challenges.* Polity Press, Cambridge, UK.

Sharifi, A. (2016) A critical review of selected tools for assessing community resilience. *Ecological Indicators* 69, 629–647.

Sharma, A. (2020) *Sustainable Tourism Development: futuristic approaches.* Apple Academic Press, Oakville, CA, USA.

Sharma, N., Ghosh, S. and Saha, M. (2021a) *Open Data for Sustainable Community: localized sustainable development goals.* Springer, Singapore, SG.

Sharma, R. (2019) *Environmental Issues of Deep-Sea Mining: impacts, consequences and policy perspectives.* Springer, Cham, CH.

Sharma, R. (Ed) (2022) *Perspectives on Deep-Sea Mining: sustainability, technology, environmental policy and management.* Springer, Cham, CH.

Sharma, R., Sharma, D.K., Bhatt, D. and Binh Thai Pham (Eds) (2021b) *Big Data Analysis for Green Computing Concepts and Applications.* CRC Press, Boca Raton, FL, USA.

Sharma, S. (2018) *Total Quality Management: concepts, strategy and implementation for operational excellence.* Sage (India) Pvt., Delhi, IN.

Shaw, B.D. (1995) *Environment and Society in Roman North Africa: studies in history and archaeology.* Routledge, Abingdon, UK.

Shaw, R. and Sharma, R. (Eds) (2011) *Climate and Disaster Resilience in Cities.* Emerald, Bingley, UK.

Sheehan, K. and Atkinson, L. (Ed) (2015) *Green Advertising and the Reluctant Consumer.* Routledge, Abingdon, UK.

Sheldon, C. and Yoxon, M. (2006) *Environmental Management Systems: a step-by-step guide to implementation and maintenance* 3rd edn. Routledge (Earthscan), Abingdon, UK.

Sheldon, C. and Yoxon, M. (2012) *Installing Environmental Management Systems: a step-by-step guide* 3rd edn. Routledge (Earthscan), Abingdon, UK.

Shellenberger, M. (2020) *Apocalypse Never: why extreme environmental alarmism hurts us.* Harper-Collins, New York, NY, USA.

Shematek, G., MacLean, P. and Lineen, P. (2016) *Health, Safety and Environmental Management Systems Auditing: design fundamentals and applications*. LexisNexis, New York, NY, USA.

Shetterley, C. (2016) *Modified: GMOs and the threat to our food, our land, our future*. Penguin (G.P. Putnam's & Sons), New York, NY, USA.

Shideler, J.C. and Hetzel, J. (2021) *Introduction to Climate Change Management: transitioning to a low-carbon economy*. Springer, Cham, CH.

Shiva, V. (2020) *Reclaiming the Commons: biodiversity, traditional knowledge, and the rights of mother Earth*. Synergetic Press, Santa Fe, NM, USA.

Shiyi Chen (2013) *Energy, Environment and Economic Transformation in China*. Taylor & Francis, London, UK.

Shmulsky, R. and Jones, P.D. (2019) *Forest Products and Wood Science: an introduction* 7th edn. Wiley-Blackwell, Hoboken, NJ, USA.

Short, M., Baker, M., Carter, J., Jay, S. and Jones, C. (2005) *Strategic Environmental Assessment and Land Use Planning: an international evaluation*. Routledge, London, UK.

Shrivastava, A.K. (2013) *Environmental Auditing*. Ashish Publishing House, Delhi, IN.

Shuanggen Jin, Cardellach, E. and Feiqin Xie (2014) *GNSS Remote Sensing: theory, methods and applications*. Springer, Dordrecht, NL.

Shukla, K. (2016) *Participatory Rural Appraisal Technique for Farm Women*. V&S Publications, New Delhi, IN.

Shunlin Liang and Jindi Wang (Eds) (2012) *Advanced Remote Sensing: terrestrial information extraction and applications* 2nd edn. Elsevier (Academic Press), London, UK.

Siegel, F.R. (2020) *Adaptations of Coastal Cities to Global Warming, Sea Level Rise, Climate Change and Endemic Hazards*. Springer, Cham, CH.

Siegle, L. (2011) *To Die For: is fashion wearing out the world?* Harper-Collins (Fourth Estate), London, UK.

Sikdar, P.K. (Ed) (2021) *Environmental Management: issues and concerns in developing countries*. Springer (with Capital Publishing Co.), New Delhi, IN.

Silberstein, J., and Maser, C. (2019) *Land Use Planning for Sustainable Development*. CRC Press, Boca Raton, FL, USA.

Simon, J.L. (1981) *The Ultimate Resource*. Princeton University Press, Princeton, NJ, USA.

Simon, T.W. (2014) *Environmental Risk Assessment: a toxicological approach* 2nd edn. CRC Press, Boca Raton, FL, USA.

Simon, Z.B. (2020) *The Epochal Event: transformations in the entangled human, technological and natural worlds*. Springer (Palgrave Macmillan), Cham, CH.

Simonen, K. (2014) *Life Cycle Assessment*. Routledge, Abingdon, UK.

Singh, B.K. (Ed) (2018) *Soil Carbon Storage: modulators, mechanisms and modelling*. Elsevier (Academic Press), London, UK.

Singh, H.P., Batish, D.R. and Kohli, R.K. (Eds) (2006) *Handbook of Sustainable Weed Management*. Haworth Press, New York, NY, USA.

Singh, J. (2009) *Tsunamis: threats and management*. IK International Publishing House, New Delhi, IN.

Singh, J.S. and Singh, C. (Eds) (2020) *Biochar Applications in Agriculture and Environment Management*. Springer, Cham, CH.

Singh, P., Hussain, C.M. and Sillanpää, M. (Eds) (2022a) *Innovative Bio-Based Technologies for Environmental Remediation*. CRC Press, Boca Raton, FL, USA.

Singh, P., Singh, S. and Sillanpää, M. (Eds) (2022b) *Pesticides in the Natural Environment: sources, health risks, and remediation*. Elsevier, Amsterdam, NL.

Singh, R. and Kumar, S. (2018) *Green Technologies and Environmental Sustainability.* Springer, Cham, CH.

Singh, R.M., Shukla, P. and Singh, P. (Eds) (2020) *Environmental Processes and Management: tools and practices.* Springer Nature, Cham, CH.

Singh, R.P. (Ed) (2022) *Asian Atmospheric Pollution: sources, characteristics and impacts.* Elsevier, Amsterdam, NL.

Singh Malyan, R. and Duhan, P. (Eds) (2018) *Green Consumerism: perspectives, sustainability, and behavior.* Apple Academic Press, Oakville, CA, USA.

Sivasubramanian, V. (Ed) (2020) *Environmental Sustainability Using Green Technologies* 2nd edn. CRC Press, Boca Raton, FL, USA.

Sjostedt, B. (2020) *The Role of Multilateral Environmental Agreements: a reconciliatory approach to environmental protection in armed conflict.* Bloomsbury Publishing, London, UK.

Skipton, N.C., Woolley, S.D., Foster, N.J., Bax, J.C., Currie, D.C., Dunn, C., Hansen, N., Hill, T.D., O'Hara, O., Ovaskainen, R., Sayre, J.P., Vanhatalo, P. and Dunstan, K. (2020) Bioregions in marine environments: combining biological and environmental data for management and scientific understanding. *BioScience* 70(1), 48–59.

Sklair, L. (2021) *The Anthropocene in Global Media: neutralizing the risk.* Routledge, Abingdon, UK.

Slatin, C. (2017) *Environmental Unions: labor and the superfund.* Routledge, Abingdon, UK.

Slocombe, D.S. (1993) Environmental planning, ecosystem science, and ecosystem approaches for integrating environment and development. *Environmental Management* 17(3), 289–303.

Slocum, S.L., Aidoo, A. and McMahon, K. (2020) *The Business of Sustainable Tourism Development and Management.* Routledge, Abingdon, UK.

Slovic, P. (2010) *The Feeling of Risk: new perspectives on risk perception.* Routledge (Earthscan), Abingdon, UK.

Smaje, C. (2020) *A Small Farm Future: making the case for a society built around local economies, self-provisioning, agricultural diversification and a shared Earth.* Chelsea Green Publishing, Hartford, VT, USA.

Smil, V. (1983) *The Bad Earth: environmental degradation in China.* Zed Press, London, UK.

Smil, V. (2001) *Feeding the world: a challenge for the twenty-first century.* MIT Press, Cambridge, MA, USA.

Smit, B., Reimer, J.A., Oldenburg, C.M. and Bourg, I.C. (2014) *Introduction to Carbon Capture and Sequestration.* Imperial College Press, London, UK.

Smit, T. (2001) *Eden.* Corgi Books, London, UK.

Smith, C.T. and Davies, E.T. (2012) *Emigrating Beyond Earth: human adaptation and space colonization.* Springer-Praxis, New York, NY, USA.

Smith, G. (Ed) (2017) *The War and Environment Reader.* Just World Books, Chicago, IL, USA.

Smith, J. and Smith, P. (2007) *Environmental Modelling: an introduction.* Oxford University Press, Oxford, UK.

Smith, K. (2013) *Environmental Hazards: assessing risk and reducing disaster* 6th edn. Routledge, Abingdon, UK.

Smith, L., Porter, K., Hiscock, K., Porter, M.J. and Benson, D. (Eds) (2017) *Catchment and River Basin Management: integrating science and governance.* Routledge (Earthscan), Abingdon, UK.

Smith, M. (2011) *Against Ecological Sovereignty: ethics, biopolitics, and saving the natural world.* University of Minnesota Press, Minneapolis, MN, USA.

Smith, W.H. (2012) *Air Pollution and Forests: interactions between air contaminants and forest ecosystems*. Springer-Verlag, New York, NY, USA.

Smuts, J.C. (1926) *Holism and Evolution* 2nd edn. Macmillan, London, UK.

Snobar, B., Oweis, T. and Nofal, H. (2011) *Microcatchment Water Harvesting Systems for Fruit Trees and Shrubs*. ICARDA, Aleppo, SY.

Soderberg, R.W. (2017) *Aquaculture Technology: flowing water and static water fish culture*. CRC Press, Boca Raton, FL, USA.

Solocova, O., Korovkin, N. and Hayakawa, M. (2021) *Geomagnetic Disturbances Impacts on Power Systems: risk analysis and mitigation strategies*. CRC Press, Boca Raton, FL, USA.

Sonwani, S. and Saxena, P. (2022) *Greenhouse Gases: sources, sinks and mitigation*. Springer, Singapore, SG.

Sookoor, T., Srinivasan, R., Houbing Song and Jeschke, S. (Eds) (2017) *Smart Cities: foundations, principles, and applications*. Wiley, Hoboken, NJ, USA.

Spangenberg, J.H. (Ed) (2019) *Scenarios and Indicators for Sustainable Development: towards a critical assessment of achievements and challenges*. MDPI, Basel, CH.

Spash, C.L. (Ed) (2017) *Routledge Handbook of Ecological Economics: nature and society*. Routledge, Abingdon, UK.

Spee, J.C., McMurray, A.J. and McMillan, M.D. (2021) *Clan and Tribal Perspectives on Social, Economic and Environmental Sustainability: indigenous stories from around the globe*. Emerald Publishing, Bingley, UK.

Speight, J. and Singh, K. (2014) *Environmental Management of Energy from Biofuels and Biofeedstocks*. Wiley (Scrivener), Beverly, MA, USA.

Spellman, F.R. (2015) *Handbook of Environmental Engineering*. CRC Press, Boca Raton, FL, USA.

Spellman, F.R. (2021) *The Science of Environmental Pollution* 4th edn. CRC Press, Boca Raton, FL, USA.

Spence, M., Annez, P.C. and Buckley, R.M. (Eds) (2008) *Urbanization and Growth*. World Bank/IBRD, Washington, DC, USA.

Spenceley, A. (Ed) (2021) *Handbook for Sustainable Tourism Practitioners: the essential toolbox*. Edward Elgar, Cheltenham, UK.

Spencer, L. (2018) *Industrial Ecology*. Larsen & Keller Education, New York, NY, USA.

Sponsel, L. (2012) *Spiritual Ecology: a quiet revolution*. ABC-CLIO, LLC (Praeger), Santa Barbara, CA, USA.

Spring, U.O. (2020) *Earth at Risk in the 21st Century: rethinking peace, environment, gender, and human, water, health, food, energy, security and migration*. Springer, Cham, CH.

Squires, V.R., Milner, H.M. and Daniell, K.A. (Eds) (2014) *River Basin Management in the Twenty-First Century: understanding people and place*. CRC Press, Boca Raton, FL, USA.

Srivastav, A.L., Madhav, S., Bhardwaj, A.K. and Valsami-Jones, E. (Eds) (2022) *Urban Water Crisis and Management: strategies for sustainable development* 6th edn. Elsevier, Amsterdam, NL.

Staplehurst, T. (2009) *The Benchmarking Book*. Elsevier (Butterworth-Heinemann), Oxford, UK.

Stead, E. and Stead, J.G. (2003) *Sustainable Strategic Management*. Taylor & Francis, London, UK.

Stefanescu, L., Stefanescu, A., Ungureanu, L., Constantinescu, M. and Barbu, C. (2011) Expert system and its applications for a sustainable environment management. *Journal of Environmental Protection and Ecology* 12(3), 1582–1591.

Steffen, A. (Ed) (2006) *Worldchanging: a user's guide for the 21st century*. Abrams, New York, NY, USA.

Steffen, W., Crutzen, P.J. and McNeill, J.R. (2007) The Anthropocene: are humans now over-whelming the great forces of nature? *Ambio* 36(8), 614–621.

Steg, L., van den Berg, A.E. and de Groot, J.I.M. (Eds) (2013) *Environmental Psychology: an introduction.* Wiley-Blackwell, Oxford, UK.

Steger, U., Fang Zhaoben and Lu We (2003) *Greening Chinese Business: barriers, trends and opportunities for environmental management.* Routledge, London, UK.

Steinbeck, J. (1939) *The Grapes of Wrath.* Heinemann, New York, NY, USA.

Steiner, F.R. (2016) *Human Ecology: how nature and culture shape our world* 2nd edn. Island Press, Washington, DC, USA.

Steiner, R. (2008) *Spiritual Ecology: reading the book of nature and reconnecting with the world.* Rudolf Steiner Press, London, UK.

Stern, N.H. (2007) *The Economics of Climate Change: the Stern Review on the economics of climate change 2006.* Cambridge University Press (HM Treasury, UK), Cambridge, UK.

Stern, N.H. and Patel I.G. (2009) *A Blueprint for a Safer Planet: how to manage climate change and create a new era of progress and prosperity.* Bodley Head, London, UK.

Sternfeld, E. (Ed) (2017) *Routledge Handbook of Environmental Policy in China.* Routledge (Earthscan), Abingdon, UK.

Stevens, S. (Ed) (2014) *Indigenous Peoples, National Parks, and Protected Areas: a new paradigm linking conservation, culture, and rights.* University of Arizona Press, Tucson, AZ, USA.

Stevenson, H. (2017) *Global Environmental Politics: problems, policy and practice.* Cambridge University Press, Cambridge, UK.

Stickney, R.R. (2017) *Aquaculture: an introductory text* 3rd edn. CABI, Wallingford, UK.

Stilwell, F.J.B. (2011) *Political Economy: the contest of economic ideas* 3rd edn. Oxford University Press Australia & New Zealand, Melbourne, VI, AU.

Stokols, D. (2018) *Social Ecology in the Digital Age: solving complex problems in a globalized world.* Elsevier (Academic Press), London, UK.

Strähle, J. (Ed) (2017) *Green Fashion Retail.* Springer, Singapore, SG.

Štreimikienė, D. and Mikalauskiene, A. (2021) *Climate Change and Sustainable Development: mitigation and adaptation.* CRC Press, Boca Raton, FL, USA.

Strieber, W. (2012) *Solar Flares: what you need to know.* Tarcher/Penguin, New York, NY, USA.

Stringer, L.C. and Reed, M.S. (2016) *Land Degradation, Desertification and Climate Change: anticipating, assessing and adapting to future change.* Routledge (Earthscan), Abingdon, UK.

Struik, P.C. and Kuyper, T.W. (2017) Sustainable intensification in agriculture: the richer shade of green: a review. *Agronomy for Sustainable Development* 37(5), 1–15.

Strydom, H.A. and King, N.D. (Eds) (2017) *Fuggle & Rabie's Environmental Management in South Africa* 2nd edn. Juta Legal and Academic Publishing, Kenwyn, ZA.

Strydom, H.A., King, N.D., Fuggle, R.F. and Rabie M.A. (2009) *Environmental Management in South Africa* 2nd edn. Juta, Cape Town, ZA.

Subrahmanian, V.S., Ovelgonne, M., Dumitras, T. and Prakash, A. (2015) *The Global Cyber-Vulnerability Report.* Springer, Cham, CH.

Subramanian, M.N. (2019) *Plastics Waste Management: processing and disposal* 2nd edn. Wiley (Scrivener Publishing), Beverly, MA, USA.

Sumi, A., Fukushi, K. and Hiramatsu, A. (Eds) (2021) *Adaptation and Mitigation Strategies for Climate Change.* Springer, Tokyo, JP.

Sun, A.Y. and Scanlon, B.R. (2019) How can big data and machine learning benefit environment and water management? A survey of methods, applications, and future directions. *Environmental Research Letters* 14(7), article No. 073001 (17 pp.).

Sungsoo Pyo (Ed) (2012) *Benchmarks in Hospitality and Tourism* 2nd edn. Routledge, New York, NY, USA.

Sunil, K. (Ed) (2020) *Southeast Asia and Environmental Sustainability in Context.* Rowman & Littlefield (Lexington Books), Lanham, MD, USA.

Surampalli, R.Y. (Ed) (2013) *Climate Change Modeling, Mitigation, and Adaptation.* American Society of Civil Engineers (ASCE), Reston, VA, USA.

Sveiby, K.-E. and Skuthorpe, T. (2006) *Treading Lightly: the hidden wisdom of the world's oldest people.* Allen & Unwin, Sydney, NSW, AU.

Svensmark, H. and Calder, N. (2007) *The Chilling Stars: a cosmic view of climate change.* Icon Books, Cambridge, UK.

Svenson, H. (2012) *The End is Nigh: a history of natural disasters.* Reaktion Books, London, UK.

Swain, A. and Ejendal, O. (Eds) (2020) *Routledge Handbook of Environmental Conflict and Peacebuilding: an introduction.* Routledge (Earthscan), Abingdon, UK.

Swearingen, W.D. and Bencherifa, A. (2021) *The North African Environment at Risk* (first published 1996). Routledge, Abingdon, UK.

Swiss Sustainable Finance (2017) *Handbook on Sustainable Investments: background information and practical examples for institutional asset owners.* CFA Institute Research Foundation, Charlottesville, VA, USA.

Sylvan, R. and Bennett, D. (1988) Taoism and deep ecology. *The Ecologist* 18(4–5), 148–158.

Symons, J. (2019) *Ecomodernism: technology, politics and the climate crisis.* Polity Press, Cambridge, UK.

Szwedo, P. (2018) *Cross-Border Water Trade: legal and interdisciplinary perspectives.* Brill Nijhoff, Leiden, NL.

Tagliaferro, A., Rosso, C. and Giorcelli, M. (Eds) (2020) *Biochar: emerging applications.* IOP Publications, Bristol, UK.

Tai-Chee Wong and Yuen, B. (Eds) (2015) *Eco-City Planning: policies, practice and design.* Springer, Dordrecht, NL.

Tait, J. and Napompeth, B. (Eds) (2018) *Management of Pests and Pesticides: farmers' perceptions and practices.* Routledge (Taylor & Francis), Abingdon, UK.

Tallis, H., Ricketts, T.H., Daily, G.C. and Polasky, S. (Eds) (2011) *Natural Capital: theory and practice of mapping ecosystem services.* Oxford University Press, Oxford, UK.

Tan, C. and Faúndez, J. (Eds) (2017) *Natural Resources and Sustainable Development: international economic law perspectives.* Edward Elgar, Cheltenham, UK.

Tanil, G. (2021) *Environmental Sustainability: water and waste management policy in the European Union and the Czech Republic.* Rowman & Littlefield, Lanham, MD, USA.

Tankha, G. (2017) *Environmental Attitudes and Awareness: a psychosocial perspective.* Cambridge Scholars Publishing, Newcastle-upon-Tyne, UK.

Tanner, C.J. and Adler, F.R. (2013) *Urban Ecosystems: ecological principles for the built environment.* Cambridge University Press, Cambridge, UK.

Tansley, A.F. (1935) The use and abuse of vegetational concepts and terms. *Ecology* 16(2), 284–307.

Tapia, M.P., Henríquez Fernández, D. and Vidal Moranta, B. (2019) Participatory environmental management: grounded theory proposals. *Revista de Gestão Ambiental e Sustentabilidade* 8(3), 489–507.

Taşeli, B.K. (Ed) (2020) *Sustainable Sewage Sludge Management and Resource Efficiency.* IntechOpen, London, UK.

Tauringana, V. (Ed) (2019) *Environmental Reporting and Management in Africa.* Emerald Publishing, Bingley, UK.

Taylor, B. (2009) *Dark Green Religion: nature spirituality and the planetary future.* University of California Press, Berkeley, CA, USA.

Taylor, M. (2015) *The Political Ecology of Climate Change Adaptation: livelihoods, agrarian change and the conflicts of development.* Routledge (Earthscan), Abingdon, UK.

Taylor, P. (2010) *The Biochar Revolution: transforming agriculture & environment.* Global Publishing Group, Jacksonville, FL, USA.

Taylor Klein, P. (2022) *Flooded: development, democracy, and Brazil's Belo Monte Dam (nature, society, and culture).* Rutgers University Press, New Brunswick, NJ, USA.

Tedim, F., Leone, V. and McGee, T.K. (Eds) (2020) *Extreme Wildfire Events and Disasters: root causes and new management strategies.* Elsevier, Amsterdam, NL.

Teebken, J. (2022) *The Politics of Human Vulnerability to Climate Change: exploring adaptation, lock-ins in China and the United States.* Routledge, Abingdon, UK.

Teegavarapu, R.S.V., Kolokytha, E. and de Oliveira Galvão, C. (Eds) (2020) *Climate Change-Sensitive Water Resources Management.* CRC Press, Boca Raton, FL, USA.

Teehankee, M. (2020) *Trade and Environment Governance at the World Trade Organization Committee on Trade and Environment.* Kluwer Law International, Alphen ann den Rijn, NL.

Tehan, M.F., Godden, L.C., Young, M.A. and Gover, K.A. (Eds) (2017) *The Impact of Climate Change Mitigation on Indigenous and Forest Communities: international, national and local law perspectives on REDD+.* Cambridge University Press, Cambridge, UK.

Teilhard de Chardin, P. (1959) *The Phenomenon of Man.* Harper and Row, New York, NY, USA.

Teilhard de Chardin, P. (1964) *The Future of Man.* Collins, London, UK.

Teisl, M. (Ed) (2007) *Labelling Strategies in Environmental Policy.* Routledge, Abingdon, UK.

Telešiene, A. and Gross, M. (Eds) (2017) *Green European: environmental behaviour and attitudes in Europe in a historical and cross-cultural perspective.* Routledge, Abingdon, UK.

Tene, V.T., Boiral, O. and Heras-Saizarbitoria, I. (2021) Internalizing environmental management practices in Africa: the role of power distance and morality. *Journal of Cleaner Production* 291 article No. 125267 (12 pp.).

Ten Have, H. (2016) *Vulnerability: challenging bioethics.* Routledge, Abingdon, UK.

Tetlock, P.E. and Gardner, D. (2015) *Superforecasting: the art and science of prediction.* Penguin Random House (Crown), New York, NY, USA.

Thabit, T.H. and Ibraheem, L.K.(2019) Implementation of environmental management accounting for enhancing the sustainable development in Iraqi oil refining companies (March 7, 2019). 3rd Scientific Conference of Administration and Economic College, University of Anbar, Anbar, IQ.

Thaler, M. (2022) *No Other Planet: utopian visions for a climate-changed world.* Cambridge University Press, Cambridge, UK.

Tharp, D.P. (2014) *Genetically Engineered Salmon: background and issues.* Nova Science Publishers, Hauppauge, NY, USA.

Theodore, L. and Dupont, R.R. (2012) *Environmental Health and Hazard Risk Assessment: principles and calculations.* CRC Press, Boca Raton, FL, USA.

Theodore, L. and Dupont, R.R. (2019) *Water Resource Management Issues: basic principles and applications.* CRC Press, Boca Raton, FL, USA.

Therivel, R. (2010) *Strategic Environmental Assessment in Action* 2nd edn. Routledge (Earthscan), Abingdon, UK.

Therivel, R. and Partidário, M.R. (1996) *Practice of Strategic Environmental Assessment.* Earthscan, London, UK.

Therivel, R. and Wood, G. (Eds) (2018) *Methods of Environmental and Social Impact Assessment* 4th edn. Routledge, New York, NY, USA.

Thiébat, F. (2019) *Life Cycle Design: an experimental tool for designers.* Springer, Wiesbaden, GR.

Thomas, D.S.K., Phillips, B.D., Lovekamp, W.E. and Fothergill, A. (Eds) (2013) *Social Vulnerability to Disasters* 2nd edn. CRC Press, Boca Raton, FL, USA.

Thomas, D.S.G. and Middleton, N.J. (1994) *Desertification: exploding the myth.* Wiley, Chichester, UK.

Thomas, I.G. and Murfitt, P. (2011) *Environmental Management: processes and practices for Australia* 2nd edn. Federation Press, Alexandria, NSW, AU.

Thomas, J.A., Williams, M. and Zalasiewicz, J. (2020) *The Anthropocene: a multidisciplinary approach.* Polity Press, Cambridge, UK.

Thomas-Hope, E. (Ed) (2013) *Environment Management in the Caribbean: policy and practice.* University of West Indies Press, Kingston, JM.

Thompson, A. and Bendik-Keymenr, J. (Eds) (2012) *Ethical Adaptation to Climate Change: human virtues of the future.* MIT Press, Cambridge, MA, USA.

Thompson, A. and Gardiner, S.M. (Eds) (2015) *The Oxford Handbook of Environmental Ethics.* Oxford University Press, New York, NY, USA.

Thompson, D. (Ed) (2014) *Tools for Environmental Management: a practical introduction and guide* (first published 2002). University of Calgary Press, Calgary, AB, CA.

Thompson, J.N. (2005) *The Geographic Mosaic of Coevolution.* University of Chicago, Chicago, IL, USA.

Thompson, S. (2021) *Green and Sustainable Finance: principles and practice.* Kogan Page, London, UK.

Thomson, D.R. and Wilson, M.J. (1994) Environmental auditing: theory and applications. *Environmental Management* 18(4), 605–615.

Thore, S. and Tarverdyan, R. (2021) *Measuring Sustainable Development Goals Performance.* Elsevier, Amsterdam, NL.

Thoreau, H.D. (1854) *Walden, or Life in the Woods.* Houghton Mifflin, Boston, MD, USA (1960 edn, New American Library, New York, NY, USA).

Thornbush, M.J. (2021) *The Ecological Footprint as a Sustainability Metric: implications for sustainability.* Springer International, New York, NY, USA.

Thornton, T.F. and Bhagwat, S.A. (Eds) (2021) *The Routledge Handbook of Indigenous Environmental Knowledge.* Routledge, Abingdon, UK.

Thunberg, G. (2019) *No One Is Too Small to Make a Difference.* Penguin, London, UK.

Tidwell, A.C. and Scott Zellen, B. (Eds) (2017) *Land, Indigenous Peoples and Conflict.* Routledge, Abingdon, UK.

Tiefenbacher, J.P. (Ed) (2022) *Environmental Management: pollution, habitat, ecology, and sustainability.* InTech Open, London, UK.

Tietenberg, T.H. (2006) *Emissions Trading: principles and practice* 2nd edn. Routledge, Abingdon, UK.

Tietenberg, T.H. and Lewis, L. (2020) *Environmental Economics: the essentials.* Routledge, New York, NY, USA.

Tiffen, M., Mortimore, M. and Gichuki, F. (1994) *More People, Less Erosion: environmental recovery in Kenya.* Wiley, Chichester, UK.

Ting, D.S.-K. and Vasel-Be-Hagh, A. (Eds) (2020) *Environmental Management of Air, Water, Agriculture, and Energy*. CRC Press, Boca Raton, FL, USA.

Tinsley, S. and Pillai, I.(2006) *Environmental Management Systems: understanding organizational drivers and barriers*. Routledge, Abingdon, UK.

Tiwari, S. and Agrawal, M. (Eds) (2018) *Tropospheric Ozone and Its Impacts on Crop Plants: a threat to future global food security*. Springer, Cham, CH.

Tobin, B. (2016) *Indigenous Peoples, Customary Law and Human Rights: why living law matters*. Taylor & Francis (Routledge), Abingdon, UK.

Tobin-de la Puente, J. and Mitchell, A.W. (Eds) (2021) *The Little Book of Investing in Nature: a simple guide to financing life on Earth*. Global Canopy, Oxford, UK.

Toensmeier, E. (2016) *The Carbon Farming Solution: a global toolkit of perennial crops and regenerative agriculture practices for climate change mitigation and food security*. Chelsea Green Publishing, White River Junction, VT, USA.

Tolba, M.K. (1982) *Development Without Destruction: evolving environmental perceptions*. Tycooly International, Dublin, IE.

Tomar, P. and Kaur, G. (Eds) (2020) *Green and Smart Technologies for Smart Cities*. CRC Press, Boca Raton, FL, USA.

Tomson, J.A. (2006) *GM Crops: the impact and the potential*. CSIRO, Collingwood, VI, AU.

Toro, F.G. and Tsourdos, A. (Eds) (2018) *UAV Sensors for Environmental Monitoring*. MDPI, Basel, CH.

Torp, S. and Andersen, T.J. (2020) *Adapting to Environmental Challenges: new research in strategy and international business*. Emerald Publishing, Bingley, UK.

Torres, P.H.C. and Jacobi, P.R. (Eds) (2021) *Towards a Just Climate Change Resilience: developing resilient, anticipatory and inclusive community response*. Springer (Palgrave), Cham, CH.

Tortajada, C. (Ed) (2015) *Integrated Water Resources Management: from concept to implementation*. Routledge, Abingdon, UK.

Tortajada, C. (Ed) (2016) *Increasing Resilience to Climate Variability and Change: the roles of infrastructure and governance in the context of adaptation*. Springer, Singapore, SG.

Tortajada, C., Altinbilek, D. and Biswas, A.K. (Eds) (2012) *Impacts of Large Dams: a global assessment*. Springer-Verlag, Berlin, GR.

Tow, P., Cooper, I., Partridge, I.J. and Birch, C. (Eds) (2011) *Rainfed Farming Systems*. Springer, Dordrecht, NL.

Townsend, A.M. (2013) *Smart Cities: big data, civic hackers, and the quest for a new utopia*. W.W. Norton & Co., New York, NY, USA.

Traer, R. (2020) *Doing Environmental Ethics* 3rd edn. Routledge, Abingdon, UK.

Tricker, R. (2016) *ISO 9001:2015 Audit Procedures* 4th edn. Routledge, Abingdon, UK.

Triggs, G.D. (Ed) (1988) *The Antarctic Treaty Regime: law environment and resources*. Cambridge University Press, Cambridge, UK.

Trigo-Rodríguez, J.M. (2022) *Asteroid Impact Risk: impact hazard from asteroids and comets*. Springer, Cham, CH.

Trigo-Rodríguez, J.M., Gritsevich, M. and Palme, H. (Eds) (2017) *Assessment and Mitigation of Asteroid Impact Hazards*. Springer, Cham, CH.

Trombley, R.B. (2006) *The Forecasting of Volcanic Eruptions*. iUniverse, Lincoln, NE, USA.

Trudgill, S. (2014) *The Terrestrial Biosphere: environmental change, ecosystem science, attitudes and values*. Routledge, Abingdon, UK.

Trumper, K., Dickson, B.K., van der Heijden, G., Jenkins, M. and Manning, P. (Ed) (2009) *The Natural Fix? The role of ecosystems in climate mitigation*. UNEP/WCMC, Cambridge, UK.

Trushenski, J. (2019) *Understanding Aquaculture*. 5M Publishing, Sheffield, UK.

Tsang, D.C.W. and Yong Sik Ok (Eds) (2022) *Biochar in Agriculture for Achieving Sustainable Development Goals*. Elsevier (Academic Press), London, UK.

Tsipursky, G. (2020) *Resilience: adapt and plan for the new abnormal of the Covid-19 coronavirus pandemic*. John Hunt Publishing, Alresford, UK.

Tundisi, J.G., Jørgensen, S.E. and Tundisi, T.M. (2019) *Handbook of Inland Aquatic Ecosystem Management*. CRC Press, Boca Raton, FL, USA.

Turner, N.J. (Ed) (2020) *Plants, People, and Places: the roles of ethnobotany and ethnoecology in indigenous peoples' land rights in Canada and beyond*. McGill-Queen's University Press, Montreal, QC, CA.

Tzotzos, G.T., Head, G. and Hull, R. (2020) *Genetically Modified Plants: assessing safety and managing risk*. Elsevier (Academic Press), Cambridge, MA, USA.

Ueasangkomsate, P. and Wongsupathai, C. (2018) *Environmental management systems in Thai small and medium-sized manufacturing firms*. 2018 IEEE International Conference on Industrial Engineering and Engineering Management (IEEM), Bangkok, Thailand, pp. 695–699.

Ulas, A. (Ed) (2016) *Handbook of Research on Waste Management Techniques for Sustainability*. IGI Global, Hershey, PA, USA.

Umetsu, C. and Sakai, S. (Eds) (2014) *Social-Ecological Systems in Transition*. Springer, Tokyo, JP.

UN (1992) *Agenda 21: programme of action for sustainable development*. United Nations, New York, NY, USA.

UN (1993) *Systems of National Accounts*. United Nations, New York, NY, USA.

UN (2014) *Big data and open data as sustainability tools*. A working paper prepared by the Economic Commission for Latin America and the Caribbean (ELAC), UN, Santiago, CL.

UN (2015) *UN Agenda 2030 -Transforming Our World: the 2030 Agenda for Sustainable Development*. UN, Paris, FR.

UN (2016) *Sustainable Development Goals Report*. UN Publications, New York, NY, USA.

UN (2017) *System of Environmental-Economic Accounting 2012: applications and extensions*. United Nations, New York, NY, USA.

UN, EEC, FAO, IMF, OECD and World Bank (2014) *System of Environmental-Economic Accounting 2012: central framework*. UN Publications, New York, NY, USA.

UNCTAD (2020) *Natural Resource Management in the Context of Climate Change*. United Nations Conference on Trade and Development, Geneva, CH.

Underkoffler, S.C. and Adams, H.R. (Eds) (2021) *Wildlife Biodiversity Conservation: multidisciplinary and forensic approaches*. Springer, Cham, CH.

UNDP (1991) *Human Development Report (for UN Development Programme)*. Oxford University Press, Oxford, UK.

UNDP (2020a) *How to Integrate Gender into Socio-Economic Assessments*. New York, NY, USA.

UNDP (2020b) *Human Development Report 2020: the next frontier – human development and the Anthropocene*. United Nations Development Program (UNDP), New York, NY, USA.

UNEP (2005) *Register of International Treaties and Other Agreements in the Field of the Environment*. UNEP, Nairobi, KE.

UNEP (2013) *Integrating the Environment in Urban Planning and Management: key principles and approaches for cities in the 21st century*. UNEP Publishing, Nairobi, KE.

UNEP (2021) *Adaptation Gap Report 2020. The gathering storm – adapting to climate change in a post-pandemic world*. United Nations Environment Programme, Nairobi, KE.

UNESCO (2016) *Drought Risk Management: a strategic approach*. UNESCO, Paris, FR.

UNESCO (2020) *Education for Sustainable Development: a roadmap*. UNESCO, Paris, FR.

Ungar, M. (Ed) (2012) *The Social Ecology of Resilience: a handbook of theory and practice*. Springer-Verlag, New York, NY, USA.

Ungar, M. (Ed) (2021) *Multisystemic Resilience: adaptation and transformation in contexts of change*. Oxford University Press, New York, NY, USA.

UN-HABITAT (2010) *Solid Waste Management in the World's Cities*. Routledge (Earthscan), London, UK.

UN-HABITAT (2018) *Climate Change Vulnerability Assessment Manual*. https://unhabitat. org/climate-change-vulnerability-assessment-manual (accessed 29/11/23).

UNIDO (2015) *Guide on Gender Mainstreaming Environmental Management Projects*. UNIDO, Vienna, AT.

Unnisa, S.A. and Rav, S.B. (Eds) (2012) *Sustainable Solid Waste Management*. CRC Press, Boca Raton, FL, USA.

Urquhart, J., Marzano, M. and Potter, C. (Eds) (2018) *The Human Dimensions of Forest and Tree Health: global perspectives*. Springer (Palgrave Macmillan), Cham, CH.

Ünsalan, C. and Boyer, K.L. (2011) *Multispectral Satellite Image – understanding: from land classification to building and road detection*. Springer, London, UK.

USAID (2007) *Adapting to Climate Variability and Change: a guidance manual for development planning*. USAID, Washington, DC, USA.

US Global Change Research Program (2019) *The Climate Report: national climate assessment-impacts, risks, and adaptation in the United States*. Melville House, Brooklyn, NY, USA.

US National Parks Service (2018) *Dutch ELM Disease and its Management* (first published 1982). FB&C Limited, London, UK.

Ussiri, D.A.N. and Lal, R. (2017) *Carbon Sequestration for Climate Change Mitigation and Adaptation*. Springer, Cham, CH.

Vaccaro, A., Parente, R. and Veloso, F.M. (2010) Knowledge management tools, inter-organizational relationships, innovation and firm performance. *Technological Forecasting and Social Change* 77(7), 1076–1089.

Vaidhyanathan, S. (2021) *Antisocial Media: how facebook disconnects us and undermines democracy* (first published 2018). Oxford University Press, Oxford, UK.

Vajravelu, R. (2009) *Ethnobotany: a modern perspective*. Kendall/Hunt Publishing, Dubuque, IA, USA.

Vakoch, D. and Mickey, S. (Eds) (2018) *Women and Nature? Beyond dualism in gender, body, and environment*. Routledge (Earthscan), London, UK.

Vale, R. and Vale, B. (Eds) (2013) *Living Within a Fair Share Ecological Footprint*. Routledge (Earthscan), Abingdon, UK.

van Bavel, B., Curtis, D.R., Dijkman, J., Hannaford, M., de Kayzer, M., van Onacker, E. and Soens, T. (2020) *Disasters and History: the vulnerability and resilience of past societies*. Cambridge University Press, Cambridge, UK.

van Beukering, P.J.H., Papyrakis, E., Bouma, J.A. and Brouwer, R. (Eds) (2013) *Nature's Wealth: the economics of ecosystem services and poverty*. Cambridge University Press, Cambridge, UK.

Van Calster, G. and Reins, L. (2017) *EU Environmental Law*. Edward Elgar, Cheltenham, UK.

Van Calster, G. and Reins, L. (Eds) (2021) *The Paris Agreement on Climate Change: a commentary*. Edward Elgar, Cheltenham, UK.

Vance, A. (2015) *Elon Musk: how the billionaire CEO of SpaceX and Tesla is shaping our future*. Virgin Books, London, UK.

Vanclay, F. (Ed) (2014) *Developments in Social Impact Assessment*. Edward Elgar, Cheltenham, UK.

Vanclay, F. (2020) Reflections on social impact assessment in the 21st century. *Impact Assessment and Project Appraisal* 38(2), 126–131.

Vanclay, F. and Esteves, A.M. (Eds) (2011) *New Directions in Social Impact Assessment: conceptual and methodological advances*. Edward Elgar, Cheltenham, UK.

Vanclay, F., Esteves, A.M., Aucamp, I. and Franks, D. (2015) *Social Impact Assessment: guidance for assessing and managing the social impacts of projects*. International Association for Impact Assessment, Fargo, ND, USA.

Van Eyk McCain, M. (Ed) (2010) *GreenSpirit: path to a new consciousness*. John Hunt Publishing, Ropeley, UK.

VanGuilder, C. (2014) *Environmental Audits*. Mercury Learning and Information, Dulles, VA, USA.

van der Land, V. (2017) *Migration and Environmental Change in the West African Sahel: why capabilities and aspirations matter*. Routledge (Earthscan), Abingdon, UK.

Vandermeer, J. and Perfecto, I. (2017) *Ecological Complexity and Agroecology*. Routledge, London, UK.

van der Ven, H. (2019) *Beyond Greenwash: explaining credibility in transnational eco-labelling*. Oxford University Press, Oxford, UK.

Van Dyke, F. (2008) *Conservation Biology: foundations, concepts, applications* 2nd edn. Springer, Dordrecht, NL.

Van Dyne, G.M. (Ed) (1969) *The Ecosystem Concept in Natural Resource Management*. Academic Press, New York, NY, USA.

van Swaaij, W.P.M., Kersten, S.R.A. and Paltz, W. (Eds) (2015) *Biomass Power for the World*. CRC Press, Boca Raton, FL, USA.

Van Triip, H.C.M. (Ed) (2013) *Encouraging Sustainable Behavior: psychology and the environment*. Routledge (Psychology Press), New York, NY, USA.

van Veenhuizen, R. (Ed) (2006) *Cities Farming for the Future: urban agriculture for green and productive cities*. IDRC, Ottawa, ON, CN.

van Zeben, J. and Rowell, A. (2020) *A Guide to EU Environmental Law*. University of California Press, Berkeley, CA, USA.

Vasconcellos, E.A. (2013) *Urban Transport Environment and Equity: the case for developing countries* 2nd edn. Routledge (Earthscan), Abingdon, UK.

Vasenev, V., Dovletyarova, E., Zhongqi Cheng, Valentini, R. and Calfapietra, C. (Eds) (2020) *Green Technologies and Infrastructure to Enhance Urban Ecosystem Services*. Springer, Cham, CH.

Vaughn, J.C. (Ed) (2011) *Watersheds: management, restoration, and environmental impact*. Nova Science Publishers, Hauppauge, NY, USA.

Verhoeven, H. (Ed) (2018) *Environmental Politics in the Middle East: local struggles, global connections*. Oxford Scholarship Online (Oxford University Press), Oxford, UK.

Verma, M.K. (Ed) (2018) *Globalisation, Environment and Social Justice: perspectives, issues and concerns*. Routledge, Abingdon, UK.

Verma, R. and Naidoo, V. (Eds) (2019) *Green Marketing as a Positive Driver Toward Business Sustainability*. IGI Global, Hershey, PA, USA.

Vezzoli, C. (2018) *Design for Environmental Sustainability: life cycle design of products* 2nd edn. Springer-Verlag, London, UK.

Vicente, J., Vercauteren, K.C. and Gortázar, C. (Eds) (2021) *Diseases of the Wildlife – livestock Interface: research and perspectives in a changing world*. Springer, Cham, CH.

Victor, D.G. (2004) *The Collapse of the Kyoto Protocol and the Struggle to Slow Global Warming*. Princeton University Press, Princeton, NJ, USA.

Viertl, R. (2011) *Statistical Methods for Fuzzy Data*. Wiley, Oxford, UK.

Vig, N.J. and Kraft, M.E. (Eds) (2021) *Environmental Policy: new directions for the twenty-first century* 11th edn. CQ Press, Washington, DC, USA.

Vigilance, C. and Roberts, J.L. (Eds) (2011) *Tools for Mainstreaming Sustainable Development in Small States*. Commonwealth Secretariat, London, UK.

Vince, G. (2022) *Nomad Century: how to survive the climate upheaval*. Allen Lane, London, UK.

Vinke, K. (2019) *Unsettling Settlements – Cities, Migrants, Climate Change: rural-urban climate migration as effective adaptation?* LIT Verlag, Zurich, CH.

Vinod Kumar, T.M. (Ed) (2020) *Smart Environment for Smart Cities*. Springer, Singapore, SG.

Virapongse, A., Brooks, S., Metcalf, E.C., Zedalis, M., Gosz, J., Kliskey, A. and Alessa, L. (2016) A social-ecological systems approach for environmental management. *Journal of Environmental Management* 178, 83–91.

Visgilio, G.R. and Whitelaw, D.M. (Eds) (2007) *Acid in the Environment: lessons learned and future prospects*. Springer, New York, NY, USA.

Visvizi, A. and Miltiadis, D.L. (Eds) (2019) *Smart Cities: issues and challenges – mapping political, social and economic risks and threats*. Elsevier, Amsterdam, NL.

Vizeu Pinheiro, M., Rojas Sánchez, L. and Long, S.C. (2020) *Environmental Governance Indicators for Latin America & the Caribbean*. Inter-American Development Bank, Washington, DC, USA.

Vlachos, E. (1985) Assessing long range cumulative impacts. In V.T. Covello, J.L. Mumpower, P.J.M. Stallen and V.R.R. Uppuluri (Eds), *Environmental Impact Assessment, Technology Assessment, and Risk Analysis: contributions from the psychological and decision sciences* (NATO ASI series G, Ecological Science; pp. 49–80). Springer-Verlag, Berlin, GR.

Vogelaar, A.E., Peat, A. and Hale, B.W. (Eds) (2019) *The Discourse of Environmental Collapse: imagining the end*. Routledge, Abingdon, UK.

Vogler, D., Macey, S. and Sigouin, A. (2017) Stakeholder analysis in environmental and conservation planning. *Lessons in Conservation* 7, 5–16.

Vogler, J. (2016) *Climate Change in World Politics*. Palgrave Macmillan, London, UK.

Vogt, W. (1948) *Road to Survival*. William Sloane, New York, NY, USA.

Vogtlnder, J.G. (2014) *A Practical Guide to LCA for Students, Designers and Business Managers: cradle-to-grave and cradle-to-cradle*. Delft Academic Press, Delft, NL.

Von Frese, R.R.B. (2020) *Basic Environmental Data Analysis for Scientists and Engineers*. CRC Press, Boca Raton, FL, USA.

Vos, R. (2015) *Technology and Innovation for Sustainable Development*. Bloomsbury, London, UK.

Wackernagel, M. and Beyers, B. (2019) *Ecological Footprint: managing our biocapacity budget*. New Society Publishers, Gabriola Island, BC, CA.

Waddington, P. (2010) *Shades of Green: a (mostly) practical A-Z for the reluctant environmentalist*. Eden Project Books, Bodelva, UK.

Wade, W. (2012) *Scenario Planning: a field guide to the future*. Wiley, Hoboken, NJ, USA.

Wagner, S.M. (2021) *Business and Environmental Sustainability: foundations, challenges and corporate functions*. Routledge, Abingdon, UK.

Wainwright, J. and Mulligan, M. (Eds) (2013) *Environmental Modelling: finding simplicity in complexity* 2nd edn. Wiley, Chichester, UK.

Walker, B. and Salt, D. (2012) *Resilience Practice: building capacity to absorb disturbance and maintain function.* Island Press, Washington, DC, USA.

Walker, I. and Schand, H. (Eds) (2017) *Social Science and Sustainability.* CSIRO, Canberra, ACT, AU.

Walker, L.R. and Shiels, A.B. (2013) *Landslide Ecology.* Cambridge University Press, Cambridge, UK.

Wall, D. (2010) *The No-nonsense Guide to Green Politics.* New Internationalist Publications, Oxford, UK.

Wallace, K.J. (2007) Classification of ecosystem services: problems and solutions. *Biological Conservation* 139, 235–246.

Wamsler, C. (2014) *Cities, Disaster Risk and Adaptation.* Routledge, Abingdon, UK.

Wang, H.-F. and Gupta, S.M. (2011) *Green Supply Chain Management: product life cycle approach.* McGraw-Hill, New York, NY, USA.

Wang, L.K., Chen, J.P., Yung-Tse Hung and Shammas, N.K. (Eds) (2009) *Heavy Metals in the Environment.* CRC Press, Boca Raton, FL, USA.

Wang, L.K., Wang, M.-H.S., Hung, Y.-T. and Shammas, N.K. (Eds) (2021) *Integrated Natural Resources Management: handbook of environmental engineering.* Springer, Cham, CH.

Wang Tianjin (2018) *China Environmental Anthropology.* North American Business Press, West Palm Beach, FL, USA.

Wani, K.A., Ariana, L. and Zuber, S.M. (Eds) (2019) *Handbook of Research on Environmental and Human Health Impacts of Plastic Pollution.* IGI Global, Hershey, PA, USA.

Wani, S.P., Rockström, J. and Sahrawat, K.L. (Eds) (2011) *Integrated Watershed Management in Rainfed Agriculture.* CRC Press, Boca Raton, FL, USA.

Ward, P.D. (2010) *The Flooded Earth: our future in a world without ice caps.* Basic Books, New York, NY, USA.

Wardlow, B.D., Anderson, M.C. and Verdin, J.P. (Eds) (2012) *Remote Sensing of Drought: innovative monitoring approaches.* CRC Press, Boca Raton, FL, USA.

Warne, K.P. (2012) *Let Them Eat Shrimp: the tragic disappearance of the rainforests of the sea.* Island Press, Washington, DC, USA.

Warner, K., Leighton, M., Xiaomeng Shen, Wendler, A. and Brach, K. (Eds) (2011) *Climate Change and Migration: rethinking policies for adaptation and disaster risk reduction.* UNU-EHS, Grand-Saconnex, CH.

Warner, R. and Marsden, S. (Eds) (2016) *Transboundary Environmental Governance: inland, coastal and marine perspectives* (first published 2012). Routledge, London, UK.

Washington, H. (2015) *Demystifying Sustainability: towards real solutions.* Routledge, London, UK.

Watson, D. and Adams, M. (2011) *Design for Flooding: architecture, landscape, and urban design for resilience to climate change.* Wiley, Hoboken, NJ, USA.

Watson, J. (2013) *The WTO and the Environment: development of competence beyond trade.* Routledge, London, UK.

Watson, R.R. and Preedy, V.R. (Eds) (2016) *Genetically Modified Organisms in Food: production, safety, regulation and public health.* Elsevier (Academic Press), London, UK.

Watson, S. (2019) *City Water Matters: cultures, practices and entanglements of urban water.* Springer (Palgrave Macmillan), Singapore, SG.

Watt, K.E.F. (Ed) (1966) *Systems Analysis in Ecology.* Academic Press, New York, NY, USA.

Watt, K.E.F. (1969) *Ecology and Resource Management: quantitative resource.* McGraw-Hill, New York, NY, USA.

Watts, J. (2010) *When a Billion Chinese Jump: how China will save mankind – or destroy it.* Simon & Schuster, New York, NY, USA.

WBGU (2009) *Climate Change as a Security Risk*. Routledge (Earthscan), London, UK.

Webb, R. (2020) The great population debate. *New Scientist*, 248(3308), 34–40.

Weber, M. (1958) *The Protestant Ethic and the Spirit of Capitalism* (English translation by T. Parsons – first published in 1904 and 1905 in 2 vols). Charles Scribner, New York, NY, USA.

Wegmann, M., Leutner, B. and Dech, S. (Eds) (2016) *Remote Sensing and GIS for Ecologists: using open source software* (Data in the Wild). Pelagic Publishing, Totnes, UK.

Wehrmeyer, W. (Ed) (2017) *Greening People: human resources and environmental management* 2nd edn. Routledge, Abingdon, UK.

Weishaar, S., Kreiser, L.A., Milne, J.E., Ashibor, H. and Mehling, M. (Eds) (2017) *The Green Market Transition: carbon taxes, energy subsidies and smart instrument mixes*. Edward Elgar, Cheltenham, UK.

Wei-Yin Chen, Seiner, J., Suzuki, T. and Lackner, M. (2012) *Handbook of Climate Change Mitigation*. Springer, New York, NY, USA.

Welford, R. (1992) *Environmental Auditing: the EC eco-audit scheme and the British standard on environmental management systems*. University of Bradford, Bradford, UK.

Welford, R. (2013) *Corporate Environmental Management 1: systems and strategies* revised edn (first published 1996). Routledge (Earthscan), Abingdon, UK.

Welti, R.C. (2012) *Satellite Basics for Everyone: an illustrated guide to satellites for non-technical and technical people*. iUniverse, Bloomington, IN, USA.

Wenk, M.S. (2005) *The European Union's Eco-Management and Audit Scheme (EMAS)*. Springer, Dordrecht, NL.

Wenshan Guo, Huu Hao Ngo, Surampalli, R.Y. and Zhang, T.C. (Eds) (2021) *Sustainable Resource Management: technologies for recovery and reuse of energy and waste materials*. Wiley-VCH, Weinheim, GR.

Westra, L. (2009) *Environmental Justice and the Rights of Ecological Refugees*. Earthscan, London, UK.

Westra, L. (2013) *Environmental Justice and the Rights of Indigenous Peoples: international and domestic legal perspectives*. Routledge, Abingdon, UK.

Wezel, A., Herren, B.G., Kerr, R.B., Barrios, E., Rodrigues-Gonçalves, A.L. and Sinclair, F. (2020) Agroecological principles and elements and their implications for transitioning to sustainable food systems: A review. *Agronomy for Sustainable Development* 40(6), 38–51.

Wheeler, S.M. and Beatley, T. (Eds) (2005) *The Sustainable Urban Development Reader*. Routledge, London, UK.

White, G. (2011) *Climate Change and Migration: security and borders in a warming world*. Oxford University Press, New York, NY, USA.

White, G.F. (Ed) (2019) *Environmental Effects of Complex River Development* (first published 1977). Taylor & Francis, London, UK.

Whitehead, M. (2014) *Environmental Transformation: a geography of the Anthropocene*. Routledge, Abingdon, UK.

Whittaker, D. (2018) *Integrated Waste Management: a sustainable approach*. Callisto Reference, Blue Ridge Summit, PA, USA.

Wich, S.A. and Lian Pin Koh (2018) *Conservation Drones: mapping and monitoring biodiversity*. Oxford University Press, Oxford, UK.

Wiersma, G.B. (Ed) (2017) *Environmental Monitoring*. CRC Press, Boca Raton, FL, USA.

Wilcox, C.P. and Turpin, R.B.(Eds) (2009) *Invasive Species: detection, impact, and control*. Nova Science Publishers, Hauppague, NY, USA.

Wildi, O. (2017) *Data Analysis in Vegetation Ecology* 3rd edn. CABI, Wallingford, UK.

Wilhite, D.A. and Pulwarty, R.S. (2017) *Drought and Water Crises: integrating science, management, and policy* 2nd edn. CRC Press, Boca Raton, FL, USA.

Wilkinson, C.R. and Buddemeier, R.W. (1994) *Global Climate Change and Coral Reefs: implications for people and reefs* (Report of UNEP-IOC-ASPEI-IUCN Global Task Team on the Implications of Climate Change on Coral Reefs). IUCN, Gland, CH.

Will, M. (2019) *An Operations Guide to Safety and Environmental Management Systems (SEMS): making sense of BSEE SEMS regulations.* Elsevier, Amsterdam, NL.

Williams, A. (2017) *China's Urban Revolution: understanding Chinese eco-cities.* Bloomsbury Publishers, London, UK.

Williams, B.K. and Brown, E.D. (2013) Adaptive management: from more talk to real action. *Environmental Management* 53(2), 465–479.

Williams, E. (2021) Environmental history of the Middle East and North Africa. In J. Hanssen and A. Ghazal (Eds), *The Oxford Handbook of Contemporary Middle Eastern and North African History* (Chapter 1). Oxford University Press, Oxford, UK.

Williams, J., Robinson, C. and Bouzaroviski, S. (2020) China's Belt and Road Initiative and the emerging geographies of global urbanization. *The Geographical Journal* 186, 128–140.

Williams, M. (2010) *Deforesting the Earth: from prehistory to global crisis, an abridgment.* University of Chicago Press, Chicago, IL, USA.

Williams, P.J. (1979) *Pipelines and Permafrost: physical geography and development in the circumpolar north.* Longman, London, UK.

Wilmsen, C., Elmendorf, W., Fisher, L. and Ross, J. (Eds) (2008) *Partnerships for Empowerment: participatory research for community-based natural resource management.* Routledge (Earthscan), London, UK.

Wilson, E.O. (1992) *The Diversity of Life.* Penguin, Harmondsworth, UK.

Wilson, G.A. (2012) *Community Resilience and Environmental Transitions.* Routledge (Earthscan), Abingdon, UK.

Winegard, T.C. (2020) *The Mosquito: a human history of our deadliest predator.* Allen Lane, London, UK.

WinklerPrins, A.M.G.A. (Ed) (2017) *Global Urban Agriculture.* CABI, Wallingford, UK.

Wiskerke, H. (Ed) (2020) *Achieving Sustainable Urban Agriculture.* Burleigh Dodds Science Publishing, Cambridge, UK.

With, K.A. (2019) *Essentials of Landscape Ecology.* Oxford University Press, Oxford, UK.

Wittfogel, K.A. (1957) *Oriental Despotism: a comparative study of total power.* Yale University Press, New Haven, CT, USA.

Wolf, E. (2019) *Can Physics Save Miami (and Shanghai and Venice, by Lowering the Sea)?* Morgan and Claypool Publishers, San Rafael, CA, USA.

Wolf, E.C. (1986) *Beyond the Green Revolution: new approaches to Third World agriculture* (Worldwatch Paper No. 73). Worldwatch Institute, Washington, DC, USA.

Wollenberg, E., Nihart, A., Tapio-Bistrom, M.-L. and Grieg-Gran, M. (Eds) (2012) *Climate Change Mitigation and Agriculture.* Routledge (Earthscan), Abingdon, UK.

Wolters, G.J.R. (1994) Integrated environmental management: the Dutch Governmental view. *Marine Pollution Bulletin* 29(6–12), 272–274.

Wondolleck, J.M. and Yaffee, S.L. (2017) *Marine Ecosystem-Based Management in Practice: different pathways, common lessons.* Island Press, Washington, DC, USA.

Wong, C.H.M. and Ho Wing-Chung (2015) Roles of social impact assessment practitioners. *Environmental Impact Assessment Review* 50(1), 124–133.

Wong, J.W.C., Surampalli, R.Y., Selvam, A. and Tyagi, R.D. (Eds) (2012) *Sustainable Solid Waste Management.* American Society of Civil Engineers, Reston, VA, USA.

Wong Shiu-Fai (2006) *Environmental Technology Development in Liberal and Coordinated Market Economies: tweaking institutions*. Palgrave Macmillan, New York, NY, USA.

Wood, E.F. and Sheffield, J.K. (2012) *Drought: past problems and future scenarios*. Routledge (Earthscan), London, UK.

Wood, M.E. (2017) *Sustainable Tourism on a Finite Planet: environmental, business and policy solutions*. Routledge (Earthscan), Abingdon, UK.

Woodcock, A., Saunders, J., Gut, K., Moreira, F. and Tavlaki, E. (2021) *Social Impact Assessment: tools, methods and approaches*. European Commission Directorate-General for Mobility and Transport Unit C.1 – Clean Transport & Sustainable Urban Mobility B-1049 Brussels, BE. 31 pp.

Woods, W.I., Teixeira, W.G., Lehmann, J., Steiner, C., WinklerPrins, A.M.G.A. and Rebellato, L. (Eds) (2009) *Amazonian Dark Earths: Wim Sombroek's vision*. Springer, Dordrecht, NL.

Woodward, G., Dumbrell, A.J., Baird, D.J. and Hajibabaei, M. (Eds) (2014) *Big Data in Ecology*. Elsevier (Academic Press), Amsterdam, NL.

Woolaston, K. (2022) *Ecological Vulnerability: the law and governance of human–wildlife relationships*. Cambridge University Press, Cambridge, UK.

World Bank (1997) *The LogFrame Handbook: a logical framework approach to project cycle management*. World Bank, Washington, DC, USA.

World Bank (2007) *World Development Report 2008: agriculture for development*. World Bank, Washington, DC, USA.

World Bank (2008) *Informal Recycling Sector in Developing Countries: organizing waste pickers to enhance their impact*. World Bank, Washington, DC, USA.

World Bank (2010) *The Cost to Developing Countries of Adapting to Climate Change: new methods and estimates*. World Bank, Washington, DC, USA.

World Bank (2013a) *Overcoming Institutional and Governance Challenges in Environmental Management: case studies from Latin America and the Caribbean region* (Resources Occasional Papers Series). World Bank, Washington, DC, USA.

World Bank (2013b) *Good Practice Handbook on Cumulative Impact Assessment and Management: guidance for the private sector in emerging markets*. World Bank (IFC), Washington, DC, USA.

World Bank (2014a) *Urban China: toward efficient, inclusive, and sustainable urbanization*. World Bank, Washington, DC, USA.

World Bank (2014b) *Climate Change and Migration: evidence from the Middle East and North Africa*. World Bank Publications, Washington, DC, USA.

World Bank (2016) *The World Bank Group's partnership with the Global Environment Facility: Vol. 1: Main report*. World Bank Publications, Washington, DC, USA.

World Bank (2022) *Impact Evaluation in Practice* 2nd edn. World Bank and Inter-American Development Bank, Washington, DC, USA.

World Commission on Environment and Development (1987) *Our Common Future (Brundtland Report)*. Oxford University Press, Oxford, UK.

Worldwatch Institute (2016) *Can a City Be Sustainable?* Island Press, Washington, DC, USA.

Wratten, R., Sandhu, H., Cullen, R. and Costanza, R. (Eds) (2013) *Ecosystem Services in Agricultural and Urban Landscapes*. Wiley-Blackwell, Oxford, UK.

Wren, C. (2012) *Risk Assessment and Environmental Management*. Maralte BV, Voorschoten, NL.

Wright, D., Camden-Pratt, C. and Hill, S. (Eds) (2013) *Social Ecology: applying ecological understanding to our lives and our planet*. Hawthorn Press, Stroud, UK.

Wright, E. (2017) *Environmental Management: past, present and future*. Nova Science Publishers, Hauppauge, NY, USA.

Wun Jern Ng, Nair, S., Jinadasa, K.B.S.N. and Valencia, E. (2018) *Saving Lakes: the urban socio-cultural and technological perspectives*. World Scientific Publishing, Singapore, SG.

Wynn, J. (2017) *Citizen Science in the Digital Age: rhetoric, science, and public engagement*. The University of Alabama Press, Tuskaloosa, AL, USA.

Xenarios, S., Schmidt-Vogt, D., Qadir, M., Janusz-Pawletta, B. and Abdullaev, I. (Eds) (2021) *The Aral Sea Basin: water for sustainable development in Central Asia*. Routledge, London, UK.

Xian, G.Z. (2015) *Remote Sensing Applications for the Urban Environment*. CRC Press, Boca Raton, FL, USA.

Xiang Cao, Congbo Teng and Jijun Zhang (2021) Impact of the Belt and Road Initiative on environmental quality in countries along the routes. *Chinese Journal of Population, Resources and Environment* 19(4), 344–351.

Xiangzheng Deng and Gibson, J. (2019) *River Basin Management*. Springer, Singapore, SG.

Xiangzheng Deng, Zhihui Li, Wang Yi, Feng Wu and Tao Zhang (2014) *Integrated River Basin Management: practice guideline for the IO table compilation and CGE modeling*. Springer, Berlin, GR.

Xiaohong Li (2018) *Industrial Ecology and Industry Symbiosis for Environmental Sustainability: definitions, frameworks and applications*. Springer (Palgrave Macmillan), Cham, CH.

Xinghua Li (2016) *Environmental Advertising in China and the USA: the desire to go green*. Routledge (Earthscan), Abingdon, UK.

Xuan Zhu (2016) *GIS for Environmental Applications: a practical approach*. Routledge, Abingdon, UK.

Xuemei Bai (2010) Industrial relocation in Asia a sound environmental management strategy? *Environment: Science and Policy for Sustainable Development* 44(5), 8–21.

Yamaguchi, M. (Ed) (2012) *Climate Change Mitigation: a balanced approach to climate change*. Springer, London, UK.

Yanarella, E.J. and Levine, R.S. (2020) *From Eco-Cities to Sustainable City-Regions: China's uncertain quest for an ecological civilization*. Edward Elgar, Cheltenham, UK.

Yang Xiaojun and Jiang Shijun (Eds) (2019) *Challenges Towards Ecological Sustainability in China: an interdisciplinary perspective*. Springer, Cham, CH.

Yanzhong Huang (2020) *Toxic politics: China's environmental health crisis and its challenge to the Chinese State*. Cambridge, Cambridge University Press, Cambridge, UK.

Yedla, S. (2013) *Urban Transportation and the Environment: issues, alternatives and policy analysis*. Springer, New Delhi, IN.

Yekini, K.C., Yekini, L.S. and Ohalehi, P. (Eds) (2021) *Environmentalism and NGO Accountability: Vol. 9*. Emerald Publishing, Bingley, UK.

Yeung, D.W.K., Yingxuan Zhang, Hongtao Bai and Islam, S.M.N. (2021) Collaborative environmental management for transboundary air pollution problems: A differential levies game. *Journal of Industrial & Management Optimization* 17(2), 517–531.

Yifei Li and Shapiro, J. (2020) *China Goes Green: coercive environmentalism for a troubled planet*. Polity Press, Cambridge, UK.

Yongseung Yun (Ed) (2017) *Recent Advances in Carbon Capture and Storage*. IntechOpen, Rijeka, HR.

Yong Sik Ok, Tsang, D.C.W., Bolan, N. and Novak, J.M. (Eds) (2018) *Biochar from Biomass and Waste: fundamentals and applications*. Elsevier, Amsterdam, NL.

Yong Sik Ok, Uchimiya, S.M., Chang, S.X. and Bolan, N. (Eds) (2016) *Biochar: production, characterization, and applications*. CRC Press, Boca Raton, FL, USA.

Young, M. and Esau, C. (Eds) (2016) *Transformational Change in Environmental and Natural Resource Management: guidelines for policy excellence*. Routledge (Earthscan), Abingdon, UK.

Young, N. (2015) *Environmental Sociology for the Twenty-First Century*. Oxford University Press, Oxford, UK.

Young, W. (2014) Rio Conventions redux: an argument for merging the trio into a single Convention on Environmental Management. *Consilience: The Journal of Sustainable Development* 12(1), 197–215.

Young, W. (Ed) (2021) *Managing Water Resources in Large River Basins*. MDPI, Basel, CH.

Yousuf, A. and Singh, M. (Eds) (2021) *Watershed Hydrology, Management and Modeling*. CRC Press, Boca Raton, FL, USA.

Yuan Su Yanni and Yu Ning Zhang (2020) Carbon emissions and environmental management based on Big Data and Streaming Data: a bibliometric analysis. *Science of the Total Environment* 73(1), article No. 138984 (11 pp.).

Yuan Xu (2021) *Environmental Policy and Air Pollution in China: governance and strategy*. Routledge (Earthscan), Abingdon, UK.

Yue Rong (Ed) (2011) *Practical Environmental Statistics and Data Analysis*. ILM Publications, St Albans, UK.

Yuhong Zhao (2021) *Chinese Environmental Law*. Cambridge University Press, Cambridge, UK.

Yung-Tse Hung, Wang, L.K. and Shammas, N.K. (Eds) (2012) *Handbook of Environment and Waste Management: air and water pollution control*. World Scientific Publishing, Singapore, SG.

Yun Tong, Haifeng Zhou, Lei Jiang and Biao Hc (2021) Investigating the factors underlying participation by the Chinese public in environmental management: an approach based on spatial heterogeneity. *Environmental Science and Pollution Research* 28, 48362–48378.

Yusuf (Wie), J.-E. and Saint John III, B. (Eds) (2021) *Communicating Climate Change: making environmental messaging accessible*. Routledge (Earthscan), Abingdon, UK.

Zaccaï, E. (Ed) (2007) *Sustainable Consumption, Ecology and Fair Trade*. Routledge, Abingdon, UK.

Zakour, M.J. and Gillespie, D.F. (2013) *Community Disaster Vulnerability: theory, research, and practice*. Springer, New York, NY, USA.

Zambon, I., Salvati, L. and Ferrara, C. (Eds) (2019) *Land Degradation: the main challenge*. Nova Science Publishers, Hauppauge, NY, USA.

Zasada, I., Piorr, A., Novo, P., Villanueva, A.J. and Valanszki, I. (2017) What do we know about decision support systems for landscape and environmental management? A review and expert survey within EU research projects. *Sustainable Ecosystems* 98, 63–74.

Zdruli, P., Pagliai, M., Kapur, S. and Cano, A.F. (Eds) (2010) *Land Degradation and Desertification: assessment, mitigation and remediation*. Springer, Dordrecht, NL.

Zebich-Knos, M. and Grichting, A. (Eds) (2017) *The Social Ecology of Border Landscapes*. Anthem Press, London, UK.

Zedilo, E. (Ed) (2008) *Global Warming: looking beyond Kyoto*. Brookings Institution Press, Washington, DC, USA.

Zehner, O. (2012) *Green Illusions: the dirty secrets of clean energy and the future of environmentalism (our sustainable future)*. University of Nebraska Press, Lincoln, NE, USA.

Zeppel, H. (2006) *Indigenous Ecotourism: sustainable development and management*. CABI, Wallingford, UK.

Zetter, R. and Watson, G.B. (Eds) (2016) *Designing Sustainable Cities in the Developing World* 2nd edn. Routledge, Abingdon, UK.

Zhang, C., Ni, G., and Fu, G. (2019) *Adaptive Catchment Management and Reservoir Operation*. MDPI, Basle, CH.

Zhang, J.Y. and Barr, M. (2013) *Green Politics in China: environmental governance and state-society relations*. Pluto Press, London, UK.

Zhang-Yue Zhou (2020b) *Global Food Security: what matters?* Routledge, Abingdon, UK.

Zhen Chen and Heng Li (2016) *Environmental Management in Construction: a quantitative approach*. Routledge, Abingdon, UK.

Zhenjiang Shen, Ling Huang, KuangHui Peng and Jente Pai (Eds) (2018) *Green City Planning and Practices in Asian Cities: sustainable development and smart growth in urban environments*. Springer, Cham, CH.

Zhifeng Yang (Ed) (2017b) *Eco-Cities: a planning guide*. Routledge, Abingdon, UK.

Zhihua Zhang (2017) *Environmental Data Analysis: methods and applications*. De Gruyter, Berlin, GR.

Zimdars, M. and Mcleod, K. (Eds) (2020) *Fake News: understanding media and misinformation in the digital age*. MIT Press, Cambridge, MA, USA.

Zolnikov, T.R. (Ed) (2019) *Global Adaptation and Resilience to Climate Change*. Springer (Palgrave Macmillan), Cham, CH.

Zommers, Z. and Alverson, K. (Eds) (2018) *Resilience: the science of adaptation to climate change*. Elsevier, Amsterdam, NL.

Zubelzu, S. and Fernández, R.A. (2016) *Carbon Footprint and Urban Planning: incorporating methodologies to assess the influence of the urban master plan on the carbon footprint of the city*. Springer, Cham, CH.

Zubrin, R. (2012) *Merchants of Despair: radical environmentalists, criminal pseudo-scientists, and the fatal cult of antihumanism*. Encounter Books, New York, NY, USA.

Zuur, A.F., Leno, E.N. and Smith, G.M. (2007) *Analysing Ecological Data*. Springer, New York, NY, USA.

Index

Milton Keynes UK
Ingram Content Group UK Ltd.
UKHW031334131124
451128UK00019BA/386